Climate Change and Agriculture

About the Editors

Dr. GSLHV Prasada Rao, had his B.Sc (Physics), M.Sc (Meteorology) and Ph.D (Agroclimatology) from Andhra University, Waltair. He commenced his service in Ground Water Department, Govt. of Rajasthan as Junior Hydrometeorologist in January 1980 and later joined as Assistant Professor (Agricultural Meteorology) in Kerala Agricultural University in December 1980. He served as Professor & Head, Department of Agricultural Meteorology, Associate Dean, College of Horticulture, Associate Director of Research and Special Officer, Academy of Climate Change Education and Research, Comptroller and Coordinator (Finance), Kerala Agricultural University before his superannuation from KAU service on 29-07-2012. Later, he took up assignment as Consultant Professor, Centre for Animal Adaptation to Environment and Climate Change Studies, Kerala Veterinary and Animal Sciences University from 01-04-2013 to 30-09-2016. Had short term assignment as International Agrometeorologist in a World Bank aided Project on Climate Resilience, MoALM&C, Govt. of Nepal in 2018.

Instrumental in establishment of "Academy of Climate Change Education and Research (ACCER)" in Kerala Agricultural University and launched unique and innovate dual degree program in Climate Change Adaptation (B.Sc.–M.Sc. dual degree program of five year duration) after plus two. It is the first of its kind in India. It is a multi-disciplinary and multi-institutional program. Instrumental in establishment of "Centre for Animal Adaptation to Environment and Climate Change Studies (CAADECCS)" in Kerala Veterinary and Animal Sciences University and launched unique and innovative Ph.D. program in "Climate Change and Animal Agriculture" during the Academic year 2015-16 in addition to Two P.G. Diploma Courses (one year duration) in Climate Services for Veterinary Students and for general science stream students.

He possesses 38 years of experience in teaching, research and extension in Agricultural Meteorology, Climate Change and Climate Change Risk Management in Agriculture. Many publications are to his credit. He is currently involved in farming and editing/writing books.

Dr. V. Uma Maheswara Rao (VUM Rao) had his post-graduation from the Department of Meteorology and Oceanography, Andhra University, Waltair. He obtained his Ph.D in Agricultural Meteorology in 1981and had a Post Doctoral Research Fellowship during 1982-83. He joined in CCS HAU service as Assistant Climatologist/Assistant Professor in 1983 and served the Department of Agricultural Meteorology in various capacities. He was the Professor and Head in CCS HAU before he joined as Principal Scientist in ICAR System and retired as Project Coordinator (AICRP on Agricultural Meteorology), Central Research Institute for Dryland Agriculture (CRIDA), Indian Council of Agricultural Research in 2016. The **prestigious Choudary Devi Lal ICAR** best AICRPAM award was obtained under his dynamic leadership as the Project Co-ordinator (Agricultural Meteorology) in 2014. He had 35 years of experience in the field of Agricultural Meteorology as researcher, scientist and teacher. He is Life Member in six scientific societies viz., "Indian Meteorological Society, The Arid Zone Research Association of India, Indian Society of Remote Sensing, Association of Agrometeorologists, International Society for Agrometeorology and The Indian Society of Dryland Agriculture. He is a Fellow of Association of Agro-meteorologists. He served as President and Vice-President of Association of Agrometeorologists. He is a reviewer for various international and national research journals, expert in several selection committees, member in Board of Studies of several universities, paper setter and external examiner and one of the renowned resource personnel in Agricultural Meteorology. Many students were trained under him with postgraduate degrees, spread across the country and settled in various positions in the field of Agricultural Meteorology. He has more than 200 publications including popular articles. He visited abroad on many occasions in his official capacity.

Dr. D.V. Subba Rao obtained B.Sc(Ag) in 1980 and M.Sc.(Ag) in Agricultural Economics in 1982 from ANGRAU. He was an ICSSR Doctoral Fellow at Institute of Economic Growth during 1988-93. He obtained Ph.D. from the Department of Economics, Delhi School of Economics, University of Delhi in 1995 under the supervision of Professor Kanchan Chopra and Professor V.N. Pandit. He has been awarded with Jawaharlal Nehru Award, ICAR, Professor Chablani Memorial Prize, University of Delhi and Vetiver Network Award of World Bank for his research contributions. He joined as research assistant in 1983 in the same university and is working for more than three decades in various capacities. At present, he is holding the NABARD Chair Professor at Agricultural College, Bapatla, ANGRAU, Andhra Pradesh. Earlier, he worked as Professor & University Head in the Department of Agricultural Economics, and Hon Director, Cost of Cultivation Scheme at College of Agriculture, Rajendranagar, Hyderabad. Further, he has worked as an Economist on projects pertaining to natural resource accounting for a year and half at Institute of

Economic Growth, Delhi during 1996-97on deputation. Subsequently, he worked as Asst. Professor, Associate Professor, and Professor at College of Agriculture, Rajendranagar, Hyderabad. He guided 20 M. Sc (Ag) and 6 PhD students in different areas such as agro forestry, minor forest products, infrastructure, water management, tank and well irrigation, institutional aspects of irrigation, potential for horticultural products, trade competitiveness of oil seeds, profitability and plant protection, land policy, women empowerment etc. in India and abroad. He attended several conferences and seminars and organized conferences on agribusiness, climate change and smart farming. He has been life member of the Indian Society of Agricultural Economics, the Indian Econometric Society and Agricultural Economics Research Association. Dr. D.V. Subba Rao has published several articles on natural resource and environmental economics.

Climate Change and Agriculture
Causes, Impacts and Interventions

GSLHV Prasada Rao
VUM Rao
DVS Rao

NEW INDIA PUBLISHING AGENCY
New Delhi – 110 034

NEW INDIA PUBLISHING AGENCY
101, Vikas Surya Plaza, CU Block, LSC Market
Pitam Pura, New Delhi 110 034, India
Phone: + 91 (11) 27 34 17 17 Fax: + 91 (11) 27 34 16 16
Email: info@nipabooks.com
Web: www.nipabooks.com

Feedback at feedbacks@nipabooks.com

ISBN: 978-93-87973-62-6

Composed and Designed by NIPA

Rupa Kumar Kolli
Chief
World Climate Applications and Services Division
World Meteorological Organization
Geneva, Switzerland

Foreword

Global warming is real and happening now, and India is not an exception, though there might be some spatial differences in warming across the country. An increasing body of scientific evidence, as brought out by the Intergovernmental Panel on Climate Change (IPCC) assessment reports and special reports, conclusively establishes that the major cause of global warming is the unabated increase in greenhouse gas (e.g., carbon dioxide, methane, etc.) concentrations in the atmosphere way beyond their natural levels. It is also abundantly clear that global warming and the consequent climate change are almost entirely due to the uncontrolled human activities such as burning of fossil fuels, deforestation, land use and land cover changes, intensive exploitation of natural resources, unsustainable agricultural practices, etc. The rapid scientific and technological advances over the past couple of centuries fed the industrial revolution and all-round development, but at an enormous cost to the entire human society and ecosystem. Rise in surface air temperature is widespread over the globe and is greater at higher northern latitudes. Land regions have warmed faster than the oceans. The latest Annual Statement of the Status of the Climate issued by the World Meteorological Organization (WMO) reports that the long-term warming trend has continued in 2018, with the average global temperature set to be the fourth highest on record. As many as 20 of the warmest ever-recorded years happened to be in the past 22 years, with the top four in the past four years, according to the statement. Other tell-tale signs of climate change, including sea level rise, ocean heat and acidification and sea-ice and glacier melt continue, whilst extreme weather left a trail of devastation on all continents, according to the WMO Statement. The HudHud cyclone over Visakhapatnam Coast in October 2014, Chennai floods in 2015, Kerala floods in August 2018 and the Titli cyclone over Orissa and North Andhra Pradesh Coast in October 2018, the

Gaja Cyclone over Tamil Nadu Coast in mid November and the Phethai Cyclone in mid December 2018 devastated the crops and infrastructure to a large extent in respective regions. In contrast, the Indian foodgrains production was adversely affected due to consecutive droughts during monsoon season in 2014 and 2015 since agricultural production is mainly dependent on monsoonal rains in India. Similarly, the occurrence of cold and heat waves in northern states of the country causes recurring damage to crops. Untimely snowfall over western Himalayan region in early November 2018 severely damaged apple production. The ill effects of climate change in the form of frequent occurrence of floods, droughts, cold and heat waves are clearly evident on various agricultural production systems, thereby posing a serious threat to food security.

Climate change is considered to be the biggest challenge confronting the global society, both to prevent future changes and to adapt to changes that are already happening and are imminent. To mitigate the climate change effects on agricultural production and productivity a range of adaptive strategies need to be considered. Changing cropping calendars and pattern will be the immediate best available option with available crop varieties to mitigate the climate change impact. The options like introducing new cropping sequences, late or early maturing crop varieties depending on the available growing season, conserving soil moisture through appropriate tillage practices and efficient water harvesting techniques are also important. Developing heat and drought tolerant crop varieties by utilizing genetic resources that may be better adapted to new climatic and atmospheric conditions should be the long-term strategy. Genetic manipulation may also help to exploit the beneficial effects of increased carbon dioxide on crop growth and water use. One of the promising approaches would be gene pyramiding to enhance the adaptation capacity of plants to climatic change inputs. There is thus an urgent need to address the climate change and variability issues holistically through improving the natural resource base, diversifying cropping systems, adapting farming systems approach, strengthening of extension system and institutional support. Latest improvements in biotechnology and information technologies need to be used for better agricultural planning and weather/climate based management to enhance the agricultural productivity of the country.

Successful adaptation to climate change requires long-term investments in strategic research and new policy initiatives that mainstream climate change adaptation into development planning. As a first step, we need to document all the indigenous practices adopted by rainfed farmers over time, for coping with climate change. Secondly, we need to quantify the adaptation and mitigation potential of the existing best bet practices for different crop and livestock production systems in different agro-ecological regions of the country. Thirdly, a long-term strategic research planning is required to evolve new tools and techniques including crop varieties and management practices that help in adaptation. In 2010, the Indian Council of Agricultural Research (ICAR) launched the National Initiative on Climate Resilient Agriculture (NICRA) as a comprehensive project covering strategic research, technology demonstration and capacity building. Targeted research on adaptation and mitigation is at nascent stage in India but based on knowledge already generated, some options for adaptation to climate variability induced effects like droughts, high temperatures, floods and sea water inundation can be suggested. These strategies fall into two broad categories viz., (i) crop based and (ii) resource management based.

Several experts have identified research areas that would reduce uncertainty and improve knowledge to face the consequences of climate change and provide improved planning. These include, but not limited to:

- Precision weather-forewarning at farm level
- Quantitative assessment of specific crop responses at different crop stages to enhanced levels of greenhouse gases, precipitation and UV-B radiation
- Breeding agricultural crops for tolerance to extreme temperatures, floods, salinity and droughts
- Innovative agroclimatic resource mapping to avoid potentially inappropriate land-use choices
- The impacts of elevated carbon dioxide levels on plant and soil-water balances
- Improved crop simulation models as tools for both research and crop management.

This publication revolves around interventions to minimize crop losses due to global warming and climate change, organized in 32 chapters collectively spanning more than 560 pages. The contributions emerged from research findings in different parts of the country in the ICAR and the State Agricultural Universities, which I consider to be a commendable compilation. I am sure this publication will be of immense use to researchers, teachers, students as well as farmers.

Rupa Kumar Kolli
Chief
World Climate Applications and Services Division
World Meteorological Organization
Geneva, Switzerland

Professor (Dr.) Kadambot Siddique
Hackett Professor of Agriculture Chair and Director
The UWA Institute of Agriculture, The University of Western Australia
Perth, Australia

Preamble

Global warming is real. Rise in temperature is likely to be around 2°C by the end of this century with regional uncertainties in rainfall. It is a threat to the society linked sectors viz., agriculture, water resources, forestry, biodiversity (both land and ocean), infrastructure and health. The adverse impact of climate change is already noticed across the world in the above society linked sectors due to weather related disasters in the form of cyclones, floods, droughts, cold and heat waves and sea level rise. The monsoon aberrations have become more frequent over **India on which rainfed Agriculture is dependant. The years 2014 and 2015 were recorded as monsoon drought years and the monsoon 2018 also ended with nine percent deficit rainfall against the long term Average. The recent years 2015, 2016 and 2017 were recorded as warmest years after 2009 and 2010 in India.** Decline in rainfall and increase in temperature are observed in the last 50-60 years across the **Country and reports indicate that the semiarid belt is increasing.** Melting of glaciers and snow from the Himalayas due to global warming is a serious threat to the seasonal water availability **across the northern states of India.** The global hydrological and energy cycles are likely to be adversely affected and thus the change in whole climate system.

This Publication deals with climate change impacts and interventions in agriculture across various regions of the Country based on the research findings emanated from ICAR institutes and State Agricultural Universities as global warming and climate change is a threat to food security. I understand that more than 60 authors are involved with 32 Chapters, containing about 560 pages focussing on mitigation and adaptation strategies in agriculture against ill effects of climate change across India. Researchers indicate that crop yields are likely to decline with increase in atmospheric CO_2 and temperature in ensuing decades and crop losses can be minimized through technologies generated in crop management, crop improvement and crop protection practices. Farmers concern to minimize crop losses against weather extremes rather than climate change as the impacts of climate change (increase in atmospheric CO_2 and temperature) on crops at the field level is not clear if water is not a constraint. The publication includes vulnerability assessment, adaptation strategies, crop choices/ combinations in different climate scenarios and preparedness to face the exigencies of climate change and emphasis is given on interventions. There is an urgent need to intensify research and development to generate technologies and location specific farm practices which can minimise the impact of climate change.

Professor (Dr.) Kadambot Siddique
Hackett Professor of Agriculture Chair and Director
The UWA Institute of Agriculture, The University of Western Australia
Perth, Australia

Preface

Weather and climate play an important role in crop distribution and production. While climate determines the adaptability of a particular crop in a region, weather determines the yield attributes of the crop. Over a period of years, changes have occurred in climate of the earth's atmospheric system mostly due to human interventions in the form of emission of large quantities of greenhouse gases (GHGs). The present level of carbon dioxide concentration in the atmosphere has crossed 400ppm level. As a result of the dumping of GHGs (CO_2, Methane, Nitrous Oxide and CFCs) the earth's atmospheric system is getting warmer at an alarming rate, which lead to a state called global warming. It is now a topic of concern to the society. The global mean atmospheric temperature has risen by 0.8°C since 1901and it is likely to be between 1.5 and 2°C by the end of 2100 A.D. This situation leads to changes in earth's climate system viz., Atmosphere, Hydrosphere, Lithosphere, Cryosphere and Biosphere. Its ill effects are evident in different society linked sectors viz., water, health, agriculture, forestry, biodiversity and infrastructure. Unprecedented floods in Kerala during August, floods in some pockets across southern states viz.,Tamil Nadu, Karnataka and Andhra Pradesh during monsoon season, cyclone Titli in October that hit Orissa and Andhra Coast (Srikakulam District), snowfall in early November in 'apple bowl' of Kashmir. Cyclone Gaja across the T.N. Coast in November and Phethai in mid-December across the Andhra Coast during 2018 led to huge damage to infrastructure including crops of the respective regions. Extreme weather events such as floods and droughts, heavy rainfall, avalanche, landslides, heat and cold waves, cyclonic storms, thunder storms, hail storms, sand storms and cloud bursts are not uncommon and likely to be frequent in ensuing decades under projected climate change scenario. The impact of drought on Indian foodgrains production is much more predominant when compared to that of flood/heavy rainfall during monsoon since large areas of cultivable land experiences drought situation if monsoon breaks or fails. Deficit monsoon during 1965, 1966, 1972, 1984, 1987, 1997, 2002, 2004, 2009, 2012, 2014 and 2015 led to drought and adversely affected the kharif foodgrains production. The unexpected rains that lashed during February, March and April, 2015 across large part of northern and central India led to extensive crop damage. March and April are the months when the rabi crop is harvested. But the heavy rainfall accompanied by strong winds dashed the hopes of many farmers. Rain induced by western disturbances usually occurs during the winter and the late precipitation close to the rabi harvest has caused widespread crop damage. Thousands of acres of wheat across the States such as Rajasthan, Madhya Pradesh, Uttar Pradesh, Gujarat and Maharashtra were destroyed. Potato crop had been damaged in U.P. In Maharashtra, the onion crop has also been hit hard, both by the rain as well as hailstorms. The unseasonal rain accompanied by high speed winds and hailstorms badly affected quantity and quality of pulses. Marathwada, Vidarbha, Northern Maharashtra and parts of Western Maharashtra were the worst affected by the unprecedented hailstorms and unseasonal rainfall. Rabi crops like wheat, harbhara, cotton, jowar, summer onion are lost, horticultural crops like Papaya, sweet lime, grapes are battered and orchards which took years to grow are ridden to the ground. The unseasonal rains also adversely affected the yield of the king of fruits this season, with many mango trees in Maharashtra being uprooted due to strong winds accompanying rain. It also affected vegetable prices as cauliflower, tomato and coriander produce suffered in a State like Maharashtra. The vegetable baskets of Junnar, Ambegaon and Khed in Maharshtra were some of the worst affected areas. Even fruits like grapes grown in parts of Maharashtra were damaged. The conditions in Gujarat were also similar; the yield of spices like jeera and coriander suffered due to unusual rains. The destruction in standing crops had resulted in vegetable prices rising in North Indian cities. Landslides and snowfall led to the closure of the Jammu & Kashmir highway leaving thousands of people stranded. Such events occur and reoccur every year somewhere or other across the World. Therefore, it is important to understand the impact of climate variability or climate change on crop production and suitable interventions are need of the hour as a proactive measure to sustain agriculture/food security in the event of global warming and climate change. Hence, an attempt has been made to compile the work done by the authors on climate change impacts and interventions across the country and published in an abridged form for the benefit of farmers and to sustain Agriculture against malevolent effects of global warming and climate change.

In introductory chapter, greenhouse effect, global warming, status of atmospheric carbon dioxide in the atmosphere and its impact on rising temperature, aerosols and their impacts on global cooling, dual effect of clouds on global warming or cooling, ozone depletion and UV radiation and climate change awareness in society linked sectors viz., water, agriculture, forestry, biodiversity and health are discussed. The need to implement crop weather insurance as a part of climate risk management is also pointed out. Global warming and its impact on food price along with climate change awareness programmes are also included. The need for conducting climate change awareness programmes on the impact of climate change is focused. Chapter two is on Rainfall Variability in India. The arid and semi-arid areas are expanding whereas the moist subhumid, humid and perhumid areas are shrinking. The semi-arid area, which occupies the large and central portion of the country, is expanding significantly. The overall effect is that the country is experiencing dry climate in the recent period with a westward shifting of rainfall. A declining tendency in monsoon rainfall can be seen from 1965 onwards and a sharp declining trend from 1995. This is essentially due to weakening of the Tibetan Anticyclone associated with cooling of the upper troposphere (600–150 hPa) over the Himalayan region and surrounding atmosphere. The recent sharp increasing trend in the surface air temperature over India is due to a declining trend in rainfall during winter and summer monsoon seasons. Climate change and food security is included in chapter 3. Zonal differences in warming across the Country are highlighted. Impact of drought on Indian food grains production, climate change impacts on Indian food grains production, climate variability and food price are discussed in detail with facts and figures in chapter 3.

Climate Change: Extreme Events, Causes and Effects on Agriculture and Farmers Economy is included in chapter four. There is a need to intensify research to develop cultivars and develop management practices which can withstand the impact of climate change. The experiences under technology demonstration component of NICRA shows that technologies developed by the National Agricultural Research System can minimize the losses under variable climates and found to stabilize the productivity and incomes. These resilient practices need to be scaled up as part of the developmental programs for large scale adoption of resilient practices and to impart resilience to the agriculture production in the country. Chapter 5 includes climate risk management in smallholder farming systems in the Semiarid Tropics. Climate risk management requires holistic solutions derived from cross-disciplinary and participatory, user-oriented research – this implies that the multiple institutions/agencies present in India are required to better integrate their efforts for the benefit of smallholder farmers. State-wise climate change, impacts and interventions in Agriculture is focused in various chapters commencing from chapter 6 to chapter 23. The States include Kerala, Tamil Nadu, Karnataka, Andhra Pradesh, Maharastra, Gujarat, Haryana, Punjab, Himachal Pradesh, Jammu and Kashmir, Assam, Bihar, Jharkhand, Chhatisgarh and Orissa. The crops include field and horticultural crops that are grown across the Country. This publication highlights more on interventions to mitigate ill effects of climate change in Agriculture. In chapter 24, climate change impacts on soil microflora and fauna are included since soil health is very important for better crop productivity and more so, in a changed climate scenario. Chapter 25 focuses on importance of monitoring weather parameters in crop insurance schemes. Weather insurance is the only one which will help to a little extent when the farmers lose everything from their farm land during weather disasters. Agroadvisory services based on medium range forecasting play an important role in mitigating the ill effects of climate variability/change, therefore its relevance is included in chapters 26 and 27. Role of satellites and remote sensing in impacts of climate change monitoring is incorporated in chapters 28 and 29. NABARD's initiatives and interventions in climate change are highlighted inchapter 30. Chapter 31 high lights on sorghum while wheet in chapter 32 in relation to climate change issues. Altogether, the publication contains more than 500 pages with Tables, Figures, Illustrations and Plates.

Successful adaptation to climate variability/change requires long-term investments in strategic research and new policy initiatives. As a first step, we need to document all the indigenous practices that farmers follow with time for coping with climate change. Secondly, we need to quantify the adaptation and mitigation potential of the existing best bet practices for different crop and livestock production systems in different agro-ecological regions of the country. Thirdly, a long-term strategic research planning is required to evolve new tools and techniques including crop varieties and management practices that help in adaptation. More recently during 2010, ICAR has launched the National Initiative on Climate Resilient Agriculture (NICRA) as a comprehensive project covering strategic research, technology demonstration and capacity building. Targeted research on adaptation and mitigation is at nascent stage in India but based on knowledge already generated, some options for adaptation to climate variability induced effects like droughts, high temperatures, floods and sea water inundation can be suggested. Like NICRA at ICAR level, SICRA (State Initiative on Climate Resilient Agriculture) should be given top priority in all the States to generate technologies to minimize crop losses during the occurrence of weather extremes such as floods, droughts, cold and heat waves for sustenance of rural livelihoods.

This publication revolves around interventions to minimize crop losses due to global warming and climate change across the Country. In that respect, the contributions emerged based on research findings from different parts of the Country in the ICAR and SAUs system are noteworthy, unique and innovative. we are sure this publication will be of immense use to researchers, teachers, students and farmers and read with zeal and enthusiasm.

GSLHV Prasada Rao
V. Umamahehare Rao
D.V. Subba Rao

Contents

1

Introduction

G.S.L.H.V. Prasada Rao

Founder Academy of Climate Change Education and Research
Kerala Agricultural University, Vellanikkara, Thrissur-680 656. Kerala, India

Climate Change has emerged as one of the most serious environmental concerns of recent times. The most imminent climatic change is increase in atmospheric temperature due to increased levels of greenhouse gases such as carbon dioxide (CO_2), methane (CH_4), nitrous oxide (N_2O) and hydrochlorofluoro carbons (HCFCs). Because of increasing concentrations of the greenhouse gases in atmosphere, there is much concern about future changes in the climate system viz., Atmosphere, Lithosphere, Biosphere, Hydrosphere and Cryosphere. Direct or indirect effects on the society linked sectors like water, forestry, agriculture, animal agriculture, fisheries, land and marine biodiversity and health due to climate change/variability are many-fold and economic impact of climate change is enormous. The recurrent drought and desertification threaten seriously the livelihood of over 1.2 billion people who depend on land for most of their needs. The global economy has adversely been influenced due to droughts and floods, cold and heat waves, forest fires, landslips and mudslips, icestorms, duststorms, hailstorms, thunder clouds associated with lightning and cloudbursts, unusual rains and the sea level rise (Fig.1.1). Increase in aerosols (atmospheric pollutants) due to emission of greenhouse gases including black carbon and chlorofluorocarbons (CFCs), hydrochlorofluorocarbons (HCFCs, hydrofluorocarbons (HFCs) and perfluorocarbons (PFCs), Ozone depletion and UV-B filtered radiation, eruption of volcanoes, the "human hand" in deforestation in the form of forest fires and loss of wetlands in the process of imbalanced development are causal factors for climate variability and change. The carbon dioxide (CO_2) concentration was in the steady state at 280 ppm till the pre industrial period (1850). It was rising since then at the alarming rate due to industrialization and burning of fossil fuels and the concentration of carbon dioxide is almost currently touching more than 400 ppm. Global increases in CO_2 concentrations are due primarily to fossil fuel use, with land-use change providing another significant but smaller contribution. The loss of forest cover and wetlands, which normally intercept rainfall and allow it to be absorbed by the soil, causes precipitation to reach across the land, eroding top soil, causing floods and droughts. Paradoxically, the lack of trees also exacerbates drought in dry years by attaining soil to dry out more quickly. For the first time the world over felt the effects of global warming and climate change during the year 1998 in the form of weather related disasters.

The year 1998 was one of the warmest and declared as the weather related disaster year, which caused hurricane havoc in Central America and floods in China, India and Bangladesh. Canada and New England in the U.S suffered heavily due to icestorm in January while Turkey, Argentina and Paraguay with floods in June 1998. In contrast, huge crop losses were noticed in Maharashtra (India) due to unseasonal and poor distribution of rainfall during 1997-98. The 1997/1998 El Nino event, the strongest of the last century affected 110 million people and cost the global economy nearly US$ 100 billion. A string of 16 consecutive months saw record high global mean temperature in 1997-98.

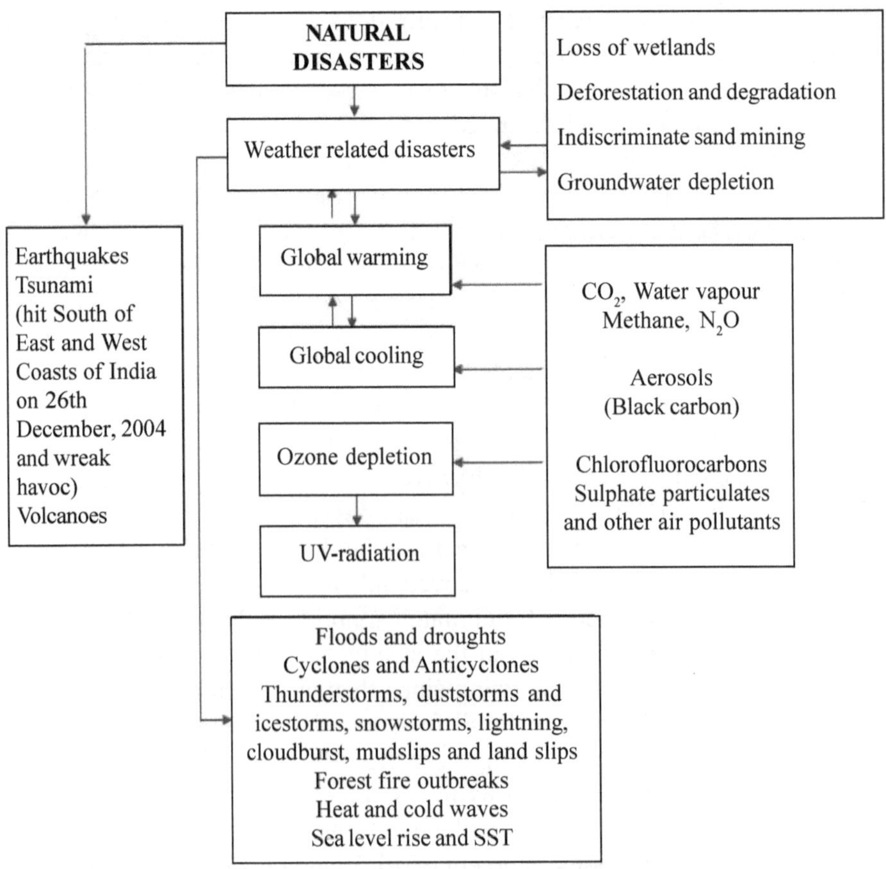

Fig. 1.1: Weather related disasters

Statistics compiled from insurance companies for 1950-1999 showed that major natural catastrophes, which were weather related, caused estimated economic losses of US$960 billion. Most of the losses were recorded in recent years since 1995 onwards as the top fifteen warmest years occurred during the last two decades. The first decade of the current century was the global warmest so far.The European Union experienced such disastrous event in the form of heat wave in 2003. The year 2005 was another historic worst warm year on record for hurricanes. The hurricane Katrina over new Orleans (USA) in August; the hurricane Rita in Texas, Central and Western Cuba and Southern Florida and typhoon over Hainan Province in South China and Vietnam during the last week of September, while early October over Mexican's Gulf coast; heavy downpour over Mumbai on 26th July, 2005 (Single-day the highest record rainfall of 944 mm) and 3rd September, 2005 over Bangalore; severe tropical storms in Andhra Pradesh in September; floods in Kerala, Karnataka, Maharashtra, Gujarat, Orissa and Himachal Pradesh during the Southwest monsoon (June-September), 2005 in India devastated cropped area to a large extent in addition to losses of thousands of human lives. In contrast, it was declared as a famine year in 24 sub-Sahara African countries due to drought and attack of locusts in 2005. Similarly, Australia experienced a severe drought in 2002 and 2006 and heavy crop damage was noticed. Again in 2006, occurrence of droughts and floods devastated rice and other crops in Andhra Pradesh and 40% cereal production was affected in Karnataka due to drought. Similar was the case during monsoon 2007, causing floods across several continents (Hurricane Dean in August slammed into Mexico) including India and Bangladesh. Torrential downpour in June, 2007 over Kerala, Karnataka, Andhra Pradesh and Maharashtra while in July and August over Gujarat, West Bengal, Orissa, Bihar, Uttar Pradesh and Assam, led to floods. Heavy rains again in September in Andhra Pradesh, Karnataka and Kerala led to floods and thus the year 2007 was one of the flood years in India. A huge crop loss was noticed in several states of the Country due to floods in *kharif*, 2007. Food shortage was reported in majority of African countries due to heavy floods, which devastated several crops in the region. Mali, a West African country more often plagued by droughts, received unprecedented rains during 2007. Similar was the case in Algeria, Uganda, Sudan, Ethiopia and Kenya. The year 2010 was the second global warm year after 1998 and the warmest in India. Pakistan was the worst affected due to unprecedented floods in August 2010. Similarly, Jammu and Kashmir (Leh in Ladakh region) in India suffered to a large extent due to floods during the same period. In contrast, Russia suffered due to wild fires and drought conditions in the same year. It led to an estimated loss of cereals up to 30 per cent. However, the European Union experienced one of the coldest winters in 2010.To add to nature's fury, sunburns and related health issues have become a hot topic in Kerala (The Humid Tropics). The sunburns were noticed in March 2010 in Palghat and Thrissur Districts. It also reveals that within the global warming, extremes of cold and heat waves, droughts and floods are experienced in different parts of the World and it is a threat to the world economy. Sea level rise is again a threat to the Coastal Regions and to marine diversity along the East and West Coasts of India. The Super Cyclone that hit Orissa Coast in October 1999, the Tsunami that hit the West and East Coasts, the cyclone "Thane' that hit Tamil Nadu Coast during the North East Monsoon season in 2011 and the disastrous HudHud cyclone in October 2014 over the Coast of Visakhapatnam and other three northern districts (East Godavari, Srikakulam and Vizayanagaram) of Andhra Pradesh are still in our memory when we look back the devastation that took place during the above natural calamities. The unusual rains including hailstorms during March and April in 2015 across the Country damaged crops like mango, paddy and other field crops to a considerable extent. The unprecedented floods across the State of Kerala in August 2018 due to heavy to very heavy monsoon rains against normal devastated in such a way that Red Alert was declared in almost all the districts except Kasaragod and human loss was in few hundreds. With less gravity, Tamil Nadu, Karnataka and Andhra Pradesh suffered due to floods during monsoon season in 2018. Floods were noticed in Maharastra and Himachal Pradesh too during the same season. After Hudhud Cyclone that hit Andhra Pradesh in 2014, the Titli Cyclone in October 2018 devastated Srikakulam District in Andhra Pradesh (Plate 1.1) and southern most districts in Orissa (Plate 1.2). Seasonal and horticultural crops (Rice, maize, banana, papaya, cashew and coconut plantations) were hit very badly and a huge loss (overall loss was estimated to more than Rs. 3000/ - crores) to farmers and rural livelihoods suffered to a considerable extent. Of course, cyclone warning now-a-days minimize human losses to a large extent unlike in the past due to technological interventions, but no way to save crops and infrastructure (Plate 1.3). Untimely snowfall over Jammu & Kashmir during early November 2018 destroyed Apple crop to a large extent (Plate 1.4). The cyclone "Gaja" in mid-November over Tamil Nadu coast and "Phethai Cyclone" in early December over the coast of Andhra Pradesh created nightmare to the farmers. Huge crop loss was noticed in both the states due to cyclones. The first decade (2001-2010) of the current century was the global warmest so far and the second decade (2011-2020) is likely to be the warmest as per the latest indications. It indicates that warming is a threat to the global economy during the ensuing decades. Therefore, in detail studies in this direction are needed at the regional scale as part of climate change adaptation and mitigation strategy in society linked sectors so as to minimize the losses that are likely to take place under the projected climate change scenario in the ensuing decades.

Plate 1.1: Titli cyclone: Damage in coconut gardens in Srikakulam District, A.P October 2018.

Plate 1.2: Titli cyclone: Flood-affected Ganjam district (Odisha) October 2018.

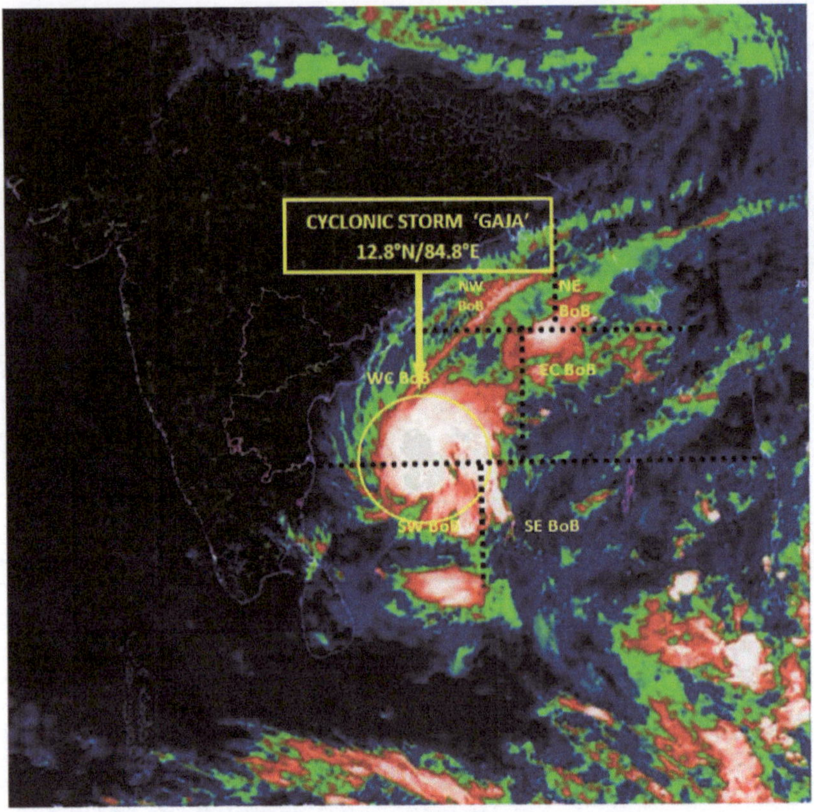

Plate 1.3: Gaja cyclone as on 14-Nov-2018

Plate 1.4: Heavy snowfall in J&K during November, 2018

1.1 Greenhouse effect

Carbon dioxide, water vapour, methane, carbon monoxide, sulphur and nitrous oxides, chlorofluorocarbons and chlorofluoromethanes are the atmospheric constituents of major importance. Incoming shortwave radiation is not absorbed by the above atmospheric constituents and the outgoing longwave radiation from the earth is absorbed by them and reradiate back to earth surface. In the process, earth warms up and its temperature raises. It is also known as the glasshouse effect as property of glass is that it allows solar radiation through it and disallows longwave radiation from earth, thereby warming takes place. The atmospheric constituents which have the property of absorbing longwave radiation and transparent to shortwave radiation are known as the greenhouse gases (GHG). Human activities like fossil fuel combustion, production of synthetic chemicals, biomass burning, deforestation, excess use of chemical fertilizers and pesticides change the chemical composition of atmosphere, thereby enhancing greenhouse effect. The frequent eruption of volcanoes in different parts of the globe in recent times is one of the predominant factors to change the chemical composition of atmosphere on which our knowledge is limited. The present growth rate of GHG is 1% in CH_4, 0.4-0.5% in CO_2 and 0.2-0.3% in N_2O as per the Intergovernmental Panel on Climatic Change (IPCC, 2007). In addition to the above, chlorofluorocarbons and chlorofluoromethanes contribute significantly to greenhouse effect. Is the greenhouse effect mitigated by regional cooling effect from aerosols like black carbon? New research showed that the aerosol effect could dominate the greenhouse effect over the continents and by and large the effect of greenhouse will be slowed down. In contrast, the ozone - friendly substitutes for chlorofluorocarbons (CFCs) like hydrofluorocarbon (HFCs), hydrochlorofluorocarbons (HCFCs) and perfluorocarbons (PFCs) are also powerful greenhouse gases that contribute to global warming. There is also evidence that faster rate of global warming since 1976 is human-induced. The impact of climate change is felt on atmospheric processes viz., pressure gradient between oceans and continents, atmospheric thickness of troposphere, cloud amount and precipitable water. These events reflect on rainfall pattern in terms of shift in rainfall and rainfall decline in recent years.

1.2 Atmospheric carbon dioxide (CO_2)

Carbon dioxide ($CO_{2)}$being an efficient absorber of long wave radiation from earth, it is of great climatic significance. Carbon dioxide has important absorption peaks at about 2, 3 and 4μ and a major absorption peak at about 15μ. This gas emits half of the absorbed heat back to earth, influencing flow of energy between earth and atmosphere. $CO_2$is a very important atmospheric constituent in the heat budget of earth and atmosphere. There is a sharp increase in carbon dioxide in recent decades due to human-made interventions in terms of burning fossil fuels like coal, oil and natural gases. The projected concentration of CO_2 in the atmosphere by 2050 is shown in Table 1.1

Table 1.1: Projected concentration of CO_2 in the Atmosphere

Period	CO_2 *concentration*
Beforepre-industrial revolution	250-290 ppm
1958	315ppm
1984	354ppm
2015	400ppm
2018	410ppm
Projection by 2050	>450ppm (?)

The concentration of carbon dioxide in atmosphere was in a steady state at 280ppm till the pre - industrial period, 1950. It is rising since then at the rate of 1.5 to 1.8 ppm/ year. As per the NOAA-ESRL, the rate of increase in carbon dioxide was 1.9 ppm per year during the decade 1995-2004 and it was 2.1 ppm per year during the decade 2005-2014. The concentration of carbon dioxide in atmosphere increased from 280 before 1950 to 410ppm in 2018 (Fig. 1.2). It is likely to increase further by the end of 21st century if preventive steps are not taken. Measurements at IARI, New Delhi showed that the concentration of carbon dioxide increased from 330 to more than 370ppm till the year 2000, indicating similar trend of global phenomena. Over the same period, atmospheric concentrations of methane and nitrous oxide increased by 151 and 17%, respectively. This resulted in an increase of global temperature by 0.72°C, causing global warming. As a result of warming, the global mean sea level rose by between 10 and 20 cm. The rise is as much as 10 times the average increase in the last 3000 years. An increase of 2.5mm/year is noticed in sea level of the Indian Ocean since 1950s. The trend appears to be higher on the East Coast when compared to that of the West Coast. It is likely that the rate and duration of warming of 20th century were larger than at any other time during the last 1000 years. In temperate soils, unexpected loss of carbon is taking place due to global warming. It leads to more carbon dioxide in the atmosphere, which means even more global warming. The findings showed that the carbon was being lost from soil at an average of 0.6% a year and richer the soils, higher the rate of carbon loss. There was no single factor other than global warming that could explain such losses in non-agricultural soils.

1.3 Global warming

Global warming is the biggest long term threat to life on earth. Rise in temperature may drive thousands of species to extinction, trigger more frequent floods and droughts and sink low lying islands and coastal areas by rising sea levels. It is the result of rising atmospheric content of CO_2 mainly owing to burning of hydrocarbons or fossil fuels like as petrol and diesel. Destruction of forests and their degradation too contribute to rise in carbon dioxide levels. The winter 2007 was the warmest and recorded 0.85°C above average of 12 °C and the previous highest was 0.71°C, which occurred in 2002 in Northern Hemisphere. The entire Europe Union recorded the warm winter, having more than 2°C above average. New York experienced the highest temperature of 21.7 °C on a day in January, 2007 and the second highest was recorded as 17.2 °C in 1950. Floods and excess rains were also noticed due to hurricanes and tropical storms worldwide in 2007. The years 2009 and 2010 were recorded as warm years including the recent warm years 2015, 2016 and 2017. The years 2014 and 2015 were also recorded as consecutive drought years after 1965 and 1966 during the monsoon season (June to September) in India. The IPCC (2006) projected the rate of warming for the 21st century to be between 0.8 and 4.4°C at various stabilized CO_2 levels in atmosphere and it is most likely to be 3°C by end of this century. It could cost global economy almost $7 trillion by 2050, is equivalent to a 20 % fall in growth if no action is taken on greenhouse gas emissions (Fig.1.3).If action is taken, it will cost only $350 billion due to climate change already taken place, just 1% of the global GDP. With all the awareness on global warming across the World and the efforts put by the political governments, the global warming is likely to be less than 2°Cby the end of this century so that less harmful to mankind to live on the Earth.

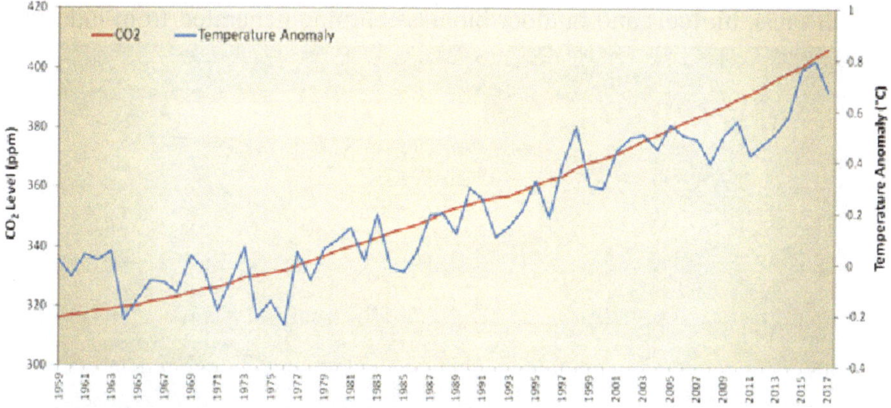

Fig. 1.2: Carbon dioxide versus atmospheric temperature
Source: CO_2 data ; NOAA ESRL (ftp://aftp.cmdl.noaa.gov/products/trends/CO2/CO2_annmean_mlo.txt)

1.4 Aerosols and global cooling

Climate change is usually associated with increasing levels of greenhouse gases, a large amount of soot and other pollutants into the atmosphere in the form of tiny particles, known as aerosols. Aerosols can, however, be produced naturally too. There is great variation in the amount of aerosols around the globe and are unevenly distributed. They change from time to time depending on type of aerosols and quantities in which they are present. Many aerosols are invisible to naked eye and are microscopic. They include sea salts from breaking sea waves, pollen, fine seeds of plants, spores, bacteria and various organisms lifted by wind, smoke and black carbon (soot) from fires, tiny sand particles, volcanic ashes and meteoric dust. Likewise, sulphur-containing compounds known as sulphates arise by natural processes and plants release organic materials. Aerosols in the process of scattering, contribute to varied colours of red and orange at sunset and sunrise. Blue of the sky is also due to selective scattering by microscopic dust particles. From global warming point of view, these tiny soot particles are very important. They absorb a part of incoming shortwave solar energy, thereby heating up the lower atmosphere and totally burning off clouds that might have formed. The combined effect of increasing greenhouse gases and human-induced soot in the warming of lower atmosphere could be as much as 0.25 degrees Celsius/ decade (positive effect). Aerosols are tiny particles, of about one micron, which scatter sunlight back to space and then cause a regional cooling effect (negative effect). Such aerosols are sulphates, soot, organic carbon and mineral dust. Black carbon absorbs solar radiation in visible spectrum and reduces amount of radiation reaching earth surface. Increase in black carbon alters large scale atmospheric circulation and hydrological cycle. Black carbon is a product of incomplete combustion of coal, diesel fuels, biofuels and outdoor biomass burning generated from industrial

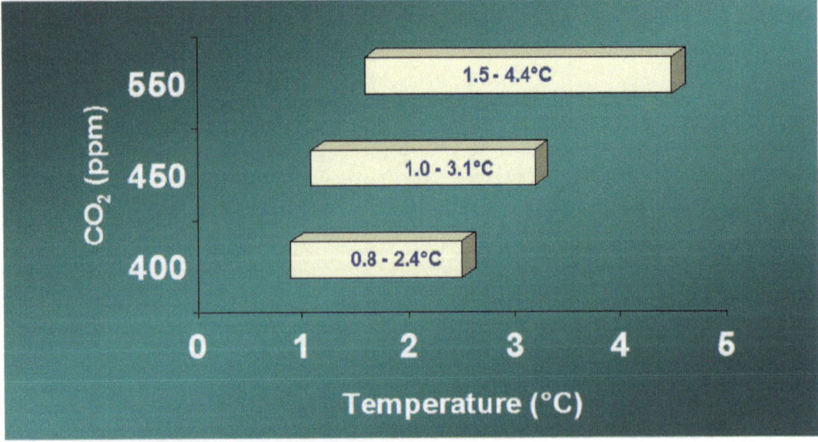

Fig. 1.3: CO_2 versus Temperature rise (IPCC, 2006) by 2100 A.D.

pollution, traffic, outdoor fires and household burning of coal and biomass fuels. The imbalance between positive (warming effect) and negative (cooling effect) aerosol forcing may slow down sometimes the global warming or may lead to global cooling. As the life span of human induced aerosols in the atmosphere is short- lived (one to three weeks) while the long-lived greenhouse gases may persist and thereby accumulate, leading to overall global warming. Many models indicate that global warming would increase rainfall over India while decrease in rainfall was noticed when the effect of aerosols was introduced. The gradual decrease in rainfall over India since last fifty years supports the theory of effect of aerosols on decline in rainfall. If that is the case, frequency of droughts in India could increase in coming decades if pollution continued unabated. The years 2002, 2009, 2012, 2014 and 2015 were the recent monsoon drought years during which Indian foodgrains was adversely affected.

1.5 Clouds and global warming/cooling

Hygroscopic aerosols act as nuclei for cloud formation. Clouds would not form if there are no fine particles in the atmosphere around which water vapour could condense into droplets. Brown clouds contain a variety of aerosols, including sulphates from coal combustion, nitrates produced by vehicular emissions, and soot resulting from fossil fuel combustion and the burning of biomasses. They enhance warming quite a bit in lower atmosphere. Clouds play a dual role in cooling and warming of earth. If the sky is totally covered with clouds, clouds act as an active surface instead of earth surface. The incoming solar radiation is reflected back to a large extent from top of clouds and thus low solar radiation reaches earth surface. This keeps the planet cool. On the other hand, they absorb long wave radiation and emit back in all directions and transmit little long wave radiation to space. In the process, atmosphere traps huge amounts of heat emitted from earth's surface and clouds, leading to green house effect. Clouds are known as warm clouds in the tropics while cold clouds in temperate zones. Clouds over the oceans differ in character and heating effects to that of clouds over the continents. Their nature and effects also vary with season and time of day. Tropical clouds trap three times as much heat as clouds trap on a global average, but also reflect so much sunlight that the heating effect is cancelled out. Since the tropics account for 20% of earth's surface and effects of tropical cumulonimbus clouds are disproportionate, the delicate heat balance they maintain is critical. A warmer ocean would likely to cause more clouds to form, increasing both their heating and cooling effect. Analysis of satellite data also showed that the clouds like stratocumulus over temperate regions reflect enough sunlight to exert a net cooling effect locally, and this appears to account for clouds overall cooling effect worldwide. The behaviour of disproportionate cooling and heating may change itself under global warming, perhaps. If it is understood, we, climatologists may be able to forecast the timing and extent of expected global warming with far more confidence than today. Cloud bursts have become more frequent with prolonged dry spells, leading to floods and droughts simultaneously.

1.6 Change in atmospheric processes

A significant overall decline in monsoon rainfall was noticed during the period of post-climate shift from 1979 to 2014 across the Country. The zonal trends in all India rainfall also indicated that the three zones viz., the East Coast (N), the East Coast (S) and the North East India (NEI) showed increase in rainfall while decline in all the remaining zones viz., Central Peninsular India, Ganga Plains, North Central India, North Mountainous India, North West India, South Peninsular India, West Coast (S), and West Coast. The Mean Sea Level Pressure has increased by 1 to 7 hPa over continental Asia like northern parts of India, Tibetan plateau, China, and Mangolia while decreasing over the South Temperate Regions by 1 to 3 hPa and over South Polar Region by 4 to 7 hPa during last 30-40 years. However, the mean sea level pressure has been lowered by 1 to 5 hPa over the continental Asia during the good monsoon years. In contrast, the composite anomaly plot of bad monsoon years showed a higher mean sea level pressure of about 1 to 3 hPa over the continental Asia during the post climatic shift. Increase in the global tropospheric temperature was 0.59 °C while 0.87 and 0.32 °C over the Southern and Northern Hemisphere, respectively during the above period. A similar pattern was observed in the upper tropospheric temperature as well as the lower tropospheric temperature. Interestingly, the troposphere was shrunk by 60m over the continental Asia and inflated by 90m over the oceanic south temperate region. The global sea surface temperature during the post climatic shift was warmer except some isolated cooling over the Northern Central Pacific, where a decrease of 0.3°C was noticed. The SST showed an increase of 0.4 to 0.6°C over the Equatorial, Southern Indian Ocean and Eastern and Equatorial Pacific. A decline in precipitable water during the summer monsoon was also noticed across the Country with cloud cover. Decline in rainfall over India could be explained due to changes that took place in land, ocean and atmosphere continuum such as the tropospheric and surface temperatures, mean sea level pressure and sea surface temperature, geopotential height, precipitable water and cloud amount in addition to deforestation.

1.7 Ozone depletion and UV radiation

The greatest concentration of ozone is at an average in height of 25 km and above in the stratosphere. The ozone molecule is made up of three atoms of oxygen. It is the most efficient absorber of ultraviolet radiation from sun and thus protects all life forms in planet earth. Ozone depletion due to industrialization in recent decades is the concern of humankind and biological activities. Release of compounds like chlorofluorocarbons (CFCs), carbon tetrachloride and methyl chloroform could significantly deplete ozone layer that shielded the planet from ultraviolet radiation. The CFCs are used in a variety of industrial, commercial, and household applications. These substances are non-toxic, non-flammable and non-reactive. They are used as coolants in commercial and home refrigeration units, aerosol profellants and electronic cleaning solvents.The global average thickness of ozone is 300 Dibson units, equivalent to 3mm. In contrast, it is about 100 Dibson units, which alarm us that we are in great danger if precautionary steps are not taken up. The ozone level fell to 90 Dibson Units, which was the lowest value recorded on 30-09-98 nearly equalled to the lowest value ever recorded 88 Dobson Units on 28-09-94 over Antarctica. The ozone losses are caused by chlorine and bromine compounds released by chlorofluorocarbons and halons. Year-to-year variation of size and depth of ozone hole depend on variations in meteorological conditions. The unusual cold temperature by 5-9°F over Antarctic zone enables greater activation of reactive chlorine that ultimately causes more ozone loss and lower ozone levels. Increase in chlorine levels should peak in the Antarctic stratosphere within a few years (Shashi et al., 2002). Among UV radiations, UV-B radiation in the range of 0.280-0.32μ is more sensitive to ozone fluctuations and reaches earth surface. A global network is created to monitor UV-B filtered radiation in terms of Minimum Erythema Dose (MED). The human-made interventions in industrial development lead to ozone depletion, thereby filtered UV radiation reaches the ground, resulting in various human, animal and crop diseases. The UV-B radiation profile recorded in this part of India indicated that it reaches to the ground beyond tolerable limits irrespective of the season between 10/11 a.m to 2/3 p.m (Fig. 1.4). The ozone loss has the potential to increase incidence of skin cancer, cataracts and damage to people's immune system, harm some crops and interfere with marine life. However, little is known on impact of ozone depletion and increasing UV-B radiation on ontogeny of tropical plants and human and animal diseases since studies in this direction are lacking. Because CFCs remain in the atmosphere for100 years, continued accumulation of these chemicals pose ongoing threats, even after their use is discontinued. However, recent NASA reports indicate that the ozone over Antarctica is recovering since 2000 onwards.

1.8 Climate change awareness

Models on global warming indicate that rise in temperature is likely to be around 2°C by end of this century. Of course, it depends upon the adaptation and mitigation measures taken up by various political governments across the World. The projected climate change scenario indicates that the frequency of weather related abnormalities like floods, droughts, heat and cold waves is likely to increase in ensuing years and food insecurity is likely to persist if corrective proactive measures are not taken up. If the sea level increase as projected, the coastal areas which are thickly populated will be in peril and for the existing population, the safe drinking water will be a great problem. The whole climate change is associated with increasing greenhouse gases and human- induced aerosols and the imbalance between them may lead to uncertainty even in year-to-year monsoon behaviour over India.Therefore, short term proactive measures play a major role along with long term strategies to tackle the issues of foodgrains production, food security, food safety in terms of quality and escalation of food price under projected climate change scenarios. There should be a determined effort from developed and developing countries to make industrialisation environment-friendly by reducing greenhouse gases pumping into the atmosphere so as to slow down the rate of increase in global warming.

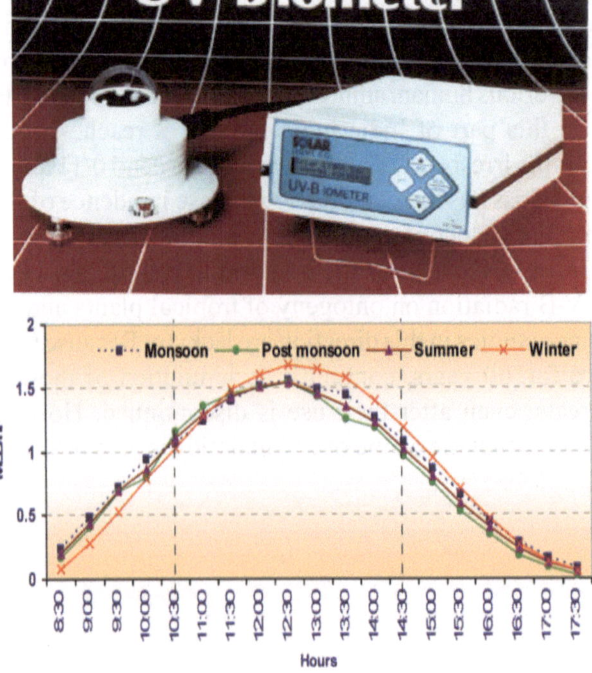

Fig. 1.4: UV-B Radiation at Vellanikkara (Thrissur)

Awareness programmes on climate change and its effects in various sectors viz., food security and safety, health, infrastructure, water, forestry, land and ocean biodiversity and sea level and the role played by human interventions in climate change need to be taken up on priority. From the agriculture point of view, effects of extreme weather events on crops are to be documented so that it will be handy to planners in such reoccurrence events for mitigating the ill effects to certain extent. Also, there is need to guide planners on projected future crop scenarios based on climate change events, which will be more realistic at field level as models always overestimate the impacts. Finally, we have to foresee these extreme events and prepare ahead to combat them so that the losses can be minimized (Plates 1.5 and 1.6). Lifestyles of people should also be changed so as not to harm the earth-atmosphere continuum by pumping greenhouse gases and CFCs into the atmosphere. Similarly attempts are to be made to forewarn local weather systems and weather extremes so as to minimize the human and crop losses. In this context, the agro-advisory based on medium range weather forecasting, which is now in operation across the Country, is of immense use to the farming community to minimize crop losses against weather abnormalities and climate change (*refer chapters 26 and 27*). In addition, weather insurance package to the farmers against weather related disasters should be made compulsory and operational in an event of their occurrence. It will help them to maintain their livelihood in an event of weather extremes those who depend solely on the income of Agriculture.

Plate 1.5: Change in cropping pattern in Punjab

Plate 1.6: Drought management practices in North Karnataka

References/Suggested Readings

Attri, S. D and AjitTyagi. 2010. Climate profile of India. Met Monograph No. Environment Meteorology-01/2010. India Meteorological Department. 122p.

IPCC (Intergovernmental Panel on Climatic Change) 2006. The Economics of Climate Change: Stern Review. The Summary of conclusions. Survey of the Environment 2007, The Hindu, pp141-145.

IPCC (Intergovernmental Panel on Climatic Change) 2007. Climate Change: The Physical Science Basis. Extracts from the IV Assessment Report. Survey of the Environment 2007, The Hindu, pp147-155.

Martinez, R., Hemming, D., Malone, L., Bermudez, N., Cockfield, G., Mushtaq, S., Diongue, A. Hansen, J., Hildebrand, A., Ingram, K., Jakeman, G., Kadi, M., McGregor, G.R. Rao, P., Pulwarty, R., Ndiaye, O., Srinivasan, G., Eh.R., Seck, White, NandZougmore, R. 2012. Improving Climate Risk Management at Local Level – Techniques, Case Studies, Good Practices and Guidelines for World Meteorological Organization Members-Chapter 21 (477-532): In Risk Management-Current Issues and Challenges (Ed: NerijaBanaitiena). In Tech Publishers. Rijeka, Croatia,584p

Ramakrishna, Y.S., Rao, G.G.S.N., Rao, S.G. and Vijayakumar, P.2006. Impact of climate change in Agriculture. In: Environment and Agriculture (eds. Chadha, K.L. and Swaminathan, M.S.). Malhotra Publishing House, New Delhi, pp. 1-30

Rao, G.S.L.H.V., Rao, G.G.S.N. and Rao, V.U.M. 2010. Climate Change and Agriculture over India. PHI Learning Pvt. Ltd., New Delhi. 328p

Rao, G.S.L.H.V.P. 2011. Climate Change Adaptation Strategies in Agriculture and Allied Sectors. Scientific Publishers (India) Ltd. Jodhpur, India. 336p

Rao, G.S.L.H.V.P. and Gopakumar, C.S. 2016. Climate Change and Plantations in the Humid Tropics. New India Publishing Agency, New Delhi. 461p

Rathore, L.S. and Parvinder Maine. 2008. Economic Impact Assessment of Agrometeorological Advisory Service of NCMRWF. Report NO. NMRF/PR/1/2008. NCMRWF, MoES, Govt. of India, New Delhi.160p

Rathore, L. S., Attri, S.D, and Jaswal, A. K.2013. State level climate change trends in India. Meteorological Monograph No. ESSO/IMD/EMRC/02/2013. India Meteorological Department. 147p.

Shukla, P.R., Sharma, S.K and Ramana, V.P. 2002. Climate Change and India - Issues, Concerns and Opportunities. Tata – McGraw-Hill Publishing Company Ltd, New Delhi, 314p

2

Rainfall Variability Across India

H. N. Singh

Indian Institute of Tropical Meteorology, Pune 411008, India

Natural disasters are of meteorological or hydrometeorological origin. Disasters such as earthquakes and tsunamis are occasional, but rainfall-related hazards, flood or drought, are very frequent and thus call for proper monitoring of large-scale rainfall variation, and better manage- ment and planning to combat the problem(s). Variation in rainfall is very large spatially and temporally compared to any other atmospheric parameter, such as pressure or temperature. The large temporal variability in the Indian rainfall has produced frequent flood/ drought and periods of above/below normal rainfall for several centuries past. Areas of high rainfall and runoff are flooded in the summer monsoon season whereas others face water stress and drought conditions. The influence of global warming on rainfall variations over India is reported in numerous observational (Singh *et al*., 1991, 1992) and modelling studies (Lal, 1994). Though considerable rainwater enters the Indian territory from the Himalayan area, the major hydrological activities and water resources of the country are due to *in situ* rainfall occurrences, mostly summer monsoon rainfall during June–September. During a normal hydrological wet season the country receives 2911.064 bcm (billion cubic metres) rainwater of which 1856.947 bcm evaporates and 1054.117 bcm enters into surface hydrological processes (surface storage, streamflow/riverflow and deep aquifer storage) (Ranade *et al*., 2008). Rainfall events display large spatial variability due to variation in intensity, frequency and geographical location of the rain-producing weather systems, such as circular systems (low, depression, tropical cyclone and mid-tropospheric cyclone), linear systems (trough, meander, squall line and convergence zone) and wave disturbances (easterly and westerly waves). Rainfall in other seasons is equally important to monitor, though restricted to some parts of the country, such as winter rainfall over the extreme north, summer thunder showers over the southwest peninsula and northeast India, and post-monsoon rainfall over the south peninsula.

Indian communities have for long faced rainfall-related disasters and have a history of traditional mechanisms of coping with such calamities by the way of irrigation, bunds, canals and several other techniques during the excess rainfall. Prabhakar & Shaw (2008) have considered the available evidence for climate change over the Indian subcontinent and assessed the existing preparedness and mitigation mechanisms for drought risk reduction in the country. Revi (2008) has considered the likely changes that climate change will bring in temperature, precipitation and extreme rainfall, drought, river and inland flooding, storms/ storm surges/coastal flooding, sea- level rise, environmental health risks and the necessary adaptation and mitigation agenda for cities in India where the urban population is rapidly growing. Real-time monitoring of the large-scale rainfall variations can provide vital information in this endeavour. Collection, compilation, analysis of long-period rainfall data and display in the form of maps and graphs on smaller spatial scales showing rainfall variability are required to decide adaptation and mitigation plans and policies. The objectives of the present study are:

- to document climatological and fluctuation features of different moisture regions and dry/wet areas during different seasons and months over the country;
- to report recent tendencies of rainfall spatio-temporal fluctuations; and
- to understand possible causes of recent trends in large-scale rainfall fluctuations.

To understand spatial variability of rainfall fields over India, sequential thematic maps have been prepared in a GIS environment (GeoMedia Professional 5.1) using data from 316 raingauge stations for the period 1871–2006. The thematic maps show interannual expansion/contraction of moisture regions (arid, semi-arid, dry subhumid, moist subhumid, humid and perhumid), and very dry, dry, wet and very wet zones of the four seasons and the twelve calendar months. The annual moisture regions and seasonal as well as monthly wet and dry zones exhibit large interannual variation but in an organized manner. The arid area shows a dry period during 1999–2004, semi-arid during 1964–2004, moist subhumid during 1999–2004, humid during 1998–2004 and perhumid 1995–2006, while the dry subhumid shows a wet period during 1984–2006. Overall during recent years (1995–2006) the climate of the country has been drier. To describe spatio-temporal features of the seasonal and monthly rainfall an integrated Rainfall Spatial Distribution Index (RSDI) has been developed by lumping the areas under very dry, dry, wet and very wet conditions. In winter rainfall, a dry period is seen during 1987–2006 and in the summer monsoon rainfall during 1972–2006, while a wet period is seen in summer rainfall during 1977–2006 and in post monsoon rainfall during1994–1999. In recent years, declining trends are seen in eight months (January, February, March, July, August, September, November and December) and increasing trends in the other four months.

2.1 Rainfall Variations

The spatio-temporal variability of annual, seasonal and monthly rainfall over India is studied using highly quality-controlled data from 316 widely distributed locations (Fig. 2.1) for the period 1871–2006. Rainfall data up to 1900 were obtained from Eliot (1902) and for the period 1901–2006 from the records of the India Meteorological Department (IMD), Pune. A digitized map of India on 1:6 million scale with the projection system of "Albers Equal Area Projection with Two Standard Parallels, 15°N and 30°N" was utilized in a GIS environment (GeoMedia Professional 5.1) for the spatial analysis of the rainfall data. To understand spatial variation of annual rainfall, expansion/contraction of different moisture regions, viz. arid (rainfall d" 560 mm), semi-arid (561–1040 mm), dry subhumid (1041–1420 mm), moist subhumid (1421–1630 mm), humid (1631–2450 mm) and perhumid (>2450 mm), are examined. This climatic classification of the different moisture regions is based on the average annual precipitation as the climatic conditions of any place in the country can be assessed with the average annual precipitation over it (Singh, 1984). Thematic maps of the country showing the normal (1901–2000) position of the moisture regions (Fig. 2.1) as well as their yearly positions (figures not shown) have been prepared using the GeoMedia GIS. The different moisture regions of the country show large variations from one year to another. On the normal annual isohyetal map the arid region occupies over 15.3% of the area of the country, semi-arid 33.8%, dry subhumid 24.3%, moist subhumid 8.9%, humid 12.4%, and perhumid 5.3%. The standardized value (actual minus mean divided by standard deviation) of the area under different moisture conditions along with its 11-year moving average are plotted in Fig. 2.2. The sequence of dry and wet periods in the fluctuations of the moisture regions is also given in Table 2.1. In the low-frequency mode fluctuations (11-year running means), notable features of the moi.sture regions are: spreading tendency in the area under semi-arid, dry subhumid and moist subhumid conditions from the early 1940s, and a contracting tendency in arid, humid and perhumid regions. Overall the climatic condition of the country shows a drier tendency in recent years/decades.

Table 2.1: Dry and wet periods in the fluctuations of the six moisture regions over India

Moisture regions	Dry periods	Wet periods
Arid	1896–1925, 1934–42, 1965–74, 1984–87, 1999–2004	1871–95, 1926–33, 1943–64, 1975–83, 1988–98
Semi-arid	1900–12, 1950–57, 1964–2004	1871–99, 1913–49, 1958–63
Dry subhumid	1914–17, 1948–83	1871–1913, 1918–47, 1984–2006
Moist subhumid	1871–83, 1952–89, 1999–2004	1884–1951, 1990–98
Humid	1899–1913, 1949–53, 1964–89, 1998–2004	1871–98, 1914–48, 1954–63, 1990–97
Perhumid	1898–1910, 1935–42, 1960–87, 1995–2006	1871–97, 1911–34, 1943–59, 1988–94

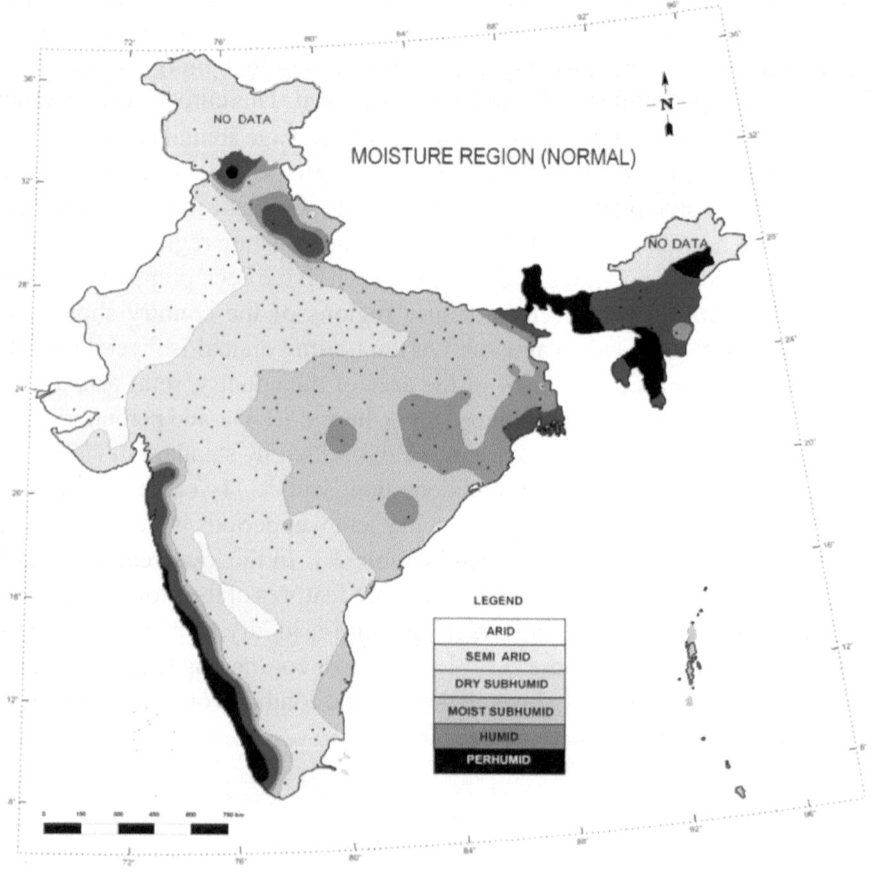

Fig. 2.1: Map of the country showing climatic moisture regions. Dots indicate the location of 316 rain- gauge stations used.

Fluctuations of Moisture Regions Over India: 1871-2006.

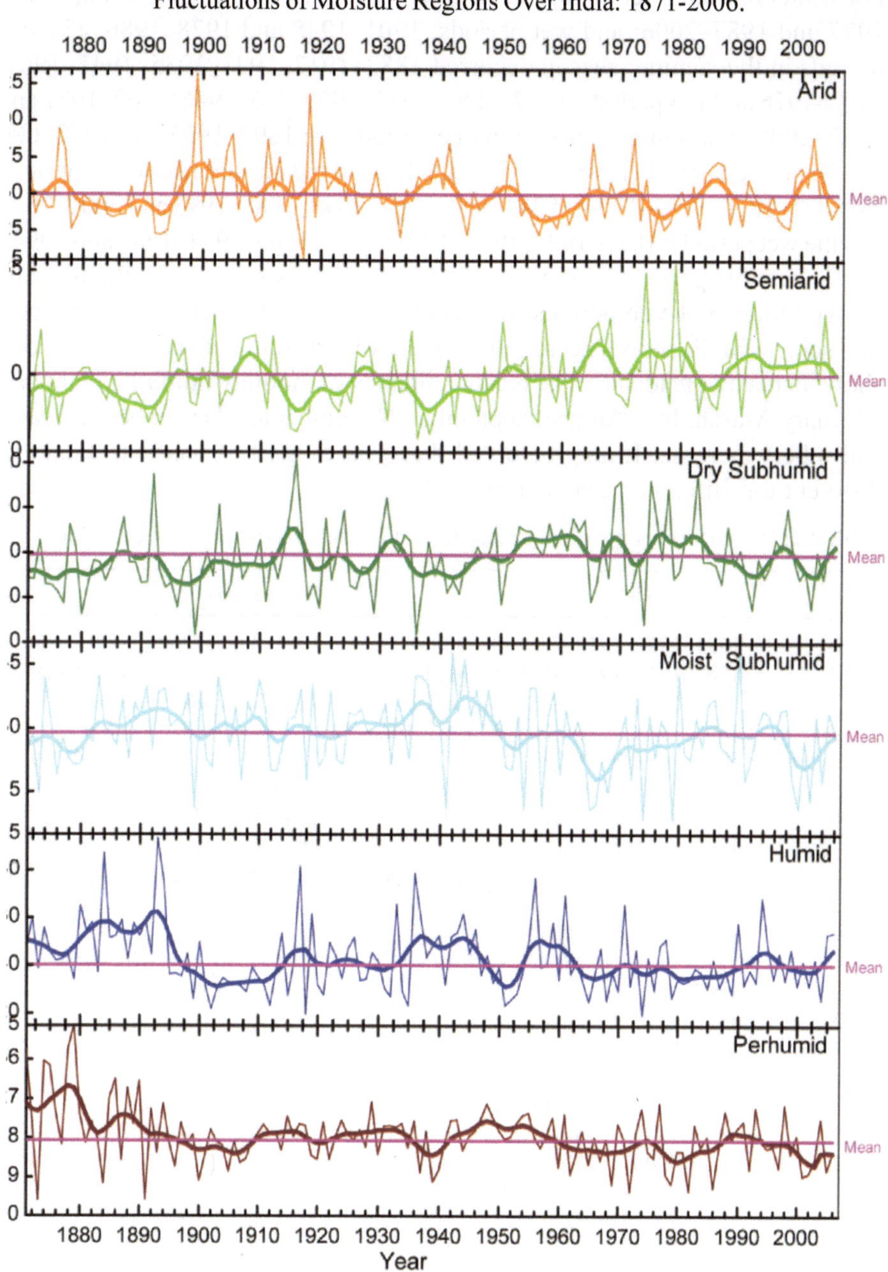

Fig. 2.2: Time series plot of the inter-annual variation of the area under six moisture conditions over the country during 1871–2006

For winter rainfall fluctuations the distinctly dry periods are 1871–1900, 1949–1977 and 1987–2006; and wet periods: 1901–1948 and 1978–1986. The dry periods in the summer season occurred 1882–1912, 1921–1935, 1947–1961, 1972–1976 and wet periods: 1877–1881, 1913–1920, 936–1946, 1962–1971 and 1977–2006. The summer monsoon dry periods are 1899–1925 and 1972–006 and wet periods 1871–1998 and 1926–1971. And the post monsoon dry periods are 1871–1876, 1895–1926, 1938–1943, 1964–1973, 1988–1993 and 2000–2006 and the wet periods: 1877–1994, 1927–1937, 1944–1963, 1974–1987 and 1994–1999. Over the period of 136 years (1871–2006) seasonal rainfall fluctuations showed four to eleven periods. In recent years/ decades, winter and summer monsoon rainfalls show a decreasing tendency while summer and post monsoon show increasing tendency. The recent tendency of monthly rainfall is: January, February, March, July, August, September, November and December decrease, and others increase. It may be noted that year of start of recent tendency is different for different months (Table 2.2).

Table 2.2: Dry and wet periods in the seasonal and monthly rainfall spatial fluctuations

Period	Dry periods	Wet periods
WIN	1871–1900, 1949–77, 1987–2006	1901–48, 1978–86
SUM	1882–1912, 1921–35, 1947–61, 1972–76	1877–81,1913–20,1936–46,1962–71,1977–2006
MON	1899–1925, 1972–2006 1871–98, 1926–71	
POST MON	1871–76, 1895–1926, 1938–43, 1964–73, 1988–93, 2000–06	1877–94, 1927–37, 1944–63, 1974–87, 1994–99
JAN	1871–86, 1896–1918, 1930–40, 1962–79, 1988–2003	1887–95, 1919–29, 1941–61, 1980–87
FEB	1871–92, 1909–26, 1949–78,2001–06	1893–1908, 1927–48, 1979–2000
MAR	1871–1903, 1921–35, 1971–78, 1983–87, 1996–2004	
APR	1871–94, 1916–24, 1947–58, 1966–75, 1987– 95	1895–1915, 1925–46, 1959–65, 1976–88, 1996–2006
MAY	1890–1912, 1921–68	1873–89, 1913–20, 1969–2006
JUN	1900–08, 1923–32, 1943–74	1871–99, 1909–22, 1933–42, 1975–2002
JUL	1895–1918, 1970–2004	1871–94, 1919–69
AUG	1898–1937, 1979–2005	1878–97, 1938–78
SEP	1899–1909, 1927–31, 1965–2004	1871–98, 1910–26, 1932–64
OCT	1871–76, 1895–1927, 1940–54, 1964–70, 1976–82	1977–94, 1928–39, 1955–63, 1971–75, 1983–2005
NOV	1871–79, 1895–1909, 1937–71, 1983–2005	1880–94, 1910–36, 1972–82
DEC	1871–79, 1891–1904, 1910–22, 1939–60, 1968–76, 1992–2006	1880–90, 1905–09, 1923–38, 1961–67, 1977–91

2.2 Rainfall Spatial Distribution Index (RSDI)

Spatial variation of the seasonal and monthly rainfall has been studied by examining interannual variation of an optimum number of four rainfall zones under different rainfall conditions, viz. very dry (VD), dry (D), wet (W) and very wet (VW), identified on the mean rainfall (1901–2000) chart of the particular period (season/month) of equal size (25%) area of the country. The criterion has been applied to yearly rainfall spatial distribution, and the area of the country under different rainfall conditions for each season and month has been computed. Based on these threshold values for the respective seasons and months, the thematic maps have been prepared showing very dry, dry, wet and very wet areas in each year from 1871 to 2006 for each season and month. In order to understand the rainfall pattern both in terms of amount and areal distribution across the country, a Rainfall Spatial Distribution Index (RSDI) was developed (Singh *et al*., 2005) by lumping VD, D, W and VW areas. A high RSDI value is indicative of enhanced rainfall over large areas, including dry areas, and *vice versa*. For a particular year, the RSDI is calculated as:

$$RSDI = \frac{1' \ \alpha_{VD} + 2'\alpha_{D} + 3' \ \alpha_{W} + 4' \ \alpha_{VW}}{A}$$

where *aVD* denotes area of the country under very dry condition, *aD* under dry, *aW* under wet and *aVW* under very wet condition, and *A* is the total geographical area of the country.

2.3 Possible reason for recent changes in monsoon rainfall

The gradient in the geopotential height (GPH) of the upper tropospheric isobaric levels from the Tibetan Anticyclone (TA) region to the Southern Tropical Indian Ocean High (STIOH) region is useful as an index of intensity of the Indian monsoon circulation (Singh *et al*., 2006; Sontakke *et al*., 2008). Besides STIOH, the North Pacific High (NPH), the Azores High (AH) and the North Sub Polar Low (NSPL) have been identified and gradients between TA and NPH, TA and AH and TA and NSPL have also been found useful as indexes of intensity of the Indian monsoon circulation. The combined area of the Iranian High (32.5°N–40.0°N; 50.0E–72.5°E) and the West Tibet (27.5°N–37.5°N; 72.5E–85.0°E) is considered as the TA region, the area between parallels 40.0°E and 110°E and meridians 5°S and 20°S as the STIOH region, the area between parallels135°E and 120°W and meridians 20.0°N and 45.0°N as the NPH region, the area between parallels 60°W and 10°W and meridians 20.0°N and 50.0°N as the AH region, and the area between parallels 40°E and 110°E and meridians 55.0°N and 75.0°N as the NSPL region.

The interannual variation (1949–2006) of the geopotential height gradient (GPHG) at 300 hPa (NCEP/NCAR reanalysis data, Kalnay *et al.*, 1996) from TA to STIOH region for the monsoon season and each of the monsoon months has been examined. Interesting to notice is that the GPHG and the rainfall over northern India (north of 18°N) are in excellent agreement during extreme years and in the latest period from 2001, which suggests the possibility of exploring cause and effect relationships between them. The two parameters GPH and rainfall for June show a rising tendency from the late 1950s, but for other three monsoon months and the season they show declining tendency. The June GPH of 300 hPa is rising over the TA region at the rate of 5.96 m/decade (significant at 1% level) and over the STIOH region at the rate of 4.93 m/decade (significant at 0.1% level). Therefore the GPHG shows a mild increasing tendency which is producing a slight increase in June rainfall over the country. During July, August, September (200 hPa level) and the whole season the GPH over the STIOH region is rising at the rate of 5.03 m, 5.47 m, 4.59 m and 5.0 m per decade, respectively, all significant at 0.1% level, but over the TA region the rising rate is 3.84 m/decade (significant at 10% level), 5.03 m/decade (significant at 1% level), 1.27 m/decade (not significant) and 4.03 m/decade (significant at 5% level), respectively. The rising tendency in the 200 hPa GPH is at a faster rate over the STIOH region compared to the TA region, therefore, the GPHG from the TA region to STIOH region shows a slight decreasing tendency (statistically not significant). The lower GPH in the upper tropospheric level contributes partially to reduction in rainfall over the country. The relationship between GPHG at NPH, AH and NSPL with TA and rainfall over the country have been examined for June, July, August, September and the whole summer monsoon season (figures not shown). In brief, the GPHG from the TA region to the STIOH region showed highest correlation with the corresponding rainfall over the Indian region, that from TA to NPH region showed correlation almost of the same order, but from TA to AH region and from TA to NSPL region showed weaker correlation of the same order.

2.4 Mechanism

During heightened summer heating of the Afro-Eurasian dry province (Arabia to Mongolia) and the land area over and around the Tibet-Himalayan highlands, the Tibetan Anticyclone intensifies. A major portion of the outflows from the Tibetan Anticyclone sinks over the North Pacific and intensifies the high over there. The intense North Pacific High produces strong easterlies over the central and western North Pacific. Part of these easterlies turns towards China, Korea and Japan, and part enters into Bay of Bengal after crossing the Vietnam-Laos-Thailand-Cambodia-Mynamar sector (Southeast Asia) where it merges with the main southwest monsoon current and the combined wind blows over the Indo-Gangetic Plains of northern India as an easterly wind parallel to the southern flank of the Himalayas. Part of this easterly wind converges over central India and the rising air flows towards the southern hemisphere via the upper troposphere. Also, the large outflows from the Tibet-Pacific Anticyclone in the upper troposphere, after crossing the equator above the western Indian Ocean, sink over southern Indian Ocean High. Part of this upper tropospheric cross-equatorial flow flows from the northern to southern tropical/subtropical Indian Ocean and while circulating anticyclonically in the lower troposphere it crosses the equator along the Somali coast and becomes southwest monsoon flow, and part merges with the southern temperate westerlies. The upper tropospheric GPHG from the Tibetan anticyclone region to the NSPL region provides a major intensification of mid-latitude upper tropospheric westerlies. While steep gradients suggest strong poleward-shifted westerlies and a well developed and intense Tibetan Anticyclone, the reverse is true during shallow gradient. In the upper troposphere, the Azores High appears as a westward extension of the Tibetan Anticyclone. Examination of geopotential height and temperature fluctuation of standard isobaric levels over the five centres of circulation revealed the following notable features:

(i) During June, over TA region the GPH at each standard isobaric level from 925 to 200 hPa is rising (non-linearly) at a faster rate compared to temperature, while over other centres of circulation positive trends in GPH and temperature fluctuations closely parallel each other;

(ii) During the other three monsoon months, over TA region the temperature showed a mild rising tendency in the lower troposphere and mild decreasing trend in the upper troposphere, but the GPH of the different levels showed rising tendency; and

(iii) Over the surrounding four centres of circulation, rising trends in the GPH and temperature closely paralleled each other from the surface to the tropopause.

From time series plots of low-frequency mode of fluctuations (9-point Gaussian low-pass filtered values) of the GPH and temperature (not shown) it is clear that within deep highs over oceanic regions, e.g. the STIOH, the NPH and the AH, both GPH and temperature showed rising tendencies at different levels from 925 to 200 hPa. In the TA region, however, the temperature showed a rising trend in the lower troposphere and falling trend in the upper troposphere. Thus the rising trend in the GPH of different levels in the TA region is essentially due to rising trend in the temperature in the lower tropospheric layers. Investigation reveals that the rising tendency in post monsoon rainfall over South Peninsular India is due to rising trends in the sea surface temperatures (SST) of the Arabian Sea and Bay of Bengal (Sontakke *et al.*, 2008). Possible causes of change in the winter and summer rainfall over the country are being investigated.

Summary

1. The arid and semi-arid areas are expanding whereas the moist subhumid, humid and perhumid areas are shrinking. The semi-arid area, which occupies the large and central portion of the country, is expanding significantly. The overall effect is that the country is experiencing dry climate in the recent period with a westward shifting of rainfall.
2. Analysis of the periodic pattern suggests that in the recent years/decades winter and summer monsoon rainfall show a decreasing tendency while summer and post-monsoon shows an increasing tendency. January, February, March, July, August, September, November and December show a decreasing trend in recent years/decades while other months show increasing tendency.
3. Gradient in the GPH of upper tropospheric isobaric levels from the Tibetan Anticyclone (TA) region to southern tropical Indian Ocean High (STIOH) region, to North pacific High (NPH) region, to Azores High (AH), and to North Sub Polar Low (NSPL) region, can be used as index of intensity of the Asian Summer Monsoon Circulation (ASMC).
4. The rising trend in post-monsoon rainfall over South Peninsular India is due to the rising trend in the SST of the oceans around the area.
5. A declining tendency in monsoon rainfall can be seen from 1965 onwards and a sharp declining trend from 1995. This is essentially due to weakening of the Tibetan Anticyclone associated with cooling of the upper troposphere (600–150 hPa) over the Himalayan region and surrounding atmosphere. The recent sharp increasing trend in the surface air temperature over India is due to a declining trend in the rainfall of winter and summer monsoon seasons (D. R. Kothawale, pers. comm.).

References/Suggested Readings

Eliot, J. (1902) Occasional discussions and compilations of Meteorological data: India and the Neighboring Countries. Indian Meteorological Memoirs vol. XIV. Published by Order of His Excellency the Viceroy and Governor General of India in Council, Calcutta, India.

Kalnay, E., Kanamitsu, M., Kistler, R., Collins, W., Deaven, D., Gandin, L., Iredell, M., Saha, S., White, G., Woollen, J., Zhu, Y., Chelliah, M., Eblsuzaki, W., Higgins, W., Janowiak, J., Mo, K.C., Ropelewski, C., Wang, J., Leetmaa, A., Reynolds, R., Jenne, R. & Joseph, D. (1996) The NCEP/NCAR 40-Year Reanalysis Project. Bull. Am. Met. Soc. 77, 437–471.

Lal, M. (1994) Water resources of the south-east Asian region in a warmer atmosphere. Adv. Atmos. Sci. 11, 239–246.

NATMO (1986) National Atlas of India Physiographic Regions of India, third edn, Plate 41. Prepared under the Direction of G. K. Dutt, Director NATMO (the National Atlas & Thematic Mapping Organization), Kolkata, India.

Prabhakar, S. V. R. K. & Shaw R. (2008) Climate change adaptation implications for drought risk mitigation: A perspective for India. Climatic Change **88**, 113–130.

Ranade A. A., Singh, N., Singh, H. N. & Sontakke, N. A. (2008) On variability hydrological wet season, seasonal rainfall and rainwater potential of the river basins of India (1813–2006). J. Hydrological Research and Development, 23, 79–108.

Revi, A. (2008) Climate change risk: An adaptation and mitigation agenda for Indian cities. Environment and Urbanization, 20(1), 207–229.

Singh, N. (1984) Fluctuations of different moisture regimes in India. Arch. Met. Geophys. Biocl. B35, 239–256.

Singh, N. & Mulye, S. S. (1991) On the relations of the rainfall variability and distribution with the mean rainfall over India. Theor. Appl. Climatol. 44(3-4), 209–221.

Singh, N., Pant, G. B. & Mulye, S. S. (1991) Distribution and long term features of the spatial variations of the moisture regions over India. Int. J. Climatol. 11, 413–427.

Singh, N., Mulye, S. S. & Pant, G. B. (1992) Some features of the arid area variations over India: 1871–1984. Pure Appl. Geophys. 138, 135–150.

Singh, N., Sontakke, N. A., Singh, H. N. & Pandey, A. K. (2005) Recent trends in spatiotemporal variation of rainfall over India – an investigation into basin-scale rainfall fluctuations. In: Regional Hydrological Impacts of Climate Change- Hydroclimatic Variability (ed. by S. Franks, T. Wagener, E. Boegh, H. V. Gupta, L. Bastidas, C. Nobre & C. de O Galvao), 273–282. IAHS Publ. 296. IAHS Press, Wallingford, UK.

Singh, N., Sontakke, N. A. & Singh, H. N. (2006) Global warming, subtropical anticyclones and the Indian summer monsoon. Proc. National Symposium Tropmet-2006 on Role of Meteorology in National Development, vol. I, d29–d31. IITM, Pune, India.

Sontakke, N. A., Singh, N. & Singh, H. N. (2008) Instrumental period rainfall series of the Indian region (1813–2005): Revised reconstruction, update and analysis. The Holocene 18: 1055–1066.

3

Climate Change and Food Security

G.S.L.H.V. Prasada Rao

Founder, Academy of Climate Change Education and Research
Kerala Agricultural University, Vellanikkara, Thrissur-680 656, Kerala, India

Food security is one of the most important issues during this century under the projected climate change scenario. The world is facing severe food crisis due to weather vagaries as a part of global warming and climate change. Many countries are under severe food crisis during the crisis of weather abnormalities since 75 per cent of world's population is dependent on agricultural activities for their livelihood. Therefore, there is an urgent need to produce more and more foodgrains at the global as well as national levels to meet the food requirement of growing population under the projected climate change scenario. There has been a major decline in world rice production since late 2007 due to various reasons including global warming and climate change in top rice producing countries. Deficit monsoon in India results to less foodgrains production including rice as the area under cultivation was less. Floods are also not uncommon in many States of India. Andhra Pradesh, one of the major rice producing states of India, was affected first by drought and then by floods. Thus rice production decline in India was a real set back in 2009. Similar was the case in 2010 due to excess rain during monsoon and floods during NE monsoon across the rice growing regions. At the same time, Philippines got hit by two major typhoons in September-October, 2009 causing damage to rice crop on ground. Approximately one million tonnes of rice in storage also got damaged. These events resulted to low world rice production including India in 2009. The immediate impacts of global warming and climate change on rice production systems and food security will be felt in the form of adverse effects of extreme weather events such as floods, droughts, heat and cold waves, hailstorms and cyclones on rice production. Less immediate but significant impacts are anticipated due to changes in mean temperatures, increasing weather extremes and sea level rising. West Bengal, Uttar Pradesh, Andhra Pradesh, Punjab, Tamil Nadu, Bihar, Orissa, Assam, Karnataka and Haryana are the major rice producing states in the Country. More than 50 per cent of the total rice production comes from West Bengal, Uttar Pradesh, Andhra Pradesh and

Punjab. Although the country exports several varieties of rice, many scientists have expressed concern that current Indian rice production techniques cannot sustain the growing domestic population. In the event of projected climate change scenario, foodgrains production is likely to be under threat. Rapid urbanization, desertification, environmental issues, decline of agricultural lands and shifting of cropping patterns led to less foodgrains production. Apart from this, monsoon behaviour, its timely onset, further advancement, distribution of rainfall and occurrences of droughts and floods determine the foodgrains production in India. Long breaks in monsoon during July-August results to drought like situation in the Country as experienced in 2002, 2009, 2012, 2014 and 2015. The saying "Indian agriculture is a gamble of monsoon" is true as it is solely dependent on monsoon. A shift is noticed in India that the area under horticultural crops is increasing at the cost of area under foodgrains. This is one way beneficial to farmers since they can withstand against droughts during monsoon season and sustain horticultural production while not so in the case of foodgrian crops.

3.1 Climate change over India

Climate change is one of the concerns to India in view of the damages that occur along the East and West Coasts from the cyclones that form in the Bay of Bengal and Arabian Sea. Any increase in the frequency or intensity of tropical disturbances due to climate change in the future could cause increased damages to life and property in the coastal regions. **The cyclone Titli** hit in September 2018 along the southern districts of Orissa and Andhra Pradesh (Srikakulam District) and the crop damage (rice, coconut, cashew and banana) was very severe in Srikakulam District. Early and untimely snowfall on 3rd and 4th November, 2018 over the Plains of Kashmir Region adversely affected apple bowl. Sea level rise is again a threat to sea foods. Studies showed greater number of high surges under climate change. Apart from this, models show an increased occurrence of cyclones in the Bay of Bengal, particularly in the post monsoon period, along with increased maximum wind speeds associated with cyclones. As far as India is concerned, the land under Agriculture is declining due to various reasons. The projected global warming is between 2 and 3^{0}C as indicated by several reports and even at the linear trends it matches with the above trends by 2100 A.D. The occurrence of floods and droughts, cold and heat weaves has adverse impact and a big threat to food security when compared to direct impacts of climate change. Of course, the frequency of occurrence of weather abnormalities is likely to be more under the projected climate change scenarios. The abnormal weather phenomena holds back the rate of increase in percentage of Indian food grains production and thereby slow growth and food security is likely to be adversely affected in such weather extreme events. The crop growth simulation models indicate that rice and wheat production is likely to decline (Fig. 3.1). The overall foodgrains production is likely to decline by 10 to 30 per cent over India by 2080 A.D.

Temperature and crop projections in India

1. Current scenario

- Temperature increase by 1°C with uncertain rainfall trends across India since last 120- 150 years
- Area under agriculture is declining while area under horticultural crops increasing (Tremendous increase in built up)
- Rate of increase is likely to decline in agriculture and horticulture production due to frequent occurrences of weather extremes like floods and droughts and heat and clod waves

2. Projections

- Temperature increase by 2100 AD will be around 2°C
- Wheat and rice production is likely to decline (?)
- The adverse Impact of drought during monsoon is more significant on foodgrains when compared to that of horticultural crops and floods as well.

Fig. 3.1: Temperature and crop projections in India

3.1 Temperature projections over India

Several reports indicate that warming is more in winter season over summer. The spatial distribution of surface warming suggests a mean annual rise in surface temperatures in north India by 3°C or more by 2050. The study also suggests that during winter the surface mean air temperature could rise by 3°C in northern and central parts while it would rise by 2°C in southern parts by 2050. In case of rainfall, a marginal increase of 7 to 10 per cent in annual rainfall is projected over the sub-continent by the year 2080. However, the study suggests a fall in rainfall by 5 to 25% in winter while it would be 10 to 15% increase in summer monsoon rainfall over the country. It was also reported that the date of onset of summer monsoon over India could become more variable in future (Table 3.1) indicates temperature projections during crop seasons in India.

Table 3.1: Temperature projections from 2020 to 2080 during Kharif and Rabi in India

Year	Season	Temperature Change (°C)		Rainfall Change (%)	
		Lowest	Highest	Lowest	Highest
2020s	Annual	1.00	1.41	2.16	5.97
	Rabi	1.08	1.54	-1.95	4.36
	Kharif	0.87	1.17	1.81	5.10
2050s	Annual	2.23	2.87	5.36	9.34
	Rabi	2.54	3.18	-9.22	3.82
	Kharif	1.81	2.37	7.18	10.52
2080s	Annual	3.53	5.55	7.48	9.90
	Rabi	4.14	6.31	-24.83	-4.50
	Kharif	2.91	4.62	10.10	15.18

(Source: Lal et al., 2001)

3.1.1 Maximum temperature

The mean annual maximum temperature over the Country has increased by 0.76°C during the study period from 1901 to 2003. A maximum increase of 1.4°C was noticed in February, followed by November (1.33°C) while minimum (0.25°) in June. Interestingly, increase in maximum temperature was noticed invariably in all the months. The highest maximum temperature was noticed in the years 2001 and 2002 (31.2°C), followed by 1966 (31.0°C), 1969 and 1941 (30.9°C each) as against the normal of 30.2°C. The maximum temperature was the least (29.0°C) in the year 1917, followed by 1933 (29.6°C), 1956, 1961 and 1971(29.7°C each). Across the Country, the increase in maximum temperature was high (1.24°C) over the West Coast of India, followed by the North East India (1.07°C) and the Western Himalayas of India (0.93°C) while low (0.14°C) over the Interior Peninsular India (Fig. 3.2). The warming was significant since last five to six decades as the decadal increase appeared to be

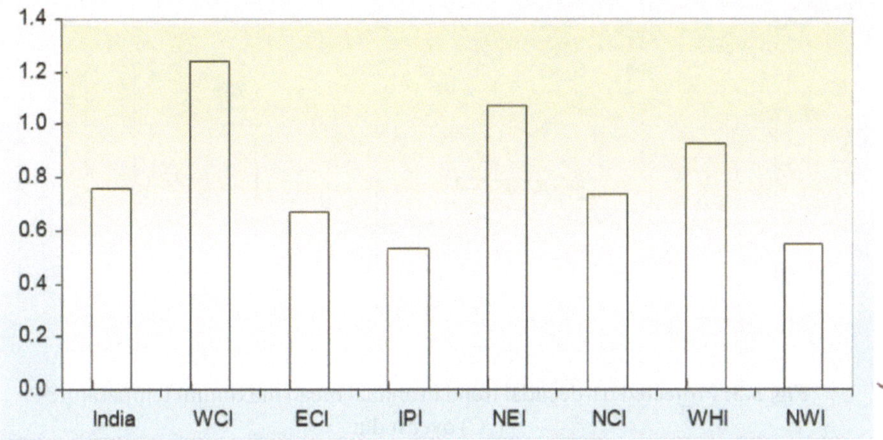

Fig. 3.2: Increase in annual mean maximum temperature (°C) across different zones of India

more since 1951-60 onwards. The tri-decadal maximum temperature over the Country also indicated that the rate of increase in the mean annual maximum temperature was high (0.4°C/13 years) in recent years. It was 0.3°C/13 years during the southwest monsoon and summer while 0.5°C/13 years in post monsoon season. It was intermediary (0.4°C/13 years) in winter and similar to that of annual. The season-wise increase in maximum temperature was 0.76°C, 1.06°C, 1.02°C and 0.65°C during southwest monsoon, post monsoon, winter and summer, respectively. It revealed that the rate of increase in maximum temperature was high in post monsoon season, followed by winter and southwest monsoon seasons. The rate of increase in maximum temperature was relatively less in summer. At this rate, the present tri-decade may be the warmest across the Country. It is clear evidence that the Country as a whole is warming and climate change in terms of maximum temperature is real. Almost all the warm years were recorded in the last two decades. The year 2010 was the warmest year over the Country, followed by 2009. It was a coincidence that the year 2009 was also an all India drought year since the monsoon rainfall was less against the Long Period Average (LPA). Probably, it might be the reason for warming in 2009. The mean annual maximum temperature is projected at the current level of increase across the Country. It is likely to be 31.2°C, 32.2°C and 33.1°C by 2020, 2050 and 2080 AD, respectively for the Country as a whole against 30.3°C (Fig. 3.3). Increase in maximum temperature over the Country from the base period of 1961-90 is projected as 0.9°C, 1.9°C and 2.8°C by 2020, 2050 and 2080 AD at the present rate of 0.4°C during 1991-2003. Of course, the rate of increase in ensuing decades may vary depending upon the emission of greenhouse gases, in particular the emission of CO_2 as it accounts up to 75% of increase in atmospheric temperature.

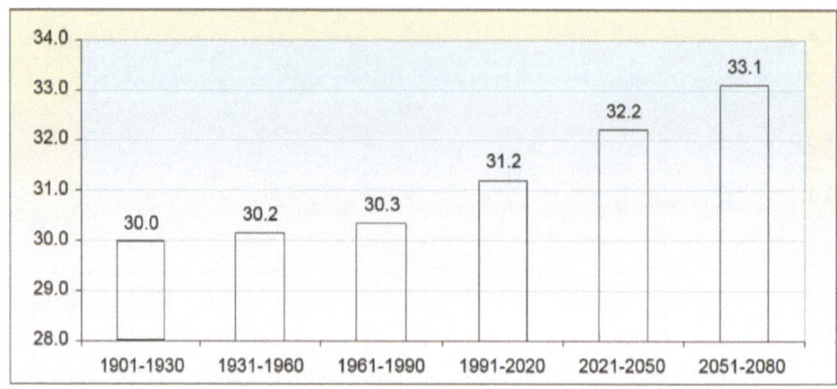

Fig. 3.3: Projected tri-decadal trend in annual mean maximum temperature (°C) over India

3.1.2 Minimum temperature

The mean annual minimum temperature over the Country has increased by 0.22°C during the study period from 1901 to 2003. Interestingly, the night temperature was low since last 40 years except in recent years when compared to the beginning of 20th century, which was evident in tri-decadal trend. A maximum increase of 1.04°C was noticed in November, followed by December (0.76°C) while minimum (0.34) in October. It was in declining trend during January, May, June, July, August and September. While the mean annual maximum temperature increased in all the months, such trend was not seen in the case of minimum temperature as half of the months showed either declining or increasing trend. The highest minimum temperature was noticed in the year 1958 (19.1°C), followed by 1988 and 2002 (19.0°C each) and 1941 and 2003 (18.9°C each) as against the normal of 18.3°C. The mean annual minimum temperature was the least (17.6°C) in the year 1971 followed by 1917 (17.8°C), 1945 and 1983 (17.9°C each), 1950 and1992 (18.0°C each) and 1989 (18.1°C). Across the Country, the increase in minimum temperature was high (0.48°C) over the Western Himalayas of India, followed by the Interior Peninsular India (0.45°C) and the East Coast of India (0.36°C) while low (0.19°C) over the North East India (Fig. 3.4). It is evident that the spatial trends were different in the case of maximum and minimum temperature. Increase in minimum temperature (warm nights) was significant in the recent decade. Interestingly, the minimum temperature was low from 1961-70 to 1981-90. It reveals that cold nights prevailed during 1981-90 when compared to that of other decades. The tri-decadal minimum temperature over the Country also indicated that the rate of increase in minimum temperature was high (0.3°C) in recent years. The highest rate of increase (0.4°C) in minimum temperature was seen in post monsoon season while it is similar (0.3°C) to that of annual in all the other three seasons. At this rate, the present tri-decade may experience warmer nights across the Country. It may be noted that warm nights are frequent in winter in recent years. The season-wise increase in minimum temperature was 0.69°C, 0.42°C and 0.18°C and post monsoon, winter and summer, respectively. It also followed the similar trend to that of maximum temperature except that a decline of 0.1°C during southwest monsoon. It is revealed that the rate of increase in minimum temperature was high in post monsoon season, followed by winter and summer. The rate of increase in minimum temperature is relatively slow in summer. Increase in both the maximum and minimum temperatures during the post monsoon and winter season is likely to be a threat to *rabi* crops in northern States of the Country. It is more so in the case of wheat, which is a predominant crop during the *rabi* season. Warmer nights adversely impact the physiological processes of seasonal crops. It is reported that increase in minimum temperature will slow down the growth of rice production in Asia. Paddy yields are likely to decline by 10 per cent in the event of 1°C rise in minimum temperature in several locations. Higher day time temperatures can increase the yield, up to a point, but future yield losses caused by higher night time temperature will likely outweigh any such gains as the temperatures are rising faster at night. If day time temperatures get too high, it too restrict the yield, causing additional yield reduction. Warmer nights can be an indicator of climate change. This will have far reaching impact on the yield of rice and wheat, causing a shortfall in an important staple crop in a crowded country already struggling with food security and inflationary issues. At the present rate of increase in minimum temperature across the Country, the minimum temperature is likely to be 18.9°C,19.6°C and 20.3 °C by 2020, 2050 and 2080 AD, respectively for the Country as a whole. The increase in mean annual minimum temperature over the Country from the base period of 1961-70 is projected as 0.7°C, 1.4°C, 2.1°C by 2020, 2050 and 2080 AD at the present rate of 0.3°C during 1991-2003.

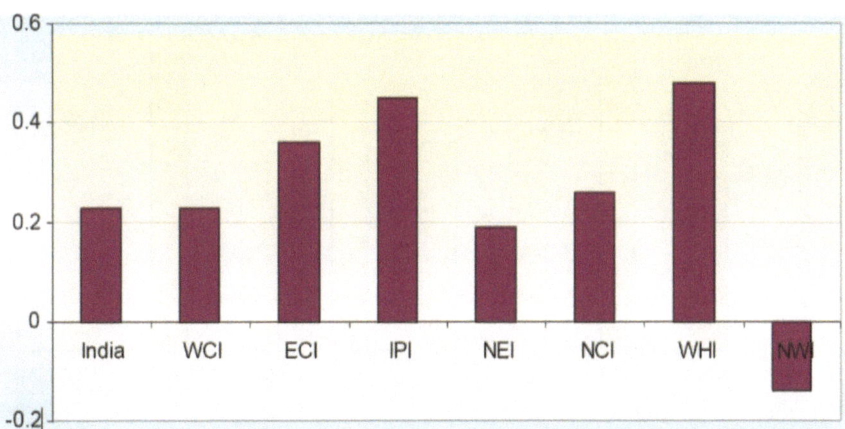

Fig. 3.4: Annual mean minimum temperature (°C) across different zones of India

3.1.3 Mean temperature

The mean annual temperature over the Country has increased by 0.49°C during the study period from 1901 to 2003. A maximum increase of 1.18°C was noticed in November, followed by December (1.03°C) and February (1.00°C) while minimum (0.02°C) in June. Interestingly, increase in mean temperature was noticed invariably in all the months. The highest mean temperature was noticed in the years 2002 and 2003 (25.1°C), followed by 1998, 1958 and 1941 (24.9°C each), 1999 (24.8°C) and 1953(24.7°C) as against the normal of 24.3°C. The mean temperature was the least (23.4°C) in the year 1917, followed by 1971 (23.7°C). It indicated that the year 2002 was the warmest year while 1917 was the coldest during the study period. Across the Country, the increase in mean temperature was high (0.73°C) over the West Coast of India, followed by the Western Himalayas of India and North East India (0.63°C) while low (0.19°C) over the Northwest India (Fig. 3.5). The increase in mean temperature was significant in the recent years. The tri-decadal mean temperature over the Country also indicated that the rate of increase was high (0.3°C) in recent years. The rate of increase was high (0.4°C) in post monsoon season when compared to that of other three seasons during which the rate of increase was similar to that of annual (0.3°C). At this rate, the present tri-decade may be the warmest across the Country. It is clear evidence that the Country as a whole is warming and climate change in terms of mean temperature is real. The season-wise increase in mean temperature was 0.19°C, 0.88°C, 0.72°C and 0.41°C during southwest monsoon, post monsoon, winter and summer, respectively. It revealed that the rate of increase in mean temperature was high in post monsoon season, followed by winter and summer. The rate of increase in mean temperature was the least in southwest monsoon. The mean annual temperature

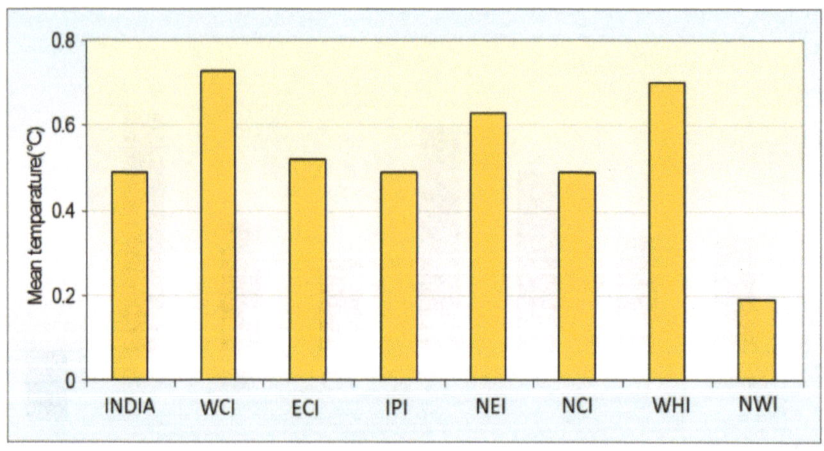

Fig. 3.5: Annual mean temperature (°C) across different zones of India

is likely to be 25.0°C, 25.7°C and 27.4°C by 2020, 2050 and 2080 AD, respectively for the Country as a whole. The increase in annual mean temperature over the country from the base period of 1961-90 is projected as 0.7°C, 1.4°C and 2.1°C by 2020, 2050 and 2080 AD, respectively.

3.1.4 Temperature range

The mean annual temperature range over the Country has increased by 0.54°C during the study period from 1901 to 2003. A maximum increase of 1.18°C was noticed in November, followed by December (1.03°C) and February (1.0°C) while minimum (0.02°) in June. Interestingly, increase in temperature range was noticed invariably in all the months. It is significant to note that the temperature range (the difference between the mean maximum and mean minimum temperature) was abnormally high in 1960s and extended to 1970s. It could be attributed to the low night temperature prevailed during the above period. The highest temperature range was noticed in the years 1972, 1974 and 1966 (12.6°C), followed by 1969, 1992, 1993 (12.5°C each) and 1965, 1967 and 1968 (12.4°C each), 1964 and 1970 (12.3°C), 1975 and 1976 (12.2°C each), as against the normal of 11.9°C. The temperature range was the least (11.2°C each) in 1917 and 1956, followed by 1948 (11.3°C), 1997(11.6°C), 1990(11.7°C) and 1978 (11.8°C). Across the Country, the increase in temperature range was high (1.0°C) over the West Coast of India, followed by the North East India (0.71°C) and the North West India (0.69°C) while low (0.07°C) over the Interior Peninsular India. The warming was significant since last five to six decades as the decadal increase in temperature range appeared to be more since 1961-70 onwards (Fig. 3.6). The increase in temperature range is very much evident since 1961 onwards unlike in maximum, minimum and mean temperature. The tri-decadal temperature range over the Country indicated that the rate of increase in temperature is insignificant and was high (0.1°C) in recent years. The season-wise increase in temperature range was 0.58°C, 0.37°C, 0.59°C and 0.47°C during southwest monsoon, post monsoon, winter and summer, respectively. It revealed that the rate of increase in temperature range was high in winter, followed by southwest monsoon season and summer. The rate of increase in temperature range is least in post monsoon. The increase in temperature range was more conspicuous during 1961-90. Increase in temperature range depends on fluctuations between the maximum and minimum temperature. The temperature range is likely to be 12.4°C, 12.7°C and 13.0°C by 2020, 2050 and 2080 AD, respectively for the Country as a whole. The increase in annual temperature range over the Country from the base period of 1961-90 is projected as 0.3°C, 0.6°C and 0.9°C by 2020, 2050 and 2080 AD, respectively. Ramakrishna (2007) studied from time to time on temperature variability across the Country. It is inferred from the above studies that the rate of increase in

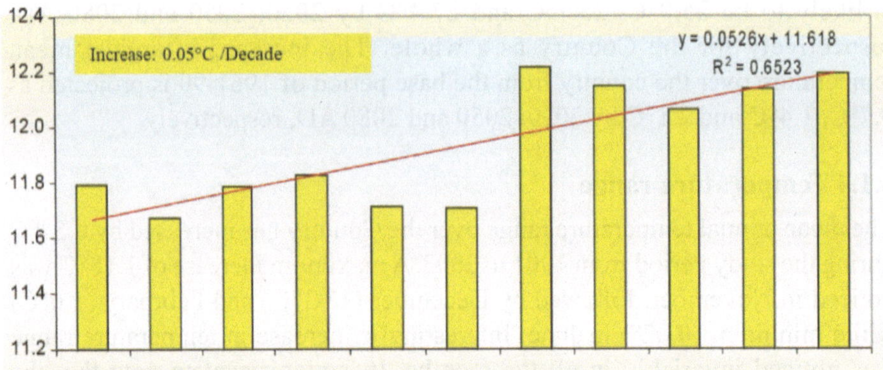

Fig. 3.6: Decadal trend in annual temperature range (°C) over India

temperature varied depending upon the data set, location, region and season. Overall, the warming trend is evident across the Country and the rate of increase in temperature was high in recent years though some locations showed cooling trend. The trend is visible in the case of maximum temperature while not so in the case of minimum temperature. The temporal and spatial variations in temperature are significant depending upon the season and the region. The spatial patterns of maximum temperature indicated that the Central India experiences more than 45°C while 35-40°C along the West Coast and above 25°C in Himachal Pradesh in north India (NATCOM Report, 2004). Some stations of the Indian Peninsular Plains also showed a falling trend in temperature. Several models indicated that the increase in temperature is likely to be around 3°C or more over the Country by 2100 A.D. It will impact the agricultural sector adversely and the crop losses will be felt much more after 2050 with increase in CO_2 and temperature. The coastal belt of the country is under threat due to sea level rise and as a whole, the food security is in threat due to global warming and climate variability / change. The present updated information also confirms the above findings.

The increase in annual maximum temperature was 0.76°C while 0.22°C in the case of minimum temperature. The increase in annual mean temperature was 0.49°C while 0.54°C in the case of temperature ranges (the difference between the maximum and minimum temperature). All the months showed warming trends though the trend in night temperature declined in half of the months. The rate of increase in temperature is relatively high in post monsoon, followed by winter. It is a concern in northern States during *rabi* season as the winter crops are adversely affected due to temperature increase. It is more so in the case of wheat. The rate of increase in maximum, mean and range of temperature was relatively high across the West Coast of India when compared to that of other zones. It was not so in the case of minimum temperature as the Western

Himalayas of India recorded high minimum temperature when compared to that of other zones across the Country. Increase in temperature might be one of the reasons for glacier melt across the Himalayas. It is again a concern and threat to the two rich hotspots of biodiversity in the Country as the rate of increase in temperature is relatively high across the Western Ghats and Western Himalayas of India. The projected annual maximum temperature from the base period of 1961-90 will be 0.9°C, 1.9°C and 2.8°C by 2020, 2050 and 2080 AD, respectively at the current rate of increase. It was 0.7°C, 1.4°C and 2.1°C in the case of minimum temperature by 2020, 2050 and 2080 AD, receptively. The rate of increase in mean temperature is similar to that of minimum temperature. Of course, the rate of increase in ensuing decades may vary depending upon the emission of greenhouse gases, in particular the emission of CO_2 as it accounts up to 70-75% of increase in atmospheric temperature. The temperature increase is likely to be 0.8-2.4°C by 2080 / 2100AD if the CO_2 level in level in the atmosphere is 400 ppm while it is 1.0 - 3.1°C if the CO_2 level is 450 ppm. The temperature increase is abnormally high of the order of 1.5 - 4.4°C if the CO_2 level in the atmosphere is 550ppm. The present study confirms with the current level of increase in CO_2.

3.2 Rainfall trends over India

The data for the study was downloaded from the IITM web site. Trend in annual as well as seasonal rainfall was worked out for different zones of the Country viz., East Peninsular India, West Peninsular India, South Peninsular India, North Central India, North East India, North Mountainous India, North West India and finally for All India.

3.2.1 Annual rainfall

Trend analysis showed that the All India annual rainfall has increased by 64.1 mm over a period of 194 years from 1813 to 2006 (Fig. 3.7). The highest maximum in annual rainfall was noticed in the year 1917 (1435.3mm) followed by 1827 (1410.9mm), 1961 (1409.9mm), 1956 (1387.9mm), 1893 (1385.2mm), 1933 (1372.5mm), 1894 (1358.6mm) and 1818 (1358.1mm) as against the average of 1150 mm. The annual rainfall was the least (868.7mm) in the year 1838, followed by 1899 (891.1mm), 1918 (895.7mm), 1965 (931.6mm), 1972 (932.7mm), 2002 (968.9mm) and 2009 (885.5mm). It also indicated that the decadal rainfall is increasing over a period of time. The increase was more evident during 1931 to 1960 while the recent decades recorded low rainfall. The highest rainfall was recorded in the decade1941-50 (1212.9mm), followed by 1951-60 (1197.6mm), 1881-90 (1187.9mm), 1991-2000 (1181.4mm) and 1931-40 (1184.3mm). The lowest decadal rainfall was recorded in the decade 1831-40 (1067.4mm), followed by 1851-60 (1081.8mm), 1841-50 (1102.1mm), 1901-

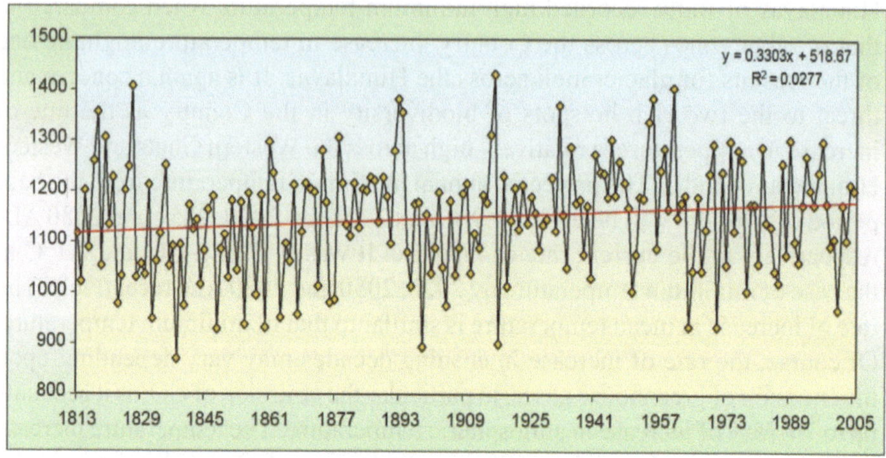

Fig. 3.7: Trend in annual mean rainfall (mm) over India from 1813 to 2006

10 (1106.8mm) and 2001-08 (1131mm). It may be noted that the current decade also comes under the group of lowest rainfall. It could be attributed to low rainfall received in 2002, 2004, 2009, 2012, 2014, 2015 and 2018. Breaks in monsoon, weak monsoon flow, and non formation of rain bearing systems over the ocean-atmosphere continuum are some of the reasons for the vagaries of monsoon over the Country. Under the projected global climate change scenario, monsoon behaviour may change and therefore, it is a threat to the economy of the Country as it is an agrarian based. Tri-decadal rainfall over the Country also indicated that the annual rainfall was steadily declining since 1960 onwards, though the overall trend was noticed from 1813-40 to 1931-1960 (Table 3.2). Increase in rainfall trend could be mainly attributed to high rainfall received from 1931-60, followed by 1871-90. The highest maximum of 1198.3mm was noticed during the tri-decade 1931-60, followed by 1871-1900 (1166.7mm) while the lowest was during the tri-decade 1841-1870 (1106mm), followed by (1813-1840). Since 1960s, tri-decadal rainfall did not show increase. Across the country, the highest increase in annual rainfall was observed in the Western Peninsular India (106.1mm), followed by the Southern Peninsular India (97.3mm) and the Eastern Peninsular India (79.9mm). The least increase was in North Central India (17.3mm) over a period of 165 years (1842-2006). In contrast, the Northern Mountainous India exhibited a declining tendency in annual rainfall by 11.2mm over the period of 162 years (1845-2006) while the decline was the least (1.8mm in 178 years) in North East India. The decline in annual rainfall over North East India, one of the areas with heavy rainfall in the world may be compared with the marginal decline in annual rainfall over Kerala, one of the wettest places in the World. The study indicated that the decline in annual rainfall is evident in high rainfall zones. The decadal annual rainfall over the Country has shown an

increase of 2.7mm while the increase was 8.9 mm when the trend was calculated during 30 year period. As in the case of annual rainfall, highest decadal and tri-decadal increase in rainfall was observed over the Western Peninsular India. Tri-decadal rainfall over the North Central India showed a marginal decline (0.79mm in 30 years) while a marginal increase (0.5mm in 30 years) over the North East India unlike annual and decadal rainfall where decline was noticed. All other trends followed the similar pattern. The tri-decadal rainfall over the Western Peninsular India is comparable to the tri-decadal rainfall of the Southern Peninsular India, comprising of Kerala, Lakshadweep, Tamil Nadu, Karnataka and Andhra Pradesh. The annual rainfall over the Southern Peninsular India is much more (1518mm) than the all India rainfall (1150mm). All India tri-decadal rainfall and the tri-decadal annual rainfall over southern Peninsular India exhibit an increasing tendency during the period from 1813 to 2006.

Table 3.2: Trend in annual, decadal and tri-decadal rainfall over different zones of India

Name of the zone	Data period	Rainfall trends		
		Annual	Decadal	Tri-decadal
All India	1813–2006 (194 year)	0.33	2.7	8.99
Eastern PeninsularIndia (EPI)	1848–2006 (159 year)	0.5	2.18	7.31
North Central India (NCI)	1842–2006 (165 year)	0.1	0.585	-0.79
North East India (NEI)	1829–2006 (178 year)	-0.01	-0.806	0.5
Northern Mountainous India (NMI)	1845–2006 (162 year)	0.069	-0.899	-13.41
North West India (NWI)	1844–2006 (163 year)	0.34	3.37	12.35
Southern Peninsular India (SPI)	1813–2006 (194 year)	0.5	2.75	17.52
Western Peninsular India (WPI)	1814–2006 (166 year)	0.64	6.18	18.39

3.2.2 Rainfall during southwest monsoon

Rainfall during the southwest monsoon over the Country has increased by 44.8mm (0.23mm/year) over a period of 194 years from 1813 to 2006. The highest southwest monsoon rainfall was noticed during the year 1961(1088.8mm), followed by 1917 (1079.1mm), 1975 (1055.5mm), 1986 (1051.2mm) and 1944 (1020.7mm) as against the normal of 894mm. The southwest monsoon rainfall was the least during 1877 (609.4mm), followed by 1918 (661.3mm), 1838 (697.5mm), 1848 (688mm), 1899 (694.6mm), 1972 (108.4mm), 2002 (715.5mm) and 2009 (885.5mm). The decadal rainfall over the Country also showed an increasing trend over the years. The highest southwest monsoon rainfall (946.0mm) was received during the decade 1941-50, followed by 1881-90 (940.1mm), 1951-60 (924.2mm), 1931-40 (921.5mm), 1881-90 (916.7mm) and 1991-2006 (910.3). Southwest monsoon was the lowest (829.6mm) during the decade 1851-1860, followed by 1831-40 (840.1mm), 1841-50 (841.2mm) and

2001-06 (850.6mm). The decline in monsoon rainfall over Northern Mountainous India is alarming. Decadal and tri-decadal analysis revealed that the rainfall trend is similar as in the case of annual rainfall except in the case of Northern Mountainous India where the decadal and tri-decadal rainfall has shown an increasing tendency while it was declining on annual basis (Table 3.3).

Table 3.3: Trend in annual,decadal and tri-decadal southwest monsoon rainfall over different zones of India

Name of the zone	Data period	SWM rainfall trends		
		Annual	Decadal	Tri-decadal
All India	1813–2006 (194 year)	0.23	2.78	8.99
Eastern Peninsular India	1848–2006 (159 year)	0.32	2.18	7.31
North Central India	1842–2006 (165 year)	0.12	0.47	-0.084
North East India	1829–2006 (178 year)	-0.13	-2.69	-3.49
Northern Mountainous India	1845–2006 (162 year)	-0.61	0.865	4.944
North West India	1844–2006 (163 year)	0.23	2.39	9.11
Southern Peninsular India	1813–2006 (194 year)	0.33	0.81	9.5
Western Peninsular India	1841–2006 (166 year)	0.54	5.18	13.99

3.2.3 Rainfall during Post monsoon

Post monsoon rainfall over India has increased marginally by 16.7mm over a period of 194 years (1813-2006) or 0.86mm per decade. The highest increase was noticed in the year 1956 (218.8mm), followed by 1955 (195.4mm), 1917 (193.4mm), 1916 (185.5mm) and 1894 (182.1mm). The post monsoon was the lowest (9.3mm) during 1828, followed by 1858 (21.5mm), 1868 (41.6mm), 2000 (59mm), 1967 (58) and 1965 (58.1). Wide inter-annual variability is noticed in the case of post monsoon rainfall over the Country. Rainfall during post monsoon season is very important to the *rabi* crops grown in the northern States of India. Decadal post monsoon rainfall also indicated an increasing trend. The highest post monsoon rainfall was received during 1951-60 (131.8mm), followed by 1911-20 (122.2mm), 2001-06 (121.5mm), 1813-20 (120.9mm), 1991-00 (120.4mm), 1971-80 (119.6mm) and 1881-90 (111.4mm). The lowest post monsoon of 91.4mm was noticed in 1871-80, followed by 1961-70 (99.5mm), 1851-60 (96.0mm), 1831-40 (97.2mm) and 1891-1900 (99.6mm). The tri-decadal rainfall also was in increasing trend. The highest tri-decadal post monsoon (120.8mm) in the latest decade 1991-2000, followed by 1931-60 (118.8mm), 1901-30 (108.4mm), and 1813-40 (105.4mm). The post monsoon rainfall was the lowest in the decade 1841-70 (100.5mm), followed by 1871-1900 (101.9mm) and 1961-90 (106.8mm). Across the country, the highest increase in post monsoon rainfall was observed in Western Peninsular India (119mm in 166years or 7.17mm per decade) followed by Southern Peninsular India (36.5mm in 194 years or 1.9 mm per decade). Except north Central India, all other zones of the country exhibited an increasing tendency in post monsoon rainfall. In the case of North Central India, a marginal decline of 8.2mm over a period of 165 years was noticed. It may be noted that an increase of post monsoon rainfall is seen in North East India and Northern mountainous India, where annual and monsoon rainfall were in declining trend. Decadal and tri-decadal analysis showed that rainfall over NMI (Northern Mountainous India) is in declining trend while all other zones exhibited trends similar to that of annual rainfall. It revealed that the tri-decadal post monsoon rainfall over the southern peninsular India has shown an increasing tendency (@7.913mm/30 year) while the rainfall over all India is increasing marginally. The Post monsoon rainfall is more or less stable over the Country as a whole irrespective of tri-decadal trends (Table 3.4).

Table 3.4: Trendinannual, decadalandtri-decadalrainfallover different zonesf India during post monsoon

Name of the zone	Data period	Rainfall trends		
		Annual	Decadal	Tri-decadal
All India	1813–2006 (194 year)	0.085	0.79	2.7
Eastern Peninsular India	1848–2006 (159 year)	0.118	0.62	5.96
North Central India	1842–2006 (165 year)	-0.049	-0.32	-1.95
North East India	1829–2006 (178 year)	0.156	1.9	4.92
Northern Mountainous India	1845–2006(162 year)	0.169	-0.54	-3.66
North West India	1844–2006 (163 year)	0.088	0.75	2.64
Southern Peninsular India	1813–2006 (194 year)	0.188	1.59	7.91
Western Peninsular India	1841–2006 (166 year)	0.072	0.64	2.87

3.2.4 Rainfall during winter

Rainfall during winter season across the country is meager and highly undependable as the coefficient of variation is high. The increase in winter rainfall is marginal over most parts of the Country. Over a period of 194 years, the increase recorded in the case of post monsoon is only 4.4 mm. The highest rainfall (96.9mm) is received in the year 1827, followed by 1998 (93.4mm), 1901 (90.1mm) and 1968 (86mm). The winter rainfall was the lowest (13.6mm) in 1876, followed by 1890 (16.4mm), 1960 (17.8mm), 1852 (18.3mm) and 1946 (19.0mm). It revealed that winter rainfall was poor in recent years. In India, major share of winter rainfall is being received from the Western Disturbances, a weather system visits Northern State's during winter from the western side of the India. This rainfall is the source for soil moisture and crop performance during the *rabi* crop season of India. Though, the share of winter rainfall is meager, the failure in the receipt of winter rainfall adversely impacts the winter season crops, especially wheat and other seasonal crops, in the northern states. Under the projected global climate change scenario, periodicity and occurrence of western disturbances may change as it s formation and travel path is beyond

India sub continent. Decadal analysis also revealed that the increase in decadal winter rainfall is marginal. During 1981-90 and 1991-00 were the latest two decades with comparatively high rainfall, while the latest decade (2001-06) has shown a decline when compared to previous decades. The highest winter rainfall was observed in decade (52.7mm), followed by 1901-10 (50.7mm), 1841-50 (50.4mm), 1981-90 (49.3mm), 1991-00 (48.3mm) and 1821-30 (48.1mm). Winter rainfall was low in the decade 1831-40 (31.2mm), followed by 2001-06 (36.3mm), 1981-70 (36.5mm), 1813-20 (36.6mm), 1871-80 (37.3mm) and 1957-60 (37.4mm). The decline in winter rainfall in the recent decade is a concern to the winter crops especially wheat. Tri-decadal winter rainfall also indicated a marginal increase mover the decades. Unlike in decadal, though not much variation is noticed in the tri-decadal rainfall, a marginal increase was noticed during 1931-60 (47.7mm) when compared to the tri-decades. It was followed by 1931-60 (44.8mm), 1991-06 (43.8mm) and 1961-90 (43.6mm). The winter rainfall was the lowest (38.9mm) in 1813-1840, followed by 1841-70 (42.2mm) and 1871-90 (42.4mm). It revealed that the tri-decadal winter rainfall followed the trend similar to that of decadal and annual. Across the Country, a marginal declining tendency was noticed in North West India. The increase in winter rainfall was the highest in southern Peninsular India (12.53mm in 194 years). It may be noted that the influence of Western disturbances is not seen across the Southern Peninsular India during winter season. Tri-decadal trend analysis indicates that winter rainfall over SPI is in increasing trend. A decline in decadal trend was noted over Northern West India while the decline was noted over Eastern Peninsular India when 30 year trend was considered. The winter rainfall across the country as a whole though shows a marginal increase, it is not significant. It revealed that the tri-decadal winter rain over India is low when compared to that of winter rainfall of Southern Peninsular India. In spite of the absence of Western disturbances, Southern Peninsular India has registered an increase in winter rainfall over the periods (Table 3.5). The highest (85mm) was recorded during 1901-1930, followed by 1901-30 (70mm) and least in 1871-1900(62mm).

Table 3.5: Trend in annual,decadal and tri-decadal winter rainfall over different zones of India

Name of the zone	Data period	Rainfall trends		
		Annual	Decadal	Tri-decadal
All India	1813–2006 (194 year)	0.022	0.25	0.71
Eastern Peninsular India	1848–2006 (159 year)	0.026	0.22	-0.22
North Central India	1842–2006 (165 year)	0.024	0.081	0.61
North East India	1829–2006 (178 year)	0.004	0.009	0.28
Northern Mountainous India	1845–2006 (162 year)	0.9	-0.158	0.37
North West India	1844–2006 (163 year)	-0.011	-0.19	-0.37
Southern Peninsular India	1813–2006 (194 year)	0.065	0.692	1.92
Western Peninsular India	1841–2006 (166 year)	0.0186	0.14	0.63

3.2.5 Rainfall during summer

Unlike annual and other three seasons, summer rainfall over the Country has shown a marginal decline on annual basis. The highest rainfall during summer was obtained in 1820 (232.6mm), followed by 1827 (204.3mm), 1852 (177.4mm), 1835(169.2 mm), 1893 (166.5mm) and 1933 (165.7mm). The summer rainfall was the lowest (62.1mm) in 1935, followed by 1854 (63.1mm), 1866 (64.2mm), 1922 (65.5mm) and 1838 (65.7mm). While annual summer rainfall declines, decadal rainfall indicated an increasing trend. The highest maximum decal rainfall during summer was recorded during the latest decade (123.3mm), followed by 1851-60 (117.3mm), 1821-30 (111.6mm), 1981-90 (109mm) and 1911-20 (108.1mm). The lowest summer rainfall was recorded during 1921-30 (91.8mm), followed by 1881-90(95.2mm) and 1901-10 (96.6mm). The tri-decadal rainfall over the Country also showed a marginal increasing trend. The latest tri-decade (1991-2006) showed a marked increase when compared to other tri-decades. The highest maximum summer rains (110.6mm) in 1991-06, followed by 1848-70(108mm), 1813-1840 (106.3mm) and 1931-60(104.3mm). The lowest summer rain was received in 1901-30 (98.8mm), followed by 1981-1900(99.8mm) and 1961-90(101.2mm). Across the Country, a decline was noticed in North East India (7.5mm in 178 years) and Southern Peninsular India (14.5mm in 194 years) in the case of summer rainfall. Summer rainfall showed marginal increase in Eastern Peninsular India, North Central India, and Western Peninsular India. Northern Mountainous India and North West India showed predominant increase in summer rainfall (52.4mm in 162 years and 53.3mm in 163 years, respectively). Tri-decadal summer rainfall over the Country as a whole has shown a tendency to increase marginally, though the annual trend is towards a marginal decline. At the same time over the Southern Peninsular India, tri-decadal summer rainfall as well as on annual basis is declining (Table 3.6). The highest increase in summer rainfall was noticed during 1931-60 (237mm) followed by 1841-70 (228mm) while low in 1901-30 (184mm).

Table 3.6: Trend in annual,decadal and tri-decadal rainfall over different zones of India during summer

Name of the zone	Data period	Rainfall trends		
		Annual	Decadal	Tri-decadal
All India	1813–2006 (194 year)	-0.008	0.097	0.14
Eastern Peninsular India	1848–2006 (159 year)	0.044	0.061	0.795
North Central India	1842–2006 (165 year)	0.006	0.295	0.585
North East India	1829–2006 (178 year)	-0.042	-0.061	-1.32
Northern Mountainous India	1845–2006 (162 year)	0.323	1.92	2.91
North West India	1844–2006 (163 year)	0.033	0.428	0.991
Southern Peninsular India	1813–2006 (194 year)	-0.075	-0.249	-1.854
Western Peninsular India	1841–2006 (166 year)	0.014	0.235	0.944

3.2.6 Monthly rainfall trends

Monthly rainfall trends over different zones of the Country were worked out on annual, decadal and tri-decadal basis to understand the monthly trends of rainfall. It is evident that rainfall during June across the country, except southern Peninsular India, is declining. July rainfall is very critical to Indian foodgrains production. Though July rainfall in majority of the zones showed an increasing tendency, it was declining marginally in North Central India and North West India. While all other zones, including All India, showed a clear upward trend in August rainfall, North East India exhibited a predominant declining tendency (32.7 mm rainfall in 178 years or 1.8mm per decade). At the same time, rainfall during September in all the zones is in increasing trend. The months of July and August are generally the most active months of the southwest monsoon season, contributing to 60% of the rainfall. Reports indicated that there exists no correlation between the monthly rainfall within the season mainly because synoptic situations of smaller time scales dominate the rainfall variability on a monthly scale. Thus the seasonal total rainfall cannot indicate the monthly distribution of rainfall. However, the highly significant correlations between the monthly rainfall and the seasonal total indicate that all the months substantially contribute to the variations in the southwest monsoon seasonal rainfall. Rainfall during October over North Central India has shown a marginal decline (10.4 mm in 165 years) while in all other zones rainfall showed a positive tendency. In Northern Mountainous India, rainfall during November and December has shown a decline (32.9 mm in 162 years and 25.8mm in 162 years, respectively). September rainfall though showed an increase on annual basis over all the zones, showed a marginal declining tendency over Eastern Peninsular India and North East India when decadal trend was examined. It was true in case of tri-decadal trend over EPI. Summer rainfall (March-April) over the Country is marginally declining. It is evident from the monthly rainfall trend. An increase of 1.9mm only was noted in April while March and May rainfall was towards decreasing. Decadal and tri-decadal trend analysis also followed more or less similar trend pattern. From the trends of monthly rainfall, it is evident that monsoon rainfall tends to be more concentrated in August, in view of positive trends over large areas in that month, particularly along the WPI, NWI and Central India. Rupa Kumar *et al.,* (2002) have also made a similar observation.

3.3 Impact of monsoon droughts on Indian foodgrains production

The Indian economy is mostly agrarian based and depends on onset of monsoon and its further behavior. The year 2002 was a classical example to show how Indian foodgrains production depends on rainfall of July. The year 2002 was declared as all India drought, as the rainfall deficiency was 19% against the long period average of the Country and 29% of area was affected due to drought. The *kharif* foodgrains production was adversely affected by a whopping fall of 19% due to all-India drought during monsoon 2002. The All India drought is defined as the drought year when the rainfall deficiency for the Country as a whole is more than 10 per cent of normal and more than 20 per cent of the Country's area is affected by drought conditions. The all India drought could be attributed to the monsoon breaks / failure during the *khairf* season. As the area under cultivation during the first crop season is mostly rainfed across the Country, the distribution of monsoon rainfall during July play a predominant role for crop spread. The Indian foodgrains production increased from 50 million tonnes in 1950-51 to more than 270 million tonnes in recent years (Fig.3.8). Recent years were the record years in terms of foodgrains production in India. Though the

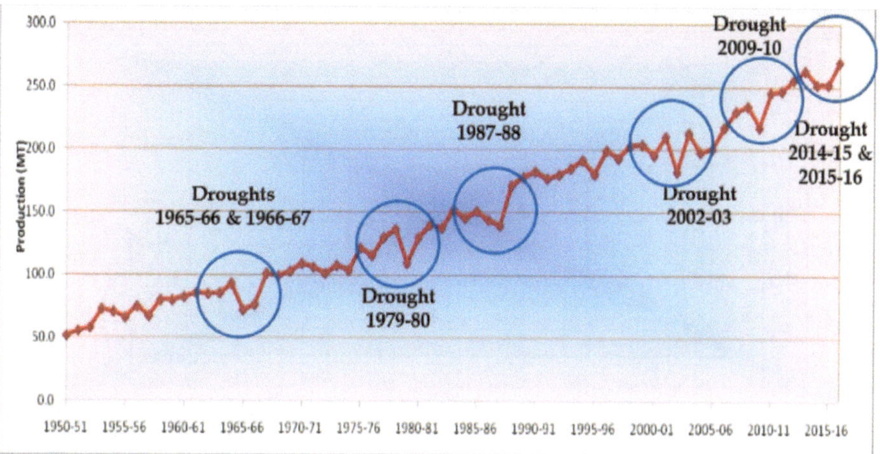

Fig. 3.8: Impact of droughts on Indian food grains productions from 1950-51 to 2017-18
Source: Foodgrain production-MoA,Govt of India, Droughts: IMD

2012-13 (257); 2013-14 (265); 2014-15 (252.02); 2015-16 (252.23); 2016-17 (274); 2017-18 (276)

increase in foodgrains production was more than five folds, the inter-annual/ intra-seasonal variations were not uncommon due to the occurrence of floods and droughts. The inter-annual dips in the production of foodgrains are mostly due to the failure of monsoon and consequent droughts. Consecutive droughts were seen across the Country in 1965 and 1966 and 2014 and 2015. The years 1979, 1987, 2002, 2009, 2012, 2014 and 2015 were declared as the drought years during which the monsoon rainfall was deficit and the country's foodgrains production was adversely affected to a considerable extent. The effect of drought on Indian foodgrains production is felt much more when compared to that of floods as large areas under cultivation is adversely affected due to drought. The years 1965, 1966, 1972, 1974, 1979, 1982, 1987, 2002, 2009 and 2012 were classified as the drought years during which the deficit of monsoon rainfall against normal varied between 7.7 and 22 per cent. When the monsoon rainfall was deficit and depending upon its distribution, the Indian foodgrains production was adversely affected. The effect of drought during monsoon on Indian foodgrains production varied between 2.9 and 28.5 per cent, depending upon the drought intensity. The average decline in Indian foodgrain production due to drought during monsoon season, which coincide with kharif season was 12.75 per cent (Fig. 3.9). As indicated, the effect of drought on Indian foodgrains production depends on the distribution of rainfall during July and August. If the deficit rainfall during July and August is more and more with erratic distribution, the effect of drought on Indian foodgrains production was also more. On pentads also, it is clearly brought out the effect of drought on Indian foodgrains production. The rate of increase in Indian foodgrains production was slowed down due to

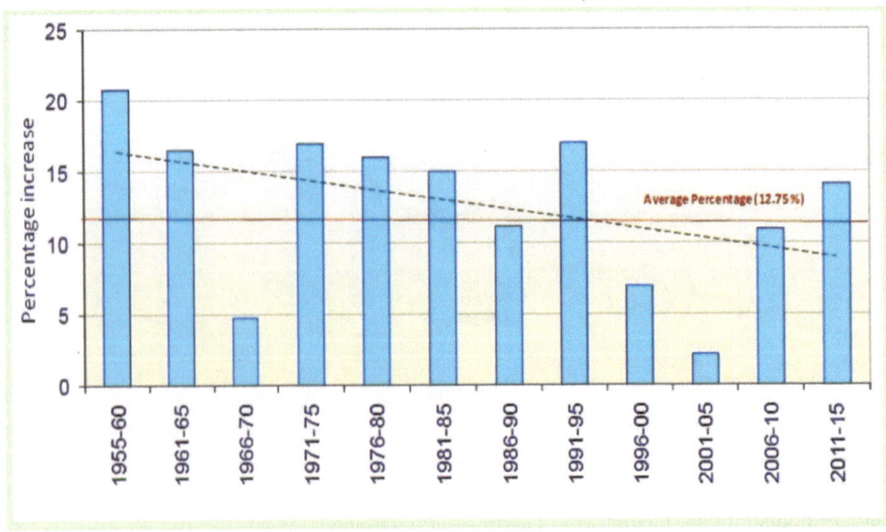

Fig. 3.9: Indian food grain productions – percentage increase

occurrence of droughts. The decline in Indian foodgrains was seen in 1966-70 (4.8%), 1986-90 (11.2%), 1996-00 (7.0%), 2001-05 (3.4%) and 2005-10 (9.3%) against the normal rate of increase of 12.75 per cent. Therefore, it is evident that the Indian foodgrains production was not in tune with the plan estimates since last 15 years. It was mainly attributed to the ill effects of droughts and floods and cold and heat waves during the crop growing seasons as the *kharif* foodgrains production is dependent on monsoon rainfall while *rabi* foodgrains is dependent on cold and heat waves. Of course, the technological interventions in recent years may be one of the factors for reducing the effect of drought on Indian foodgrains production. At the same time, the effect of drought during monsoon on Indian foodgrains production would be much more if the production during *kharif season* alone was considered. Therefore, it is possible to predict the *kharif* foodgrains production based on summer monsoon rainfall as majority of the foodgrains crops are grown during the *kharif* under rainfed conditions and depend on monsoon rainfall and its distribution. Such an attempt was made by Parthasarathy *et al.,* (1998) and developed a statistical equation to predict the Indian foodgrains production based on Indian summer monsoon rainfall.

The frequency of weather aberrations is likely to increase under the projected climate change scenario across the Country in ensuing decades and the Indian foodgrains production is adversely affected while the demand for the same will be more due to increase in human population. As per the ICMR estimates, the requirement of foodgrains is 260 million tonnes by 2010 and about 310 million tonns by 2020 according to the increase in human population (Table 3.7). Interestingly, the rate of increase in Indian foodgrains production was only about 3 million tonnes per year and at this rate, the increase in foodgrains production by 2020 is unlikely to be projects requirement of 310 million tonnes. To achieve the above target is in doubt due to slow rate of increase in Indian foodgrains production under weather uncertainties as a part of global warming and climate change. No doubt, it may lead to deficit in Indian foodgrains production due to occurrence of floods and droughts and heat and cold waves. Studies indicate that by 425 ppm carbon dioxide concentration and an increase of 2°C, wheat area at different productivity levels in India is likely to decline. Similar was the case in the case of rice also (Sinha and Swaminathan, 1991). Therefore, pro-active measures are the need of the hour to increase the Indian foodgrains production against the ill effects of weather aberrations for which location/region/crop specific technologies are to be evolved as a part of climate change adaptation in agriculture.

Table 3.7: Production requirements by 2020 to meet the balanced diet norms prescribed by ICMR

Crops	Requirement per day in gms	Requirement in million tonnes		
		2000	2010	2020
Cerealsandmillets	420	198.7	237.4	280.99
Pulsesandlegumes	40	18.92	22.61	26.76
Foodgrains	460	212.6	260.01	307.75
Rootsandtubers	75	35.48	42.39	50.18
Vegetables	125	91.66	109.52	129.62
Fruits	50	36.66	43.81	51.85
Milk	150	70.96	84.79	100.35
Fatsandoils	22	10.41	12.44	14.72
Sugar	30	14.19	16.96	20.07
Egg	45	21.29	25.44	30.11
Meat	25	11.83	14.13	16.73
Fish	25	11.83	14.13	16.73
Population(millions)	-	1005	1200.2	1420.5

3.4 Climate change impacts on Indian foodgrains

Studies revealed that the maximum, minimum, mean and temperature range over the Country has increased by 0.76°C, 0.22°C, 0.49°C and 0.54°C, respectively during the period from 1901 to 2003. Based on this, temperature projections were worked out for 2020, 2050 and 2080 to understand the impact of increased temperature on rice production in India under the projected climate change and global warming scenario. Projections indicated that the maximum temperature across the Country is likely to increase by 0.9°C by 2020, 1.9°C by 2050 and 2.8°C by 2080. Of course, the rate of increase is determined by the emission of GHGs, predominantly by carbon dioxide, as it contributes lion-share to global warming. The minimum and mean temperature projections follow the similar trend. The increase in minimum temperature is likely t be 0.7°C, 1.4°C and 2.1°C, respectively by 2020, 2050 and 2080, respectively. The temperature range across the Country is likely to increase by 0.6°C-0.9°C by 2050 and 2080 (Fig. 3.10). Moreover, the rate of increase in temperature is relatively high in post monsoon, followed by winter. It is a concern in northern States during *rabi* season as the winter crops are adversely affected due to temperature increase. It is pointed out that wheat crop will be badly affected in such an event. Warmer nights adversely affect the physiological processes of seasonal crops. It is reported that increase in minimum temperature will slow down the growth of rice production in Asia. Increase in day temperature can increase the yield up to a certain level, but future yield losses caused by higher night time temperature will likely outweigh any such gains as the temperatures are rising faster at night. If the day time temperatures get too high, it too restrict the yield, causing additional yield reduction (Science Daily, October, 30, 2010). Warmer nights can be an indicator of climate change. This will have far reaching impact on the yield of rice and wheat, causing a shortfall in an important staple crop in a crowded country already struggling with food security and inflationary issues. It is quite evident from the projected temperature that the mean annual temperature by 2050 is likely to be 1.4°C, and as a result the paddy and wheat yield are going to be adversely affected. As per the studies, yield reduction is likely to be around 10 per cent for every 1°C rise in temperature. Further increase in temperature during the ensuing years after 2050 will witness for a deficit foodgrains production, in the event of climate change and global warming under the projected increase in temperature. On all India level, annual rainfall, southwest and post monsoon and winter rainfall has shown an increasing trend, while annul summer rainfall showed a marginal declining trend. Global climate models also projected increase in rainfall over the country, though some models project rainfall deficit across some regions of the country. During the first crop (*kharif) season* July rainfall is critical. As such, the projections in rainfall indicate that monsoon rainfall over India is in increasing trend. In a country, where the agricultural production is solely dependent on monsoon rainfall, its timely onset, distribution and quantum of rainfall are very important for reaping a better yield. Uncertainties in the onset of monsoon, weak monsoon flow, and long spells of break are serious issues as far as the Indian agriculture is concerned. In the event of monsoon breaks, the nation often faces drought like situation. Drought could be caused by weak/deficit monsoon rainfall as happened in 2015 and 2018 in recent years. Since the monsoon is a global scale phenomenon, it is a complex one to be understood properly. As majority of the agricultural land is under rainfed, monsoon failure will definitely push the nation to low foodgrains production. Post monsoon rainfall is also in increasing trend across the Country.

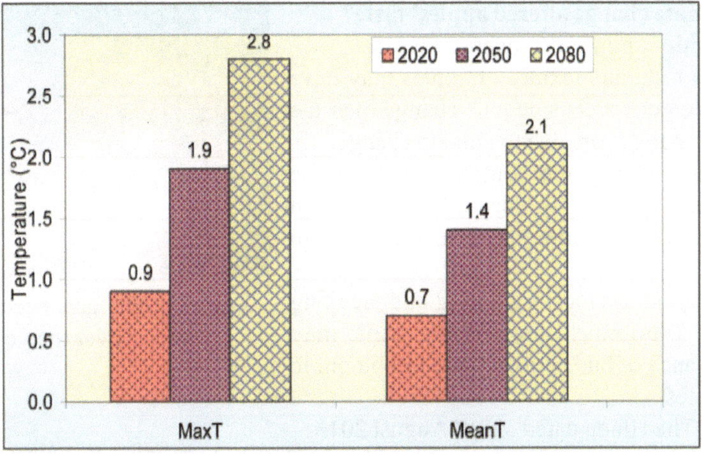

Fig. 3.10: Temperature projections based on the current trends

Post monsoon rainfall is important for *rabi* crop, mainly to wheat. Post monsoon rainfall, though lesser in quantity is essential for the second crop in the northern States of the Country as it is the source of soil moisture to the seasonal crops. Under the projected climate change scenario, it is the uncertainties in monsoon, occurrences of extreme weather events like droughts and floods, heat and cold waves, hail storms, cloud bursts, will adversely affect the crops to a large extent. It indicates that the foodgrains production is more vulnerable to climate variability and it is an immediate threat when compared to climate change. Therefore, it is high time to go for proper adaptation strategies to evolve suitable varieties in the case of wheat and rice and other foodgrains crops to withstand against the extreme weather events for sustenance of agricultural production in India.

3.5 Climate change and quality produce

The quality of fruits, vegetables and foodgrains is deterioating in recent years due to various reasons and health is adversely affected. The quality of intake is one of the reasons for illness. Few reports indicate that the global warming is one of the reasons for poor quality apples due to exposure to high atmospheric temperature in absence of soil moisture. The tad taste in apples is explained due to global warming (Plate 3.1). The milling percentage is also decreasing in rice and wheat. The plantation crops like coconut, cashew, cardamom, coffee, cocoa, black pepper, tea, areca nut, rubber and oil palm play a major role in export economy at the national level. In fact, the quality produce is the one which plays a pivotal role in plantation crops in deciding the marketability outside. For the first time in 2002 when the monsoon failed across the Western Ghats, the effect of monsoon and its behaviour on the quality of 'Monsoon Malabar Coffee' was highlighted in its processing. If the coffee beans are processed in

Has climate change altered apples' taste?

- Verities : Fuji and Tsugaru in Japan
- The taste and texture of apples may have undergone a considerable change over the last many years due to climate change
- Increase in global temperature over the years, specifically during the maturation period of the fruits, may lead to tad taste in apples
- Temperature increase during bud breaking and blooming times of the fruit, may advance of bud breaking and full bloom in apples

Source: The Hindu dated : 22TM August 2013

Apples might have become tad sweeter over the years

Plate 3.1: Tad taste in apple due to global warming

July when the active phase of monsoon is noticed along the West Coast, the quality of black coffee is superior in terms of less acidity and better sugar content. In the case of coconut also, coconut size is likely to be adversely affected when its active development phase (4–7 months) coincides with the hot summer during which the surface temperature is high. Probably, it is one of the reasons, why, nut size is relatively small in Tamil Nadu, Karnataka and Andhra Pradesh in non-traditional coconut growing areas though the number of nuts harvested per palm is high. The mean summer temperature across the coconut plantations is always between 40 and 45° C over non-traditional areas in the South while traditional coconut growing regions experience between 30 and 35° C. In the case of cashew also, nut size and kernel weight are much influenced by the maximum temperature during the nut development period. Also, a good summer shower during the nut development stage in sandy soils is enough to obtain better nut size in cashew. In contrast, unusual rains during the nut development stage may lead to immature fruit drop, thereby decline in nut quality along the West Coast. The unusual summer rains in 1984 and 2008 adversely affected the nut quality in cashew. The inferior nut quality of cashew due to unusual summer rains adversely affects the farmers' income and future market to a greater extent. The nut size and kernel weight of the same variety also vary significantly depending upon the agroclimatic zone. All these observations indicate that the climate variability is one of the factors deciding the quality produce to a greater extent, which is one of the components of food safety. Therefore, further studies in this direction need to be taken up on priority basis under the projected climate change scenario.

3.6 Climate variability and food price

One of the reasons for the global food price increase is due to frequent occurrence of weather abnormalities like floods and droughts. The foodgrains production is not in tune with the demand whenever weather related disasters take place. The foodgrains production is stagnated overall since last 15 years as several parts of the world experienced floods and droughts or heat and cold waves. It was attributed to global warming and the eight warmest years was seen during the decade of 2001-10. In addition, it appears that some pieces of the land cultivatable under foodgrains are shifted to bio-fuels across the world. It aggravated the availability of foodgrains, resulting in escalation of food price the world over. The crop loss varied between 40% and 90% in West African countries depending upon the severity of droughts in 2004-05. World cereal output in 2005 fell due to adverse hot and dry weather, and drought that hit crops in parts of the European Union. Australia and other wheat growing countries suffered heavily in 2006 due to unprecedented drought, and the wheat production was only half of the mark in 2006 as against 20-25 million tonnes normally produced in Australia. India, Bangladesh, China and other countries in Southeast Asia, too suffered due to frequent floods and droughts since the last one decade, where rice is produced largely. Floods in Pakistan, and China, cloud bursts in Leh, Ladakh region of Jammu and Kashmir, forest fires and heat waves in Russia during August 2010 and heavy rains in southern States of the Country from October to December due to depressions and cyclones were as a result of global warming and climate change. Forest fires and heat waves in Russia devastated thousands of hectares of wheat. As a result, cereal price escalated the world over. At the regional level, unusual summer rains in southern states of India resulted in heavy loss of paddy crop and vegetables during 2008. The price of both the commodities was very high locally. Similarly, extended rains and cyclones during northeast monsoon in southern states like Andhra Pradesh, Tamil Nadu and Kerala affected several crops adversely in 2010. These states were the worst hit in 2018 due to floods and cyclones. The paddy crop was the worst hit. On an average, the current price of these food items is more by 50-70% when compared to prices in 2007. It was attributed to the impact of climate change. Even in 2007, prices were soaring due to scarcity of foodgrains the world over and more demand due to increase in population and their purchasing power. On the one hand the farmers agitate for better price and on the other; consumers need to pay heavy price for food. These states were the woo... Phi in 2018 due to flood and cyclones. This scenario becomes acute when weather related disaster takes place. Therefore, there is an urgent need to tackle the issue on a war-footing and to take steps for sustenance of foodgrains production and food security so as to stabilize food price under the projected climate change scenario. The projected climate change scenario indicates that the frequency of weather related abnormalities like floods, droughts, and heat and cold waves is likely to increase in ensuing years and food insecurity and escalation in food price are likely to persist if corrective proactive measures are not taken up by the various political governments to sustain foodgrain production against weather related disasters.

3.6.1 Climate variability and rice in coastal Andhra Pradesh (Rice Bowl of India)

Marginal increase in annual rainfall and temperature is noticed across the State of Andhra Pradesh though slight decline in monsoon rainfall was the trend since last 60-70 years as per the report of the India Meteorological Department (Table 3.8). It is one of the predominant States in India in producing rice and maize. The years 2014 and 2015 were the consecutive drought years since last 50 years during *kharif* across the Country after 1965 and 1966. The summer 2017 was the hottest globally, followed by 2016 and 2015. The coastal belt of Andhra Pradesh experiences invariably heat wave conditions during May, June and July if monsoon breaks/fails. Early rains in June and July, followed by no rains during September and October across the coastal belt of Andhra Pradesh resulted to grain chaffing in paddy during *kharif* due to high maximum temperature (between 35 and 40° C) in absence of rains and standing water in paddy fields though the crop appearance was good in the year 2015. Majority of farmers obtained only between 1.5 and 2.5t/acre even less than that during *kharif* 2015. The area cultivated under paddy was also relatively less due to poor rainfall in *kharif* 2015.

Table 3.8: Rainfall and temperature trends in Andhra Pradesh (Source: IMD)

Season	Rainfall (mm/year)	Temperature (°C/year)	Remarks
Annual	1.31	0.01	
Winter	0.29	0.01 (0.02-Max)	Flowering in fruit crops is likely to be adversely affected (Mango)
Summer	0.35	0.01	Pulses (Black gram, green gram and green manure crops can be taken up
Monsoon	-0.14	0.01	Rice yields can be improved if irrigation is assured – Examples of 2015 and 2016
Post monsoon	0.46	0.01 (0.02-Max)	Flowering in fruit crops is likely to be adversely affected (Mango)

The first time release of Pattiseema water during September 2015 was of no use and not availed properly by rice farmers along the Krishna Delta (Plate 3.2). In contrast, proper utilization of water from Pattiseema and extended rains in September and October resulted to better harvest (2.5-3.0 t/acre) in 2016.

Table 3.9 Rice yields in Andhra Pradesh (*Source*: DoA, GoAP)

Year	Area in lakhs (ha)	Yield (kg/ha)	Remarks
2014	16.35	4677	-
2015	13.99	4403	• Rains in August and September and insignificant rains during October and November – crop was good and chaffing grains due to high temperature in absence of optimum soil moisture. • Farmers obtained 3500-6000 kg/ha
2016	15.21	5197	• Poor monsoon –Poor rains – rains in September and October resulted to good crop. • Farmers obtained 6000-8000 kg/ha. Similar paddy yields are obtained during 2017 too. • However, 2016 was the best. Similar yields are expected in *kharif* 2018.

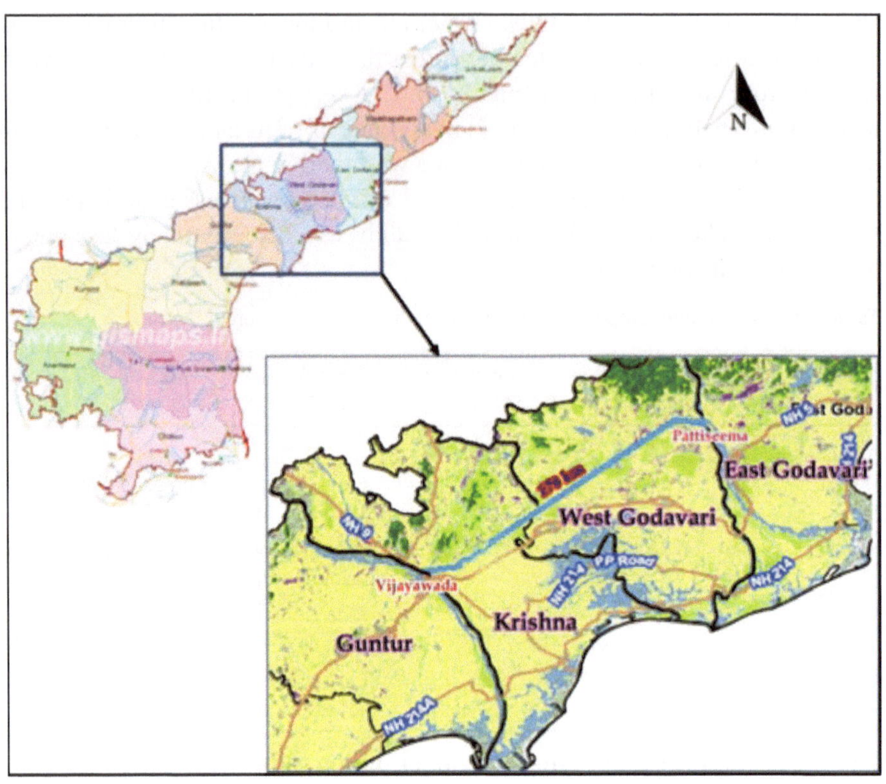

Plate 3.2: Pattiseema- A Canal connecting between Godavari and Krishna rivers for the benefit of farmers

The year 2016 was a record year during which paddy production was the highest in Andhra Pradesh (Table 3.9). Reports indicated that more than 5t/ha was recorded in *kharif* 2016 against 4.4 t/ha in 2015 and 4.6t/ha in 2014 across the rice belt of Andhra Pradesh. In terms of *kharif* rice productivity, identical situation was noticed in 2017 and 2016. However, the best overall rice yields were obtained in 2016 only. *Kharif 2018* is likely to be again a better year for rice along the coastal belt (Krishna Delta) though rainfall was scanty during reproductive and ripening phases. BPT 5204 is popular among farmers in this part of Andhra Pradesh. Fortunately, this region escaped from cyclones during the harvest time since last five years except in 2018. Due to change in monsoon rainfall pattern, the area under paddy was declining while increase in productivity was noticed since last three years (2016, 2017 and 2018) despite high temperature prevailed in June and July, coincides with dry sown of paddy. One of the reasons for better grain yield in paddy during the above years was attributed to timely utilization of fertile canal water of Pattiseema. Pattiseema is a canal connected between Godavari and Krishna rivers for utilization of flood water in Godavari

River. Crop season was well maintained due to intervention of Pattiseema Canal water in Krishna and Guntur Districts of Andhra Pradesh in all the three years (2016, 2017 and 2018) even under uneven and poor distribution of rainfall (Fig. 3.11). Therefore, timely intervention of Pattiseema Canal water across the Krishna Delta is a boon to rice farmers for timely cultivation of paddy, followed by maize/pulses/vegetables as well summer pulses without missing crop season/ calendar unlike late release of water due to uncertain rainfall and climate change in the past. No release of canal water was possible during *kharif* 2016, 2017 and 2018 but for Pattiseema in Krishna Delta due to insufficient water in Krishna River. In the past, farmers of this region know only transplanted paddy and dry sown paddy is gaining momentum in recent years against climate risks and shortage of farm workers (Plate 3.3). Water saving, high density of plant population, intact of crop calendar against monsoon aberrations and saving in farm labourer are the advantages of Pattiseema Canal brought from Godavari river to Krishna river for the benefit of rice farmers (Plate 3.4, 3.5 and 3.6). No cyclone hit in November/December across the Krishna Delta was another fortune in obtaining record rice yields in this part of Andhra Pradesh from 2016 to 2018. This type of intervention from the Govt. is a big boon to the farmers of Krishna Delta as part of climate risk management since crop calendar is well maintained with water assurance during *Kharif* season.

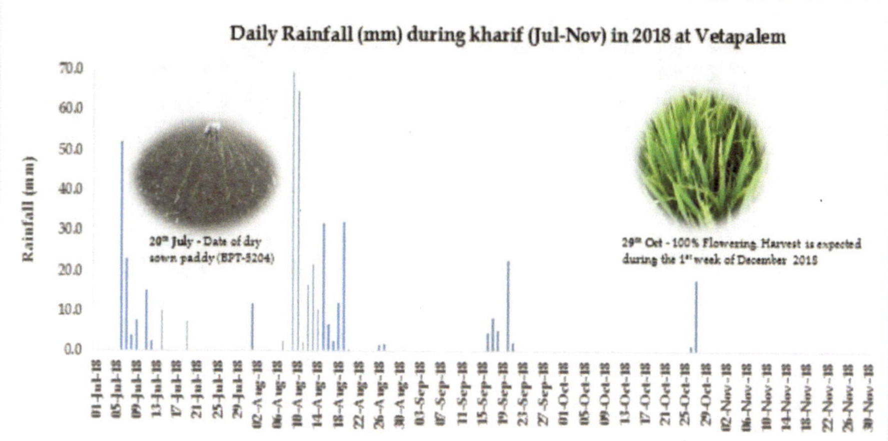

Fig. 3.11: Daily rainfall (mm) during kharif (Jul-Nov) in 2018 at Vetapalem (Guntur district, Andhra Pradesh)

Plate 3.3: Dry sown paddy during second fortnight of July-water and farm labour soving mechanism

Plate 3.4: Weeding in dry sown paddy

Plate 3.5: Rice is in reproductive stage-better yields are expected in Kharif, 2018

Plate 3.6: One of progressive farmers obtaines better rice yields in 2016-17 and 2018 (R. Abbaiah, Vetapalen)

Summary

The global food price increase, one of the reasons, is due to frequent occurrence of weather abnormalities like floods, droughts cold and heat waves. The Indian foodgrains production is not in tune with the demand whenever weather related disaster takes place. The rate of increase in foodgrains production is stagnated overall since last 10-12 years as several parts of the World experiences floods and droughts or heat and cold waves. In addition, it appears that some pieces of the land cultivatable under foodgrains are shifted to bio-fuels and horticultural crops across the World. It aggravated the availability of foodgrains, resulting in escalation of food price the World over. The crop loss during drought years varied between 40% and 90% in West African countries depending upon the severity of droughts. World cereal output in 2005 fell due to adverse hot and dry weather, and drought that hit crops in parts of the European Union. Australia and other wheat growing countries suffered heavily in 2006 due to unprecedented drought, and the wheat production was only half of the mark in 2006 as against 20-25 million tonnes normally produced in Australia. India, Bangladesh, China and other countries in Southeast Asia, too suffered due to frequent floods and droughts since the last one decade, where rice is produced largely. As a result, the food price escalated the World over. At the regional level, unusual summer rains in southern states of India resulted in heavy loss of paddy crop and vegetables during 2008. The price of both the commodities was very high locally. On an average, the current price of these food items is more by 50-70% when compared to prices in 2007. Even in 2007, prices were soaring due to scarcity of foodgrains the World over. It was attributed to rising demand for foodgrains and the impact of climate change. On the one hand the farmers agitate for better price and on the other hand, consumers need to pay heavy price for food. This scenario becomes acute when weather related disaster takes place. The projected climate change scenario indicates that the frequency of weather related abnormalities like floods, droughts, and heat and cold waves is likely to increase in ensuing years and food insecurity is likely to persist if corrective proactive measures are not taken up by the various political governments against weather related disasters. Therefore, short term proactive measures play a major role along with long term strategies to tackle the issues of foodgrains production, food security, food safety in terms of quality and escalation of food price under projected climate change scenarios. The scenario of climate change and food security was highlighted by FAO, WFO and WHO is presented in plate 3.7 which is self explanatory. This plate is just highlighted to focus on relavance of climate change in food security.

Climate Change and Food Insecurity

(FAO, WFO and WHO)

- One in nine people, still, suffer from hunger - 805 million people A staggering two billion people are affected by iron deficiency
- More than 250 million children suffer from Vitamin A deficiency - which is a major public health challenging in more than half the countries on the planet - with half a million going blind each year. Half of these children die within 12 months of going blind 1.3 billion of us are classified as overweight or obese, fuelled by a food system that is damaging not just our bodies but the environment too
- We continue to destroy tropical forests for agricultural expansion and this contributes 12 per cent to the total warming of the planet today
- We waste much of the food we produce. What is a crisis for many now could become a catastrophe for more in the future because of the effects of climate change
- Climate change is already making people hungry, by disrupting crop yields, pushing prices up and increasing food insecurity for large number of the World's population. And it is not food but nutrients that are becoming scarcer as the climate changes
- Smallholder farmers - they are able to grow, sell and eat more nutritious foods and it is crucial - a food system for the future

Plate 3.7: Climate change and food security

Acknowledgement

Thanks to Dr. C.S. Gopakumar, Scientific Officer, ACCER, KAU, V. M. Sandeep and V.P. Promod (Ph.D students) from CRIDA who assisted me in preparing this chapter with good illustrations, tables and plates.

References/Suggested Readings

Food and Agricultural Organization. 1996. Declaration on world food security. World Food Summit, FAO, Rome.

Ilangovan, R., Beevi, P., Rao, GSLHVP and Moosa, P.P. 2011. Impact of climate change on soil degradation and rice productivity at Pattambi, Kerala in the humid tropics. In: Climate Change Adaptation Strategies in Agriculture and Allied Sectors. Ed: G.S.L.H.V. Prasada Rao. Scientific Publishers (India), 2011. pp. 220-225.

IPCC. 2007(Fourth Assessment Report of the Intergovernmental panel on Climate Change: The Impacts, Adaption and Vulnerability (WG III). Cambridge University press, United Kingdom and New York, NY, USA.

IPCC (Intergovernmental Panel on Climatic Change) 2006. The Economics of Climate Change: Stern Review. The Summary of conclusions. Survey of the Environment 2007, The Hindu, pp141-145.

IPCC (Intergovernmental Panel on Climatic Change) 2007. Climate Change: The Physical Science Basis. Extracts from the IV Assessment Report. Survey of the Environment 2007, The Hindu, pp147-155.

Lal, M., Nozawa, T., Emori, S., Harasawa, H., Takahashi, K., Kimoto, M., Abe-Ouchi, A., Nakjima, T., Takemura, T. and Numaguti, A. 2001. Future climate change: Implications for Indian summer monsoon and its variability. *Current Science*. 81: 1196-1207.

Mathews, R.B., Krofpff, M.J., Horie, T and Bachelet, D. 1997. Simulating the impact of climate change on rice production in Asia and evaluating options for adoption. *Agric. Syst.* 54: 399-425.

Parthasarathy, B., Moonot , A and Kothawala, D. 1988. Regression model for estimation of Indian foodgrains production from summer monsoon rainfall. *Agri forest Meteorol*, 42: 167-182.

Prasad, R and Rana, R. 2006. A study on maximum temperature during March 2004 and its impact on rabi crops in Himachal Pradesh. *J. of Agrometeorology*, 8(1): 91-99.

Ramakrishna, Y.S., Rao, G.G.S.N., Rao, S.G. and Vijayakumar, P.2006. Impact of climate change in Agriculture. In: Environment and Agriculture (eds. Chadha, K.L. and Swaminathan, M.S.). Malhotra Publishing House, New Delhi, pp. 1-30.

Rao, G.S.L.H.V.P. 1990. Seasonal and weather influence on growth, duration and grain yield of wetland rice at Pilicode. Rice in wetland ecosystem. Proceedings of the National Symposium in rice in wetland Ecosystem held at Kottayam, Kerala during December 19-21, 1990.

Rao, G.S.L.H.V.P. 1993. Rice and weather. In sustaining rice productivity through integrated nutrient management. Published by KAU, Trivandrum. Pp.36-46.

Rao, G.S.L.H.V.P. 1990. Effect of temperature on the duration of vegetative phase of rice varieties in low lands of Kasaragod District. *Madras Agric. J.*, 81(3):134-136.

Rao, G.S.L.H.V.P. 1994. Effect of date of planting on the duration and grain yield of rice cultivar Allikkannan. *J. of Tropical Agriculture* 32(1994):96-97.

Rao, G.S.L.H.V.P. 1999. Crop growth simulation models of rice under humid climates. Advanced Technologies in Meteorology. Tata McGraw-Hill Publishing Company Limited. New Delhi. Pp.436-439.

Rao, G.S.L.H.V.P. and Subash, N. 1998. Monsoon rainfall and *kharif* rice production of Kerala . Proc. of the Tenth Kerala Science congress. Kozhikode. Pp.43-46.

Rao, G.S.L.H.V.P. and Nair, R.R 1983. Water balance and rice production in Cannanore district of North Kerala. In Proc. Of National Symposium on water balance and national development, 19-21 December, AU, Waltair. Pp.157-159.

Rao, G.S.L.H.V.P., Krishnakumar, K.N and Xavier Tony. 2001 Occurrence of meteorological droughts during monsoon and its impact on paddy yield in tropical monsoon climates. In Proc.of the Thirteenth Kerala Science Congress. January 2001, Thrissur pp.504-507

Rao, G.S.L.H.V.P, Rao, G.G.S.N. and Rao, V.U.M. 2011. Climate Change and Agriculture over India. PH India (Pvt.) Ltd. New Delhi India, 320p.

Rao, G.S.L.H.V.P. 2010. Impact of Climate Change on Agriculture and allied sectors. Scientic Publishers, Jodhpur, India, 316p.

Rao, G.S.L.H.V.P. 2018. Agricultural Meteorology (Fourth Printing). PHIndia (Pvt.) Ltd. New Delhi India, 364p.

Rice and Weather 1975. World Meteorological Organization. Technical Note No.144. pp.41.

Samra, J.S., Singh, G and Ramakrishna, Y.S. 2004, Jan.10 cold wave during 2002-03 over north India and its effect on crops. *The Hindu*.p.6.

Samui, R.P 1999. A note on the weather and rice yield relationship at some stations in India. Proc. Indian Acad. Sci.(Earth Planet. Sci.) , 108, No.4, December 1999, pp.309-316.

Sehu, D.V and Cady , F.D. 1984. Response of rice to solar radiation and temperature estimated from international yield trails. *Crop Sci.* 24:649-654.

Shukla, P.R., Sharma, S.K and Ramana, V.P. 2002. Climate Change and India - Issues, Concerns and Opportunities. Tata – McGraw-Hill Publishing Company Ltd, New Delhi, 314p

Sreenivas, G., Raji Reddy, D and Narasimba Rao., S. B. S. 2005. Influence of weather parameters on yield of lowland rice (Oryza sativa L.). *J. of Agrometeorology* 7(1):69-75 (June 2005).

Venkataraman S. 1987. Agrometeorological aspects of growth, yield and water relations with special reference to rice. Weather and Climate. : Proc. International workshop on the impact of weather parameters on growth and yield of rice, April , 1986, IRRI, Philippines pp.47-54.

Rao, G.S.L.H.V.P., Mohan, H.S.R., Gopakumar, C.S and Krishnakumar, K.N. 2008. Climate change and cropping systems over Kerala in the humid tropics. *J. of Agrometeorology* (Special issue-Part 2): 286-291.

Rao, G.S.L.H.V.P., Rao, A.V.R.K., Krishnakumar, K.N and Gopakumar, C.S. 2010a. Impact of climate change on food and plantation crops in the humid tropics of India. ISPRS Archives XXXVIII-8/W3. Workshop proceedings: Impact of climate change on Agriculture. pp.127-130.

Saseendran, S.A., Singh, K.K., Rathore, L.S., Singh, S.V and Sinha, S.K. 2000. Effects of climate change on rice production in the tropical humid climate tof Kerala, India. *Climate Change*. 44(4). 495-514.

4

Climate Change Causes and Effects on Agriculture and Farmers Economy

Ch. Srinivasa Rao

Central Research Institute for Dryland Agriculture
Santoshnagar, Hyderabad - 500 059, India

The Intergovernmental Panel on Climate Change (IPCC) in its fifth assessment report (AR5) stated that warming of the climate system is unequivocal and is more pronounced since the 1950s. The atmosphere and oceans warmed, the amounts of snow and ice have diminished, and rise in sea level is noticed. Each of the last three decades has been successively warmer at the earth's surface than any preceding decade since 1850 and the globally averaged combined land and ocean surface temperature data show a warming of 0.85°C over the period 1880 to 2012. The recent years viz., 2015, 2016 and 2017 were the hottest years on record. The warming trend in India since 1961 has indicated an increase of 0.9°C, which is likely to impact many crops, negatively impacting food production. There are already evidences of negative impacts on yields of wheat and paddy in northern India due to increased temperature, water stress, and reduction in number of rainy days. Coastal agriculture has the onus of providing livelihood to nearly 20 % of Indian population. The water logging and coastal salinity are the major factors affecting the agricultural output from the regions. During the past decade, frequency of the extreme events such as droughts, cyclone and hailstorms increased, with 2002, 2004, 2009, 2012, and 2014 being severe droughts. Frequent cyclones and severe hailstorms in various central and southern Indian states and drought prone areas have become common. Eastern part of the country is affected by sea water intrusion. Reduced food grain productivity, loss to vegetable and fruit crops, fodder scarcity, shortage of drinking water to animals during summer, forced migration of animals, severe loss to poultry and fishery sectors were registered due to the climatic stresses, threatening the livelihoods of rural poor particularly in the coastal regions of the country.

Climate change projections for Indian subcontinent up to 2100 indicate an increase in temperature by 2-4^{0}C, with different regions expected to experience differential changes in the amount of rainfall. India will experience an increase in mean and extreme precipitation during monsoon. This will increase both floods and drought. Freshwater resources will be affected due to combination of climate change and unsustainable practices. It is projected that Indian agriculture will lose over US $7 billion by 2030. There will be large reductions in wheat yield in the Indo-Gangetic plain; and substantial increase in heat stress for rice, affecting yield in the country. Coastal flooding will affect people and agriculture and also affect tourism in India. Some of the fish and other marine animals will face extinction by 2050, affecting fishing community. Glaciers in Himalaya continue to shrink affecting run-off and water resources downstream. Impacts of global climate change will be felt among the populations, predominantly on subsistence and smallholder farmers. Since agriculture currently contributes about 17 % of India's gross domestic product (GDP), a negative impact on production implies cost of climate change to roughly range from 0.7% to 1.35% of GDP per year especially with rural India. The Indian agriculture production system faces the daunting task of feeding 17.5% of the global population with only 2.4% of land and 4% of water resources at its disposal. India is more vulnerable to climate change in view of the different vulnerabilities such as heat stress, cold waves and seasonal flooding in northern India; seasonal flooding, sea water intrusion, soil salinity and severe drought in eastern India; drought, salinity, severe cyclones and scarcity monsoon in Southern India.

4.1 Extreme Events

The IPCC Fifth Assessment Report observed that the coming years will see more extreme weather events (floods, cyclones, cloud bursts, unseasonal excessive rains and drought, etc.) in most parts of the globe and India will be among the most affected countries. In recent years the occurrence of extreme weather events is on rise. Studies based on the observed precipitation records of India Meteorological Department (IMD) have shown that the occurrence of extreme precipitation events and their variability has already gone up in many parts of India (Goswami *et al.*, 2006). Chaturvedi *et al.* (2012) projected consistently increasing trends in frequency of extreme precipitation days (e.g. >40 mm/day) for decades 2060s and beyond. Cold waves have shown significant impact on crop production in northern India. Cold waves during December 2002 and January 2003 caused considerable damage to horticultural crops such as mango, guava, papaya, brinjal, tomato and potato crops. Maximum number of cold waves generally occurs in Rajasthan followed by Jammu and Kashmir and Uttar Pradesh. The frequency of events over different time periods indicates that in recent years the state of Rajasthan is experiencing more cold waves and

Jammu and Kashmir is experiencing a few (Sikka *et al.*, 2016). Hail formation requires environments with strong upward motion of air and/or lowered heights of freezing level. In the middle latitudes, hailstones are formed near the interior of continents while in tropics, they tend to be confined to higher levels of freezing (Rao *et al.*, 2014). The maximum number of heat waves occurs over East Uttar Pradesh followed by Punjab, East Madhya Pradesh, and Saurashtra and Kutch in Gujarat (Raghavan, 1967).

4.2 Causes of Climate Change

The agriculture contributes to global warming primarily through the emission of green house gases (GHGs) such as CH_4, N_2O and CO_2. CH_4 is produced in soil during microbial decomposition of organic matter under anaerobic conditions. Rice fields submerged under water are the potential source of CH_4 production. Continuous submergence, higher organic C content and use of organic manure in puddled soil enhance CH_4 emission. Burning of crop residues contributes to the global methane budget. The enteric fermentation in ruminants is major source of CH_4 emission.Nitrous oxide emissions from soils through the processes of nitrification and denitrification is another important source of GHGs. The main source of carbon dioxide production in agriculture is the soil management practices such as tillage which triggers emission of this gas through biological decomposition of soil organic matter. Tillage breaks the soil aggregates, increases the oxygen supply and exposes the surface area of organic material resulting in decomposition of organic matter. Use of fuel for various agricultural operations and burning of crop residues are other sources of carbon dioxide emission. An off-site source is the production of carbon dioxideduring manufacturing of farm implements, fertilizers and pesticides (Pathak *et al.*, 2010). In India, the GHG emissions from the agriculture sector, including crop and animal husbandry together are 3,90,165Gg of CO_2 eq., which constitutes about 18% of the total GHG emissions of the country during 2014. The enteric fermentation constituted 58% of the total CO_2 eq. emissions and the emissions from the paddy system contributes to 18% of the total emissions from the agriculture sector. Agricultural soils emitted 21% of the total CO_2 eq. whereas burning of crop residues in field contributed to 2% of the total GHG emissions from the agriculture sector (Table 4.1)

Table 4.1: Greenhouse gas emissions from the agriculture sector in India

S.No	Sector	Emissions from sector	Percentage of the total
	Agriculture sector as a whole	3,90,165Gg of CO2 equivalent emissions	18% of the total emissions of the country
A	Enteric fermentation	227033	58%
B	Manure management	2768	0.7%
C	Paddy cultivation	71368	18%
D	Nitrogen	81081	21%
e	Burning of crop residues	7915	2%

Source: Biennial Update Report of the Ministry of environment and forests published in 2016

4.3 Impacts of Climate Change on Agriculture

Indian agriculture is predominantly monsoon dependent; climate has been playing a major role in agricultural production. There have been several instances and events where adverse climate has affected the crop production. Past events include, for instance, drought of 2002 which led to reduced coverage of more than 15 million hectares of the rainy-season crops and resulted in a loss of more than 10% in food production. Higher temperatures during March in 2004, in the Indo-Gangetic plains, caused wheat crop to mature earlier by 10-20 days resulting in reduction in wheat production by more than 4 million tonnes in the country. Similarly, irregular monsoon in 2009 caused significant reduction in rice production. Apple productivity declined to the tune of 40-50% in Himachal Pradesh up to 1500 m mean sea level (msl), partly due to warmer climate resulting in lack of chilling requirement during winter and warmer summers. This is also one of the causes for shifting of apple production to higher elevation (2700 m msl). The projected climate change will not only influence the production and productivity of crops but is also likely to influence the quality of produce. Studies indicated that protein content of wheat, green gram and chickpea grains increased marginally with rise in temperature, whereas it decreased with rise in CO_2 level. Studies conducted by National Dairy Research Institute, Karnal indicated that global warming is likely to lead to a loss of 1.6 million tonnes in milk production by 2020 and 15 million tonnes by 2050, if no adaptation measures are taken. Lactating cows and buffaloes have higher body temperature and are unable to maintain thermal balance. Body temperature of buffaloes and cows producing milk is 1.5-2 °C higher than their normal temperature, therefore more efficient cooling devices are required to reduce thermal load of lactating animals. A rise in temperature could have important and rapid effects on the health of fish and their geographical distributions. With warming of sea surface, the oil sardine is able to find temperature to its preference especially in the northern latitudes and eastern longitudes, thereby extending the distributional boundaries and establishing fisheries in larger coastal areas. In recent years the phenomenon of Indian Major Carps maturing and spawning as early as March is observed and as a result it has been possible to breed them twice in a year at an interval ranging from 30-60 days.

The Inter-governmental Panel on Climate Change (IPCC), in its 5th Assessment Report (AR5) warned of dire consequences of climate change on human health, settlements and natural resources, if no measures are taken to curb the ill-effects of global warming. The region-wise findings of the report have some key messages for agriculture and food security of India. It observes that the coming years will see more extreme weather events (floods, cyclones, cloud bursts, unseasonal excessive rains and drought, etc.) in most parts of the globe and India will be among the most affected countries. According to IPCC report, India will experience decrease in seasonal mean rainfall and an increase in mean and extreme precipitation during monsoon. This will increase both floods and drought. There will be large reductions in wheat yield in the Indo-Gangetic plain; and substantial increase in heat stress for rice affecting yields in the country. Coastal flooding will affect people and agriculture. The Report suggested that severe impacts of climate change in future and can be avoided or minimized provided countries act collectively and quickly. To deal with these consequences and ensure food security and livelihoods, adaptation to climate change is essential.

4.4 Impacts of Climate Change on costal ecosystems

The major natural disaster that affects the coastal regions of India is cyclone and as India has a coastline of about 7516 km; it is exposed to nearly 10% of the world's tropical cyclones (TC). On an average, about five or six Tropical Cyclones form in the Bay of Bengal and Arabian Sea, and hit the coast every year. Out of these, two or three are severe (Sikka *et al.*, 2016). When a cyclone approaches to coast, a risk of serious loss or damage arises from severe winds, heavy rainfall, storm surges, and river floods. Most cyclones occur in the Bay of Bengal followed by those in the Arabian Sea and the ratio is approximately 4:1. The incidence of cyclonic storms, with wind speeds between 65 and 117 kmph and severe cyclonic storms with wind speeds between 119 and 164 kmph, reaching Tamil Nadu and Andhra Pradesh is high during the northeast monsoon season (October–December), whereas the highest annual number of storms and severe storms occur in the Odisha-West Bengal coast. Flash floods occur due to high rate of water flow as also due to poor permeability of the soil. Most of the floods occur during the monsoon period and are usually associated with tropical storms or depressions and active monsoon conditions (Sikka *et al.*, 2016). Coastal zones contain some of the most densely populated areas in the world as they generally present the best conditions for productivity. However, these regions face many hydrological problems like flooding due to cyclones and wave surge, and drinking water scarcity due to problem of saltwater intrusion (Bhosale and Kumar, 2002).

4.5 Development of climate resilient agriculture in Andhra Pradesh

Climate resilient agriculture involves integration of adaptation, mitigation, and other practices in agriculture which increases the capacity of the system to respond to various climate-related disturbances by resisting or tolerating the damage and recovering quickly. Such perturbations and disturbances can include events such as drought, flooding, heat/cold wave, erratic rainfall pattern, long dry spells, insect or pest population explosions, and other perceived threats caused by changing climate (NAAS, 2013). Climate resilient agriculture includes an inbuilt property in the system for the recognition of a threat that needs to be responded to, and also the degree of effectiveness of the response and focuses on judicious and improved management of resources in an integrated way through adoption of best bet practices (Srinivasarao *et al.*, 2016). To meet the challenges of sustaining domestic food production in the face of changing climate and to generate information on adaptation and mitigation in agriculture to contribute to global fora like UNFCCC, the Indian Council of Agricultural Research (ICAR) launched a flagship network project '*National Initiative on Climate Resilient Agriculture*' (NICRA) during XI Plan in February 2011, and during XII Plan it is referred as '*National Innovations in Climate Resilient Agriculture*' (NICRA). The major objective of NICRA is to enhance the resilience of Indian agriculture in crops, livestock and fisheries to climatic variability and climate change through development and application of improved production and risk management technologies; to demonstrate site specific technology packages on farmers' fields for adapting to current climate risks; and to enhance the capacity (knowledge, skill and management) of scientists and other stakeholders in climate resilient agricultural research and its application.

To address the issues related to climate variability and to enhance the adaptive capacity of communities, extensive farmer participatory, demonstrations of location-specific climate resilient technologies were initiated on farmers' fields in each of the 121 climatically vulnerable districts of the country as part of National Innovations in Climate Resilient Agriculture (NICRA) under Technology Demonstration Component (TDC) during 2011. The climatically vulnerable districts of the country were identified based on a scientific analysis of climate related problems; exposure, sensitivity and adaptive capacity in relation to climate change (Rama Rao *et al.*, 2013).The adaptation and mitigation strategies were identified following a bottom-up approach in participatory mode. The objective of the programme is not only to demonstrate the climate resilient agriculture technologies but also to institutionalize the mechanism at the village level for continued adoption of such practices in a sustainable manner. As part of TDC, demonstrations of climate resilient technologies are being taken up in five districts of Andhra Pradesh. The districts involved, the climatic vulnerability addressed and significant interventions are furnished in Table 4.2 and Fig. 4.1.

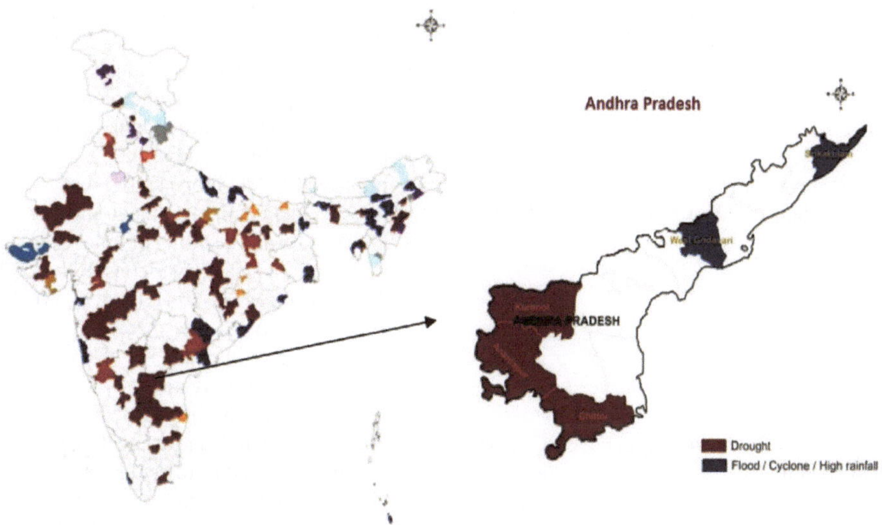

Fig. 4.1: Map showing the districts involved in Technology Demonstration in the country and Andhra Pradesh

Table 4.2: Districts involved in TDC of NICRA and the climatic vulnerability addressed in the State of Andhra Pradesh

District	NICRA village	Climatic vulnerability	Major interventions
Anantapur	Chamaluru, Chakrayapeta	Drought	• Water harvesting structure (farm pond, Check dam) • in situ moisture conservation practice for groundnut • Tank silt application • Introduction of drought resistant and high yielding varieties in groundnut • Crop diversification through foxtail millet • Introduction of fodder CoFS-29
Kurnool	Yagantipalle	Drought	• Conservation furrows in pigeonpea and seteria • Desilting of village tank • Tank silt application • Introduction of drought tolerant varieties in Pigeon pea var.ICPL-87119

			• Introduction of Seteria var SIA-3085 as a alternate crop to Desi Cotton • Intercropping System of Setaria+ Redgram(5:1) • Preservation of green fodder in silage bags • Hydroponic fodder production systems
Chittoor	Chittecherla	Drought	• Renovation of water tanks and drainage channel • Plastic mulching in horticulture crops • Short duration and drought tolerant paddy varieties of paddy, groundnut, horsegram • Direct seeding in paddy • Improved fodder crops cultivation
Srikakulam	Sirsuwada	Flood	• Improvement in field drainage channel • Renovation and desilting of village tank • Flood Tolerant paddy varieties RGL-2537, PLA-1100, MTU-1061 (Indra), MTU-1064 (Amara) • Direct seeding of paddy • Zero till maize cultivation • Captive rearing of Fish seed
West Godavarai	Matsyapuri	Cyclone and Flood	• Deepening of drainage channels • Soil health based fertiliser application· Direct sowing in paddy • Short duration paddy varieties (MTU-1121 & MTU 1156) • Flood tolerant varieties MTU 1061, MTU 1064 • Management of Water quality in fish pond

To address the climate vulnerabilities in the selected villages, various interventions were planned in four modules; however, the specific intervention under each module for a particular village was need based and also based on climatic vulnerability and resource endowments of the selected village. The four modules being implemented are (1) Natural resource management viz., *in-situ* moisture conservation, biomass mulching, residue recycling, manure management, soil health card–based nutrient application, water harvesting and recycling for supplementary or life saving irrigation, improved drainage in high rainfall/flood prone villages, conservation tillage, and water saving irrigation methods. (2) Crop production module consisting of introduction of short-duration and drought/ heat-tolerant varieties, modifications in planting dates for post rainy season crops to cope with terminal heat stress, water saving paddy systems viz., System of Rice Intensification, aerobic, direct seeding, staggered community nursery to meet delay in onset of monsoon, energy-efficient farm machinery through village custom hiring centers for timely completion of farm operations in limited sowing window, location specific intercropping systems, and suitable agroforestry systems. (3) Module III covers livestock and fisheries interventions through augmentation of feed and fodder production, fodder storage methods, and improved shelter for reducing cold/heat stress in livestock, management of fish ponds/tanks during water scarcity and excess water and promotion of integrated farming systems as adaptation strategy. (4) Module IV consists of village level institutions for developing enabling mechanisms and support system. This involves collective marketing groups, introduction of weather-based insurance and climate literacy though establishment of automated weather stations. Inputs from the relevant development departments of the state, zonal agricultural research stations of agricultural universities, progressive farmers are being and used in finalizing the interventions and action plans in a participatory manner (Srinivasa Rao *et al.* 2016).

4.6 Impacts of important climate resilient interventions taken up in frequently drought prone regions

The NICRA villages of the districts of Anantapur, Kurnool and Chittoor receive low rainfall and crops being affected by the prolonged dry spells, emphasis was laid on in-situ moisture conservation and harvesting runoff and its efficient use.

In-situ moisture conservation technology by subsoiling with chisel plough was demonstrated in groundnut and pigeonpea fields in Chamaluru village of Anantapur district. The subsoiling breaks the hard pan and helps in infiltration of the rainwater into the lower layers of soil. Sub soiling resulted in increased yields of groundnut pod and haulm by 11% and 24%, respectively. An additional income of Rs. 6300 was recorded with sub soling as compared to farmers' practice in groundnut crop.

Subsoiling for in-situ moisture conservation in at Anantpur, AP

In Yagantipalle Village in Kurnool district of Andhra Pradesh, desilting of village pond (Burrakunta) was taken up in the year 2012, wherein 1260 m^3 of silt was excavated and applied to farmers' fields. As a result, ground water was recharged and there was an increase in water table in the open wells and bore wells located around the village pond during the months of July to November during 2015-16. The water was used to irrigate an area of 107 ha in *kharif* and 96 ha in *rabi*season that resulted in increase in average in yields of sorghum (24.5 q/ ha) and maize (47 q/ha) in the village by 28% even in deficient monsoon conditions. During the year 2015-16, two check dams were desilted in NICRA village of Anantapur with a water storage capacity of 6000 m^3. Nearly 8 bore wells and 3 open wells were recharged in the vicinity of check dams benefitting farmers in the village. Water was utilized for irrigating high value crops, such as, sweet orange, pomegranate, Lilly flowers 2-3 times, growing fodder crops and for drinking water for livestock. The water was also used for growing leafy vegetables.

In village Chittecherla of Chittoor, demonstration of direct seeding method was taken up in paddy using drum seeder. About 9% yield increase was observed in comparison to the traditional transplanted method. Farmers opined that crop duration was reduced by 10-15 days and cost of cultivation was reduced by Rs.9723/ha. To demonstrate climate resilient crops and to minimize the risk of crop failure due to drought, short duration Seteria SIA-3085 and Suryanandi were takenup. Farmers got on an average yield of 19.54 q /ha of grain besides fodder and the returns are almost equal to the traditional cotton crop and the fodder will be the additional. Moreover the improved varieties mature 10-15 days early providing opportunities for rabi cropping of chickpea. The yield level and income of farmers could be substantially increased by the adoption of this technology and the improvement in yield was around 17.82 %. Seteria performed well and farmers got net returns of Rs.11,820/ha under rainfed situations.

Demonstration of pigeonpea + setaria (1:5) intercropping system was taken up, which resulted in higher gross income (Rs.57417/ha) than the sole crop of setaria (Rs. 28942/ha). This practice was proved to be a promising drought coping strategy and resulted in higher yields per unit area and time.

Short duration Seteria (SIA-3085) and Intercropping of Seteria + Pigeonpea in Kurnool

In village Chittecherla, Chittoor demonstrations on drought tolerant variety of groundnut Dharani was taken up in 1.4 ha area covering 7 farmers during rabi season. The results showed that about 11.3% increase in yield in Dharani variety over farmer's variety (Kadiri-6). The net returns obtained in the demonstration were Rs.53169/ha as compared to farmers' practice of Rs.41438/ha. Continuous drought in recent years Kurnool district has reduced the availability of green fodder, forcing the milk producers to depend on concentrates for optimum milk production which enhanced cost. To overcome this problem, low cost hydroponic fodder production model was introduced in the NICRA-village Yagantipalle. In this method, 6-8 kg of highly nutritious green fodder was produced from every kg of maize grain, within only 7 days, which solved the problem of green fodder scarcity to a great extent. An increase in milk yield (7%) and milk fat (8.97%) was observed by feeding 12 kg hydroponic maize fodder along with 7 kg dry roughages. Net income was increased by Rs.1580 due to reduction in concentrate feeding with increased hydroponic fodder. This resulted in higher B:C ratio of 8.98 as compared to farmers' practice which is 5.71.

Hydroponic fodder cultivation in Kurnool, AP

4.7 Impacts of important climate resilient interventions taken up in frequently flood prone districts of Andhra Pradesh

NICRA village Sirusuwada, Srikakulam is prone to floods during monsoon season. A drainage channel of 800 m long was dug to drain out the excess water from the fields, due to either floods or intense rainfall in order to reduce the period of water stagnation. The threat of complete inundation due to heavy rain was overcome in area of 45 acres and the period of complete inundation got reduced by 2-3 days during August 2015 when 109 mm of rain was received in one day.

Paddy is the main crop in the NICRA village of Srikakulam and farmers are growing varieties like BPT 5204 (Samba mashuri), MTU 7029 (Swarna), MTU 1001 (Vijetha) which are susceptible to flood situation though having high yield potential and market demand in the district. Hence, the flood tolerant paddy

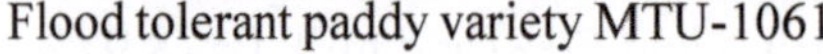
Flood tolerant paddy variety MTU-1061

Flood affected paddy varieties in Srikakulam

varieties like PLA 1100 (Badava Mashuri), RGL 2537 (Srikakulam sannalu), MTU 1061 (Indra) and MTU 1064 (Amara) were introduced and demonstrations were conducted in farmers' fields, which recorded higher yields of 5.9 q/ha compared to the farmers' practice of growing traditional varieties.

In Matsypuri village, West Godavari during *kharif* 2015 heavy rains were received during the cyclonic period in November 2015 (111.2 mm) in four days (17-11-2015 to 20-11-2015) and lodging tolerant varieties of paddy like MTU-1064 (Amara), MTU-1061 (Indra) and MTU-1140 were partially lodged whereas swarna (MTU-7029) was completely lodged due to heavy rains and high speed winds. It was observed that the yield increase was 39% in MTU 1061 and 31% in MTU 1064 compared to MTU 7029. Recommendations were also made to drain out the excess water from the fields and spray of validamycin or hexaconozole @ 2ml/lit as prophylactic measure to prevent sheath blight. Further, farmers were suggested to spray 5% salt solution to prevent grain discoloration and germination.

Damage due to lodging and flood tolerant paddy variety MTU 1061 in West Godavari, AP

In Matsypuri village during 2015-16 in *rabi*season, direct seeding of rice with drum seeder was demonstrated in 16 hectares area. Higher yields were obtained with paddy sown with drum seeder (81q/ha) as compared to farmer practice (74 q/ha) of manual transplanting. The cost of cultivation reduced in drum seeder sowing method by Rs. 4422/ha than manual transplanting and the crop also matured one week earlier than manual planting.

Summary

Climate change is considered as the biggest threat the human race has ever experienced. With the present resource use, particularly the non renewable resources and with the growing economies in several countries, the threat is likely to intensify and there is a need to adapt and make our production systems resilient against climatic variability which is the main concern in the short term. There is a need to intensify research to develop cultivars and develop management practices which can withstand the impact of climate change. The experiences under technology demonstration component of NICRA shows that technologies developed by the National agricultural research system can minimize the losses under variable climates and found to stabilize the productivity and incomes. These resilient practices need to be scaled up as part of the developmental programs for large scale adoption of resilient practices and to impart resilience to the agriculture production in the country.

References/Suggested Readings

Chaturvedi R. K, Jaideep J, Mathangi J, Bala G, Ravindranath, N H. 2012. Multi-model climate change projections for India under representative concentration pathways. *Current Science*, 103 (7): 10.

Goswami B. N, Venugopal V, Sengupta D, Madhusoodanan M. S, Prince K. X. 2006. Increasing Trend of Extreme Rain Events Over India in a Warming Environment. *Science*, 314: 1442-1445.

NAAS, 2013.Climate Resilient Agriculture in India. 26p

Pathak H, Bhatia A, Jain N and Aggarwal PK (2010) Greenhouse Gas Emission and Mitigation in Indian Agriculture – A Review, In Bulletin on Regional Assessment of Reactive Nitrogen, Bulletin No. 19, New Delhi, 1-34p.

Raghavan, K., 1967. Climatology of severe cold waves in India.Indian *J. Meteorol. Geophys.*18 (1), 91–96.

Rama Rao, C.A., Raju, B.M.K., Subba Rao, A.V.M., Rao, K.V., Rao, V.U.M.,Ramachandran, K., Venkateswarlu, B., Sikka, A.K., 2013. Atlas on Vulnerability ofIndian Agriculture to Climate Change.Central Research Institute for Dryland Agriculture, Hyderabad 116 p.

Rao V. U. M, Bapuji Rao B, Sikka A. K, Subba Rao A. V. M, Ranbir Singh and Maheswari M. 2014. Hailstorm threat to Indian agriculture: A historical Perspective and future strategies. Central Research Institute for Dryland Agriculture, 44p.

Sikka, A.K., Bapuji Rao, B., Rao, V.U.M., 2016. Agricultural disaster management andcontingency planning to meet the challenges of extreme weather events. Mausam 67(1), 155–168.

Srinivasa Rao, Ch., Gopinath, K.A., Prasad, J.V.N.S., Prasannakumar, Singh, A.K. (2016). Climate Resilient Villages for Sustainable Food Security in Tropical India: Concept, Process, Technologies, Institutions and Impacts. *Advances in Agronomy*. Vol 140: 101-214 p.

5

Climate Risk Management in Smallholder Farming Systems in the Semiarid Tropics

[1]Anthony Whitbread*, [1]Peter Carberry, [1]KPC Rao, [1]Shalander Kumar,[1]Dakshina

Urthy Kadiyala, [1]Nedumaran Swamikannu and [2]Uday Nidumolu
[1]ICRISAT, Patancheru –502324 Telangana, India
[2]CSIRO Agriculture, Agriculture & Global Change, Waite Campus, Glen Osmond South Australia.

Climate risk management in the semi-arid tropics (SAT) is one of the major challenges to achieving food security and development in India and large parts of sub-Saharan Africa and also in the case of Australia. Climate-induced production risk associated with the current season-to-season variability of rainfall is a major barrier in making rainfed agriculture sustainable and viable farm business. Since season outcomes are uncertain, even with the best climate information, farmers have limited flexibility in applying management with confidence. In fact in risky environments, farmers most often respond by adapting a risk averse strategy and are reluctant to invest in even risk reducing measures (Leathers and Quiggin 1991). In the SAT agro-ecologies, there are a limited range of enterprise or crop options to consider which may be further restricted by cultural traditions, food preferences or market opportunities.While there are fundamental differences between large scale commercial farms in Australia compared to the predominantly smallholder resource poor farms found in India, when it comes to climate risk management in the SAT, there are many commonalities. The purpose of this paper is therefore to (i) establish a framework for managing climate variability and transforming farming systems to be more resilient and sustainable for future climates; and (ii) provide some case study examples from climate risk management in low rainfall cropping system in Australia and consider how they may be applied in smallholder systems of the SAT.

In the smallholder farming systems of the semi-arid, the majority of rural livelihoods are based on the production of cereals under rainfed mixed crop-livestock systems. Under these systems, yield gap between potential and achieved productivity is large, water and nutrient use efficiencies are generally low and land degradation is widespread and severe. There are many commentaries as to why the productivity gains achieved in other more favourable agro-ecologies have not positively impacted on agricultural production in the drylands. While most conclude that lack of government support, infrastructure, poor management and governance of natural resources, i.e. the enabling environment, are the major barriers, we argue that the effects of climate variation which act as a powerful disincentive to investment across all scales should not be overlooked as a major contributing factor. This paper argues that research, development and extension efforts in the drylands pay inadequate attention to climate risk encompassing both present variability and future climate changes. The effect of extreme climatic events such as severe and long term droughts, high intensity rainfall events and heat waves leading to crop loss and land degradation are significant disincentives to investment by small holders. Non availability of customized climate information, a lack of the 'enabling environment'and the lack of knowledge on how to respond to climate information with confidence are the real constraints to the further sustainable development of semiarid smallholder farmers. While there are a number of possible solutions to manage climate risk in small holder farming systems, we explore if some successful examples of climate risk management from low rainfall commercial cereal systems in Australia can provide insights that are applicable to smallholder farms in India to manage climate risk.

5.1 Climate now and in the future

Climate in the SAT in India is driven primarily by the summer southwest monsoons (June to Sept) and the winter northeast monsoons (Nov-Feb) which contribute to the vast majority of rainfall. The El Nino Southern Oscillation (ENSO) phenomenon influences inter-annual variability in climate with strong correlations found with rainfall in these monsoon periods (Krishna Kumar *et al.* 1999, Geethalakshmi *et al.*, 2009).In SAT regions, in addition to inherent climate variability, climate change is likely tobe harsher with frequent extreme events (droughts, floods), increasing temperatures and shifting rainfall patterns.Analysis by Singh *et al.* (2014) suggests increasing variability inrainfall in the sub-continent with significant increases in the frequency of dry spells and intensity of wet spells.

5.2 Planning for now and the future- a framework

To remain profitable and food secure, farmers must cope effectively with current variability in climate as a first step to adjust their farming systems to cope better with future climate. For near term and current decision making, this can be termed 'tactical' where a flexible risk management strategy is adopted that uses multiple information sources to make decisions(Table 1). This may include pre-season enterprise planning guided by seasonal climate forecasts, a set of criteria or 'triggers' for sowing and variety selection and a range of in-season responses to the prevailing weather, market signals or other factors. The longer term perspective, where a farming system is redesigned or adjusted to be more resilient to the current and future climate patterns, can be termed 'strategic' planning (Table 5.1). Farm design should consider what mix of enterprises, crop types, and farming systems are most resilient to current and future climate also considering market and cultural factors. This requires significant analytical efforts to understand historical and projected climate, model based scenario analysis, co-design of farming systems that are more resilient to extreme events and reduce the damage of such events on the natural resource base. In some landscapes and environments, this may suggest transformational changes in landscape design. In both strategic and tactical planning, climate information that is based on both historical observations and current forecasts or projectionsare used as an underpinning tool. Such approaches are not deterministic and must be interpreted and communicated as probabilistic.

Table 5.1: A framework for planning for resilience to impacts of current and future climate

Strategic	Tactical (pre- and in-season responsive management)
Historical and future climate analyses	[a]Climate forecasting (long, medium and short term)
Modelled scenario analysis: e.g. farming systems modelling, climate change prediction.	Pre-season enterprise planning including the contingency planning
Co-design of the farm system for resilience (extreme events/ food security) and market opportunities (commercialization)	Decision making triggers for sowing and other operations
Infrastructure and institutions to enhance adaptive capacity (irrigation, NRM, markets, enhanced logistics).	In-season responses to prevailing weather

[a]*Note:* Short-term refers to a lead time 2-3 day forecast which is useful for tactical or operational purposes, medium range forecasts are 7-10 days, potentially for tactical and long term may mean extended range (10-30 days) /seasonal climate forecasts for 1-3 months in advance

5.3 The enabling environment

Rainfed agriculture plays a crucial role in India's food security by contributing about 45% of the total food grain production and around 90% of coarse cereals and pulses production (Sharma, 2011). However, rainfed agriculture faces multiple challenges like low cropping intensity, low productivity, uncertainty in output, high cost of cultivation, poor adoption of modern technology, lack of institutional credit and inadequate public investment with significant social impacts such as high incidence of rural poverty (Singh *et al.*, 2010). Policy and infrastructure development: To increase the productivity of rainfed crops and improve the profitability of rainfed farmers, the Indian government initiated a number of targetedprogrammes such as the Integrated Scheme of Oilseeds, Pulses, Oil Palm and Maize (ISOPAM), and increased minimum support price (MSP) for pulses to incentivize the farmers to increase the area under pulses to attain self-sufficiency in pulses. It also constituted a National Rainfed Area Authority (NRAA) as an advisory body for policy and programme formulation withrainfed crops of pulses and coarse cereals as major components of its agenda. These initiatives have been underpinned by the National Food Security Mission (NSFM). To realize the full potential of rainfed agriculture, higher strategic research investments to develop biotic (pests and diseases) and abiotic (weather and soil fatigue) stress-tolerant cultivars, short-duration varieties, incremental increases and arrangements to get MSP forrainfed crops (coarse cereals, pulses and oilseeds) on par with wheat and rice, sustainable seed systems to increase seed replacement rate (SRR), warehouse facilities and scientific storage methods that reduce post-harvest losses and information and communications technology (ICT) innovations to provide real time information to improve farmers decision will be required.

5.3.1 Development programs and investments

There is growing evidence on the importance of water management and investments in rainfed agriculture to reduce the risk and to face the impacts of changing climate (Reddy and Chiranjeevi 2016; Kerr *et al* 2002, Kumar *et al* 2016). To increase water use efficiency and to realize "*per drop more crop*", water conservation technologies and water saving irrigation methods should be promoted in the rainfed regions by providing customized information, low cost institutional credit and investment subsides to increase the adoption of these technologies. To realize the potential of R&D investment in rainfed agriculture, there is need of enabling policy environments that promote sustainable seed systems to increase seed replacement rate (SSR);sustains infrastructure initiatives such as modern weather and pest proof.The incentives to promote ICT innovations to provide information for real time farm decision are critical where traditional extension services are being over stretched. To enable farmers and agricultural industries to adapt to current and future climates, investment by government and the private sector will be required. In general, in the SAT regions adequate investment in infrastructure (roads, storage and marketing facilities,) and human capacity have not been made compared with more favorable agro-ecologies. Such investments are fundamental to creating agribusiness opportunitiesthrough adoption of next generation and emerging technologies.

5.3.2 Transformative technologies

The use of digital technologies is an exciting opportunity for enabling the communication of real time and context specific information to stakeholders. In Australia, for example, precision agriculture has been enabled by technology and has led to transformational change incrop management. Technology is being applied by actors in various ways. For example, in the delivery of information from various sources (downscaled market, weather, farming system, and soil) as locally specific advisories direct to farmers on smart phones or tablets; in the application of risk management intelligence on markets or weather in the banking, or insurance sectors; in business intelligence through harnessing information to a dashboard to monitor project outputs. Case study #2 provides a model for how this was successfully achieved in a commercial setting in Australia.

5.3.3 Responsive farming using decision triggers: Case study 1

In Australian low rainfall cropping regions (growing season rainfall < 300 mm) face major challenges associated with high yield variability and its effect on profitability, with production risk historically much more important than price risk. To combat variability in cropping yields, farm businesses adopt a range of practices such as diversification (mainly involving livestock) to provide more reliable cash flow during difficult seasons. The use of responsive farming systems allows flexibility in crop area and in crop type between years, and is an important component of risk management. In the absence of reliable seasonal climate outlooks, other indicators are sought to provide robust trigger points to adjust decisions about crop type and area. It is widely accepted that seasons which allow earlier seeding times with higher initial plant available water (PAW) usually results in enhanced yield outcomes. Scenario analyses using crop models (APSIM) and historical weather data compute the probability of favorable outcomes under different starting conditions of plant available water and seeding opportunity. This approach offered a means to make more informed decision making around cropping intentions and to determine appropriate trigger points at the individual farmer level.For example, even a site with a mean rainfall of about 220mm / year (Port Germein, South Australia) is responsive to the storage of out of season soil water with relatively high simulated fallow efficiencies (average of 24%).The effects of different ranges of PAW and seeding opportunity on the probability of different yield outcomes (divided into terciles) show that the interaction of plant available water at seeding and seeding opportunity is a strong indicator of final crop yield (Fig. 5.1). The combination of low PAW and late seeding rarely produces a favourable outcome with most yields in the lower tercile. At the other extreme, poor crop yields (in the lowertercile, (bottom 33% of the yields) are rare when PAW at seeding is categorised as high.Decisions around modifying sowing intentions to limit exposure in poor seasons and capitalize on better years rely on reliable indicators to "trigger" appropriate changes to the farm programs. While farmers have lamented the lack of more reliable seasonal outlooks(1-3 monthly rainfall forecasts)to provide such information, usingmultipleand integrated information sources will enable more robust decision making.

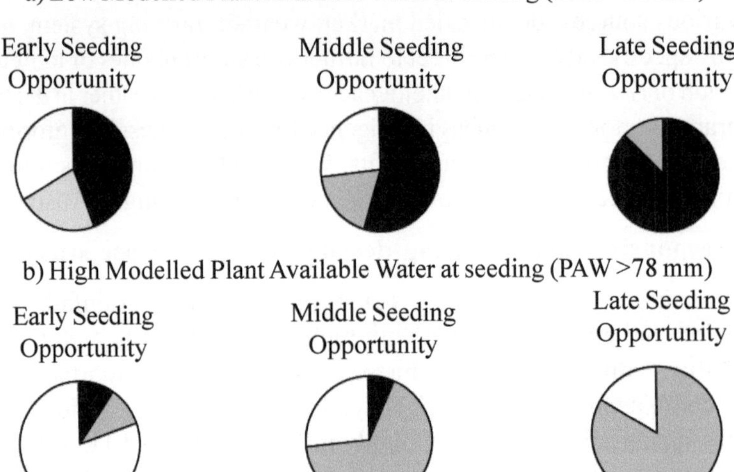

Fig. 5.1: Effect of variations in PAW and seeding opportunity on percentage of modelled yields in upper tercile (white), middle tercile (grey) and lower tercile (black)

5.3.4 In-season prediction of crop performance and benchmarking: Case study 2

Yield potential has long been used in Australian cropping systems to benchmark crop performance at a field level against a water limited potential yield target. Onewidely used method in the southern Australian wheat belt was devised by French and Schultz (1984) based on a boundary function derived from a large field data set. Predicted wheat grain yield is the result of in-season rainfall minus a fixed evaporation of 110 mm, which is multiplied by a transpiration efficiency factor of 20 kg/mm/ha. While this empirically based method was simple, a linear rainfall-yield relationship cannot address yield differences caused by soil variability, in season rainfall deficit at critical times. Since about 2005, the prediction of crop performance with crop models which simulate important process and interactions have been widely applied by researchers to analyse the seasonal-spatial dynamic nature of crop productivity (Hoffmann *et al.*, 2015; Whitbread *et al.*, 2015; Sadras and Angus 2006; Sadras and Rodriguez

2010).According to McCown *et al.* (2006) however, the effectiveness of decision support systems in influencing farmer behaviour has been disappointing. A program led by CSIRO in Australia explored whether researchers and advisors could benefit from simulation and decision support aid to risk management (Hochman *et al.* 2009). This led to the development of a highly successful commercial on-line (http://www.yieldprophet.com.au/) system where farmers and their consultants could enter site specific details about their soil, crop and location and generate reports which predicted crop growth and possible season finishes using local weather information and driven by the APSIM model (Fig. 5.2.) A range of other 'system' information (frost and heat risk, crop stage dates, soil water and mineral N status) is also reported to further inform decision making.

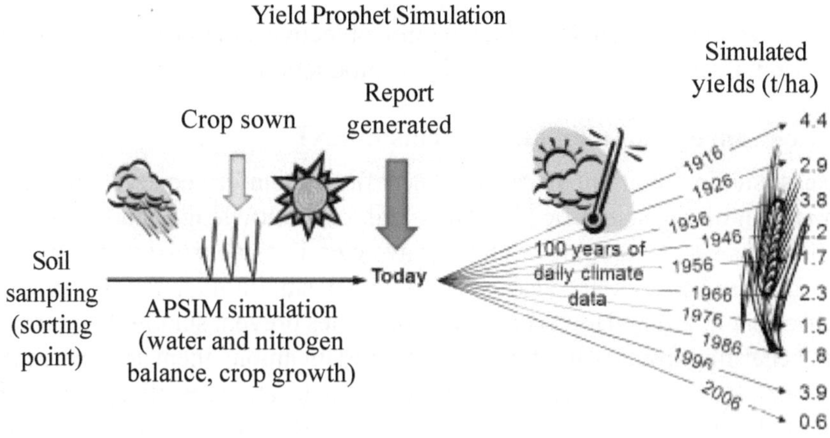

Fig. 5.2: A pictorial display of a Yield Prophet report generated at a point in the growing season with APSIM then simulating yield outcomes using historical weather data from a local station

5.3.5 Climate forecasting: Case study 3

Climate forecasting has long been an important component of climate risk management in Australian agriculture. According to Hayman (2007), while 30-50% of farmers use seasonal climate forecasts when making farm management decisions, most farmers and advisers are quick to point out that they place less weight on the information than they would like to if the forecast was more reliable. Increasing the skill of forecasting using dynamic climate modelling is the focus of much current research (Franzke *et al.*, 2015).Following Meinke and Stone (2005) some speciûc Australian examples of the use of forecasting in decision-making include:

1. Cotton growers in Queensland scheduling the timing of their cotton harvests based on the expected passing of the next Madden Julian Oscillation (MJO);
2. Farmers in northeastern Australia who use ENSO-based information to tailor their rotations and crop management based on local conditions at the time and rainfall probabilities for the coming months;
3. Sugar growers and millers in north Queensland who use targeted climate forecasts in management decisions involving harvest strategies, planting decisions, and mill throughput operations;
4. Bulk handling and marketing agencies, which require accurate regional commodity forecasts to assist them in storage and transport logistics and export sales well before harvest;
5. Government agencies, which require objective assessments of the effect and severity of climate variability on production;

5.4 Relevance to smallholder systems in SAT

The case studies highlight the potential benefits from informed decision making that is built on the scientific rationality and local knowledge and needs. Pilot studies in many developing countries have clearly established that similar or even greater benefits are also possible under smallholder faming systems (Hansen *et al.*, 2011). However, several obstacles prevent smallholder farmers from benefiting from such information. Important among them are poor access to information especially in a user friendly format, location specificity, timeliness and reliability or credibility of the information. Recent developments in ICT have opened up new opportunities to develop and deliver data and information that is location specific in real time to decision makers. However, access to information alone is not sufficient. The capacity of end users (smallholder farmers and their support agents) to understand and utilize the information is equally important. We believe that by developing appropriate information systems and by strengthening the capacity of the farmers, significant benefits can be harnessed from the experiences and successes of large scale commercial farmers in the developed world.

Summary

This paper has presented a short review of current climate risk management strategies and enabling environments in commercial dryland cereal systems in Australia and explored how these may provide insights for managing climate risk on to smallholders farming systems in semi-arid India. Three case studies have shown (i) how the analysis of historical weather records combined with systems information and a well validated crop-soil model provides powerful indication of a season outcome at the start of the season; (ii) a commercially available web based system based on the APSIM model providing farmers and their consultants with in-season predictive reports; and (iii) examples of the use of climate forecasts by agricultural industries in decision making and a range of scales. In India, enhancing climate risk assessment and management will occur through a combination of approaches, many of which are already underway to varying degrees. Of highest priority will include enhancing the use of public seasonal climate services which are location and system specific and which may be increasingly delivered through private sector led ICT delivery with support from public and other actors. An important aspect in the context of resource poor farmers, will be equity i.e. making sure access to information and inputs are socially inclusive particularly in poorer SAT regions. The building of capacity in multiple stakeholders as a pre-requisite for effective climate risk management in semiarid regions, especially around information translation to action i.e. access to inputs (seeds, fertiliser, labour etc) to apply to take advantage of the climate information. Meinke *et al.* (2006) concluded that climate risk management requires holistic solutions derived from cross-disciplinary and participatory, user-oriented research – this implies that the multiple institutions/ agencies present in India are required to better integrate their efforts for the benefit of smallholder farmers.

References/Suggested Readings

French RJ, Schultz JE (1984) Water use efficiency of wheat in a Mediterranean- type environment. I. The relation between yield, water use and climate. *Australian Journal of Agricultural Research* 35, 743-764.

Franzke, C. L. E., O'Kane, T. J., Berner, J., Williams, P. D. and Lucarini, V. (2015), Stochastic climate theory and modeling. *WIREs Clim Change,* 6: 63–78. doi:10.1002/wcc.318

Geethalakshmi, V., Palanisamy, K., Yatagai, A., Umestsu, C. (2009).Impact of ENSO and the Indian Ocean Dipole on the north east monsoon rainfall of Tamil Nadu State in India.Hydrological Processes. DOI: 10.1002/hyp.7191.

Hansen, J.W., Mason, S.J., Sun, L. and Tall, A. 2011. Review of seasonal climate forecasting for agriculture in Sub-Saharan Africa. *Experimental Agriculture*. 47: 205-240.

Hayman, PT, Crean J, Mullne J, Parton K. 2007. How do probabilistic seasonal climate forecasts compare with other innovations that Australian farmers are encouraged to adopt? *Australian Journal of Agricultural Research*, 58, 975–984.

Hochman Z, van Rees H, Carberry PS, Hunt JR, McCown RL, Gartmann A, Holzworth D, van Rees S, Dalgliesh NP, Long W, Peake AS, Poulton PL (2009) Re-inventing model-based decision support with Australian dryland farmers: 4. Yield Prophet® helps farmers monitor and manage crops in a variable climate. *Crop and Pasture Science* 60, 1057-1070.

Hoffmann MP, Llewellyn R, Davoren B, Whitbread AM (2015) Assessing the potential for zone-specific management of cereals in low rainfall south-eastern Australia: Combining on-farm results and simulation analysis. *Journal of Agronomy and Crop Science*. doi:10.1111/ jac.12159.

Holzworth DP, Huth NI, deVoil PG, Zurcher EJ, Herrmann NI, McLean G, Chenu K, van Oosterom E, Snow VO, Murphy C, Moore AD, Brown HE, Whish JPM, Verrall S, Fainges J, Bell LW, Peake AS, Poulton PL, Hochman Z, Thorburn PJ, Gaydon DS, Dalgliesh NP, Rodriguez D, Cox H, Chapman S, Doherty A, Teixeira E, Sharp J, Cichota R, Vogeler I, Li FY, Wang E, Hammer GL, Robertson MJ, Dimes J, Whitbread AM, Hunt J, van Rees H, McClelland T, Carberry PS, Hargreaves JNG, MacLeod N, McDonald C, Harsdorf J, Wedgwood S, Keating BA,(2014) APSIM - Evolution towards a new generation of agricultural systems simulation. *Environmental Modelling & Software* 62: 327-350.

Kerr, J., G. Pangare, and V.L. Pangare (2002), Watershed Development projects in India An Evolution, Research Report 127, Environment and Production Technology Division, International Food Policy Research Institute (IFPRI), Washington D.C., U.S.A.

Kirkegaard JA, Hunt JR, McBeath TM, Lilley JM, Moore A, Verburg K, Robertson M, Oliver Y, Ward PR, Milroy S, Whitbread AM (2014) Improving water productivity in the Australian Grains industry—a nationally coordinated approach. *Crop and Pasture Science* 65: 583–601. DOI10.1071/CP14019.

Krishna Kumar K, Rajagopalan B, Cane M A. 1999. On the weakening relationship between the Indian monsoon and ENSO. *Science*, 284: 2156-2159.

Shalander Kumar, Ramilan Thiagarajah, Ramarao C.A., Rao Ch. Srinivasa, Whitbread A. 2016. Farm level rainwater harvesting across different agro climatic regions of India: Assessing performance and its determinants. *Agricultural Water Management* 176 (2016) 55–66

Leathers H.D., and Quiggin J.C. (1991). Interactions between agricultural and resource policy: the importance of attitudes toward risk. *American Journal of Agricultural Economics* 73, 757-764

McCown RL., Brennan LE., Parton KE., (2006) Learning from the historical failure of farm management models to aid farm management practice: 1. The rise and demise of theoretical of farm economics. *Australian Journal of Agricultural Research* 57, 143-156.

Meinke H, Nelson R, Kokic P, Stone R, Selvaraju R, Baethgen W. (2006). Climate Research 33:101-110 doi:10.3354/cr033101.

Meinke, H. & Stone, R.C. (2005). Seasonal and inter-annual climate forecasting: The new tool for increasing preparedness to climate variability and change in agricultural planning and operations. *Climatic Change* 70: 221. doi:10.1007/s10584-005-5948-6

Monjardino, M., McBeath, T.M., Brennan, L., Lewellyn, R.S. 2013. Are farmers in low-rainfall cropping areas underfertilising with nitrogen: A risk analysis. Agricultural Systems

Mudge, B. and Whitbread, A. (2010). Making better decisions about crop rotations in low rainfall environments: should stored moisture and the timing of the seeding opportunity influence this decision? In "Food Security from Sustainable Agriculture."Proceedings of the 15th Australian Agronomy Conference. November 2010, Christchurch, New Zealand (http://www.regional.org.au/au/asa/2010/crop-production/sequence/6994_mudgeb.htm# Top of Page

Sadras, V.O., Angus, J.F., 2006. Benchmarking water-use efficiency of rainfed wheat in dry environments. *Aust. J. Agric. Res.* 57, 847-856.

Sadras, V.O., Rodriguez, D., 2010. Modelling the nitrogen-driven trade-off between nitrogen utilisation efficiency and water use efficiency of wheat in eastern Australia. *F. Crop. Res.* 118: 297-305.

Sharma K.D. (2011). Rain-fed agriculture could meet the challenges of food security in India, Current Science 100, 1615-1617.

Singh, S. and Rathore, M. S., (2010) Rain-fed Agriculture in India – Perspectives and Challenges, Rawat Publications, Jaipur.

Singh, D., Tsiang, M., Rajaratnam, B. &Diffenbaugh, N.S. (2014).Observed changes in extreme wet and dry spells during the South Asian summer monsoon season. *Nature Climate Change* 4, 456–461.

Sivakumar, M.V.K., Das, H.P. and Brunini, O. Climatic Change (2005) 70:31. doi:10.1007/s10584-005-5937-9.

Reddy VR and Chiranjeevi, T. (2016). The Changing Face of Rainfed Agriculture: Need for Integration of Science, Policy and Institutions, Vicissitudes of Agriculture in the Fast Growing Indian Economy: Challenges, Strategies and the Way Forward, Publisher: Academic Foundation in Association with the Indian Society of Agricultural Economics, Mumbai, Editors: C. Ramasamy and K. R. Ashok (ed, pp.281-312).

Whitbread AM, Davoren C, Gupta V.V.S.R. Llewellyn R, Roget D. (2015). Long-term cropping system studies support intensive and responsive cropping systems in the low rainfall Australian Mallee. *Crop and Pasture Science* 66, 553-565.DOI10.1071/CP14136

6

Climate Change and Agriculture Experiences in Kerala

G.S.L.H.V.Prasada Rao and C.S.Gopakumar

Academy of Climate Change Education and Research
Kerala Agricultural University, Vellanikkara., K.A.U.(P.O) Thrissur, Kerala, India

The State of Kerala has witnessed an unprecedented and unusual flood which devastated the entire State during August 2018 after the notorious 1924 flood, popularly known as the flood of 1099. Heavy rainfall from 14to18 August 2018 led to flash floods across the State. This was aggravated by the dam opening and subsequent water level in the downstream of the rivers. 33 dams were opened across the State. This has caused the river water to flow into the inland areas. Rivers overflowed and inundated the river side villages and towns. It was reported that the water levels at many places were of the order of 20 feet. Landslides were the major threat that claimed many human lives. A chain of mud slip/land slide events took place during the heavy rainfall period at the time of flood across the mid and high lands of the State, which are ecologically sensitive areas. Even before the heavy rainfall events started on 14th August, 2018 there were major rainfall events on August 8thand 9th. On 9thAugust 2018, Nilambur (Malappuram District) recorded extremely heavy rainfall of 400 mm which is a record rainfall in 24 hours as far as the State is concerned. The monsoon season from June to September 2018 ended with 23.4 per cent excess rainfall, with August rainfall was the highest (96 per cent of the normal) from 1stAugust to 19thAugust (excess of 164 per cent against the normal). The Govt. of Kerala has estimated the overall huge loss to the State due to the unprecedented flood was of the order of more than Rs.30000/- crore. Agricultural sector was the worst affected. It is estimated that the loss in the agricultural sector was Rs.5000/- crore. Paddy, vegetables and banana were the worst affected due to unprecedented floods. In contrast, low monsoon rainfall, followed by weak post monsoon led to severe hydrological droughts as experienced in 1983,1987,1999,2002,2004,2012,2015 and 2016. It is a threat to plantation crops' production in Kerala. The monsoon rainfall deficits in recent years were 2002(-32.8%), 2003(-17.8%), 2004(-27.7%), 2012(-21.8%), 2015(-21.8%) and 2016(-29.6%) against the long period average of 1922mm. *Kharif* droughts are common in India due to monsoon breaks while summer droughts in Kerala. In the absence of northeast monsoon and summer showers, prolonged dry spell results to summer drought, which adversely affects plantation crops' production to a large extent in Kerala.

The State of Kerala under the Humid Tropics is one of the wettest places in the Humid Tropics where annual rainfall is of the order of 3000 mm, ranging from less than 1000 mm in the South to greater than 5000 mm in the North and High Ranges. About 68 per cent of the rainfall is obtained during southwest monsoon while 16 per cent in post monsoon and the rest from summer (14 per cent) and winter rainfall (2 per cent). Bi-model rainfall pattern is noticed towards southern districts due to influence of both southwest and northeast monsoons while uni-model rainfall pattern towards northern districts of Kerala. Floods during monsoon adversely affect paddy production in the State while prolonged droughts during summer in the absence of post monsoon rainfall adversely affect plantation crops' production (not all plantations) to a considerable extent. The wetlands in Kerala are rich sources of water during summer and act as sink during monsoon season. Such wetlands are fast declining in Kerala and converted as garden lands. Decreasing wetlands might be one of the reasons for frequent floods and droughts in Kerala in recent years. The forest cover of Kerala also declined from 70 per cent to 24 per cent over a period of one- hundred- and - fifty years due to deforestation and forest fires. Consecutively deficit monsoon rainfall was noticed in 2015 (26 %), 2016 (34 %) and 2017 (9%) across Kerala while the deficit was high in Wayanad District. Such deficit rainfall was noticed in 2012 (26 %) and 2002 (33 %) as well. Excess rainfall was noticed only in 2013 (26 %), 2007 (43%) and 234 % in 2018 since 2001 onwards. The deficit monsoon rainfall from June to September together with failure of northeast monsoon rains in recent years is a crisis in water and agriculture sectors of Kerala. Frequent failure or break in monsoon across the State led to low water levels in major reservoirs and adversely affected hydro power generation too. The years 1983, 2004, 2013 and 2017 experienced prolonged summer droughts in Kerala, led to hydrological droughts. Sun burns (2010) and heat burst (2015) were also noticed in recent years in addition to wild life and human conflicts (Zoonosis), leading to viral diseases in animals and human beings.

6.1 Weather Extremes and crops

The lull in monsoon over Kerala during 2002 followed by floods in October 2002 due to cyclonic storm over the Karnataka Coast devastated seasonal crops and plantations to a considerable extent. Kannur received 370 mm of rainfall on 14th October 2002 and Malappuram recorded 400 mm on 9th August in 2018 which were the highest and not received since 1924. Heavy monsoon rainfall in 2007 followed by unusual summer showers in March 2008 together adversely affected the paddy production of the State to a considerable extent during 2007-08. The State received record rainfall during post monsoon 2010. The widespread post monsoon rainfall during 2010 also devastated the paddy fields especially in upper Kuttanad and Kole lands of Thrissur. Low lands of southern districts were flooded due to the prolonged rainfall during November, 2010. In contrast, the prolonged dry spell during summer 2013 adversely affected the water resources and several plantation crops' production affected. Floods in August/September 2014 across Wayanad District adversely affected paddy and arecanut fields. Yellowing in areca nut devastated large number of palms in Wayanad District in 2014. The heavy rain that lashed the capital city for over six hours from Tuesday (21.4.2015) evening represents a record rainfall (over a period of 24hrs) for the month of April. The city received 16.1cm of rainfall till Wednesday Morning, against the previous high of 13.41cm recorded on April 5, 2005. "Such high intensity rain is typical of the summer season in Kerala. Are these frequent weather aberrations like monsoon uncertainties, floods and droughts, sunburns, atmospheric heat load during summer and UV radiation filtered to ground due to ozone depletion could be as a part of global warming and climate change? There were reports that the heat burst might be the reason for scorching leaves across the Coast of Kerala in June 2015. Sunburns and Heat bursts are new phenomena noticed in Kerala, the science behind is yet to be studied. The weather phenomenon like heat burst was noticed in Kerala when the entire south (except Kerala) was reeling under severe heat wave in June 2015. The climate change related issues over Kerala may be decline in rainfall, wetlands, land and ocean biodiversity, increase in air and sea surface temperature and sea level, floods and droughts, landslides, groundwater depletion and saline water intrusion, decline in forest area, frequent forest fires, unusual rains and hailstorms. As a whole, the State of Kerala was moving from wetness to dryness within the Humid Climate (B_4 to B_3). It is believed that almost all the crops are likely to be under threat due to climate change/variability in the ensuing decades. The threat from global warming and climate change could be seen in the form of decline in cropped area, production and productivity, quality of crop produce and food price. The rate of increase in temperature is more across the High Ranges, followed by low lands due to increase in deforestation and sea surface temperature, respectively. There is a significant change in cropping pattern too over a period of time. Rice and cashew area was declining while area under rubber and coconut increasing. Vanilla and cocoa were introduced and in the same fashion vanilla disappeared and area under cocoa declined. Increase in area under black pepper was noticed across the State and now declining. There was a time that oranges were plenty in Wayanad District and now disappeared almost. The paddy lands in Wayanad District are mostly converted to areca nut and banana gardens. The quality of copra is better in recent years in case of coconut while the quality of coffee beans is declining in Wayanad District. Coconut productivity is better towards south despite dreaded disease of root-wilt when compared to that of northern districts due to less dry spells during summer. Cardamom is predominant in South across the High ranges while coffee is predominant in Wayanad. Cashew is predominant in mid and high lands across northern districts (Kannur and Kasaragod) of Kerala with better nut size and quality. It could be attributed mostly due to rainfall and temperature distribution. As a whole, the index of food crops was declining while increasing non-food crops due to various socio-economic factors. The altitudinal sequence of crops within the Humid Tropics is quite interesting (Plate 6.1) since the plantation crops like coffee, cardamom and tea that are grown across the highranges are not at all seen in mid and lowlands, indicating that they cannot withstand to high temperature and the crop distribution is a reflection of prevailing temperature as influenced by altitude and lapse rate.

Plate 6.1: Altitudinal sequence of crops in Kerala

6.1.1 Trends in onset of monsoon

The normal onset of monsoon over Kerala is on 1st June with +/- 7 days. The earliest onset of monsoon was on 11th May in 1918 while belated monsoon on 18th June in 1972. The onset of monsoon is said to be early if it sets on or before 25th May while late if it is on or after 8th June when one standard deviation from the mean date of onset of monsoon is taken into account. The factfile in recent years during this decade showed that the onset of monsoon was on 29th May in 2011, 5th June in 2012, 1st June in 2013, 6th June in 2014, 5th June in 2015, 8th June in 2016, 30th May in 2017 and 29th May in 2018 (Fig. 6.1), indicating that the belated monsoon was seen only in 2016 during this decade so far. It reveals that inter-annual variations of the onset of monsoon are expected within one standard deviation to the mean onset of monsoon in majority of the years. The trend analysis on onset of monsoon also indicated that it is likely to be on or before 1st June over Kerala. All these reveal that there is no deviation in onset of monsoon over Kerala from the normal date (June 1st +/- 7 days) due to global warming and climate change. Therefore, the onset of monsoon over Kerala is stable since last 149 years and impact of climate change is not seen on the onset of monsoon.

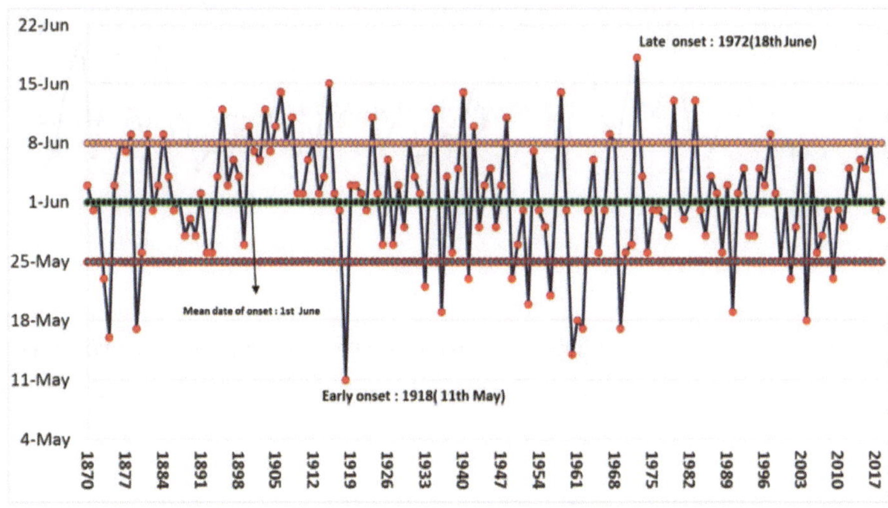

Fig. 6.1: Onset of monsoon over Kerala from 1870 to 2018 (149 years)

6.1.2 Rainfall Trends

Kerala is the land of monsoons. It is also one of the wettest places in the world, where annual rainfall is of the order of 3000 mm. About 68 per cent of the rainfall is obtained during southwest monsoon while 16 per cent in post monsoon and the rest from summer (14 per cent) and winter rainfall (2 per cent). Rao, *et al.*, (2009) reported that there exists a cyclic trend in annual rainfall with a declining trend in southwest monsoon rainfall and an increasing trend in post monsoon rainfall since last 50-60 years (Fig.6.2a & b). Rains during the monsoon period also appear to be erratic in recent years though the onset of monsoon was within normal dates of 1st June +/- 7 days. It is also understood that the monsoon set may be early (before 25th May) or late (after 8th June) occasionally during which the monsoon rainfall is likely to be below normal or normal, indicating that the chances of excess rainfall in such years (early or late monsoon years) are likely to be less. Decline in years of excess rainfall and increase in deficit rainfall years during the summer monsoon from June to September in recent years reveal that the State of Kerala was moving from wetness to dryness within the humid climate (B4 to B3). Rainfall during the first half of the monsoon (June –July) is declining while the contribution from August to September to the total southwest monsoon is increasing. It may be an indication that the monsoon is shifting its phase. Increase in monthly rainfall from August to September may not contribute significantly to annual rainfall while rainfall

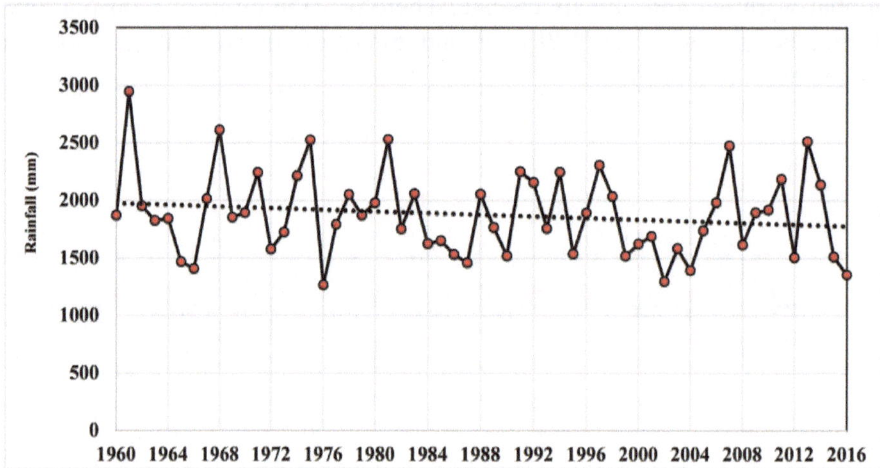

Fig 6.2 a: Trend in monsoon rainfall over Kerala from 1960 to 2016

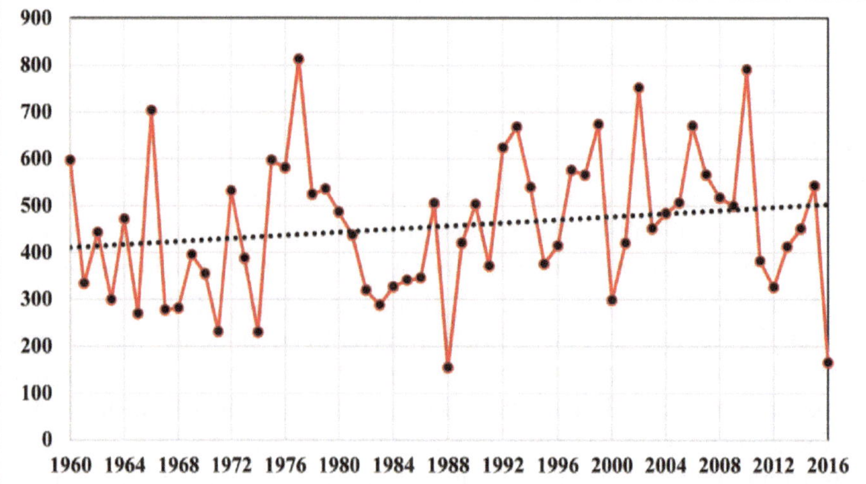

Fig. 6.2 b: Trend in post monsoon rainfall over Kerala from 1960 to 2016

decrease in June and July may contribute significantly and lead to deficit annual rainfall since June and July are the rainiest months across the State of Kerala. Failure of monsoon as well as post monsoon contributed to the low water levels in the surface water reservoirs and open wells during 2016 in which the post monsoon rainfall was deficit by 63 per cent across the State.

6.1.3 Temperature Trends

Warming Kerala is real as the trend in atmospheric temperature was increasing significantly at the rate of 0.011°C/year since 1956 onwards in average temperature (Fig. 6.3). Increase in maximum temperature was 0.99°C while 0.31°C in the case of minimum temperature from 1956 to 2014 and increase in mean temperature was 0.65°C. Rate of increase in mean temperature was high (0.814°C in 59 years @0.0138°C per year) across the Coastal belt when compared to midlands (rate of increase in mean temperature is 0.33°C in 59 years @0.0056°C per year) and high ranges (the rate of increase in mean temperature was 0.44°C in 33 years @0.0134°C per year). Seasonal increase in mean temperature was high (0.76°C over a period of six decades since 1956) during the post monsoon season (October–November), followed by southwest monsoon (June – September: increase of 0.732°C over a period of six decades since 1956). Increase in mean surface air temperature is likely to be less than 1°C while 1.5°C increase in maximum temperature and 0.3-0.4°C in the case of minimum temperature by 2100 A.D (Fig. 6.4). The decade 1981-90 was the warmest and driest decade in Kerala. The year 1987 was the warmest, followed by summer 1983 and 2014. The current decade 2011-2020 is likely to be warmer when compared to that of 1981-90. According to India Meteorological Department (IMD), during the last 43 years, the mean maximum temperature over Kerala has risen by 0.8°C, the minimum by 0.2°C and the average by 0.5°C, indicating that the temperature trends in Kerala followed the trends of West Coast. Palakkad recorded the highest of 41°C on 26th April 1950. Similar temperatures were recorded over the Palakkad region in February and March in recent years. Day time temperatures during February and March often cross 40°C at many places in Palakkad and Thrissur Districts. For the first time in the history of Kerala, sunburn events were reported during March, 2010 in Palakkad District. Heat burst like situation was noticed in June 2015.The year 1987 was the warmest year over Kerala. For the first time in the history of Kerala, heat waves were declared in Palakkad and Kozhikode districts by the India Meteorological Department during the summer 2016. Of course, rate of increase in temperature was alarming across the Highranges (where cardamom, coffee and tea are grown) due to deforestation. It is also true to some extent along the Coast (low land) due to increase in sea surface air temperature.

6.1.4 Climate change

In the first report on "Impact of climate change in four regions of the country" submitted to the Government of India by the Indian Network for Climate Change Assessment (INCCA), it has been pointed out that reduced rainfall, increased atmospheric temperature and flooding due to sea level are the climate change scenarios for the Western Ghats and Kerala in the next 20 years. The minimum surface air temperature in the Western Ghats region may rise by 2 to 4.5°C. The average temperature in the region bordering Kerala is likely to rise by 1 to 3°C. The number of rainy days is likely to decrease along the entire Western Coast, including the Western Ghats. The rice fields in the northern districts of Kerala may witness an increase in production. If the sea level rises by one metre, 169 sq km of the coastal region surrounding Kochi will be inundated.

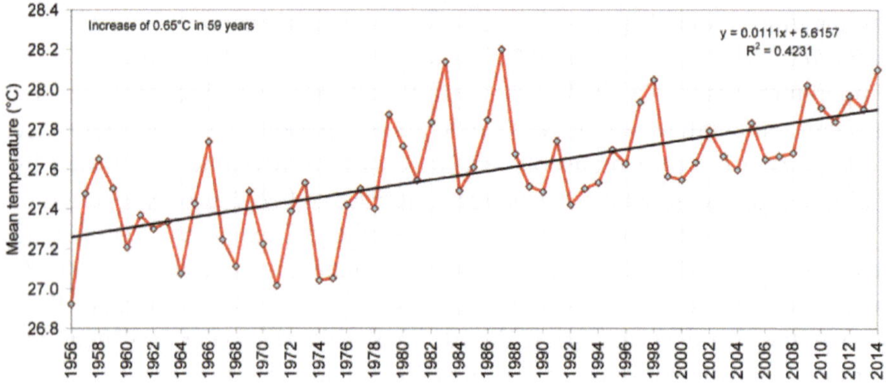

Fig. 6.3: Increase in mean temperature across Kerala

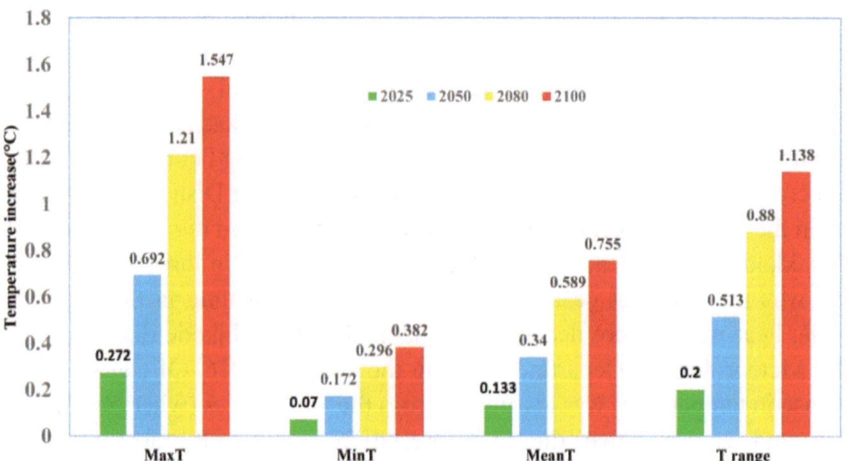

Fig. 6.4: Temperature projections over Kerala based on current trends

Under the projected climate change scenario, it is certain that the temperature is likely to increase by 2°C by 2050. An increase of surface air temperature by 2°C would have tremendous adverse influence on the crops growing across the State. Even now, the thermo-sensitive crops across the high ranges are under threat as the temperature range is widening. Climate change/variability and deforestation over a period of years have changed the typical agro-ecosystems of Wayanad. Paddy fields have been converted into banana gardens. Excessive use of agrochemicals and pesticides, large-scale sand mining from rivers have been identified as the causal factors for severe drought in summer 2004 in addition to decline in rainfall. Groundwater depletion in Wayanad was so much (25%) due to over exploitation. Rise in maximum temperature and fall in minimum temperature are the trends in Idukki agro-ecosystem, and thus the natural habitat of cardamom is also in peril. Widening in temperature range along with deforestation may be detrimental to thermo-sensitive crops like cardamom, coffee, tea, cocoa and black pepper across the high ranges of Kerala. The forest cover of Kerala declined from 70 per cent to 24 per cent over a period of one- hundred- and - fifty years. An increased demand and unregulated exploitation are threatening to accelerate depletion of groundwater resources in Kerala. Because of the heavy extraction of groundwater, the water level has fallen alarmingly. The destruction of forests in the catchment areas has led to the drying up of reservoirs and the extensive sand dredging has drastically reduced rivers' water holding capacity and percolation to groundwater. Groundwater is depleting at a faster rate than that of its recharge. Some of the reasons being decline in rainfall, more run off, over use of water for irrigation, more land is brought under irrigation for cultivation; increased demand for drinking water for the growing population, deforestation, and riverbed based sand mining, decline in wetland area and disappearance of lakes and ponds. In short, the climate change related issues over Kerala may be summarized as below:

1. Decline in rainfall, wetlands, land and ocean biodiversity
2. Increase in temperature (air and sea) and sea level
3. Floods and droughts, landslides, groundwater depletion and saline water intrusion
4. Decline in forest area and frequent forest fires

6.2 Climate change and rice

Under the projected climate change scenario, monsoon uncertainties are expected in the State of Kerala on which rice production is dependant to a large extent during the first crop season (*Virippu*). As already seen in 2007-08, prolonged monsoon coinciding with post monsoon rains and unusual summer rains in the form of floods is a threat to paddy production. There was a decline of 8.5 per cent on an average, ranging from 2 to 17 per cent depending upon the stage of crop coinciding with the occurrence of floods. Therefore, abrupt monsoon uncertainties are likely to influence the paddy production adversely in ensuing decades rather than the influence of climate change on long term basis. The rate of increase in mean temperature at the current level in ensuing decades is likely to be less than 1°C across the low lands of Kerala, where paddy is cultivated mostly. Increase in temperature of this nature during the first crop season is beneficial to paddy crop under better crop improvement (saline and flood tolerant varieties need to be introduced), crop management and crop protection measures since standing water in paddy fields is not a limiting factor. However, inter annual / seasonal variations were noticed in rice production and its productivity due to weather aberrations in the form of climate variability such as floods during the first crop season (Fig. 6.5). Under such conditions, the rice yield is bound to decline. Therefore, rice yields in Kerala during *Virippu* are unlikely to decline due to long term climate change such as increase in temperature, but bound to decline to some extent through the abrupt short term changes as noticed in 2008, 2009 and 2010 in the form of monsoon uncertainties. Of course, there was a drastic decline in rice area and production since last 60 years and at the same time increase in rice productivity (1.37t/ha in 1960-61 to more than 2-2.5t/ha since last one decade) is noticed despite increase in atmospheric temperature. However, rice-milling percentage has come down

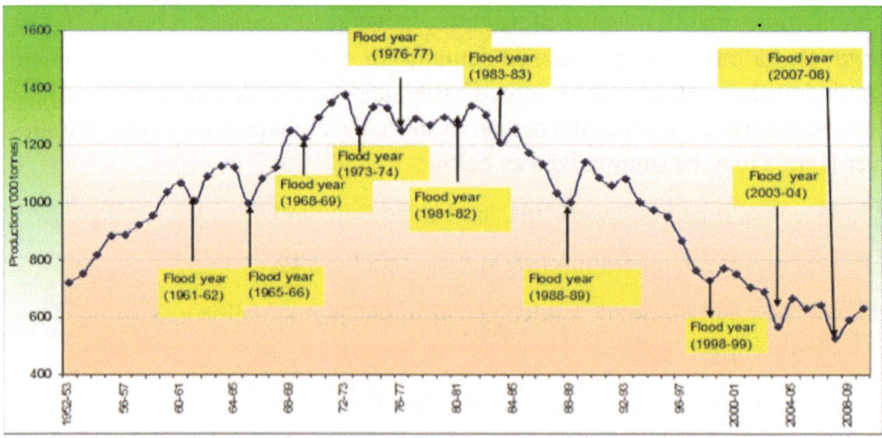

Fig. 6.5: Impact of floods on rice production in Kerala

significantly in recent years. It is attributed to global warming.

6.3 Climate change and coconut

Coconut productivity in Kerala is relatively low though the total coconut production of the State is high. Of course, total coconuts produced are high in Tamil Nadu and Kerala is in second position as per the latest reports. With increase in area under coconut, Kerala contributes major share in terms of coconut production of the country while there was no breakthrough in nut productivity of coconut. This could be attributed to several factors *viz.*, high rainfall during monsoon period, insignificant rainfall in several parts o f Kerala during summer, nutrient status of soil, lack of agronomic practices and prevalence of root wilt disease in southern districts of Kerala. The price of coconut, copra and oil content are also subjected to wide fluctuation depending upon the total coconut production of the State. However, coconuts are not seen on coconut crown due to various socio-economic factors in addition to abiotic and biotic stress conditions in recent years. Coconut is relatively tolerant to weather aberrations, unlike seasonal crops. It tolerates high rainfall and temperature to some extent. Coconut production under rainfed conditions is influenced significantly by the length of dry spells at critical stages. The primordium initiation, ovary development and button-size nut development in coconut are sensitive to soil moisture availability. The crop performs better under irrigated conditions during summer in traditional and non-traditional areas.

6.3.1 Summer drought and coconut yield

Heavy monsoon rainfall with prolonged dry spell during summer influences the ontogeny of coconut and seasonality of nut production in coconut to a large extent. The coconut production of Kerala suffered to a great extent during 1983-84 due to unprecedented drought that occurred during summer 1983, followed by 2004 and 2013. The coconut production in Kerala during 2013-14 was declined by 10 per cent as per the survey reports. Almost all the coconut growing states experienced decline in coconut production in 2013-14. It was attributed to prolonged summer drought in 2012-13 across the coconut growing areas. The effect of high rainfall from June to September and prolonged dry spell during summer adversely affect the coconut production in the following year under the Kerala's situation. The decline in coconut productivity due to severe drought during summer could be seen in the following year under rainfed conditions. Similarly, summer showers influence coconut yield positively in the subsequent year. The effect of summer drought on monthly nut yield at various locations across Kerala indicated that the effect commences in the seventh, eighth or ninth month after the drought period is over by May or June, depending upon the receipt of pre-monsoon showers or onset of monsoon. The decline in monthly nut yield due to drought continued for 12 months, reaching to its maximum decline in $12^{th}/13^{th}$ month after the drought period is over. The average decline in annual coconut yield varied from 20.6% at CDB Farm , Vellanikkara, followed by 23.4% at RARS, Kumarakom to 34.3 % at RARS, Pilicode. The minimum effect due to summer drought varied from 1.5 % at Aralam Farm, Kannur, followed by 4.2% at CRS, Balaramapuram to 16.4% at CDB Farm, Vellanikkara.The maximum nut decline was noticed between 82.6% at CPCRI Farm, Kasaragod and 16.2% at RARS, Kumarakom (Table 6.1). The percentage decline in the subsequent year due to summer drought varies depending upon the yield group too. In higher nut yield group (greater than 100 nuts per palm per annum), it revolves around 29.3 % and 30.4% (between 80 and 100 nuts per palm per annum) while around 10% in poor yielders (yield group of 40 and 60 nuts per palm per annum). It is intermediary (24.5%) in group III, producing 60 to 80 nuts per palm per annum. On an average, the decline in nut yield due to drought in the subsequent year is 23.7% and it varies based on the yield group (Table 6.2). Good planting material, coconut hybrids, good agronomic practices, floor crops, intercropping and scarce water management practices and integrated pest management alleviate the ill effects of abiotic and biotic stresses to a large extent in coconut. There was a marginal decline in coconut productivity during this tri-decade (5670 nuts/ha) when compared to previous tri-decade (5762 nuts/ha). It can be attributed to rise in temperature in tune to global warming. The negative trend in coconut productivity was much more (17.9%) in the decade 1981-90, which was the warmest and driest decade. The decline in coconut productivity was maximum during 1983-84 (19.2%), followed by 1986-87(6.2%) and 4 per cent in 1987-88. All the above three low coconut productivity years fell in the decade 1981-90 and there is a threat to coconut productivity in the ensuing decades under the projected climate change scenario. The annual coconut production in Kerala during 2013-14 was only 5,200 million nuts as against 5,799 million nuts in 2012-13. The low coconut yield in 2013-14 could be attributed to the severe summer drought that occurred in 2012-13 across the coconut growing areas of the State. Nut size in coconut under rainfed conditions is likely to decline if temperature increase is noticed and oil content is affected. Hence, there is urgent need for pro-active measures to sustain coconut productivity in Kerala as part of climate change adaptation.

6.4 Climate and cashew

The cashew productivity over the decades has shown a steep decline. It is quite alarming that the cashew productivity declined from the ever highest value of 1600 Kg/ha during late 50s (1952-60) to the lowest productivity of 659 Kg/ha during 1981-90. Slight improvement in cashew productivity was noticed during 1991-00 and 2001-10 (785 Kg/ha and 773 Kg/ha, respectively). The percentage decline in cashew productivity was 51 per cent over 1950s. In terms of tri-decadal area, an increase of 27 per cent was registered from 1951-80 (83.8 thousand ha) to 1981-10 (106.3 thousand ha). During the same period, tri-decadal production of raw cashew nuts has declined by 14 per cent from 1951-80 (89.6

Table 6.1: Effect of drought on coconut yield at different locations in Kerala

Sl.No	Name of the station	Drought period and duration	Effect of drought on monthly nut yield			
			Effect of drought on monthly coconut yield	Minimum decline in monthly yield	Maximum decline in monthly yield	Average decline in yield (%)
1	RARS, Pilicode	December 1982 to May, 1983(6 months)	Feb (8th month), 1984 to Jan (19th month), 1985	12.3%(January, 1985)	51.8% (July1984) 13th month	33.2
		December 1988 to May 1989(6 months)	Feb (8th month),1990 to January 1991 (19th month)	12.8%(March,1990)	56.4% (June 1990) 12th month	34.3
2	CPCRI, Farm, Kasaragod	Januaryto April,1998 (4 months)	Jan (8th month) 1999 to March 2000 (22nd month)	11.2%(March 2000)	82.6%(July199) 14th month	25.6
3	Aralam, Farm, Kannur	December 1997 to May1998(6 months)	March (9th month) 1999 to March 2000 (21st month)	1.5%(June 1999	51.5%August 1999 14th month	24.7
4	CDB, Farm, Vellanikkara	December 2001 to April 2002(5 months)	Jan, 2003 (8th month) to Dec, 2003	13.2%April 2003	31.5%June 2003 13th month	20.6
		December 2003 to April 2004(5 months)	Jan, 2005 (8th month) to Dec, 2005	16.4%March 2005	32.9%June 2005 13th month	24.4
5	RARS, Kumarakom	December 1982 to May, 1983(6 months)	Jan, 1984(7th month) to Nov, 1984	13.6%November 1984	16.2%July 1984 13th month	23.4
6	CRS, Balaramapuram	December 1982 to May, 1983(6 months)	Jan - Feb,1984 (7/8th month) to Mar - Apar,1985 (22/23th month)	4.2%Mar - Apr 1985	56.3% Nov-Dec 1984 18/19th month	31.8
		December 1988 to May, 1989(6 months)	Jan-Feb, 1990 (7/8th month) to Nov-Dec, 1990 (18/19th month)	10.3%May-Jun, 1990	51.3% Nov-Dec, 1990 18/19th month	26.9

Table 6.2 Percentage decline in nut yield of different yield groups of coconut

Yield group	Decline in nut yield over previous year				Decline in nut yield due to alternate bearing (B)	(A)-(B)
	1984	1986	1990	Mean (A)		
Group I (>100 nuts per palm per year)	44.3	47.6	69.5	53.8	24.5	29.3
Group II(80-100 nuts per palm per year)	54.2	24.5	41.0	39.9	9.5	30.4
Group III(60-80 nuts per palm per year)	38.9	48.6	33.6	40.4	15.9	24.5
Group IV(40-60 nuts per palm per year)	15.1	39.3	43.4	32.6	22.0	10.6
Mean	38.1	40.0	46.9	41.7	18.0	23.7

thousand tonnes) to 1981-10 (977.1 thousand tonnes). The decline in tri-decadal productivity was 42 per cent during the period from 1951-80 (1200 Kg/ha) to 1981-10 (700 Kg/ha). Farmers' reluctance to cultivate cashew crop due to poor returns and switching over to other profitable crops like rubber is one of the reasons for the decline in cashew area in the State. 1997-98 was a poor yielding year due to unfavourable weather conditions prevailed during the flowering phase of cashew. The yield recorded was only 56890 tonnes. In contrast, incidence of pest complex devastated the cashew plantation along the West Coast during 1998-99 though profuse flowering was noticed due to favourable weather. It was more so in Kannur and Kasaragod districts of Kerala, which account for more than 50 per cent of the total cashew production of the State. In 1999-2000 and 2000-01, a combination of adverse weather and incidence of pest complex adversely affected cashew production of Kerala. The percentage decline in cashew productivity was the highest (59%) in 1981-90 due to the warmest and drought conditions in the form of decline in rainfall and increase in temperature. There was also a drastic decline in cashew yield during the recent years. Therefore, the decline in cashew productivity over a period of time can be attributed to climate variability and climate change.

6.5 Climate change and coffee

The blossom and backing showers are very important in the case of coffee. Coffee yield is very poor in the absence of either of them. The growth of coffee appears to be affected when exposed, and hence it requires a certain degree of shade for better performance. Arabica is more sensitive to weather when compared to that of Robusta. Of course, the quality of Arabica appears to be better. Arabica prefers high altitudes when compared to that of Robusta. October and November rains augment black pepper yields in the next crop season. It is understood that crop in recent years matures early and berry weight declined. It is attributed to global warming and climate change. In Wayanad District, where coffee is predominantly grown, decline in rainfall and rise in temperature was noticed. Overall, the productivity of coffee in recent decades increased though inter-annual variation in coffee production is significant due to influence of blossom and backing showers. Sprinkler irrigation is practiced in the case of coffee to meet the blossom and backing showers in the absence of rains. The coffee growers are also of the opinion that drip irrigation during summer can be a substitute to sprinkler irrigation and beneficial to the crop.

Monsooned Malabar Coffee is a speciality coffee preferred by the people who like black coffee and it fetches high premium in the international market. If monsoon is weak during the monsoon season in which coffee beans are processed, the quality of coffee appears to be poor. The quality of 'Monsooned Malabar Coffee' was very poor for the first time due to bad monsoon in 2002. The quality of "Monsooned Malabar Coffee" is better if monsoon is heavy and continuous during July and August over the Malabar Region, where the coffee beans are processed. Continuous clouds with heavy rainfall and its better distribution, high atmospheric water vapour content, moderate surface air temperature and gentle sea breeze would lead to better moisture content in coffee beans if processed during the monsoon period. Probably, active monsoon period within the monsoon season may be more conducive for such processing of coffee beans. It is one of the factors leading to superior quality in addition to other genetical and environmental factors.

6.6 Climate change and black pepper

Whenever summer showers are received, the yield of black pepper declines. Interestingly, if the same showers coincide with blossom and backing showers, coffee flourishes. Of course, mortality rate is high under poor management conditions in prolonged dry spell conditions in the case of young black pepper vines. It was the reason, why, black pepper gardens were wiped out in Wayanad District during summer 2004 due to hydrological drought. Another important aspect is that whenever black pepper yield is low the coffee yield is better across the high ranges. It is true in 2003 and 2008 when dried coffee beans were high the yield of dry black pepper was low. In 2006 and 2000 when dried coffee beans were low the yield of black pepper was high (Table 6.3). Therefore, mixed cropping of coffee with black pepper is better against prolonged dry spell during summer rather than to allow as mono crop. Cashew, coffee and black pepper appear to be better across the mid-and-high lands of northern region of Kerala since the prolonged dry spell from November to May may not be detrimental for better yield in terms of quality and quantity depending upon the nature of the crop under rainfed conditions. Area under black pepper and arecanut is declining in recent years in Wayanad District. Heavy monsoon rains in 2014 across Wayanad District led to waterlogging and due to yellowing, many arecanut gardens were adversely affected. The wetlands in the District are converted into arecanut gardens and banana plantations. Both demand large quantity of water for irrigation. It may the reason for frequent floods and summer droughts in Wayanad District.

Table 6.3: Mean yield of coffee and black pepper from 2000 to 2008 at RARS, Ambalavayal

Year	Dried coffee (kg/ha)	Yield of dry pepper (kg/ha)
2000	656	387
2001	802	398
2002	639	326
2003	739	285
2004	690	324
2005	660	334
2006	598	318
2007	705	357
2008	703	198

6.7 Climate change and cocoa

Cocoa is grown as an intercrop in coconut or part of homestead garden. Both the crops cocoa and coconut respond identically with reference to weather and climate under the Humid Tropics. Of course, the phenology of both the crops is totally different. In case of cocoa, the difference in monthly yield is high during the monsoon season between good and bad yield years when compared to other seasons. The number of cocoa (cacao) pods is high during summer, but pod weight is low. A harvest of five pods during October/November may be equal to eight/nine pods harvested in summer, showing superiority in cocoa pods harvested in October/November. There is a lag period of 4 to 6 months between weather effects and cocoa yield. The maximum temperature from January to March influences cocoa yield to a considerable extent. A prolonged dry spell from November to May with high maximum temperature adversely affects the pod yield to the tune of 40 per cent depending upon the crop management. High rainfall versus cocoa yield showed inverse relationship, indicating that high rainfall during rainy season may not be conducive in the case of cocoa due to waterlogging and lack of soil aeration. Overall, it reveals that high maximum temperature during summer, followed by heavy rains during the Southwest monsoon may not be conducive for obtaining good yield in cocoa under the Humid Tropics. High rainfall has a malevolent effect in places where the crop is grown under waterlogged conditions, while summer rainfall has a benevolent effect on pod yield. Cocoa needs equitable climate with well distributed rainfall. It requires an annual rainfall of 1500-2000 mm with a minimum of 90-100 mm rainfall per month. It is also grown under uneven distributed rainfall with fairly well defined dry and wet seasons. Such seasonal changes exert marked effects on the growth of the cocoa tree, and on its cycle of flushing, flowering and fruiting. Studies at Vellanikkara, Thrissur indicated that high maximum temperature followed by heavy rainfall is detrimental to cocoa yield. Mean maximum temperature during summer was high (34.7°C) in poor yield years while less (34.2°C) in good yield years (Table 6.4). A mean maximum difference of 1.1°C was noticed in April between good and poor yield years. A decline of 30-40 % in annual cocoa yield was noticed in 2004 when compared to that of 2003 due to rise in maximum temperature of the order of 1-3 °C from 14th January to 16th March, 2004 along with prolonged dry spell (Fig. 6.6). Such trend was noticed whenever summer temperature shot up by 2-3 °C when compared to that of normal maximum temperatures of 33-36.5 °C. The adversity of weather aberrations on rhythm of normal growth of cocoa is reflected in yield after a lag period of 4-5 months. Flowering behaviour in cocoa depends on rainfall distribution as flower dip is noticed under prolonged wet spell with heavy rains during the rainy season. Under the projected climate change scenario, the cocoa production is likely to be adversely affected as it is influenced by rise in maximum temperature and uncertainties in monsoon rainfall pattern in the form of occurrence of floods and droughts. In cocoa, mango and cashew the flowering period is very much influenced by the geographical co-ordinates such as latitude, longitude and altitude of the crop growing regions and appears to follow the famous Hopikins Bio-climatic Law. It states that "A biotic event in North America will, in general, show a lag of four days for each one degree of latitude, five degree of longitude and 400 feet of Altitude, northward, eastward and upward in spring and early summer". Hence, there is need to develop climate adaptation strategies at the plantations/farm level to mitigate the ill effects of climate change/climate variability.

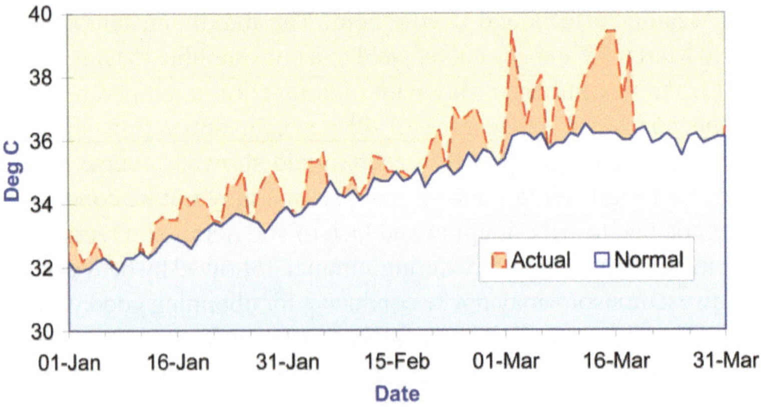

Fig. 6.6: Increase in maximum temperature at Vellanikkara in 2004

6.8 Wayanad and Idukki agro-ecosystems

The highranges of Kerala comprising of Idukki and Wayanad districts are vulnerable to the projected climate changes. Normal average annual rainfall over the Wayanad district is 3417 mm, having 2760 mm in the southwest monsoon, 340 mm in northeast monsoon and the rest in other months. Since the last 16

Table 6.4: Monthly maximum temperature during summer in good and bad yield years of cocoa

Month	Max. temp. in good yield years (°C)					Max. temp. in poor yield years (°C)				Normal (°C)
	1992	1993	1996	1999	Mean	2004	1998	1995	Mean	
January	32.6	32.7	33.1	32.4	32.7	33.4	33.1	32.9	33.1	32.8
February	34.4	34.1	34.7	34.5	34.4	35.2	34.4	35.4	35.0	34.8
March	36.9	35.4	36.4	35.5	36.1	36.5	36.2	37.6	36.8	36.1
April	36.3	35.6	34.6	33.4	34.9	34.8	36.5	36.6	36.0	35.4
May	33.8	34.4	32.8	30.7	33.0	30.4	34.2	33.5	32.7	33.8
Mean	34.8	34.4	34.3	33.3	34.2	34.1	34.9	35.2	34.7	

years, the annual and monsoon rainfall were declining in the Wayanad district. The district had a deficit (186 cm) of 55 per cent in 2003 and 35 per cent (191 cm) in 2002. Continuous decline in rainfall and rise in temperature in addition to severe summer drought in 2004 led ecosystem of Wayanad in peril. Rainfall decline and temperature increase are projected over Ambalavayal, Wayanad district. Several black pepper gardens were wiped out during summer 2004. There was a time when oranges were plenty in Wayanad district. At present such agro-ecosystems are extinct. It was attributed to change or variability in climate and deforestation. Conversion of large tracts of paddy fields into banana garden in addition to excessive use of chemical fertilizer and pesticides and large-scale mining of sand from rivers which began in the 1980s, has now been identified as the causal factors for severe drought with continuous decline in annual rainfall. Introduction of banana in paddy lands has deprived the district of large areas of wetlands needed to store water. With banana cultivation so extensive, nature lovers said the very name of the district (Wayanad means land of paddy fields) has lost its meaning. Groundwater depletion is so much (25%) due to over exploitation. Widening temperatures may be detrimental to cardamom and black pepper in addition to deforestation. A weather scenario which is not conducive to coffee may be favourable to black pepper and vice-versa. Absence of backing showers in April / May 2009 led to low bean production in the case of coffee in Wayanad district.Therefore, mixed cropping coffee plantations with black pepper may sustain in Wayanad district as a strategic approach to mitigate the ill-effects of climate change / variability.

6.9 Climate change and cardamom

The natural habitat of cardamom under the forest ecosystem is also in peril over Idukki district due to climate variability. Rise in maximum temperature and fall in minimum temperature are projections over Pampadumpara, Idukki. Widening temperatures may be detrimental to cardamom and black pepper in addition to deforestation.The production and productivity of small cardamom show an increasing trend though a sharp decline was noticed in cardamom area across the Western Ghats. This could be attributed to technological interventions since last three decades. However, the inter-annual fluctuations in cardamom production were common due to weather aberrations. For example, the cardamom production was badly hit during 1983 due to unprecedented drought that occurred from November 1982 to May 1983 across the cardamom tract of the Western Ghats. Similar was the case during 2003-04 over Kerala. It is noticed that the cardamom production over Kerala in recent years was also badly hit due to dry spells that occurred in recent years. Rainfall from Dec-May, temperature range during southwest monsoon and annual temperature range could explain the variability in cardamom production by 78 per cent. It revealed that monsoon uncertainties, declining summer rainfall, increase in annual and summer temperatures, widening the temperature difference between the maximum temperature in summer and minimum temperature in winter are the concern at several locations across the cardamom tract and it is detrimental to cardamom production under the projected climate change scenario. A sharp decline in area under cardamom was noticed across the cardamom tract of the Western Ghats in recent years. However, the production and productivity of small cardamom were increasing. The important physiological stages like panicle initiation and subsequent growth (forwarding) depend on receipt of showers from January to May. The failure of showers during this period results in poor growth and crop yield. The unprecedented drought in summer 1983 adversely affected the growth and yield of cardamom in Coorg district, and the same trend prevailed in other cardamom-growing tracts of South India. The drought that prevailed during 1982–83 resulted in as high as 50% mortality in some of the cardamom estates. In the Palney hills of Tamil Nadu, both panicle and flower production are seen in October and November, during which northeast monsoon rains are received. It indicates that both panicle and flower production in cardamom are commensurate with the rainfall pattern of the cardamom tract. There is a close relationship between cardamom production and rainfall distribution during summer. Summer rains during March, 2008 have benefited the cardamom crop of 2008–2009 up to 20–30%. Therefore, summer rains have significant positive influence on cardamom. There was a strong negative relationship between soil moisture stress and cardamom yield. Climate risk is more in the case of cardamom in Karnataka region, followed by Ambalavayal and neighbouring areas in northeast of Kerala while climate risk is less in Thandikudi and Idukki regions in south of cardamom tract.

6.10 Climate change and rubber

Area, production and productivity of rubber have been increasing over a period of time. The rate of increase in rubber productivity was affected during 1959-63 (-24.3%), followed by 1979-83 (-8.3%), 1984-88 (2.2%), 1999-03 (13.5%) and 2004-08 (18.9%). Despite increase in temperature, aridity index, number of droughts, decline in rainfall and moisture index (index of climate shifts) the rubber productivity was in increasing trend. However, decline in rate of increase in rubber productivity was noticed. At the same time, the prolonged dryspells during summer due to absence of rainfall during northeast monsoon and failure of summer showers is likely to affect rubber productivity adversely as seen in recent years. Abrupt changes in weather may affect the productivity of the crop as noticed in 2007-08, during which the yield declined (1471 Kg/ha) due to continuous rainfall and more number of rainy days when compared to that of previous year 2006-07 (1554 Kg/ha). Climate change in the form of climate variability, especially unpredictable, irregular and deficient rainfall pattern adversely affects the growth and productivity of natural rubber. Changes in rainfall pattern triggers fungal infection of leaves, which adversely affects productivity. It indicated that the long term increase in temperature is likely to affect rubber yield adversely and the rate of increase is likely to decline.

6.11 Climate Change and Bio-diversity

The Economic Review prepared by the State Planning board, 2003 warns that one-third of the State's biodiversity would vanish or would be close to extinction by 2030 unless steps were taken to check extinction of species. Of the 300 rare, endangered ot threatened species in the Western Ghats, 159 are in Kerala. Of these, 70 are herbs, 23 climbers, eight epiphytes, 15 shrubs and 43 trees. Besides, 10 species of fresh water fish are identified as most threatened. Kerala has a flora of 10035 species, which rep[resents 22% of Indian flora. Of these, 3872 are flowering plants of which 1,272 are endemic. As many as 102 species of mammals, 476 birds, 169 reptiles, 89 amphibians and 262 species of fresh water fish are reported from Kerala. Many of these are endemic. The review recalls that during the 20th century, at least 50 plant species became extinct in the country. Three species of birds-Himalayan mountain quail, forest spotted owlet and pink-headed duct-have become extinct. Besides, as many as 69 bird species have been categorised as extinct. The mammals, Indian cheetah and lesser one-horned rhinoceros have also perished. The Malabar civet is on the threshold of extinction and 173 species have been listed as threatened. Among flowering plants, about 1500 species come under threatened categories. It may be noted that nearly 23 % of the total endemic flora species of the country are in Kerala. Of 1272 such species, 102 species occur exclusively in Kerala. Describing a conservation strategy, the review says that ecologically sensitive areas have to be identified with reference to topography, hydrological regimes and this has to be networked with species diversity.

6.12 Climate Resilient Agriculture

The State Action Plan on Climate Change (SAPCC) indicated that Alappuzha, Palakkad, Wayanad and Idukki are the 'Climate Change Hot spots' in Kerala while Alappuzha and Palakkad are the districts most vulnerable to climate change. Idukki, Wayanad, Thiruvananthapuram and Kannur are highly vulnerable to Climate Change. Kerala is predominantly a plantation state, falls under the Tropical Humid Climate, moving from wetness to dryness within the Humid Climate ($B_{4\ to}\ B_3$ as per the Thorthwaite's climate classification). Monsoon rainfall across the State was declining since last 60 years (cyclic trend of 40-60 years is also noticed with annual and monsoon rainfall) while rise in temperature is evident. Of course, rate of increase in temperature was alarming across the Highranges (where cardamom, coffee and tea are grown) due to deforestation.

It is also true to some extent along the Coast (low land) due to increase in sea surface air temperature. The decade 1981-90 was the driest and warmest decade and 2011-2020 is likely to be still warmer, therefore warming Kerala is true since last three decades. 1987 was the warmest year, followed by 1983. 2015, 2016 and 2017 were also warmer during the current decade. The decadal and tri-decadal changes in climate will lead to weather extremes and thereby, effects crop productivity and crop production. Hence, it is high time to take initiatives to minimize crop losses in the ensuing years/decades against weather extremes.

Rice productivity was increasing while not so in cashew and coconut and to some extent in the case of rate of increase in rubber too. In contrast, the crop productivity was increasing in cocoa, cardamom, coffee and black pepper and not so in the case of tea. At the same time, all the plantation crops showed low productivity or low rate of increase in productivity during the warmest and driest decade 1981-90. Therefore, the adverse affect of climate variability on plantation crops is evident and at the same time the effects of climate change cannot be quantified though decline in crop productivity could be seen in coconut, cashew, tea and rubber. However, crop growth simulation models indicate that crop yields including cropped area are likely to decline in the ensuing decades under the projected global warming and climate change scenario. Climate change adaptation strategies and awareness are particularly important for managing the risks posed by climate variability and change, not only on plantation crops but also across all the society linked sectors that are sensitive to weather and climate. Various agroclimatic techniques have been used to effectively manage some of the risks posed by climate change to plantation crop productivity. Expansion and further development of such techniques will be vital for the continued sustenance of plantation crops' production in the Humid Tropics. As pointed out by Prof. M.S. Swaminathan "India's strength lies in its ability to manage monsoons" instead of saying" Indian agriculture is a gamble of the monsoon". Climate smart agriculture, climate resilient agriculture, precision farming, hi-tech farming, conservation agriculture, traditional agriculture, protected cultivation, mixed farming and integrated farming are some of the technologies/tools are coined to minimize the ill effects of climate change and drought on crops. These technologies may be sound in the case of field crops/ seasonal crops, but not in the case of plantation crops since the plantation crops are in the field round the life time, having economic yield age of 20 to 60 years depending upon the type of crop. In the case of coconut depending upon the type of cultivar or hybrid, the first flowering commences between fourth and eighth year after planting under better management conditions and coconut yield is stabilized by its 15/20th year and yields economically up to eighty and hundred years. Therefore, integrated farming as done traditionally or crop combinations in the case of coffee and black pepper or multi-tier cropping systems in coconut gardens may be strategic approach to withstand against the ill effects of climate change/variability rather than looking for mono-crops. The farm-level weather insurance may be another tool which should be seriously examined for the benefit of the farmers in the event of crop losses due to weather extremes. Unlike seasonal crops, the effects of climate change may be standing many more subsequent years depending upon the phenology of the crop. It is also observed that crop mix/integrated farming may withstand relatively better against the ill effects of climate variability/change as seen in the case of coffee and black pepper. When black pepper performs better coffee suffers and vice-versa. Recent findings indicate that the climate change effects not only crop output but also commodity quality and price. The current scenario in coconut price fluctuations could be attributed to low coconut production in southern states since last three years. One of the factors for overall price escalation in recent years could be attributed to global warming and climate change. Therefore, it is high time that the government agencies and policy makers to be proactive with short term and long term strategies to mitigate the ill effects of weather related disasters with the people's participation and minimize the losses against the weather related disasters. Need of the hour is expect the unexpected guests in the form of weather related disasters under the projected climate change scenario and tune to the system to minimize the losses with well preparedness at various levels.

Summary

The National Action Plan on Climate Change (NAPCC) identified 'Sustainable Agriculture' is one of the thrust areas to tackle the climate change related issues. The mission intends making Agriculture more resistant to climate change by identifying and developing new varieties of crops that are thermal resistance and capable of withstanding extreme weather. These issues are being tackled as a part of National Innovations on Climate Resilient Agriculture (NICRA). When it comes to the State of Kerala such initiatives need to be taken up on priority basis in developing technologies against the weather extremes such as saline/flood resistance and drought resistance/tolerance rather than thermal resistance, for which a definite State Action Plan on Climate Change (SAPCC) in Agriculture Sector need to be highlighted and tackled with concerted efforts. To tackle the climate change related issues in Agriculture, a definite policy framework must be chalked out and related institutes should look into. To me it appears, we need have an action plan similar to that of NICRA at the State level with technical people in the form of 'State Initiative on Climate Resilient Agriculture'. It can be known as SICRA at the State level similar to that of NICRA at the national level (Fig. 6.7). It may be related to plantation crops or horticulture or medicinal plants or seasonal or as a whole integrated/homestead faming. Such system should be made operational and in vogue at the earliest possible time since climate change/variability is a big threat in Agriculture Sector.

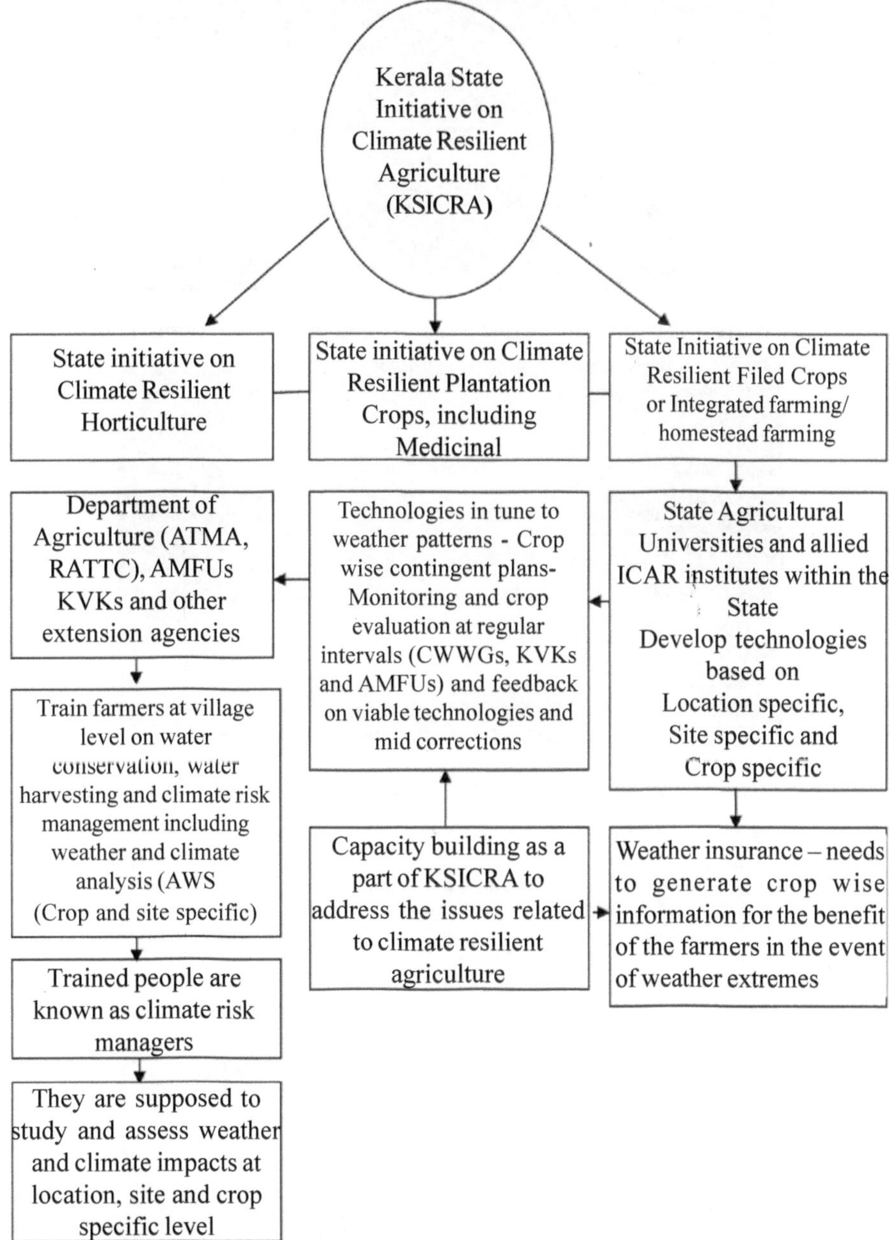

Fig. 6.7: Concept of SICRA

Plate 6.2: Crop damage due to 2018 flood in Kerala (August)

Plate 6.3: Landslides on National Highway at Kuthiran, Thrissur (August 2018)

Plate 6.4: River overflow due to flood in 2018 (August)

Plate 6.5: Aerial view of August floods in Kerala, 2018

References/Suggested Readings

Ananthakrishnan R and Soman M.K. 1988. The onset of southwest monsoon over Kerala for the period 1801-1980 *J. Climatol.* 8: 283-296.

Gopakumar CS and Rao, GSLHVP 2018. Climate change over Kerala. . Presented in the National Seminar on Recent Trends in Climate Change and Carbon Research. Organized by the Department of botany, University of Calicut during 26-28, February 2018.

Kannan, K., Devadas, V.S and Thomas, C.G. 1987 Effect of weather parameters on the productivity of coffee and pepper in Wayanad. In: Proc. of National Seminar on Agrometeorology of plantation crops held at RARS, Pilicode during 12-13, March 1987. P.187.

Krishnakumar, K.N., Rao, G.S.L.H.V.P. and Gopakumar C.S. 2000. Rainfall trends in twentieth century over Kerala, *India. Atmospheric Environment* 43: 1940-1944.

Krishnakumar., K.N and Rao, G.S.L.H.V.P, Gopakumar, C.S and Ram Mohan, H.S 2007. Rainfall variability during southwest monsoon over Kerala (India: 1871-2005). *Indian Journal of Polar Sciences,* Vol 1(1) .pp 81-87.

Martinez, R., Hemming, D., Malone, L., Bermudez, N., Cockfield, G., Mushtaq, S., Diongue, A. Hansen, J., Hildebrand, A., Ingram, K., Jakeman, G., Kadi, M., McGregor, G.R. Rao, P., Pulwarty, R., Ndiaye, O., Srinivasan, G., Eh.R., Seck, White, NandZougmore, R. 2012. Improving Climate Risk Management at Local Level – Techniques, Case Studies, Good Practices and Guidelines for World Meteorological Organization Members-Chapter 21 (477-532): In Risk Management-Current Issues and Chanllenges (Ed: Nerija Banaitiena),584p.

Monteith, J.L. 1977. Climate. In: Ecophysiology of Tropical Crops. (Eds. P.de T. Alvin and T.T.Kozlowski). Academic Press. New York, pp.1-27.

Nelliat, E.V. and Padmaja, E.K. 1978. Irrigation requirement of coconut and response to levels of fertilizer under irrigated condition during the early bearing stage. In: Proc. PLACROSYM – I, Indian Society of Plantation Crops. Kasaragod. pp 196-199.

Nybe, E.V. and Rao, G.S.L.H.V.P. 2005. Drought management in plantation crops. Kerala Agricultural University. Thrissur, Kerala, India. 176p.

Rajagopal, V. Shivashankar, S and Jacob Mathew. 1996. Impact of dry spells on the ontogeny of coconut fruits and its relation to yield. Agro Physiology. 251-255.

Ramakrishna, Y.S., Rao, G.G.S.N., Srinivasa Rao, G. and Vijayakumar, P. 2006. *Impact of climate change in agriculture*. Environment and Agriculture (eds. Chadha, K.L. and Swaminathan, M.S.). Malhotra Publishing House, New Delhi, pp. 1-30

Rao, G.S.L.H.V.P. 1991. Agricultural droughts with reference to coconut. Journal of Plantation crops. 18 (supplement): 166-170.

Rao, G.S.L.H. V.P. 1994. Drought management in coconut. *Indian Coconut J.* Vol. XXIV(I 1): pp 6-8.

Rao, G.S.L.H.V. and Gopakumar, C.S. 1999. Latitudinal and altitudinal influence on cashew flowering over Kerala. In proc. of XI Kerala Science Congress held at CPCPRI, Kasaragod during February 27-March 1, 1999. pp 242-246.

Rao, G.S.L.H.V.P and Krishnakumar, K.N. 2005. Monsoon onset over Kerala (India): 1870-2004. *J. of Agrometeorol.*(2) : 161-167.December 2005.

Rao, G.S.L.H.V.P., Mohan, H.S.R., Gopakumar, C.S and Krishnakumar, K.N. 2008. Climate change and cropping systems over Kerala in the humid tropics. *J. of Agrometeorol.* (Special issue-Part 2): 286-291. pp. 286-291.

Rao, G.S.L.H.V.P and Gopakumar, C.S. 2009. Agroclimatology of cashew. In: Cashew Research and Development in Humid Tropics with emphasis on cashew apple processing. p. 26-36. (Eds). Mathew, J., Mini, C and Abraham, M. 2009.

Rao, G.S.L.H.V.P and Gopakumar, C.S. 2009c. Agroclimatology of cashew. In: Cashew Research and Development in Humid Tropics with emphasis on cashew apple processing. p. 26-36. (Eds). Mathew, J., Mini, C and Abraham, M. 2009.

Rao, G.S.L.H.V.P. and Manikandan, N. 2009. Climate Variability and small cardamom across the Western Ghats. Kerala Agricultural University. Thrissur, Kerala, India. 117p.

Rao, G.S.L.H.V.P. and Manikandan, N. 2009. Crop weather relationships of cocoa in the Humid Tropics. Kerala Agricultural University. Thrissur, Kerala, India. 117p.

Rao, G.S.L.H.V.P., Rao, G.G.S.N. and Rao. V.U.M. 2010. Climate change and agriculture over India. PHI Learning Private Limited, New Delhi, India. 320p.

Rao, G.S.L.H.V. P. 2011. Climate change adaptation strategies in agriculture and allied sectors. Scientific Publishers (Jodhpur), India, 336p.

Rao, G.S.L.H.V.P. and Gopakumar, C.S. 2011. Climate change and plantation crops in Kerala. Kerala Agricultural University. Thrissur, Kerala, India. 255p.

Rao, G.S.L.H.V.P., Gopakumar CS and Krishnakumar, K.N.2011. Climate variability and Indian foodgrains production. In: Proc. of International Symposium on Geographical Congress held at CWRDM, Kozhikode held during 6-8, May 2011.

Rao, G.S.L.H.V.P. 2011. Climate change adaptation strategies in agriculture and allied sectors, Scientific Publishers (India), 336p.

Rao, G.S.L.H.V.P, Gopakumar, C.S and Krishnakumar, K.N. 2013. Impacts of Climate Change on Horticulture Across India. In: Climate-Resilient Horticulture: Adaptation and Mitigation Strategies. H.P.Singh *et al* (eds). Springer Publication. DOI 10.1007/978-81-322-0974-4_2, pp 13-20.

7

Practicing Smart Agriculture in Tamil Nadu

T.N.Balasubramanian*, R. Gowtham** and V.Geethalashimi ***

**Professor(Agrl. Met)-Rtd, TNAU*
***Ph.D. Scholar, Agro Climate Research Centre, TNAU, Coimbatore*
**** Professor (Agronomy), Agricultural College and Research Institute*
Tamil Nadu Agricultural University, Madurai, Tamil Nadu, India

Tamil Nadu has unique physiography with western and eastern Ghats on the west and northwest, Bay of Bengal in the east and Indian Ocean to the south. Tamil Nadu has high dependence on natural resources and hence faces the threat of climate change and its impacts. Available evidence shows that there is high probability of increase in the frequency and intensity of climate related natural hazards (Stephen, 2012). Out of 13 million hectares of geographical area, about 7 million hectares of land is under cultivation. Of this 55 per cent is irrigated and rest is under rain fed / dry lands. In Tamil Nadu, ground water has been harnessed fully and dry land productivity is found oscillating with ruling variability in seasonal monsoon rain and hence agricultural productivity is always under stake. To improve agricultural productivity, Tamil Nadu has been divided into seven agro climatic zones viz., North Eastern zone, North Western zone, Western Zone, Cauvery Delta zone, the Southern zone, High rainfall zone and High Altitude and Hilly zone. Tamil Nadu is vulnerable to increase in temperature and sea level rise with change in climate as a result of accumulation of green gas gases beyond the threshold level in the atmosphere at the global level (TNSAPCC, 2014). United Nations Organization (UNO) defines vulnerability as the extent to which a natural or social system is susceptible to sustaining damage from climate change process and vulnerability is a function of the sensitivity of the system to change in climate, adaptive capacity of the said system (natural or social) and the degree of exposure of the system to climate hazards (UNFCCC, 2008). Climate change is one of the biggest challenge to food security, but it is also an opportunity for the scientists and farmers to innovate and adapt. More than 70 per cent of the population in Tamil Nadu are engaged in agriculture and are facing the threat of climate change. The food grain productivity need to be increased by 130 per cent by 2050 to meet the demand of growing population. The Government of Tamil Nadu is moving towards the goal of doubling the food production and tripling the income of the farming community in a sustainable way through effective management of natural resources including soil and water. This could be achieved by practicing climate smart agriculture that aims to help find new ways to adapt to changing climate. This process would increase agricultural productivity and farm income sustainably and equitably and simultaneously helping food system pillars and farmer's livelihoods and at the same time minimizing agriculture's green house gas emissions. Each and every farm can become climate smart unit for promoting sustainable weather based farming practices, using inputs more efficiently and effectively to grow more and building market where farmers can access what they need and sell any surplus of what they produce.

7.1 Climate trends

Normal annual average of rainfall of Tamil Nadu is 958.4 mm, mainly distributed in three seasons, viz., south west monsoon, north east monsoon and pre monsoon season. About 50 per cent of the total annual average rainfall is received during north east monsoon, while about 31 per cent is received during southwest monsoon and the rest is received in the other seasons. Spatial distribution of the rainfall received over Tamil Nadu is highly variable with both inter seasonal and intra seasonal variability. Rainfall over coastal areas is more and decreases in the inland areas, also the rainfall over northern part is more than the southern parts. Jain and Kumar (2012) has reported that the annual rainfall has increased by (+)8.5 per cent and (+4.4) per cent in the Cauvery river basins and the river basins north to Cauvery river basin in Tamil Nadu respectively in the last 100 years with respect to the average rainfall during this period. The river basins that are in the south of the Cauvery river basin have experienced decrease in annual rainfall by (-)9.8 per cent. An analysis of annual rainy days indicates that there is no change in the Cauvery basin in the last 100 years period. However, the river basins north and south of the Cauvery basins have experienced decreasing trend by (-)3.6 per cent and (-)32.3 per cent respectively. In general, the maximum temperature rarely exceeds 43°C and the minimum temperature rarely falls down below 18°C. The mean annual temperature is 28.2°C in the plains and 15.2°C in the hills. The temperature is minimum in the month of December with 24.7°C and maximum in May with 37.3°C. Annamalai *et al* (2011), based on their study on temperature over Cauvery basin of Tamil Nadu reported that the average year to year variation in surface temperature lies in the range about 0.4° C with few years warmer or cooler by 0.8°C. Based on the technique of deducting the long time forced component (trend), the temperature series in both the seasons clearly indicates a warming tendency. For the period from 1951 to 2008, the warming is of the order of 0.7 to 0.8°C and this is above the natural variability. One difference is that, while the warming is gradual during south west monsoon season, it appears to occur abruptly during northeast monsoon season. Balasubramanian *et al* (1994) based on their analysis of the prevailed temperature in Coimbatore from 1962 to 1992 found that there was decadal variability in maximum and minimum temperature and this was on the rising level up to 0.1 to 2.7°C.

7.2 Future climate projections

Future climate projections for Tamil Nadu for RCP 4.5 and RCP 8.5 at a spatial resolution of 0.25 x 0.25 degree was done by Gowtham *et.al.* (2017) for the mid century time slice (2041 - 2070) using 29 GCMs prescribed in the IPCC AR-5. The results indicated that both maximum and minimum temperature over Tamil Nadu are projected to increase irrespective of the models and scenarios studied and the rate of increase in minimum temperature is higher than that of maximum temperature. Maximum temperature is expected to increase up to 3.7°C and 4.7°C and minimum temperature up to 4.9°C and 5.7°C by the end of mid century as projected through stabilization (RCP 4.5) and overshoot emission (RCP 8.5) pathways. As far as rainfall is concerned, annual rainfall over Tamil Nadu is expected to vary between a decrease of (-)17.1 per cent to an increase of (+)34.6 per cent and (-)33.0 per cent to (+)45.1 per cent as projected through RCP 4.5 and RCP 8.5 scenarios respectively. In another study conducted by Anna University based on UK Met Office Hadley Centre regional climate model PRECIS with boundary data inputs from 6 out of 17- member perturbed-physics ensemble (HadCM3Q0-Q16, known as 'QUMP') at a spatial resolution of 25 km x 25 km and the GHG emission drivers are generated by the IPCC A1B SRES scenario, the maximum temperature over Tamil Nadu is projected to increase by 1.0°C, 2.0°C and 3.1°C and minimum temperature by 1.1 °C, 2.4 °C and 3.5 °C for the periods 2010-2040, 2040-2070, 2070- 2100 respectively with reference to the baseline of 1970-2000. The rainfall projection indicates a slight decrease of about 50 mm by end of the century (2070-2100) with reference to the baseline. South west and north east monsoons being principal rainy seasons, analyses have been carried out for these two seasons. North east monsoon may experience more intense rainfall when compared to south west monsoon by end of the century (TNSAPCC,2014). The Climate Change 4X4 assessment for India at a sectoral and regional analysis for 2030's report published by the MoEF&CC, GoI in 2010, concludes that in future, the number of cyclones hitting the eastern Indian coast including Tamil Nadu is likely to reduce, however, the intensity i.e., the wind speed of the cyclones may increase. Using CLIM SYSTEM - a model used for predicting sea level rise, with climate inputs from various GCMs(Global Climate Model), projected that by the end of the century i.e. by 2100 the sea level may rise in Tamil Nadu coast is likely to range from 0.19 m to a maximum of 0.73 m.

7.3 Status of Agriculture

Tamil Nadu is one of the most urbanized and industrialized States in India and only 22 per cent of its income comes from the agriculture and allied sectors and the share has shown a declining trend over the years. The sector, which grew at 15.28 per cent during 2006-'07 declined to (-) 4.46 per cent during 2007-'08 due to the crop damages caused by natural calamities. However, till date, 40 per cent of the State population is dependent on this sector for their livelihoods. Hence the growth in agriculture is important not only to ensure food security, but also for reducing poverty. A comprehensive package combining several components to revitalize the sector should be designed to enhance the productivity working within the water constraints and stabilizing or enabling inter and intra seasonal risk proofing of rain fed production systems (TNSAPCC, 2014). In Tamil Nadu, millet crops such as jowar, bajra, maize, ragi and other minor millets; pulses such as black gram, red gram, green gram, horse gram and cowpea; oilseed crops such as groundnut, gingelly, sunflower and castor are best suited for dry land as they are mostly grown under rain fed condition. In addition, the frequent and recurrent hydro-meteorological events such as droughts, extreme rainfall events and cyclones impact agriculture in the State negatively and the growth of this sector is further constrained by a number of other factors such as reduced availability of water and declining cropped area, which has declined from 33 per cent of available land area in 2000-'01 to 31 per cent of total land area in 2010-'11. Further, small land holdings, deterioration in soil health due to depletion of topsoil, decline in organic content, decrease in cropping intensity and shortage of labor besides reluctance to work in the farms, and poor adoption of crop management practices etc. are adding to the already existing pressure on agriculture.

7.4 Impact of climate change on agriculture

Climate change may exacerbate the impacts and thus limit agricultural production. In respect of Tamil Nadu agriculture, with change in climate there would be water shortage against its demand, increase in temperature would affect biomass production in agricultural crops, sea level rise would reduce the quality of the fertile lands, extreme rainfall events would increase the frequency of droughts and floods. Under this situation, the sustainable agriculture that is being practiced at Tamil Nadu presently would get disturbed negatively. This is partly being observed at now. Hence it is necessary to recommend and subsequently practice smart agriculture at Tamil Nadu to enhance the resilience capacity in terms of adaptive capacity to reduce the degree of exposure of the system to the anticipated climate hazards. Geethalakshimi and Dheebakaran (2008) studied the impact of climate change on rice and maize crop duration and its yield for Tamil Nadu using INFOCROP model and concluded that the yield reduction in rice would be more during *kharif* season (June -Sep) than in *rabi* season(Oct -Feb). The reduction would be 10 to 15 per cent during *kharif* season in 2020 with increase in temperature and change in rainfall. The expected yield reduction in rice would be 30 to 35 per cent in 2050, while it would be around 80 per cent during 2080. The interesting facts from the study is that, during *rabi* season, the rice yield would increase by 10 per cent. This might be due to increase in air and water temperature by 1-2 °C in 2020 and 2050. With further increase in temperature, the rice yield would be reduced drastically to 25 per cent during 2080. The rain fed maize crop duration would be reduced with 1°C rise in both maximum and minimum temperature and the reduction would be around six days during good years and four days in bad and normal years. Increase in rainfall by 10 per cent did increase the crop duration by four days during bad and normal years. Geethalakshimi *et al.* (2008), also reported that there would be reduction in sorghum grain yield by 4.5, 11.2 and 18.7 per cent respectively at 2020, 2050 and 2080 because of changes in climate. It is also a concern to note that crops productivity would fall down with change in temperature and rainfall including cotton and red gram. These statement would indicate that food insecurity would prevail in the coming years. Increase in temperature both maximum and minimum do likely to affect fodder and feed production and there by decrease in milk production in cattle and egg and meat productivity in poultry. The expected Thermal Humidity Index would be >72 with increase in temperature and relative humidity. This will affect livestock comforts and put the animals and birds under environmental stress. The gap between demand and supply for water during 2020 would be around 2211MCM (11%) and it is likely to go up to 17 per cent by 2050. As per the analysis done under normal climate situation out of five years cycle, one year each would be under drought and floods respectively and the remaining three years would be under normal. But under climate change scenario, the cycle will be broken and three would be continuous visit of either droughts or floods.

7.5 Management strategies

A special package for climate change adaptation should be developed both for irrigated and rain fed / dry land areas for minimizing production risks. Steps towards this could include Integrated farming including crops, livestock, fisheries, poultry and agro forestry based on farming situation; promotion of homestead gardens supported by nurseries; promotion of farm ponds, legume trees and biogas plants; selection of thermal and water stress resistant varieties; soil test-based INM practices for improving soil health; improving the seed production mechanism for timely supply of quality seeds; efficient use of water through micro irrigation; rain water management and introduction of farm mechanization. Some of the reasons for not implementing latest technology in Tamil Nadu are non availability of crop specific/ location specific technologies, high investment cost that is not affordable by small and marginal farmers despite Government support, mind set of the farmers in favour of traditional agriculture, non-availability of quality inputs required for high tech agriculture, non availability of inputs in time, less availability of credit support to most of the small and marginal farmers. Some of the general management strategies for reducing the climatic risks are given below:

7.5.1 Avoiding exposure to risk

Area under high water requiring crops like rice, sugarcane and banana must be reduced and cultivated only in the efficient cropping zone identified across Tamil Nadu through scientific means (Kanwar, 1972). Practicing weather based agriculture will help in reducing the input cost, maximize the output and profit. Establishing strong net work on crop insurance and making 100 per cent participation of farmers by law in crop insurance programme will compensate the crop loss in the event of extreme weather events. Developing bio-shield in coastal villages by growing mangroves that would not only protect the coastal communities from cyclones and other sea related disasters but will also provide nutrients for marine species that thrive in the mangrove soils

7.5.2 Crop diversification

Popularization of minor millets, vegetables and fruit crops that are resilient to climatic hazards may be encouraged to grow by the farmers. Inter-cropping and mixed farming will also help in stabilizing the farm income. Alternate cropping strategies for "agro climatic zone based cropping pattern" may be developed to ensure maximum utilization of available land and water with an ultimate aim to increase cropping intensity.

7.5.3 Adoption of advanced cropping techniques

To maximize the input use efficiency, farmers can adopt innovative agricultural technologies. The Integrated Crop Management (ICM) technology including Integrated Nutrient Management (INM), Integrated Water Management (IWM), Integrated Pest Management (IPM) could be practiced. Under this umbrella, some of the technologies are cultivating resistant varieties under modified cultivation practices like SRI and aerobic system of rice cultivation, SSI for sugarcane cultivation, changing the sowing window to put the crop under comfortable weather during the critical crop growth stage, switching to micro irrigation for improving water use efficiency, fertigation for increasing fertilizer use efficiency, use of bio-pesticides environmentally safe for pest control, etc.

7.5.4 Maintaining the soil health

Minimum Tillage is recommended to retain soil carbon content and also to reduce the top soil erosion. Soil health and nutrient content could be managed through soil test based integrated nutrient management.

7.5.5 Diversification of income source

Encouraging development of Integrated Farming Systems suitable for each agro climatic zone which encompasses various agricultural enterprises viz., crop husbandry, animal husbandry, fishery, forestry, etc. to facilitate changes in the farming techniques for optimizing the use of resources, maximizing production, supplementing the income of the farmers and thereby improving the economic status of the poor farmers. Establishment of Framers Producer Company across Tamil Nadu and strengthening its activities will reduce the market risks.

7.5.6 Capacity building

Capacity building through training on latest feasible and effective farm mechanization practices will improve the uptake of new technologies by the farming community

7.6. Climate smart agricultural practices for rice – A case study from Clima Rice program

In Tamil Nadu, the availability of water is declining with projected water supply-demand of about 21 billion cubic metre in 2025. Rice is the staple food and vital for realizing local food security of resource poor farmers of Tamil Nadu. The area cultivated with rice is about 2 million ha of which 70 per cent is irrigated. Rice cultivation requires water from 1300 to 2000 mm depending upon the local climatic conditions (Geethalakshmi *et al.*, 2011). Alternatively, rice needs about 3000 to 5000 liters of water to produce 1 kg of grain. The alarming water crisis and water-intensive nature of rice cultivation drive to search for modified management method to increase water productivity in rice cultivation. Tamil Nadu Agricultural University under ClimaRice and ClimaAdapt projects identified the climate smart technologies for rice cultivation under changing climate and those are as follows.

7.6.1 Community nursery

Rice crop is mainly grown in the Cauvery Delta Zone of Tamil Nadu. Farmers in the recent days are unsure about the release of water from Mettur Dam. Farmers' raise nursery of rice varieties after the release of water from the dam in mid June. Seedlings of 21-25 days age are transplanted in July and harvested in September end or October first week. Delay in harvest of *Kharif* rice normally affects the planting of subsequent northeast monsoon rice crop in the cropping system. Hence, community nursery was promoted. Community nursery was established at the village level in one of the farmers' field where there is availability of underground source of irrigation water (two weeks prior to anticipated date of the release of water from the DAM). As soon as the water is released from the dam, the farmers undertake preparatory cultivation and plant the seedlings. This helped in timely plating and saving of irrigation water.

7.6.2 Direct sowing of rice

In Cauvery delta zone, rice is predominantly cultivated by transplantation methods. Transplanting requires at least 200 mm of water for puddling operation, which creates a dense clay layer in the sub-soil to prevent seepage losses. The crop requires about 1300 mm of water to complete its life cycle. Under delayed release of water Direct Seeded Rice (DSR) cultivation is recommended. Direct seeding of drought tolerant varieties of rice in dry soil is done with pre-emergence herbicide application under sufficient soil moisture condition followed by a post-emergence herbicide application at 25-35 days after sowing or hand weeding at 35-45 days after sowing to effectively manage weed problem. Some advantages of DSR technology are: saving in water and energy up to 25 per cent, saving of labour by 40 man-days, increased fertilizer use efficiency as result of root zone placement of fertilizer, early maturity of crops by about a week and these help in timely planting of *rabi* season rice crop in time, reduction in methane emission and less disturbance to soil structure.

7.6.3 Drought tolerant early maturing cultivars

In the Cauvery Delta zone rice variety CO 51 was promoted as it proved to be early maturing, less water requiring and drought tolerant. This variety had yield advantage compared to normal cultivars.

7.6.4 Changing the cultivation method to SRI

Farmers attempt to develop a different environment for rice under System of Rice Intensification (SRI) method by which the rice plants can exhibit their full potential. Major principles of SRI includes perfect land leveling, lower seed rate, transplanting young seedlings of 12 days old, single seedling / hill with wider spacing (25 x 25 cm), careful water management (applying a minimum of water during vegetative growth and at later stages maintaining only one inch water) and weed management using cono weeder thrice in the vegetative stage. Researchers and farmers' experience in the Cauvery basin in Tamil Nadu from ClimaRice have shown that yields of rice could be increased considerably (18–27%) with reduced water consumption (22- 25%) and less expenditure, just by following SRI method of cultivation. Thus SRI is beneficial for small and marginal farmers under changing climatic conditions.

7.6.5 Azolla and cyanobacterial systems in supplementing nitrogen to rice

Rice fields are to be considered as a significant source of green house gases (CH4 and N2O) as per the report of Bronson *et al.,* (1997) and emission from rice field eco systems account for about 60 Tg CH4 per year or about 12 per cent of the global annual CH4 emission (IPCC, 1996). The emission of green house gases (CH4 & N2O). from rice ecosystem is likely to increase in the days to come due to intensification of rice cultivation. Much attentions have been paid on the influence of fertilization, organic matter amendment, water management and rice varieties on green house gas (CH4 and N2O) emission from rice fields. The utilization of cyanobacteria (blue green algae) as a bio fertilizer for rice in supplementing nitrogen is highly promising as the rice field ecosystem provides congenial environment for this self supporting diazotroph. Azolla is a floating water fern that also fixes atmospheric nitrogen. Azolla and Cyanobacteria have been identified as ecofriendly natural nitrogen fixers in the rice field ecosystem. A judicious use of these blue green algae could provide entire rice acreage of India as much nitrogen as obtained from 15-17 lakh tonnes of urea. Blue green algae was applied at the rate of 4 litres of composite blue green algal inoculum diluted in 200 litres of water per acre(4000m^2). It is sprayed around the root zone of the standing rice crop during 10-15 days after transplanting. Azolla could be applied at the rate of 100 kg of wet azolla fronds per acre in the standing water during 10-15 days after transplanting of rice. The combination of azolla and cyanobacterial systems recorded a dissolved oxygen content of 7.4 ppm as against 4.3 ppm in the un inoculated plot. Cyanobacteria and azolla that grow on the soil surface and also as a floating mass act as live aerators in rice field ecosystem and oxygen released during the photosynthetic activity got liberated as minute air bubbles and consequently aerate the water impounded in rice field that resulted in increased dissolved oxygen content, which ultimately decreased the methane flux. In addition to the aeration of water, cyanobacteria and azolla might alleviate the toxicity due to accumulation of reduced iron and sulphites in rice fields as observed under continued submergence. Hence, these nitrogen fixing biological systems can be used to reduce methane flux from flooded rice ecosystem besides their ability in supplementing nitrogen to the paddy crop.

7.6.6 Other climate smart production technologies

a. Soil management

Off season tillage : These are scheduled during un -cropped seasons for leveling to the desirable gradient, leaching to remove salts, lowering of a seasonal water table & reducing the population of harmful flora and fauna in the soil.

Summer ploughing: Deep summer ploughing (off season tillage) with pre-monsoon showers (during May) to recharge the soil profile. It facilitates to sow the crops immediately after onset of southwest monsoon. Off season tillage increases water content of soils and reduces runoff. It also reduces pest and weed infestation.

Conservation tillage: Conservation tillage is any method of soil cultivation that leaves the previous year's crop residue (such as corn stalks or wheat stubble) on fields before and after planting the next crop to reduce soil erosion and runoff, as well as other benefits such as carbon sequestration .

b. Nutrient management

Site-Specific Nutrient Management (SSNM): It is a plant-based approach for managing the nutrient requirements of crops. It provides principles and tools for supplying crops with nutrients as and when needed to achieve high yields while optimizing use of nutrients from indigenous sources.

Soil test based nutrient application: DESIFER is the user friendly software developed in Visual Basic "Stand Alone" version. Farmers realize nutrient potential of organic manures, crop residues, compost and bio fertilizers. Soil analytical reports and nutrient requirement for a targeted yield are taken into consideration for fertilizer recommendation under DESIFER.

Integrated Nutrient Management: INM refers to the maintenance of soil fertility and of plant nutrient supply at an optimum level for sustaining the desired productivity through optimization of the benefits from all possible sources of organic, inorganic and biological components in an integrated manner.

Leaf Colour Chart (LCC) : It is used for deciding the time of topdressing of N fertilizer. This will help in minimizing the loss of nutrients.

c. Water management

Micro-irrigation that has the potential to save water and nutrients by allowing water to drip slowly to the roots of plants, either from above the soil surface or buried below the surface. The goal is to place water directly into the root zone and minimize evaporation. Drip irrigation systems distribute water through a network of valves, pipes, tubing, and emitters. Depending on how well designed, installed, maintained, and operated it is, a drip irrigation system can be more efficient than other types of irrigation systems, such as surface irrigation or sprinkler irrigation. Efficient management of nutrients can be achieved through fertigation

d. PPFM (Pink Pigmented Facultative Methyl bacterium) spray

Bacteria often live in environments containing diverse substrates. One of such habitat is the phyllosphere, where facultative methylotrophic bacteria are found to be ubiquitous and abundant. These methylotrophic bacteria belong to the genus *Methylobacterium* and are known to metabolize methanol but also a limited number of alternative carbon substrates, such as organic acids and alcohols. Plant leaf surfaces release diverse carbon sources, mainly sugars and organic acids, at low amounts (µM range), and these sources are heterogeneously located and result of leaching through the cuticle. In addition to these substrates, volatile carbon substrates, particularly the plant cell wall metabolism byproduct methanol, are released via the stomata. Methanol emission peaks in the morning, when the stomata first open. Spraying of PPFM during flowering found to increase the yield.

7.7 Management strategies

Comprehensive management technologies including some important listed above specific to Tamil Nadu to alleviate the impact of climatic risk are listed below:

- High water requiring crops like rice, sugarcane and banana areas must be reduced to 50 per cent of the present area under cultivation and these crops have to be raised at efficient cropping zone only under efficient drip irrigation system .The efficient cropping zone have to be identified across Tamil Nadu through scientific means as reported earlier. In addition, for rice, both aerobic and System of Rice Intensification could be recommended.
- For irrigated condition, dry irrigated crops may be introduced
- Minor millets, vegetables and fruit crops may be encouraged to grow
- The Integrated Crop Management (ICM) technology must be introduced and this includes Integrated Nutrient Management(INM), Integrated Water

Management(IWM), Integrated Pest Management(IPM), Importance may be given to bio-pesticide.

- Practicing weather based agriculture
- Establishing strong net work on crop insurance and making 100 per cent participation of farmers by law in crop insurance programme
- Encouraging for the establishment of Framers Producer Company across Tamil Nadu and strengthening its activities
- Modified Training and Visit system must be introduced to revitalize present agricultural extension system
- Seed production and seed supply programme must be strengthened
- Modified agronomic practices must be recommended like change in variety, date of sowing/ planting, spacing etc.,
- Minimum Tillage to retain soil carbon content
- Promotion of use of drought and flood tolerant varieties
- Managing soil erosion through minimum or no tillage,
- Managing soil health and nutrient content through soil test based integrated nutrient management.
- Encourage development of Integrated Farming Systems suitable for each agro climatic zone especially for poor farmers to augment their incomes
- Undertake capacity building through training on latest feasible and effective farm mechanization practices
- Alternate cropping strategies for "agro climatic zone based cropping pattern" to ensure maximum utilization of available land and water with an ultimate aim to increase cropping intensity
- Promote leguminous feed production in degraded pasture lands in villages
- Develop bio-shield in coastal villages by growing mangroves that would not only protect the coastal communities from cyclones and other sea related disasters but will also provide nutrients for marine species that thrive in the mangrove soils.

Conclusion

There is no meteorological boundary between countries in the atmosphere like that of geographical boundary that exists presently between countries. In addition to this, the Mother earth rotates at its own axis and hence any over accumulation of Green House Gases in any part of the global atmosphere would spread to entire mass of atmosphere and thus universal negative impact would occur in all parts of the world. Against this, the strategies like mitigation and adaptation are recommended at each **CoP** held at different countries of the globe. Strong agricultural research to develop weather based technologies, developing tolerant crop varieties against environmental stresses, enforcing farmers friendly agricultural extension services and enhancing farmers skill and knowledge in scientific agriculture would enhance the resilience capacity of the region against climate hazards. Let us lead the world including India for 'Be Weather wise-Otherwise-Not wise.

Reference/Suggested Readings

Annamalai, H., Prasanna, V., Hafner, J., Nagothu, U. and Kitterod, N. O., 2011. Modeling the current and future climate over the Cauvery river basin. In Sustainable Rice Production in a Warmer Planet, Macmillan Publishers, New Delhi, 2011, pp. 59–81.

Balasubramanian. T.N., Chinnamuthu, C.R., Geethalakshimi.V., Purusothaman.S and SP.Palaniaapn.1994. Climate Variability of Coimbatore. National Symposium on climate variability, TROPMET 94. Feb. 8-11 IITM, Pune.

Bronson, K.F., H.U. Neue, U. Singh and E.B. Abao Jr. 1997. Automated chamber measurement of methane and nitrous oxide flux in s flooded rice soil. *Soil Sci. Soc. Am.J.*, 61: 981-987.

Geethalakshimi, V and Ga. Dheebakaran 2008. Impact of climate change on agriculture over Tamil Nadu (pp 80-92): In Climate change and Agriculture over India (Eds.G.H.L.H.V. Prasda

Rao., G.G.S.N.Rao., V.U.M.Rao and Y.S.Ramakrishna). AICRP on Agro meteorology and Kerala Agricultural University. P 256.

Geethalakshmi,V., S. Kokilavani, R. Nagarajan, C. Babu and S. Poornima. 2008. Impacts of Climate Change on rice and ascertaining adaptation opportunities for Tamil Nadu state. In the proceedings of the International symposium on Agrometeorology and Food security conducted by CRIDA, Hyderabad between 18 – 21, February, 2008 : pp: 21 - 22

Geethalakshmi1,V., A. Lakshmanan, D. Rajalakshmi, R. Jagannathan, Gummidi Sridhar, A. P. Ramaraj, K. Bhuvaneswari, L. Gurusamy and R. Anbhazhagan. 2011. Climate change impact assessment and adaptation strategies to sustain rice production in Cauvery basin of Tamil Nadu. *Current Science*, VOL. 101 (3): 342 – 347.

Gowtham Ramasamy, S. Panneerselvam and V. Geethalakshmi . 2017. Rainfed maize productivity under changing climate in Tamil Nadu. AgMet paper present at HAU Hissar.

Intergovernmental Panel on Climate Change (IPCC). 1996. The science of climate change, Cambridge, Cambridge University Press,P.572.

Jain,S.K.. and Kumar, V. 2012. Trend analysis of rainfall and temperature data for India Current Science (Bangalore) 102 (1), 37-49

Kanwar,J.1972. Cropping pattern, scope and concept. In Proceeding of symposium on cropping pattern in India. ICAR. New Delhi.PP.11-32

MoEF&CC,2010.http://www.moef.nic.in/downloads/public-information/fin-rptincca.pdf),

Stephen, A. 2012. Natural disasters in India with special reference to Tamil Nadu. *Journal of Academia and Industrial Research* (1): 59 - 67

TNSAPCC, 2014. Tamil Nadu State Action Plan for Climate Change. http://www.moef.gov.in/sites/default/files/Tamilnadu%20Final%20report.pdf

UNFCCC 2008. Climate Change: Impacts, Vulnerabilities and Adaptation in Developing Countries. UNFCCC Secretariat, Bonn.

8

Climate Resilient Agriculture in Karnataka

H.S. Shivaramu and H.S. Padmashri

Agrometeorology Section, GKVK, UAS, Bangalore, Karnataka, India

The State of Karnataka is located between 11.50° to 18.50° N latitudes and 74.25° to 78.50° E longitudes and covers an area of 19.1 m ha which accounts for 5.8 percent of the total geographical area (TGA) of the country. Out of the total 19.1 m Ha in Karnataka the net cultivated area (2010-11) was 10.5 M Ha, net irrigated area 3.49 m Ha and net rainfed cultivated area 7.01 m ha. Karnataka has the largest area under rain-fed agriculture (Fig. 8.1). Nearly 55% of total food grain production and 74% of oilseeds production come from rainfed agriculture in Karnataka. Therefore rain-fed agriculture plays an important role in total food grain production in the state. Further the rain-fed agriculture has substantial untapped potential, which can be brought to use by increasing the crop yields in the dry land areas through the adoption of proper dry-land production technologies. The climate of the state is determined mainly by the geographical location with respect to the Sea, monsoon winds and physiography. Karnataka State has sub-humid to humid climate on the west coast and western ghats and semi-arid to arid (very warm) climate in central and northern districts of plateau region. The year is divided into four seasons viz., winter (December to February), summer (March to May); South-West monsoon (June to September) and North-East monsoon (October to November). The occurrence of rainfall and its spatial distribution is highly variable and not dependable. Rainfall contribution is very high, from South-west monsoon season. It is seen that the annual rainfall is highest (5051mm) over the western ghats and lowest (408mm) in the eastern parts of Chitradurga district. Taluk wise annual rainfall variation is increasing over most of the taluks in south interior Karnataka. More than 2/3rd of the state receives less than 750 mm of rainfall. The atmospheric temperature in the state ranges from 23°C to 43°C in summer and 9 °C to 27 °C in winter. Hailstorms are common in plateau region of the state during pre-monsoon period.

The State has been divided into four physiographic regions namely 1) South Deccan Plateau 2) Western Ghats 3) Eastern Ghats and 4) West Coast plains. There are different kinds of soils in these regions of the state. Because of complex topographic, geologic, vegetation and climatic conditions, there are different kinds of soils in the state. These soil resources are characterized, (Shivaprasad, *et al*, 1996) and grouped taxonomically under Alfisols (27% of TGA), Inceptisols (25% of TGA), Entisols (16% of TGA), Vertisols (15% of TGA), Ultisols(8% of TGA), Aridisols(5% of TGA) and Mollisols (1% of TGA). Traditionally, these soils can be grouped under five categories Viz., red soils, lateritic soils, black soils, alluvial soils, forest soils (Table. 8.1). Karnataka State is divided into ten Agro-Climatic Zones by UAS, Bengaluru under NARP program. These zones could be grouped in to three categories namely:

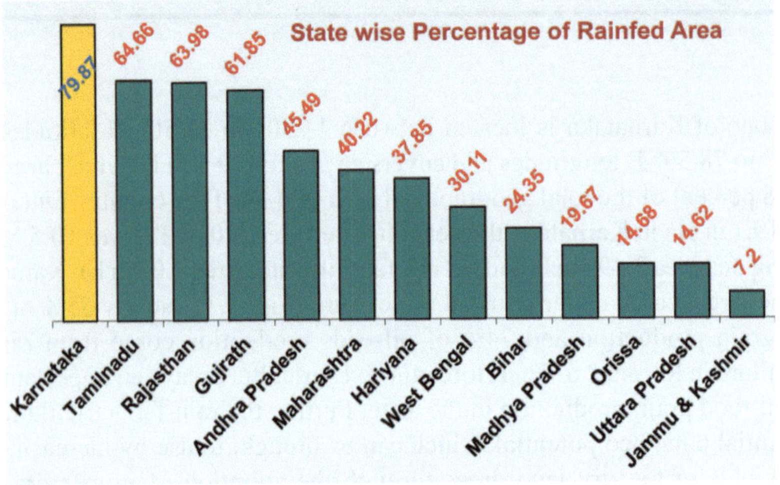

Fig. 8.1: State wise Percentage of Rainfed Area

a) **Dry Zones:** North Eastern Dry Zone (2), Northern Dry Zone (3), Central Dry Zone (4), Eastern Dry Zone (5) and Southern Dry Zone (6)

b) **Transition Zones:** North-Eastern Transition Zone (1), Southern Transition Zone (7) and Northern Transition Zone (8)

c) **Hill and Coastal Zones**: Hilly Zone (9) and Coastal Zone (10)

8.1 Spatial and Temporal distribution of Annual Rainfall and Rainy days

During the period of 1980-2013, inter-annual variability has been greater during the past 15 years in interior Karnataka in the period of analysis (Venkatesh *et al.*, 2016). Trends of rainfall and rainy days are almost the same. The significance of the trends resulted from Mann Kendall test is presented in Fig. 8.2 for annual rainfall and in Fig. 8.3 for annual rainy days. 82 % of the 392 stations in the state showed no significant trend (322 stations) in annual rainfall while 71 % showed no significant change (279 stations) with respect to annual rainy days. Annual rainfall of among the remaining 70 stations, 51 stations showed positive

Table 8.1: Soils of Karnataka

Soil	Region	Characters	Constraints for Crop Production	Crops Grown
Red soils - 35.15% of TGA	almost all physiographic regions of the state	Deep to very deep, well drained and generally with a clay enriched sub surface-horizon	Shallow rooting depth, coarse texture, high gravel, low moisture retentivity, low fertility, surface crusting & compaction, soil erosion, low CEC and base saturation, low nutrient status	agricultural, horticultural and plantation crops underefficient soil, water and nutrient management practices
Lateritic soils - 12.1% of TGA	commonly found in area of highaltitude and heavy rainfall in Karnataka (Western Ghat region)	Under humid and warm condition (in the hilly terrain) such soils have high content of iron and magnesium but are well drained and porous	deficiency of phosphates accentuated by high phosphate fixing capacity as Fe and Al phosphate, high acidity, toxicity of aluminum and manganese, deficiency of potassium, calcium, zinc, magnesium and boron	paddy, banana, coconut, arecanut, cocoa, cashew, coffee, tea, rubber
Black soils-6.9% of TGA	Deccan plateau region-covering large parts of North Interior Karnataka	Generally calcareous, Highcation exchange capacity, high base status, higher contents of exchangeable calcium and magnesium	These soils have impeded drainage and low permeability, neutral to slightly alkaline in reaction, swelling and shrinkage, plasticity and stickiness, poor in organic matter, nitrogen and available phosphorus.	cotton, millet, sorghum, soybean etc. in the state. Under irrigated conditions sugarcane
Alluvial Soils - 16.1% of TGA	lower topographic situations	fairly sufficient in phosphorous and well supplied with potassium and lime	Due to the injudicious use of irrigation water some such soils remain waterlogged and pose problems to soil salinity. Some of these soils show deficiency in zinc, Sulphur and sodium because of intensive cultivation problems,	These are the best agricultural soils and used for growing most agriculturalCrops
Forest soils - 6.1%of TGA	Western ghat region of the state	Vary from deep alluvium in the river basins and lower slopes to highly immature residual gravelly on higher altitudes. Areas of steep relief are mostly devoid of soil. Various types of crops are grown in different regions like rice in valley, orchards on slopes in almost all areas.		

trend (13% of total stations), while 19 stations negative trend (6% of total stations). Out of the stations with positive trend, 50 % showed significance at 10 per cent, 28 % shown significance at 5 per cent and 22 % stations shown significance at 1 per cent. This indicates that the increasing trend was slower at 50 % of stations (PT-10%), and it was greater at other stations (PT- 5% & 1%). Out of the stations with negative trend, 38% shown significance at 10%, 48 % of stations shown significance at 5% and 14% of stations shown significance at 1 per cent. Thus, the negative trend observed was greater at more number of stations. With respect to the annual rainy days, of the remaining 113 stations, 89 stations showed positive trend (23% of total stations), while 24 stations showed negative trend (6% of total stations) for annual rainy days. An interesting point was noticed during the north east monsoon that none of the stations showed decreasing trend in annual rainy days while only 9% of the stations showed increasing trend indicating the shift in the wetness in October, November and December. Interestingly, none of the 392 stations of the state had decreasing trend in rainy days of North eastern monsoon. State as a whole experience increasing trend with respect to rainy days in north east monsoon (Fig. 8.4) with SIK (20 stations) being major contributor followed by malnad (7stations), NIK (4 stations) and coastal (4 stations) region.

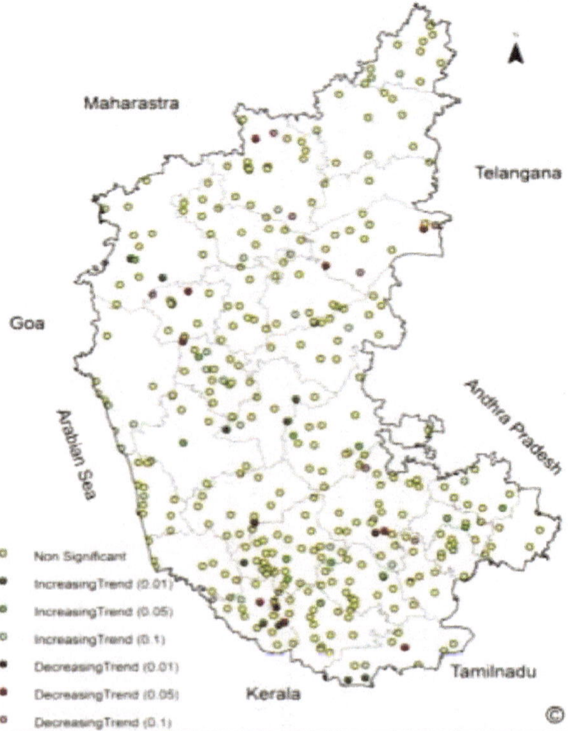

Fig. 8.2: Trends in annual rainfall in Karnataka

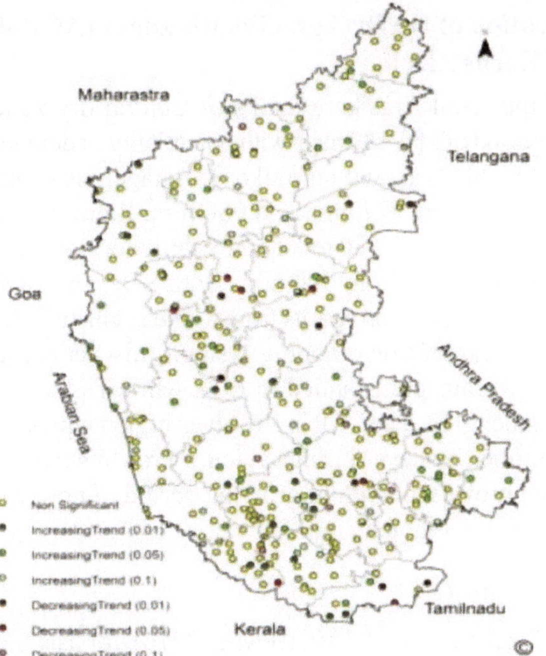

Fig. 8.3: Trends in annual rainy days in Karnataka

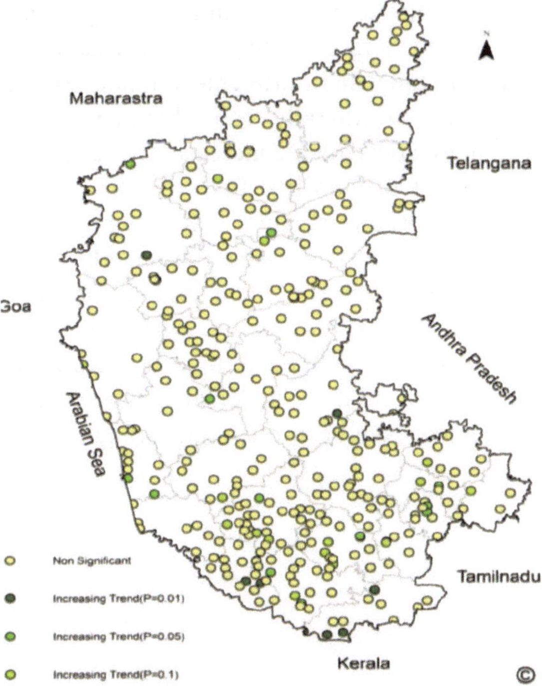

Fig. 8.4: Trends in rainy days of North east monsoon in Karnataka

8.2 Re delineation of NARP Agroclimatic zones (ACZ-4, ACZ-5 &ACZ-6) of Karnataka

Koratagere, Tiptur and Arasikere taluks of Central dry zone (CDZ: ACZ-4) have growing period of 150-180 days with potentiality of double cropping system. Average (1980 to 2013) annual rainfall of Koratagere is around 780 mm while Tiptur and Arasikere have 770mm and 960 mm. Rainfall of Eastern dry zone (EDZ: ACZ-5) ranges from 680 to 880 mm while for Southern dry zone (SDZ: ACZ-6), it ranges from 700 to 890 mm. Both SDZ and EDZ has hot, moist, semi arid climate. Considering results from K-mean cluster analysis (Shivaramu, et.al, 2017) and field data on cropping systems, district borders and other soil characteristics, Koratagere is added to EDZ while Tiptur and Arasikere have been grouped under SDZ. The LGP in any given region represents the climatically determined number of days during which a crop receives enough moisture from soil for its growth. Length of growing period of newly formed Zone 4, 5 and 6 are provided in Fig. 8.5.

8.3 Drought Analysis

Taluk wise drought analysis (AICRPAM, 2016) for last seventeen years is presented in Fig. 8.6. Several taluks of the State are growing increasingly vulnerable to drought conditions every year. Between 2001 and 2017, all the taluks in the state have enjoyed normal rainfall for only three years. Of the 176 taluks in the state, 109 have faced drought for at least 4 years between 2011 and 2017. According to the data of drought-hit taluks over the past 17 years, only coastal districts which generally record heavy rainfall are exempt from drought. However, these areas too are increasingly becoming vulnerable. The taluks with the highest probability for drought (86 per cent) include Chikkaballapur, Aland and Jewargi taluks in Kalaburagi district and Shirahatti taluk in Gadag district. Shockingly, 135 of the 176 taluks have a drought probability of over 50 per cent. A large number of them have suffered droughts for at least 8 years of the past 17 years while situation might get worse in the next few years. The worst hit regions are parts of North Interior Karnataka. Some districts in South Interior Karnataka, such as Chikkaballapur, Kolar and parts of Mysuru and Chamarajanagar too have been recording drought years at an alarming frequency.

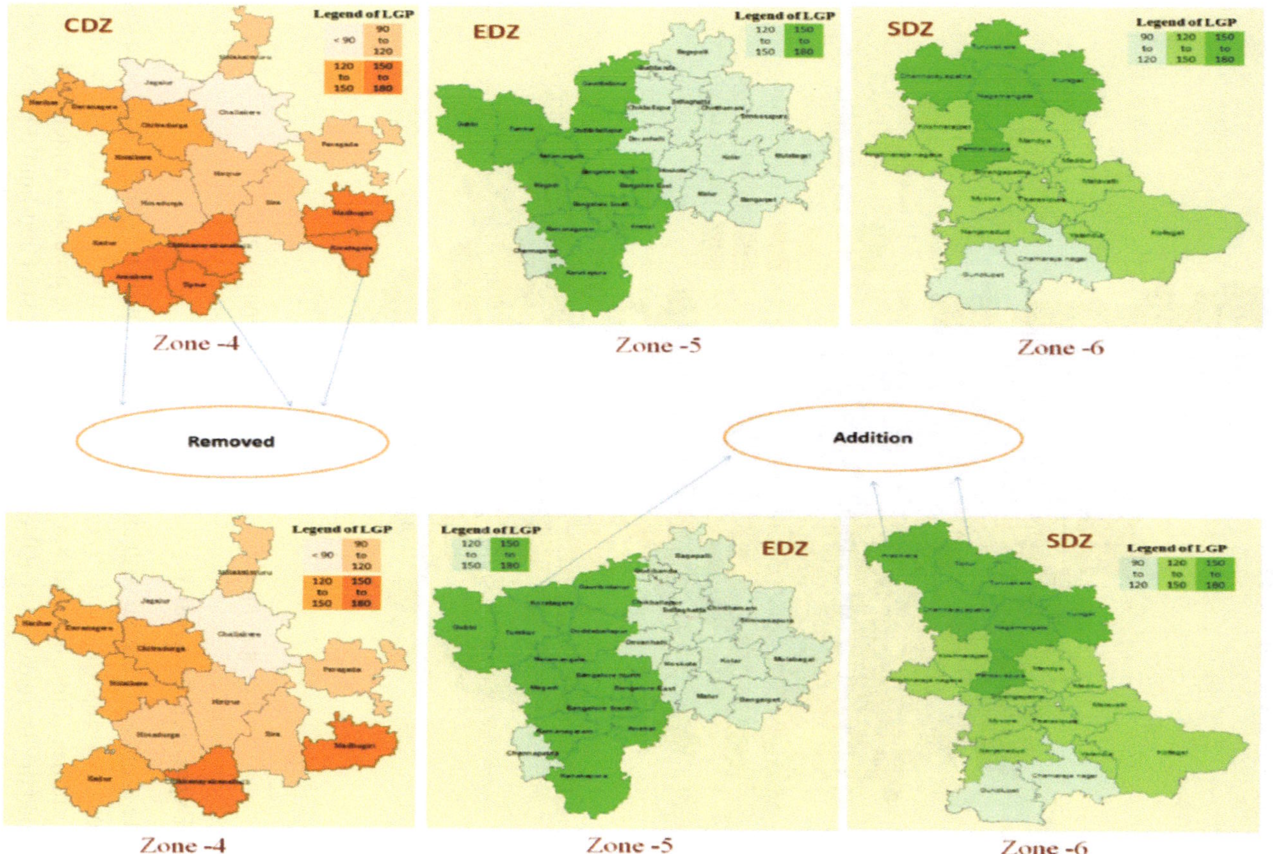

Fig. 8.5: Old and New delineation of LGP of None-4, Zone-5 and Zone-6

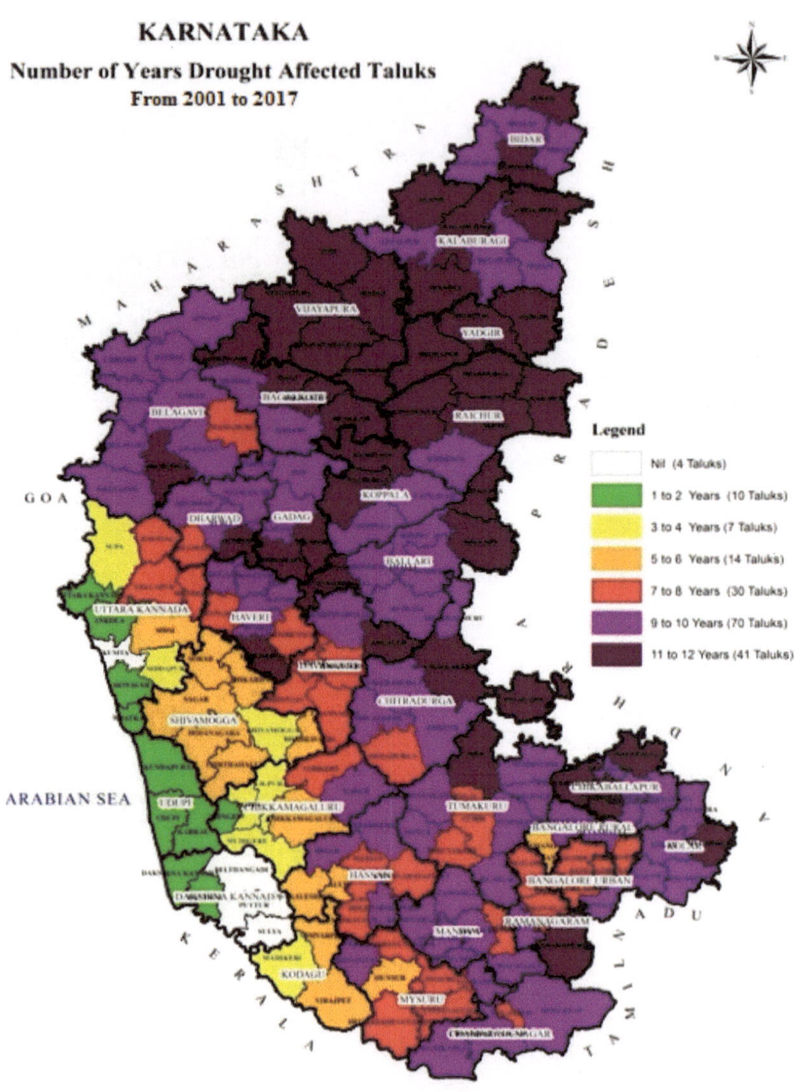

Fig. 8.6: Frequency of drought from 2001 to 2017

8.3.1 Expansion of Aridity

Climate change associated with changes in precipitation and PET would lead to changes in aridity (i.e., P / PET) and thus the areal extent of drylands. Our recent studies (Figs. 8.7 and 8.8) suggested that the area and under arid and semi arid dry increases (AICRPAM, 2017) at the cost of sub-humid moist and semi arid moist. New inclusions under arid climate are: Chitradurga part, Chikkaballapur part, Tumkur part, and Kolar part, Aland and Jewargi of Kalburgi, Shirahatti of Gadag, Ranebennur of Haveri.

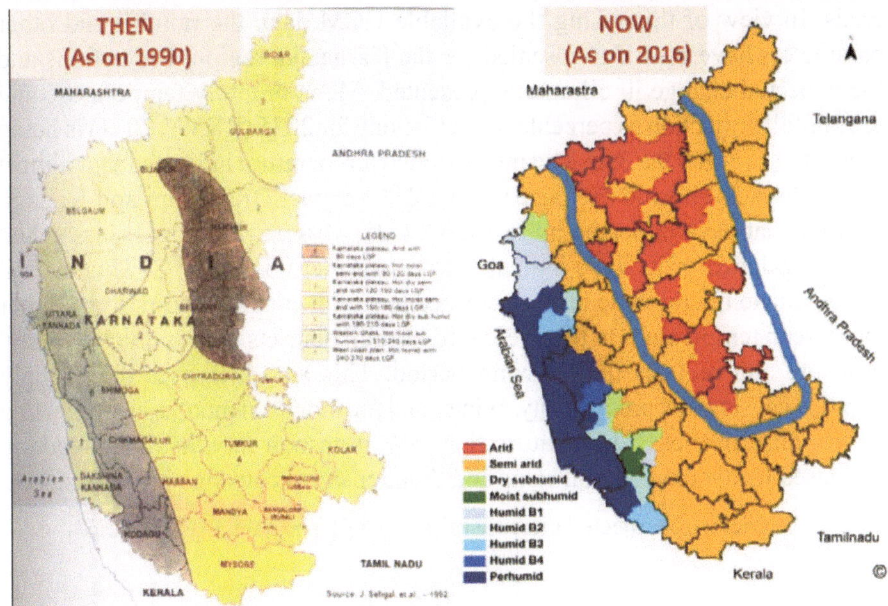

Fig. 8.7: Expansion of Arid zone in Karnataka

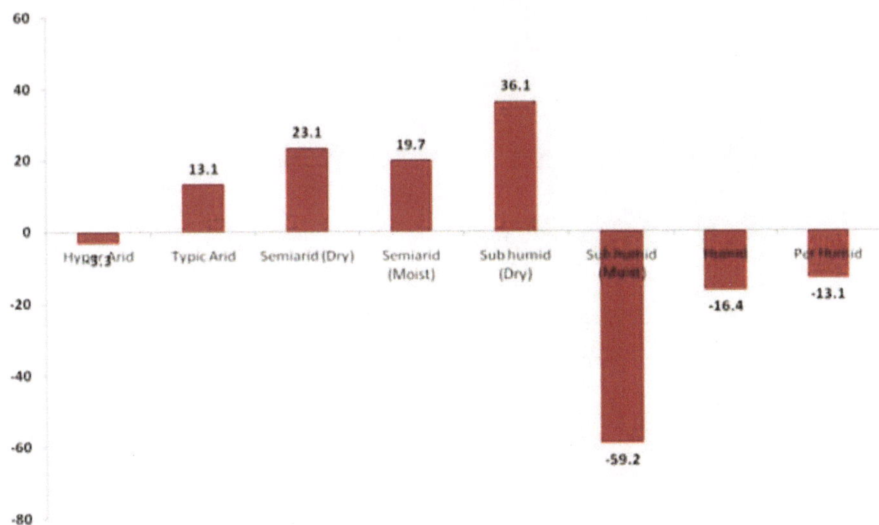

Fig. 8.8: Revised area under Bioclimate type from 1992 to 2015

8.4 Climatic Projections

The economy of the given region, arrangements for transportation of the food from the high production area to the low production area or preparedness to cope up with the expected trend in the food production helps in solving the food crisis. In view of this, using the available GCM data, the rainfall and other parameters have been down scaled for the Karnataka region. Projections and the expected change in climate is presented. Mean monthly temperature and the rainfall projection as percentages deviations for 2035 (RKVY, 2011) is being presented in Fig. 8.9 a & b. The mean annual temperature indicated that almost all the districts showed the raise up to 2.2 ^{D}C, expect DK, Udupi and Kodagu districts that have indicated up to 1.8 ^{D}C. U.K, Mysore, Shimoga, Hassan and Chikmagalur districts have indicated to be warmer up to 1.9 ^{D}C. This overall increase in temperature leads to increased potential evapotranspiration. Higher increase in temperature in the decreased rainfall districts causes quick loss soil moisture and reduces the growing period. This also increases the drought duration causing low productivity. Winter and pre monsoon months are indicating higher raise in mean temperature than SW monsoon months. This leads to increased local convectional rains and thunder storm causing natural disaster.

District wise projected increase in precipitation in 2035 A1B scenario

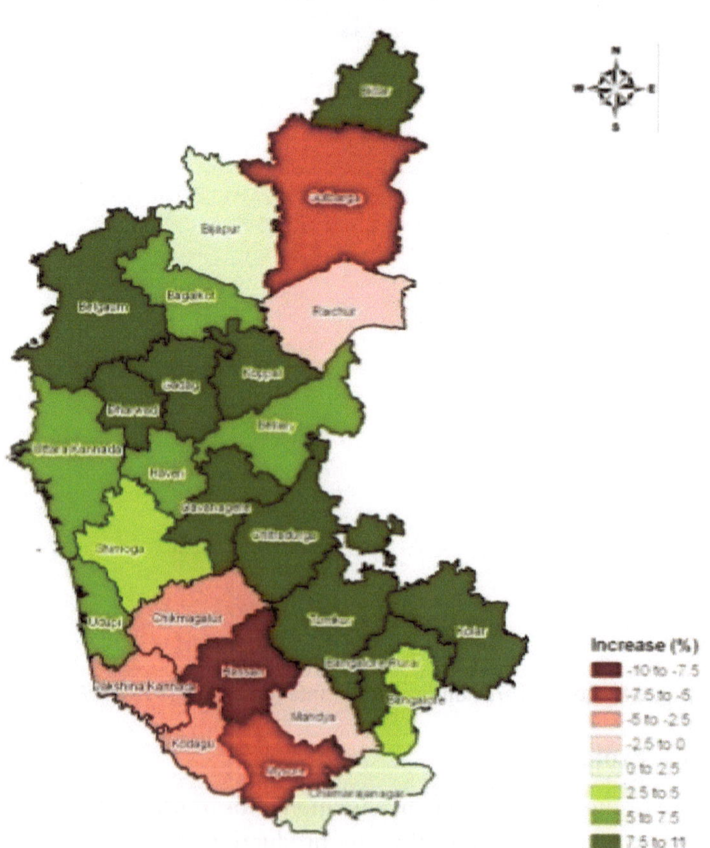

Fig. 8.9a: Rainfall projections as percentage deviations for 2035

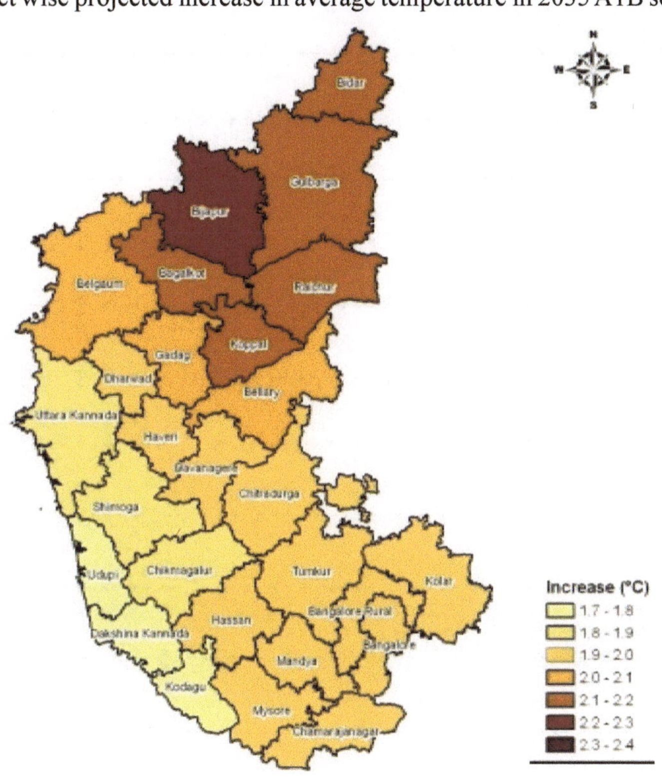

Fig. 8.9b: Temperature projections as percentage deviations for 2035

8.5 Trends in crop yield under current rainfall variability

Trends in area and production of different crops grown in all districts of Karnataka for the period from 1951 to 2008 presented in Table 8.2 and Table 8.3 indicated the shift in yield of major crops of Karnataka. Trends of both rainfall and crop status clearly indicates lesser incidences of "No change" category.

8.5.1 Crop yield projections

Climate projection models output on temperature and rainfall from Alb-2035 scenario, projected carbon dioxide levels as per Bern CC model, crop coefficients and crop management for 10 years (1991 – 2000) were used as input for Infocrop model to forecast the 2035 yield estimates of major crops of Karnataka (Rajegowda, *et al.* 2013). Results say that yield from major rice, pigeonpea and cotton growing districts will be reduced by 1.5 to 9, 8.3 to 47 and 6.5 to 11.7 per cent (Fig. 8.10). Maize and sorghum yields are likely to be increased up to 1.2 and 2.5 per cent where as potato and soybean yields are expected to decline by 13.5 and 2.8 per cent (Table 8.4)

Table 8.2: Area, production and productivity trends of major crops

Districts	Rice		Maize		Jowar		Ragi		Cotton		Pigeonpea		Groundnut		Sunflower		Sugarcane	
	A	P	A	P	A	P	A	P	A	P	A	P	A	P	A	P	A	P
Bagalkote	↑	↓	↑	↓	↑	↑	↑	↑	↓	-	↑	↓	↓	-	↓	-	↓	↑
Bangalore (U)	↑	↓	↑	↑	-	-	↓	↓	-	-	↑	↓	↓	↓	↑	↓	-	-
Bangalore (R)	↓	↑	↑	↑	-	-	↓	↑	-	-	↔	↔	-	↓	↑	↓	-	-
Belgaum	↔	↓	↑	↑	↔	↑	↓	↑	↓	↑	↓	↓	↓	↓	↑	↔	↑	↓
Bellary	↑	↓	↑	↑	↓	↑	↑	↓	-	↓	↓	↓	-	↓	↑	↔	↓	↑
Bidar	↓	↓	↑	↑	↓	↑	↓	↑	↓	↑	↑	↓	↓	↓	↑	↑	↓	↑
Bijapur	↓	↓	↑	↑	↓	↑	↑	↑	↓	↓	↑	↔	↓	↓	↑	↑	↑	↓
Chamarajanagar	↔	↓	↑	↑	-	↑	↑	↑	↓	↓	↔	-	↓	↓	↑	↑	↓	↑
Chikkamagalur	↔	↑	↑	↑	-	↑	↑	↑	-	↑	↔	↔	↔	↔	↑	↓	↓	↑
Chitradurga	↓	↑	↑	↑	↔	↑	↓	↑	↓	↑	↓	↓	↑	↑	↑	↑	↓	↑
Davanagere	↑	↓	↑	↑	↔	↑	↓	↑	↓	↑	↓	↓	↑	↑	↑	↑	↓	↑
Dharwad	↑	↓	↑	↓	-	↑	↓	↑	↓	↑	↑	↓	↑	↓	↑	↓	↓	↑
Gadag	↔	↑	↑	↑	↓	↑	-	↑	↓	↑	↑	↑	↓	↓	↓	↓	-	↑
Gulbarga	↓	↑	↑	↑	↓	↑	-	↑	-	↑	↑	↑	↓	↓	↑	↑	↓	↑
Hassan	↓	↓	↑	↑	-	↑	↓	↑	↓	↓	↓	↔	↔	↔	↑	↑	↓	↑
Haveri	↓	↓	↑	↑	↓	↓	↓	↑	↔	↑			↓	↓	↓	↑	↓	↑
Kodagu	↓	↑	↑	↑	-	-	↔	↔	↓	↑	-	-	↓	↓	↑	↑	↓	↑
Kolar	↑	↓	↑	↑	↓	↓	↓	↓	↓	↓	↓	↓	-	-	↑	↑	↓	↑
Koppal	↑	↓	↑	↑	↓	↓	-	↑	↓	↓	↓	-	↓	↔	↑	↑	↓	↑
Mandya	↔	↑	↑	↑	↓	↑	↓	↑	↓	↑	↓	↔	↔	↔	↑	↑	↑	↑
Mysore	↑	↑	↓	↑	↓	↓	↓	-	↑	↑	↓	-	”!	↔	↔	↔	↑	↑
Raichur	↑	↑	↑	↑	↓	↔	↓	↓	↑	↑	↔	↔	↓	↓	↑	↑	↑	↑
Shimoga	↓	↑	↑	↑	↓	↓	↓	↓	↑	↑	↓	↓	-	-	↔	↔	↑	↑
Tumkur	↓	↓	↑	↑	↑	↓	↓	-	↓	↑	↑	↓	↑	↑	↑	↑	-	-
Udupi	↓	↓	↑	↑	-	-	-	-	-	-	-	-	↓	↑	-	-	↓	↓
Uttara Kannada	↔	↑	↑	↑	↑	↓	↓	↓	↓	↓	↔	↔	↑	↑	↓	↓	↓	↑

Note: A: Area, P: Production, “!: Decreasing, ‘!: Increasing, ”!: No Change

Table 8.3: Trends in crop yield and rainfall variability

Particulars		Increasing	Decreasing	No Change	Particulars		Increasing	Decreasing	No Change
Rice	Area	9	11	6	Cotton	Area	7	15	1
	Prdn	12	14	0		Prdn	17	6	0
	Prdty	11	13	2		Prdty	16	6	1
Maize	Area	25	1	0	Pigeonpea	Area	9	11	4
	Prdn	24	2	0		Prdn	6	11	7
	Prdty	18	8	0		Prdty	12	5	7
Jowar	Area	8	12	3	Groundnut	Area	9	13	4
	Prdn	16	6	1		Prdn	8	13	5
	Prdty	14	8	1		Prdty	10	11	5
Ragi	Area	7	16	1	Sugarcane	Area	7	16	0
	Prdn	17	6	1		Prdn	20	3	0
	Prdty	20	4	0		Prdty	18	5	0
Trend		Pre monsoon		S-W monsoon		N-E monsoon		Annual RF	
Increasing		17		12		16		12	
Decreasing		8		14		11		15	
No Change		4		3		2		2	

Table 8.4: Yield predicted (q/ha) and yield percentage variation with respect to base yield

District	Rice			Maize			Sorghum			Pigeonpea			Cotton		
	Current	Predicted	%	Current	Predicted	%	Current	Predicted	%	Current	Predicted	%	Current	Predicted	%
Bagalkote	36	39	8.3	28.9	29.3	1.4	19.5	20.6	5.4	6	5.5	-9	15	14.7	-2.3
Bangalore- u	35	35.6	1.7	28.1	35.8	27.6	19.3	21.3	10	17.6	20.1	14.1	13.5	16.6	22.6
Bangalore- r	33	38	15.2	25.8	31.8	23.3	18	20	11	18	21.2	17.9	13.3	15.6	16.7
Belgaum	27	26	-3.7	30.6	34.6	13.2	17.7	18.5	4.9	10	9.8	-2	16.8	18.6	10.4
Bellary	28.7	27	-5.9	10	9.8	-2	13	15	15	6	5.5	-8.3	16.5	14.9	-10
Bidar	40.2	33.2	-17.5	12	13	8.3	14.8	17	15	10.2	10.4	1.5	11	12	9.1
Bijapur	39.7	41.1	3.4	27.8	25.3	-9.1	25.3	26	2.9	10.2	9.8	-4.2	13	15	15.4
Chamarajanagar	38	36	-5.3	25	28	12	20	22	10	5.5	6	9.1	10.9	11.2	2.9
Chikmagalur	38	35	-7.9	4.2	3.5	-17	15	14	-7	6.4	7.5	17.6	12	10.2	-14.8
Chitradurga	38	40	5.3	25	27	8	20	23	15	5	6.7	34	12.8	12	-6.3
Dakshina kannada	40.1	45.1	12.5	9.5	8	-16	10	9	-10	5	4	-20	9.5	8	-16
Davangere	36	37	2.8	28	30	7.1	23	21	-9	8	9	12.5	14.2	15.3	7.7
Dharwad	34.9	35.6	2	28.9	29.3	1.4	19.5	18.6	-5	7.5	8.3	11.1	15	13.3	-11.7
Gadag	35.9	37	3.1	27	29.6	9.6	18.9	21	11	6.5	8	23.1	15.5	14.5	-6.5
Gulbarga	28	24.6	-12.3	12.6	13.3	5.8	13	15.2	17	5.7	5.7	-0.5	16.6	17	2.4
Hassan	38.9	40	2.8	25	24	-4	18	16.5	-8	8.8	9.5	8	12	11.6	-3.7
Haveri	36	38	5.6	28	30	7.1	23	24.3	5.7	8	9	12.5	15	16.2	8.2
Kodagu	38.1	44	15.5	6.5	5.5	-16	9.4	10.6	13	10.2	7	-31	9.3	8.5	-8.6
Kolar	34.3	36.6	7	18	19	5.6	17	15	-12	6.4	7.5	17.6	11.2	10.2	-8.9
Mandya	41.7	40.9	-1.9	13	14	7.7	18	20	11	6	7	16.7	10	11	10
Mysore	40.6	42	3.6	15	16.5	10	20	22	10	8.5	9.5	11.8	10.6	10.9	3.3
Raichur	36.6	40.9	11.8	35.9	36.4	1.4	21.3	23.4	9.9	5.5	4.7	-15	15.6	17.2	10.7
Shimoga	39.5	36	-9	15.1	13	-14	15	13	-13	8.5	10.2	20.7	11.4	10.2	-10.1
Tumkur	37.9	40	5.5	13.5	12.9	-4.4	16	13.1	-18	8.9	7	-21	10.6	13.6	28.4
Udupi	38	50.1	31.9	10.2	8.2	-19	15	14	-7	9.4	6	-36	10	9.8	-2
Uttara kannada	39	48	23.1	12	10	-17	14	12.3	-12	13.2	7	-47	9.6	8.2	-14.1

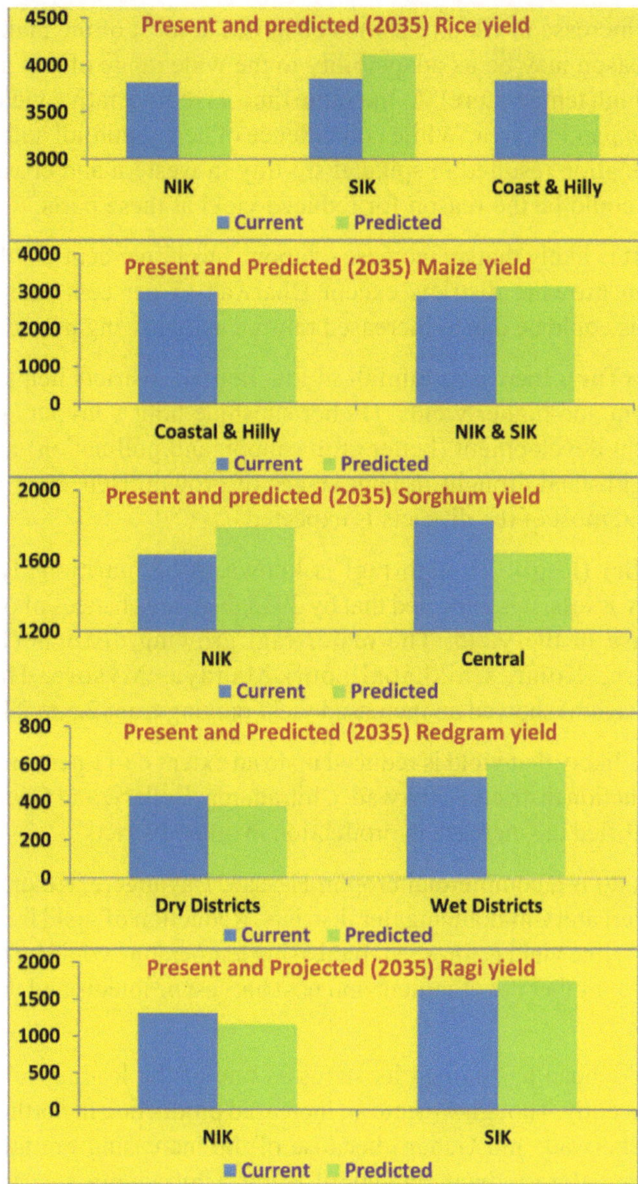

Fig. 8.10: Present and Projected (2035) yield

Rice: It can be noted that rice yield is decreasing (-17.5%) in major rice growing area of Karnataka. While districts like Raichur, Udupi, Uttara Kannada, Kodagu, Bangalore rural and Bangalore Urban have noted an increasing rice production. Changing cropping pattern could be the reason for reduced yield. Higher cloud cover, higher rainfall and low sunshine hours are other major causes for reduction in yield.

Maize: An increase in yield (2.2%) is observed in most of the maize growing districts. Reason may be its adaptability to the wide range of soil and climate (withstand high temperature). At the same time its remunerative yield has made farmers to expand its area. While coincidence of heavy rainfall and cloudiness during pollination resulted in spikelet sterility in western and coastal parts of Karnataka, could be the reason for reduced yield at these parts.

Sorghum: It is likely that an increase of about 3 to 17 per cent yield in most of the sorghum growing districts except Dharwad (5 per cent reduction) was noticed. This could be due to increased rainfall and growing period.

Pigeonpea (Tur): Increased rainfall in low rainfall districts helps good crop establishment and higher yields. Higher sunshine hour's favours pigeonpeas physiological development (better crop canopy and pollination) and its deep roots can withstand drought in later stages of growth. Hence, an increase in production at most of the districts is expected

Finger millet (Ragi): Though ragi is known to be nutritionally rich, still acceptability is less. It is expected that by 2035 there are chances of overall 2 per cent increase in the yield. The major ragi growing districts (Bengaluru, Ramanagara, Kolar, Chikkaballapur, Mandya, Mysore, Hassan and Chamarajanagara) showed an increased yield ranging from 2.5 to 24.7 per cent.

Cotton: It is likely that yield is reduced up to an extent o f 11 per cent in leading cotton production districts (Dharwad, Chitradurga, Bellary, and Gadag). This in turn has nullified the increase in production in other districts.

Potato: Potato is a commercial crop in Hassan, Davangere, Kolar, Bangalore Urban, Haveri and Chikkamangalur districts. Reduction of yield by 20 per cent from the base line yield is foreseen. Increasing temperature could be the possible reason for poor tuber development and post harvest problems and thus reduced productivity.

Soybean: Soybean has gained its status as one of the leading oilseed in the state very recently. Though we note an increased production in northern districts (Raichur, Dharwad, and Gadag) because of the increasing rainfall, southern districts (Bangalore Urban, Bangalore Rural, Chitradurga and Tumkur) are expected to lose the production.

8.6 Implementation of agro met advisories

An agriculturally relevant forecast is not only useful for efficient management of farm inputs but also leads to precise impact assessment (Gadgil, 1989). Studies were conducted on implementation of agro met advisories by the farmers (AAS farmers), regular estimation of benefit/loss on each item of the advisory realized at the farmers level, economic benefit/loss on adoption of the agro advisory issued by the agro met division compared with the non-AAS farmers. For this purpose, villages near the nodal centre have been identified to know the economic benefit obtained by the farmers adopting the agromet advisories. Weather forecast and Agro advisories were provided to AAS farmers during *Kharif* and Rabi season and no information was provided to non AAS farmers. A field survey of the study area was conducted and feedback from farmers from each village were chosen and the feedbacks are collected and summarized by recording the yield of the crops from two situations viz., recommended practices with agro-advisory and recommended practices without agro advisory.

8.6.1 Reliability of forecasts

To assess the reliability of weather forecasts of rainfall, temperature and cloud cover, error structure have been calculated Singh *et al.*, 1999) for forecasts issued from 2011 to 2016 (GKMS, 2011 to 2016). Trend line is also fit for the usability per cent. The overall mean usability (Table 8.5) indicates that the rainfall usability varied 39.4 to 76.6 percent over the years (Fig. 8.11). The annual usability of temperature ranged from 63.4 to 89.4 per cent while for cloud cover it was from 43.6 to 72.8 per cent. An estimate made by agribusiness a community in western countries indicates that the forecast can be put to an economical use if it is 50-60 per cent realized (Seeley, 1994). Trend lines presented in Table 8.2 shows that except for rainfall forecasts, reliability of all other weather parameters viz. maximum, minimum temperature and cloud cover is increasing over the years. Temperature and cloud cover trend lines have higher R^2 values indicating positive.

Table 8.5: Average usability (%) of rainfall, temperature and cloud cover

Year	Rainfall	Max. Temp	Min. Temp	Cloud cover
2011	76.6	64.0	63.4	43.6
2012	39.4	73.4	75.4	59.4
2013	42.8	78.4	66.2	63.8
2014	50.2	81.6	67.0	66.0
2015	49.0	74.4	71.2	72.8
2016	46.4	89.4	87.8	67.8
Average	50.7	76.9	71.8	62.2
Trend line	$y = -3.28x + 62.21$	$y = 3.805x + 63.54$	$y = 3.148x + 60.81$	$y = 4.668x + 45.89$
R^2	0.213	0.693	0.44	0.741

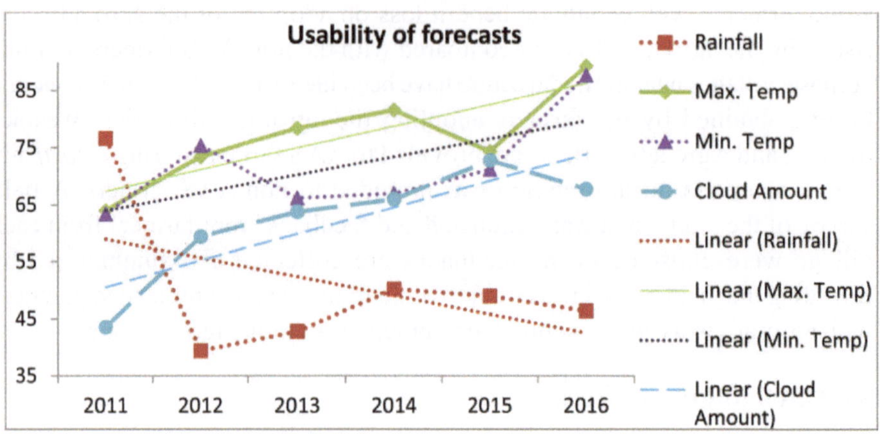

Fig. 8.11: Usability of Forecasts

8.6.1.1 Success of AAS

Every tuesday and friday agromet advisories are dispersed both in English and regional language (Kannada) to Bangalore (Urban), Bangalore (Rural), Chikkaballapur, Kolar and Ramanagara at district level. An average of around 62 per cent (Table 8.6) of the advisories dispersed advisory to 3 villages (Nayanahalli, Patrenahalli and Myalapanahali) of Chikkaballapur district and Kuthanagere watershed area of Magadi Taluk, Ramanagara district from 2011. Among the 616 AAS issued, 379 AAS had recorded direct economic benefit by saving Rs. 37 lakhs to the farmers from 2011 to 2016 (NICRA, 2011 to 2016).

Table 8.6: Success rate of Agromet advisories

Year	AAS issued	Successful AAS	Percent	Amount Saved
2011	104	61	58.7	90,599
2012	103	68	66.0	1,60,320
2013	101	72	71.3	9,06,203
2014	103	59	57.3	7,98,550
2015	104	56	53.8	13,47,885
2016	101	63	62.3	4,41,015
Total	616	379	61.5	37,44,572

Benefit: Cost Ratio of AAS and non-AAS farmers over 5 years (Annual report, NICRA, 2011-16): Among the different crops of adapted villages, vegetable crops recorded higher considerable gain in terms of B : C ratio (1:2.62), followed by fruit crops (1:2.61), flower crops (1:2.59) and field crops (1:2.2) against non-AAS farmers (1.61, 1.72, 1.42 and 1.57 respectively), Fig. 8.12).

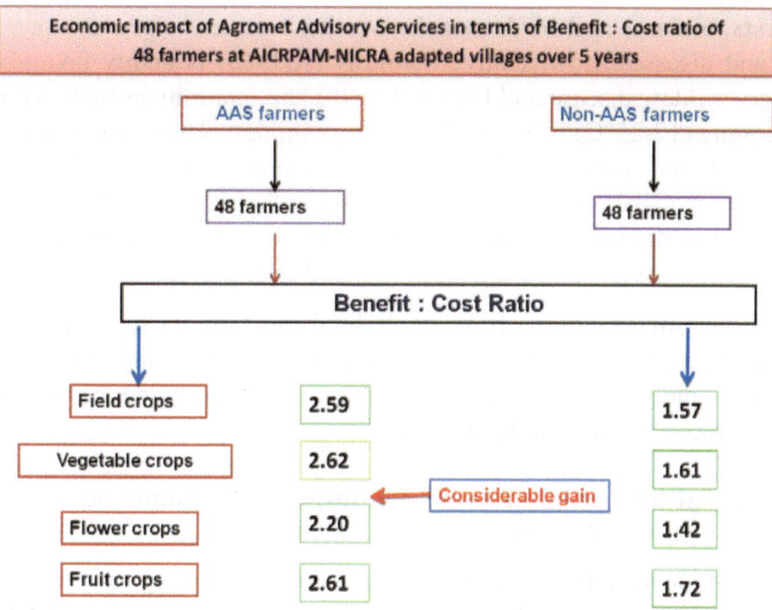

Fig. 8.12: Benefit: Cost Ratio of AAS and non-AAS farmers over 5 years

8.7 Climate Resilient Practices – Successful Stories

8.7.1 Productivity enhancement in grapes due to agromet advisories in Chikkaballapur district

Water table of Chikkaballapur– Kolar belt is highly affected beyond rejuvenation. To address this problem, meticulously worked out Agromet advisories (AAS) based on weather forecast twice a week is dispersed to the farmers of Patrenahalli and Nayanahalli villages of Chikkaballapur taluk. Partial budgeting of grapes cultivation through AAS has been done for two years (NICRA, 2015 and 2016).

Area covered: Nearly 200 farmers were monitored under the project out of which 150 were grape growers in an area of 90 hectares.

More crop per drop: All the grape growers were advised for drip irrigation through synergy from horticultural department, watershed department and KVK etc. The farmers were given subsidy for drip irrigation through this line department and advisories were given from NICRA project with respect to schedule of irrigation based on rainfall forecast. This resulted in the best water use efficiency besides the quality of produce. Among the advisories given, timely pruning was the best advice given based on the rainfall forecast and most of the AAS farmers followed this advice, thus resulted in higher yield, quality bunches and premier market rates due to appealing color and less disease infestation.

No pests, diseases and birds attack: Weather forecast helped in forecasting pests and diseases. Low relative humidity (RH<80) is highly favorable to powdery mildew disease and high RH (>90) and intermittent rains with less temperature (17-25°C) is favorable to downy mildew. These two diseases are problematic and cause significant yield loss in grapes. These two diseases are effectively managed through advisories. Grape farmers were also advised to use the nylon mesh around the grape plantation to restrict the entry of birds so that damage and yield loss could be minimized effectively.

Higher income: Timely advisories helped the farmers to decide timely cultivation practices viz., irrigation, pruning, weeding, spraying and harvesting etc. These timely practices reduced the labour wages, pesticides costs and also lead to achieve better yield by the farmers.

8.7.2 Partial budgeting to assess the impact of agromet advisories in grapes revealed following points (Fig. 8.13)

- Additional returns due to agromet advisories from sample survey of AAS farmers was Rs.2,48,000 per ha.
- Reduced cost due to savings in irrigation, pesticides spray and timely pruning was to the extent of Rs.20,000 per ha
- The total net gain including the total additional return and reduced cost was Rs.2,68,000 per ha

Impact of weather based agro advisories in grapes during 2015-16 and 2016-17 {average of two years, n=10).

Groups of farmers	Productivity (t/ha)	Cost of production (Rs/ha)	Gross income (Rs /ha)	% increase in income	Reduced cost (Rs/ha)	Additional return (Rs/ha)	B:C ratio
Non-AAS farmers	28.50	1,15,000	6,27,000	-	-	-	5.45
AASfarmers	35.00	95,000	8,75,000	149	-20,000	2,68,000	9.21

Note: Market Rate of grapes ranged from Rs. 22,000 to Rs. 40,000 per ton

Fig. 8.13: Partial budgeting in Grapes

- Overall benefits from the project due to Agromet advisories from grape growers alone was Rs.2.41Crores, including the synergy from line departments, KVKs, Agricultural Research Stations and NGOs in the district

8.7.3 Field bean- Horse gram Double Cropping System in Ramanagara district

Ramanagara is one of the south interior districts of Karnataka famous for finger millet, field bean, Pigeonpea, ground nut and mango. Besides these, mulberry cultivation for silk rearing is an age old livelihood enterprise dwindling in the recent years. Due to the intervention by the University of Agricultural Sciences, Bangalore under watershed development programme, mango based Agri-horti system was recommended and now has become a very popular practice throughout the district. Due to its stabilized income under abnormal rainfall scenario, based on rainfall pattern of 2016, instead of only finger millet mono cropping, field bean- finger millet double cropping or field bean-horse gram double cropping systems were advised to the farmers under NICRA project in operation at Kuthanagere micro watershed (NICRA, 2016). Farmers were informed of the weather condition prevailing in the next five days besides long range weather forecast and extreme weather events during the cropping season through block level agromet advisories twice a week.

Area covered: Nearly 100 farmers were monitored under the project out of which 40 had adopted the double cropping systems based on our advice instead of mono cropping of finger millet only. The total area covered under double cropping systems of field bean-horse gram was around 50 hectares in the project area of Kuthanagere micro watershed.

Bumper crop of field bean as a vegetable: Based on the advice under NICRA project, farmers who have sown field bean in May making the best use of pre monsoon showers got good yield of the crop to the extent of 20 quintals of vegetable pod per hectare. At the market rate of Rs. 40 per kg. The profit realized from field bean alone was Rs.80000 per hectare.

Second crop after field bean: Finger millet is the staple food crop of the farmers and invariably the farmers transplant the crop immediately after the harvest of field bean in the month of August. Unfortunately this year, there was no soaking rains of more than 20 mm to undertake the transplant of finger millet in the month of August. Therefore the farmers were advised horse gram+ niger intercropping after field bean instead of finger millet as a second crop. The farmers of Ramanagara consider horse gram as poor man's meat and Niger as poor man's ghee. Therefore they readily accepted to take up the second crop of horse gram intercropped with Niger in August. These two crops yielded better than finger millet since failure of rains coincided with tillering and flowering stages of finger millets. The dry spell continued in September (-77.3 %), October (-52.4%) and November (-92.9%) also. Thus, income from niger + horse gram intercropping was nearly Rs.20,000 per ha as against Rs.3,500 per ha from finger millet.

Higher Income from field bean- horse gram + niger: Timely advisories helped the farmers for the successful double cropping system of field bean followed by horse gram intercropped with niger.

Partial budgeting: Productivity enhancement (Fig. 8.14, Table. 8.7) through climate resilient system of field bean- horse gram double cropping system against single cropping of finger millet during 2016 drought year (n=10)

- The income enhancement through this climate resilient practice was Rs.1,00,500 as compared to Rs.35,500 per ha with single cropping of finger millet. The net gain from the double cropping of field bean followed by horse gram plus niger was Rs.34,500 per ha.
- The Overall benefit from the project was Rs.34,000 X 50 ha = 17.25 lakhs in the project area of Kuthanagere micro watershed even during the drought year of 2016 with deficit rainfall (-26.2 %).

Sl.no	Debit column (Rs./ha)		Credit Column (Rs./ha)	
1	Additional cost for seeds, labour and pesticides	Rs.33,000	Reduced cost	Rs.O
2	Reduced return	Rs.O	Additional returns from vegetablefield bean, niger and horse gram	Rs 67,500
3	Total additional cost	Rs 33,000	Total additional gain	Rs 67,500
4	Net gain = Rs.67,500 - Rs.33,000 = Rs.34,500 /ha			

Fig. 8.14: Partial budgeting in Double cropping system

Table 8.7: Partial budgeting in Double cropping system

Cropping system	Productivity (q/ha)	Income (Rs/ha)	% increase in income	Additional cost(Rs/ha)	Additional return(Rs/ha)	Additional return to additional cost
Finger millet	Grain 7.00 Straw 12.00	35,500	-	-	-	
Field bean-finger milletdouble cropping system	Field bean harvestedfor vegetable 20.30 Finger millet grain 0.00 Finger millet Straw 2.30	84,650	138	31500	53,150	1.69
Field bean-niger+horsegram double cropping system	Field bean harvestedfor vegetable 20.30 Horse gram 3.00 Niger 1.10	1,00,500	280	33000	67,500	2.04

Note: Market rates used for calculation - Vegetable field bean Rs.4,000/q, finger millet grain= Rs.2,500/q, finger millet straw= Rs.1,500/q, horse gram = Rs.3,500/q, niger= Rs.8000/q,

* Input cost =Fingermillet = Rs 35/kg, horsegram= Rs 60/kg, field bean = Rs 140/kg, niger = Rs 100/kg

** Farmers who have harvested field bean for seeds instead of vegetable pods got 5.1 quintal per ha (the rate of field bean seeds = Rs.8,000/q)

- The expected horizontal spread of this climate resilient system is over 200 ha during 2017-18.The additional employment created due to the climate resilient system was 75 man days per ha. Cumulative overall benefit for the project period was expected upto one crore in the next year.

8.7.4 Tamarind- A Resistant crop to hail storm in Tumkur district

Duragadanagenahalli is a NICRA village under technology demonstration centre (TDC) of KVK Hirehalli in Korategere taluk of Tumkur district. The village receives around 746 mm of rainfall in 46 rainy days. Besides low rainfall, the soil of the NICRA village is gravelly sandy loam with low moisture holding capacity on undulated terrain. Among the dry land orchard crops, tamarind is suitable in terms of soil- site and climatic suitability criteria. In addition to this, tamarind is less susceptible to hail storm as compared to mango, another prominent dry land orchard crop of area. Hail storm in May is a common extreme weather event of the area and it occurred once in 2017. Due to this, mango farmers often suffer due to hail storm resulting in heavy fruit damage and fruit drop. Often, the crop will be sold at a lower price compared to tamarind. An economic analysis (NICRA, 2017) was made among five farmers each under mango and tamarind orchard. An average mango farmers, mango farmers got Rs 52500 per ha per annum as compared to Rs 65250 per ha per annum in tamarind. Due to the fact that tamarind is less susceptible to hail storm and relatively more suitable to the locality based on soil site parameters, farmers prefer tamarind than any other dry land orchard crops. Therefore, tamarind (Fig. 8.15) is recommended under NICRA project due to its sustainable yield and income for rainfall abnormalities associated with hail storm.

Fig.8.15: Tamarind- sustainable income generating agri-horti system against hail storm and drought even on gravelly sandy loam soils

8.7.5 Millet based farming: Coping with weather extremities in 2017, Southern Karnataka

2017 was characterized by normal onset with early season drought. June and July experienced deficit rainfall in Kuthanagere NICRA village to the extent of -9 % and -68 %. Hence, weather based contingent crop plan was suggested for early season drought. Short duration finger millet variety (GPU-28, Fig. 8.16) was suggested for August 1st week sowing. Rainfall from August onwards was surplus to the extent of 55.4 % in August, 49.1 % in September and 62.4 % in October. Early exit of rain with deficit rainfall in November and December resulted in poor seed filling in long duration varieties. However, GPU-28 early maturing in 110 days was a tailor made variety for 2017 rainfall distribution pattern. Our timely advise of short duration variety in NICRA villages could reduce climate change vulnerability besides addressing the problems of malnutrition in the villages (NICRA, 2017). The heavy down pour in August, September and October would have eroded the soil without our timely advise of across the slope cultivation, strengthening of bunds, repeated Intercultivation, surplus rainwater harvesting in farm ponds, nala bund and check dams.

Fig. 8.16: Bumper crop of finger millet in 2017 due to the resilient variety- GPU-28

8.7.6 Revival of Millet based systems

Small millets particularly finger millet is a staple food crop of southern Karnataka. However, use of traditional varieties and poor knowledge of high yielding and short duration varieties resulted in low yields with heavy weed infestation due to excess rain in August. With a broad objective of reducing the climate change vulnerability and for promoting balanced nutritive millet based inter cropping and double cropping systems were also suggested to the farmers of NICRA village. If one crop fails due to poor coincidence of favorable rainfall with critical stages of the crop, another crop may give good yield is the principle of intercropping /double cropping. There were nearly 30 farmers (Table 8.8 and Table 8.9) who followed our timely advice of short duration variety besides soil conservation measures and rain water harvesting which brought added income (Rs. 15 to 30 thousand).

Table 8.8: Cost of cultivation and income of climate resilient variety (GPU-28) as compared to long duration varieties (n=10)

Finger millet variety	Productivity (q/ha)	Income (Rs)	Cost of cultivation per ha	Net Income (Rs)	Additional return per ha
Long duration varieties (MR-6, Indoaf -8, L-5)	15.62	62,500=00	43,500=00	19,000=00	-
Short duration varieties (GPU-28)	25.00	77,500=00	42,000=00	35,500=00	16,500=00

Note: Market values: Finger millet grain Rs 2300/- per quintal, straw = Rs 1500 per quintal

Table 8.9: Double cropping/intercropping systems with finger millet adopted by the NICRA farmers

Intercrop/Double cropping	Area in ha.	No. of NICRA farmers	Average Income (Rs/ ha)
Finger millet + pigeonpea (8 :1)	3	7	75,200
Finger millet + groundnut (8 :1)	3	5	70,100
Cowpea - finger millet	2	5	51,500
Sesamum - finger millet	2.5	6	55,400
Field bean - horse gram	1.5	7	60,000
Total/Average	12	30	62,440
Farmers without NICRA(Traditional system)	12	30	50,500

**Note:* Traditional system include 9 crops mixed with finger millet and is called locally as akkadi/ nava danya system

8.7.7 Impact of agromet advisory in coffee plantation in Malnad region (Hilly zone)

Karnataka is India's largest coffee producing state, contributing to nearly 70 per cent of the total production. The most commonly used coffee beans are Arabica and Robusta grown in the hills of Karnataka (Kodagu, Chikkamagaluru and Hassan), Kerala (Malabar region) and Tamil Nadu (Nilgiris District, Yercaud and Kodaikanal). Coffees gain aromatics in high altitudes between 1000-1500 mts above MSL, besides sparse shading through inter planting with spices such as cardamom, cinnamon, clove, and nutmeg, storage and handling functions. Growing altitudes range between 1,000 m (3,300 ft) to 1,500 m (4,900 ft) above sea level for *Arabica* (premier coffee), and 500 m (1,600 ft) to 1,000 m (3,300 ft) for *Robusta* (though of lower quality, it is robust to environment conditions). Ideally, both *Arabica* and *Robusta* are planted in well drained soil conditions that favour rich organic matter that is slightly acidic (pH 6.0–6.5). However, India's coffees tend to be moderately acidic which can lead to either a balanced and sweet taste, or a listless and inert one. Slopes of *Arabica* tend to be gentle to moderate, while *Robusta* slopes are gentle to fairly level.

8.7.8 Climatic conditions

Ideal climatic conditions to grow coffee are related to temperature and rainfall; temperatures in the range of 23 °C to 28 °C with rainfall incidence in the range of 1500–2000 mm, followed by a dry spell of 2–3 months suit the Arabica variety. Cold temperatures closer to freezing conditions are not suitable to grow coffee. Where the rainfall is less than 1000 mm, providing irrigation facilities is essential. In the tropical region of the south Indian hills, these conditions prevail leading to coffee plantations flourishing in large numbers. Relative humidity for Arabica ranges 70–80% while for Robusta it ranges 80–90%.

Among the climatic conditions,

- Stress during December to January extending up to February is important for flowering.
- The flower buds (spikes) burst out on each and every branch nodes due to the moisture stress and cold climate during December to February.
- After the flower bud initiation, coffee plant needs rain for full blooming. It is called as blossom showers. If the rain does not occur, farmers resort to artificial raining through rain gun/ sprinklers.
- When coffee plantations are in full bloom it is a delightful sight to watch with scenic landscape and appealing aroma. However, well distributed pre monsoon showers during March to May, play very important role for full blooming and maturity of the fruits.

Fig. 8.17: Rain gun/ sprinkler for simulating rain as blossom shower from seepage pond in coffee

- Some of the planters, have dug the seepage ponds (Fig. 8.17) based on our advise and used the same for artificial raining as and when required. Fig. 8.17 depicts blossoming of coffee through artificial blossom showers.
- Blooming of the spikes takes place immediately during night hours followed by pollination and fertilization. Insects and bees contribute to 33% (Fig. 8.18) of the pollination where as remaining 67% is through wind. For full blooming 4-5 days is needed. Rainfall during this period, play very important role for full blooming. Therefore, Agromet advisory during this period, based on rainfall and relative humidity forecast play a crucial role on the yield and income of the farmers. In the event of no rainfall forecasted during February and March, planters will resort to artificial raining through sprinklers. Seepage ponds (17.a) in the valleys serve the water needs of coffee plantation on the

Fig. 8.18: Close up view of coffee flower clusters and pollination through insect

adjoining hill slopes. There is lot of demand for digging seepage ponds by the planters through the watershed and NAREGA schemes.

- For this purpose, planters take every step to conserve the soil moisture on plantation hill slopes through contour, trenching and mulching and prepare for artificial raining in the event of no natural rain.
- Abnormal rainfall impacts on coffee noticed are presented in Fig. 8.19. Most observed impacts are:
  - (a) Unseasonal flowering due to unseasonal rains
  - (b) Damage to coffee cherries caused by drought in the final stage of ripening
  - (c) Bean damage as a result of low rainfall between weeks 14 and 20 of fruit development
- Finally, agromet advisories based on weather forecasts and land resources advisories for soil moisture conservation (trenching and mulching across the slope between coffee rows) goes a long way in proper flowering, fruiting, disease control, yield and income of the farmers in 2017 (GKMS, 2017).

Fig. 8.19: Abnormal rainfall impact on coffee

References/Suggested Readings

AICRPAM, 2016, Annual report of Agrometeorology (2015-2016). Univ. Agric. Sci., Bengaluru, p. 10-14.

AICRPAM, 2017, Annual report of Agrometeorology (2016-2017). Univ. Agric. Sci., Bengaluru, p. 5-10.

GADGIL, S., 1989, Monsoon variability and its relationship with agricultural strategies. Paper presented In: *Int. symp. on climate variability and food security in developing country,* 1989, New Delhi. February 5-7, p. 249-267.

GKMS, 2011, Annual report of Gramin Krishi Mousam Seva (GKMS), Univ. Agric. Sci., Bengaluru, p. 26-30.

GKMS, 2012, Annual report of Gramin Krishi Mousam Seva (GKMS), Univ. Agric. Sci., Bengaluru, p. 24-29.

GKMS, 2013, Annual report of Gramin Krishi Mousam Seva (GKMS), Univ. Agric. Sci., Bengaluru, p. 27-32.

GKMS, 2014, Annual report of Gramin Krishi Mousam Seva (GKMS), Univ. Agric. Sci., Bengaluru, p. 40-49.

GKMS, 2015, Annual report of Gramin Krishi Mousam Seva (GKMS), Univ. Agric. Sci., Bengaluru, p. 42-49.

GKMS, 2016, Annual report of Gramin Krishi Mousam Seva (GKMS), Univ. Agric. Sci., Bengaluru, p. 41-48.

GKMS, 2017, Annual report of Gramin Krishi Mousam Seva (GKMS), Univ. Agric. Sci., Bengaluru, p. 45-47

NICRA, 2011, Annual report of AICRPAM - NICRA (2011-2012), Univ. Agric. Sci., Bengaluru, p. 29-37.

NICRA, 2012, Annual report of AICRPAM - NICRA (2012-2013), Univ. Agric. Sci., Bengaluru, p. 49-67.

NICRA, 2013, Annual report of AICRPAM - NICRA (2013-2014), Univ. Agric. Sci., Bengaluru, p. 31-47.

NICRA, 2014, Annual report of AICRPAM - NICRA (2014-2015), Univ. Agric. Sci., Bengaluru, p. 31-40.

NICRA, 2015, Annual report of AICRPAM - NICRA (2015-2016), Univ. Agric. Sci., Bengaluru, p. 23-30

NICRA, 2016, Annual report of AICRPAM - NICRA (2016-2017), Univ. Agric. Sci., Bengaluru, p. 27-37

NICRA, 2017, Annual report of AICRPAM - NICRA (2017-2018), Univ. Agric. Sci., Bengaluru, p. 25-35

RAJEGOWDA, M.B., PADMASHRI, H. S., PAVITHRA.B.V. SHILPA.C.N, SOUMYA.D.V. and

RKVY, Annual report of Characterization of Climate of Karnataka State, Contingent Crop Plan and Hobli wise Agromet Advisories for Karnataka State, 2011, Univ. Agric. Sci., Bengaluru, p.10-25

Seeley, M. W., 1994, The future of serving agriculture with weather/climate information and forecasting: some indications and observations. *Agric. Forest Meteorol..* 69:p. 47-59.

Shivaprasad, C.R., Reddy, R.S., Sehgal, J and Velayutham,M., 1996, Soils of Karnataka for optimizing land use., NBSS. Publ. 47:p. 98.

Shivaramu, H.S., Pradmashri, H.S. and Soumya, D.V., 2017, Regrouping of NARP zone of Karnataka based on changing climate trends., Climate Smart Agriculture, XIII Agricultural Sciences Congress, February 21-24, UAS, Bengaluru, p.12.

Singh, S.V., Rathore, L. S. and Trivedi, H. K. N., 1999, Verification of medium range weather forecasts, Guide for Agrometeorological advisory services, NCMRWF, Department of Science and Technology, Government of India, p. 73-81.

Vemkatesh, H., Rajegowda, M. B., Shivaramu, H. S and Rao, V. U. M., 2016, Agroclimatic atlas of Karnataka, All India Coordinated Research Project on Agrometeorology, Vijayapura and Bengaluru., Univ. Agric. Sci., Dharwad.

9

Climate Change and Agriculture – Experiences in North Karnataka

H. Venkatesh

Principal Scientist (Agrometeorology), AICRPAM
Regional Agricultural Research Station, Vijayapura (UAS Dharwad), India

The extreme weather events of the present day and the expected future climate scenarios at global and regional scales have been attributed to the changes in the atmospheric chemical composition and the massive amounts of greenhouse gases (GHGs) emitted by humans since the industrial revolution, bringing about the global warming and resulting in Climate Change. Further, the natural disasters like flash floods are attributed to the highly conspicuous human interventions with nature – the earth and its atmosphere. The parody of the situation is that, while the GHGs sourcing countries are mainly the developed ones, the sink of the result are the agrarian-dominated and eco-sensitive economies of the tropics. Invariably, tropical regions are to be affected the most, because, on the one hand the general circulation of the atmosphere directly and naturally causes divergence of the polluted air from extra tropical regions towards tropics, while on the other, the industrialization within the tropical regions too would affect the local atmosphere. Karnataka State, being located well within the tropics, can be truly impacted by the climate change caused by anthropogenic sources and it is demarcated into 10 agroclimatic zones (Fig. 9.1) based on crops, soils and climate. Climate experts have warned that Karnataka is likely to be more vulnerable to climate change than other states in India. The specific impact would be in terms of warming, to the extent of 1.5-2.0° Celsius by 2030s under the likely high-emissions scenario, compared to the pre-industrial period (1880s) levels. Dependable flow in many reservoirs would be 75 per cent to 90 per cent by the middle of the present century. Reduced availability of surface +water, higher water stress for crops, floods and drought are expected to make, particularly, Kalaburgi, Raichur and Yadgir districts, vulnerable to climate change. Assessment of the impact of climate change showed that the yield of rice, maize, sorghum, red gram and ragi is likely to decline by the 2030s, anticipating setting off a food crisis. Changes in Forest and bio-diversity types imply that

1. North Eastern Transit io 11 Zone Pulses,/"euvwr, oilseeds, *bajra,* cotton and sugarcane
2. North Eastern Dry Zone Rabi./ownr, *bajra,* pulses, oilseeds and cotton
3, Northern Dry Zone Maize, *bajja,* ground nut, cotton, wheat, sugarcane and tobacco
4, Central Dry Zone Ragi./buw, pulses and oilseeds
S- Eastern Diy Zone (Ragi, rice, pulses, maize and oilseeds
6. Southern Dry Zone Rice, ragi, pulses, *jowar* and tobacco
7. Southern Transition Zone Rice, ragi, Pulses; Jewar and tobacco
8. Northern Transition zone Rice,/oh'ftr, groundnut, pulses, sugar-cane and tobacco
9. Hilly Zone Rice and pulses
10. Coastal Zone Rice and sugarcane

Fig. 9.1: Agroclimatic zones of Karnataka

future climate may not be suitable for the existing forest types and bio-diversity and can result in forest dieback. The consequences would therefore be severely adverse for sustenance of the ecosystem. From an agricultural point of view, in short, the climate change suggests an increase of extreme events, both in magnitude and frequency of primary weather variables like rainfall, temperature, dry spells, off-season rainfall including hailstorm events. Since the turn of the twenty first century, northern parts of Karnataka have experienced one extreme event or the other, and repeatedly too, which can be considered is a precursor to the events to come later. The state has been alerted that it needs to develop and implement coping strategies to deal with the current climate variability and build resilience to long-term climate change. This has been duly attended to by the state government from time to time. It is pertinent to note that Karnataka is the first state in India to have "Natural Disaster Monitoring Centre". However, considerable ground work is needed to be performed towards identification and implementation of mitigation strategies. In this chapter, we focus on thirteen districts of North Karnataka.

9.1 Climate of North Karnataka

There are 13 districts in North Karnataka, out of which, Uttara Kannada district is located in the Hilly (Zone 9) and the Coastal (Zone 10) zones. Haveri, Dharwad and Belagavi districts comprise both the Northern Transition Zone (Zone 8) and the Northern Dry Zone (Zone 3). Bidar district is in Northeast Transition Zone (Zone 1), whereas Kalaburgi, Raichur and Yadgir districts are in Northeastern Dry Zone (Zone 2). All the other districts, viz., Bagalakote, Ballari, Gadag, Koppala and Vijayapura are located exclusively in the Northern Dry Zone, which is the largest Agroclimatic Zone in the country. The northern region of Karnataka is endowed with all types of climate – per-humid to arid. It experiences four traditional seasons – winter (January to February), summer (March to May), monsoon (June to September), and post-monsoon (October to December). The climate of the region is extremely diverse – from arid to semi-arid in the plateau regions through sub-humid in the foothills of the Western Ghats to perhumid along the Ghats and the Coastal region. Rainfall situation in each year is different, which could be – delayed onset of southwest monsoon, or mid-season dry period caused by low rainfall sequences during July and/or August. Most of north Karnataka is dominated by *rabi* crops that are grown in receding soil moisture situation. As such, rainfall in September and October is also very critical; either deficiency in its quantity or its early cessation can cause agricultural drought. even the double cropping, which is a sort of subsistence farming, is adversely affected in some years due to poor rainfall in both the seasons, resulting in greater hardship to the farmer. Therefore the role of agro-technologists, the Government and the NGOs becomes crucial in alleviating the adverse effects of drought on year to year basis.

The first extensive study on the rainfall patterns of Karnataka was performed by Dr. A. Krishnan (Krishnan, 1984), whose name is synonymous with Agricultural Meteorology and Droughts in Karnataka. Later many scientists have attempted the same with different objectives. North Interior Karnataka meteorological subdivision, which covers 12 of the 13 districts of north Karnataka (except Uttar Kannada) has mean annual rainfall of 659.3 mm of which 67 percent is received in the southwest monsoon season, 21 percent in northeast monsoon season, and 11 percent in summer season. The coastal region, which comprises Uttar Kannada district, has the mean annual rainfall of 3943.9 mm of which 87.6 % is received in the southwest monsoon season, 7.9 percent in northeast monsoon season and 4.4 percent in summer season. Rainfall of 4000 mm is received in the coastal belt and on the Western Ghats. The sudden drop in altitude (down to 500 m) in the transitional belt along the eastern portion of the Western Ghats has led to rain-shadow effect resulting in vast area further east, with annual rainfall of just about 500-600 mm, which is ill distributed both temporally and spatially. There exist pockets of extremely low annual rainfall of less than 500 mm along the border of Vijaypura and Bagalkot districts as well as in Ballari district.

Temperature over North Karnataka is lowest in the months of December and January, and is highest in May. Majority areas of North Interior Karnataka the mean daily temperature is about 30°C, and the maximum temperature reaches 43°C in Kalaburgi, Yadgir, Raichur and Bellary districts. It remains considerably lower (20°C -24°C) over the Ghats and coastal area. North interior Karnataka experiences incidences of heat waves also during May and June.

9.2 Trends in climate

In a multi-institutional effort involving experts from Indian Institute of Science (IISc), Institute for Social and Economic Change (ISEC), University of Agricultural Sciences (UAS), and Centre for Study of Science, Technology and Policy (CSTEP), the Bangalore Climate Change Initiative – Karnataka (BCCI-K), studied Karnataka's climate trends and made projections, in most cases, based on a coupled atmosphere-ocean general circulation HadCM3 model. The study observed a general decline in annual rainfall over Northern Karnataka (Table 9.1). In another recent publication, where the micro level rainfall data analysis was performed, Venkatesh *et al.* (2016) pointed out that, pockets in Vijayapura, Dharwad and Raichur districts have experienced decreasing trend of annual rainfall since 1980. BCCI—K (2011) reported a warming trend particularly in Bidar, Vijayapura, Kalaburgi, Yadgir and Raichur districts, with both the minimum and the maximum temperature increase by at least 0.6°C over the last 100 years, which is 0.1°C more than the 0.5°C observed over most parts of the country.

Table 9.1: Projected Rainfall by 2021-50 from the baseline of 1961-90

District		Projected change				
		Winter (%) (%)	Summer season (%)	Monsoon season (%)	Post monsoon	Annual (%)
1.	Bagalkote	-35.42	43.65	-10.47	16.87	1.38
2.	Belagavi	-6.25	20.34	-1.32	20.00	6.15
3.	Ballari	-14.47	36.16	-8.11	6.87	0.52
4.	Bidar	-71.30	22.29	6.89	50.76	27.03
5.	Vijayapura	-50.81	39.41	-9.81	12.54	0.60
6.	Dharwad	0.00	30.55	0.06	9.94	3.71
7.	Gadag	0.00	37.48	-6.79	12.42	1.93
8.	Kalaburgi	-63.11	12.77	-1.44	9.49	4.26
9.	Haveri	0.00	12.28	1.91	10.87	5.80
10.	Koppal	-2.08	45.19	-12.35	8.81	-1.56
11.	Raichur	-4.17	20.50	-10.99	-7.59	-6.79
12.	Uttara Kannada	0.00	10.62	4.87	6.80	5.40
13.	Yadgir	-65.04	19.32	-7.14	-5.22	-4.07

9.3 Projection of climate Change

As per BCCI-K (2011), the change in annual rainfall is positive for most of the districts, but negative for the southwest monsoon season, and indicating increase-situations in northeast monsoon season. Looking back to the climatology of north Karnataka (Venkatesh, 2003), had revealed opposing epochs of rainfall during *kharif* (June-August) and *rabi* (September-October) season as well as September and October on monthly scale. The data of BCCI-K could in a way represent the same fluctuations, as it is possible that the 1961-90 period could be of high rainfall epoch in *kharif* season and the 2021-2050 period could be the opposite. Hence, in addition to just the trend, it would be far more important to analyse the variability patterns in different periods. However, the important information for the BCCI-K report is the increase in summer rainfall, which could more of less-utility for crops, but harmful for horticultural crops like mango, grapes and pomegranate. The study of BCCI-K also projects further warming by 1.7°C to 2.2°C by the 2030s in continuation of the trends observed at the beginning of the century, resulting in increase by 2.0°C to 2.4°C from the base period, in case of both maximum and minimum temperatures. In another study, Anandhi (2010) reported that there would be an increase in wet season length, but no change in length of cold and warm seasons at the end of 21st century. Such reporting for staggered periods by different sets of scientists would be of less use for the planners, and would be of only academic interest. It is therefore better to provide customised projections as per the needs of the stake holders including the governments, giving specific applications.

9.3.1 Climate change and monsoon characteristics

AICRP on Agrometeorology, Vijayapura Centre and the Institute for Agriculture Research on Climate Change, UAS Dharwad studied the impact of climate change projections of rainfall on the monsoon characteristics and drought situations. Anonymous (2015) reported on the monsoon characteristics of onset of rainy season, cessation of rainy season and duration of rainy season. It was noticed that the 30 year mean of onset time of monsoon during 2021-2050 would not be much different from that of the base period of 1961-1990. However, there would be increase in standard deviation in onset of monsoon in northeast transition and northeast dry zones, but reduced in northern dry zone. In both cases, the standard deviation is expected to be between 1.6 to 2.0 weeks. This suggests that crop planning would be riskier in northeast transition and northeastern dry zones. End of rainy season would be delayed in the northernmost districts of Bidar, Vijayapura and Belagavi as also the Uttara Kannada district on the western side. However, the standard deviation of end of rainy season would reduce in Bidar, Ballari and Koppal districts, but would enhance in Kalaburgi, Yadgir and Raichur districts. This calls for rabi crop planning particularly with respect to varietal selections in the riskier districts. Analysis of duration of rainy season indicated that, it would reduce in Kalaburgi, Yadgir, Gadag, Bagalkot, Koppal and Ballari districts, whereas there would be increased variability in terms of standard deviation in Bidar, Kalaburgi, Vijayapura, Raichur and Koppal districts (Fig. 9.2). When analysing the meteorological drought situations projected for 2021-50 with the base period, it was noticed that frequency of mild category of drought is expected increase in

Districrt Average:
Duration of rainy season (Mean)

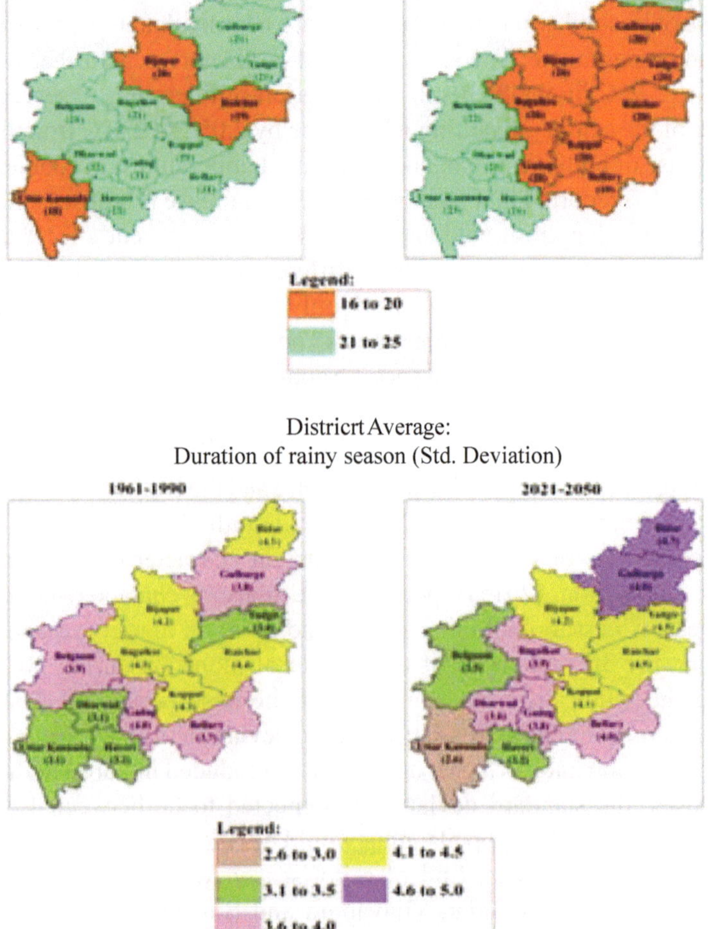

Fig. 9.2: District wise characteristics of duration of rainy season (weeks) in north Karnataka

Vijayapura, Bagalkot, Yadgir, Raichur, Koppal, Gadag, Dharwad, Haveri and Ballari districts. Frequency of moderate droughts is expected to decrease in Bidar district. Severe droughts are not expected to occur in the projected climate change period in any of the districts of North Karnataka.

9.3.2 Climate change vulnerabilities

The projected levels of temperature and patterns of precipitation might cause more extreme weather situations in both magnitude and frequency, thereby causing shifts in cropping patterns, the loss of traditional species and increased vulnerability of crops and animals to vector borne diseases. Further, it is anticipated that these weather events would reduce disease resistance capability in crops and animals. In totality, all these point to great decrease in agricultural productivity, affecting the livelihood of farmers. On the other hand, the projected sea level rise and surface temperature increase would impact the coastal communities dependent on coastal agriculture and fisheries. The cup of vulnerability is full.

9.4 Cropping pattern

Rainfed Agriculture during *kharif* (for crops sown during June-August) and *rabi* (for crops sown during September-November) is mostly based on rainfall received in southwest monsoon (June-September) and northeast monsoon (October-December). Wide spectrum of crops ranging from high-water requirement crops like paddy and sugarcane to very low water requirement crops like millets is grown. Short to long duration pulses, oilseeds, fibre crops, spices, and seasonal and perennial horticulture crops also are being grown in the region. In red and shallow black soils, *kharif*-dominated cropping is in vogue, whereas in medium and medium to deep black soils, *rabi*-cropping or double cropping is adopted. Rainfed *rabi* crops thrive on receding soil moisture situation. Irrigated paddy, chillies, maize, cotton, soybean, sorghum, groundnut, pigeon pea, pearl millet, little millets, potato and wheat dominate the cropping area in north Karnataka. Vegetable and fruit crops are also abundantly grown in north Karnataka. Mango, banana, grape, pomegranate, lime and papaya are the most important fruit crops grown. Unless there is drastic change in rainfall pattern – either a jump or decrease – is encountered under climate change, not much change in the cropping system can be foreseen except in case of changes in irrigated areas. By superposing the recommended crops at different sowing windows on the rainfall distribution, Crop Selection Calendars (Fig. 9.3) have been prepared for weather based cropping (Venkatesh *et al.*, 2008). The charts have been prepared for shallow and deep soil types for all districts north Karnataka. From the figure, it can be known as to which crops or cropping systems need to be taken up for sowing at different times of onset of monsoon.

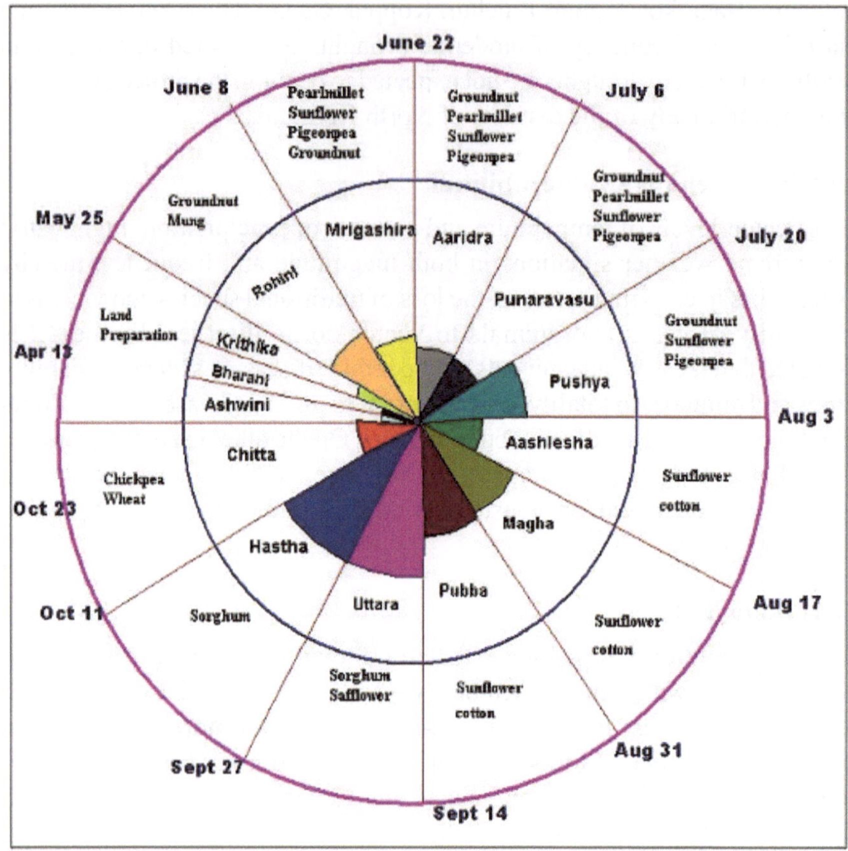

Fig. 9.3: Crop selections in different sowing windows

9.4.1 Cropping pattern changes

There was a decrease in cultivation area of Jowar from 29.7 lakh ha to around 12.6 lakh ha from 1961 to 2011. Over the past five decades, The net cropped area decreased marginally from 102.3 lakh ha (1961) to 101.7 lakh ha (2009), while the gross cropped area increased from 105.9 lakh ha to 123.7 lakh ha in the same period. The trends in cropping pattern analysed by BCCI-K (2011) are given in Table 9.2. As per the report, introduction of high yielding varieties (HYVs) has progressively reduced the cultivated area for traditional varieties in the state. This resulted in the extinction of few local varieties. Traditional varieties of paddy, banana, mango and vegetables with distinct taste and aroma have high nutritional values and disease resistance. Landraces of many crops have provided the genes needed for pest and disease resistance or for the crops to adapt to poor soils, drought and cold temperatures, and hence there is need to preserve them.

Table 9.2: District wise trends in crops grown in North Karnataka: 1961 - 2008

District	Rice			Maize			Jowar			Cotton			Pigeon pea			Groundnut			Sunflower			Sugarcane		
	Area	Production	Productivity	Area	Production	Productivity	Area	Production	Productivity	Area	Production	Productivity	Area	Production	Productivity	Area	Production	Productivity	Area	Production	Productivity	Area	Production	Productivity
Bagalkot	⇧	⇩	⇩	⇧	⇩	⇩	⇧	⇧	⇩	⇩	⇧	⇧	⇧	⇩	⇧	⇩	⇧	⇩	⇩	⇧	⇧	⇩	⇧	⇩
Belagavi	⇔	⇩	⇩	⇧	⇧	⇩	⇔	⇧	⇧	⇩	⇧	⇩	⇩	⇩	⇔	⇩	⇩	⇧	⇧	⇔	⇔	⇧	⇩	⇩
Ballari	⇧	⇩	⇩	⇧	⇧	⇩	⇩	⇧	⇧	⇧	⇩	⇩	⇧	⇩	⇧	⇧	⇩	⇧	⇧	⇔	⇔	⇩	⇧	⇧
Bidar	⇩	⇩	⇩	⇧	⇧	⇧	⇩	⇧	⇧	⇩	⇧	⇧	⇧	⇩	⇧	⇩	⇩	⇔	⇧	⇧	⇩	⇩	⇧	⇧
Vijayapura	⇩	⇩	⇧	⇧	⇧	⇧	⇩	⇧	⇧	⇩	⇩	⇧	⇧	⇔	⇔	⇩	⇩	⇔	⇧	⇧	⇔	⇧	⇩	⇩
Raichur (+ Koppal)	⇧	⇧	⇧	⇧	⇧	⇧	⇩	⇔	⇔	⇧	⇧	⇧	⇔	⇔	⇔	⇩	⇩	⇧	⇧	⇧	⇧	⇧	⇧	⇧
Dharwad	⇧	⇩	⇩	⇧	⇩	⇧	⇧	⇧	⇧	⇩	⇧	⇧	⇧	⇩	⇩	⇧	⇩	⇩	⇧	⇩	⇩	⇩	⇧	⇧
Gadag	⇧	⇧	⇧	⇧	⇧	⇧	⇩	⇧	⇧	⇩	⇧	⇧	⇧	⇧	⇧	⇩	⇩	⇧	⇩	⇩	⇩	⇧	⇧	⇧
Kalaburgi (+ Yadgir)	⇔	⇧	⇧	⇧	⇧	⇧	⇩	⇧	⇧	⇧	⇧	⇧	⇧	⇧	⇧	⇩	⇩	⇧	⇧	⇧	⇧	⇩	⇧	⇧
Haveri	⇩	⇩	⇩	⇧	⇧	⇩	⇩	⇩	⇩	⇔	⇧	⇧	-	-	-	⇩	⇩	⇩	⇩	⇧	⇧	⇩	⇧	⇧
Uttara kannada	⇔	⇧	⇧	⇧	⇧	⇩	⇧	⇧	⇧	⇧	⇧	⇧	⇔	⇔	⇧	⇧	⇧	⇧	⇩	⇩	⇩	⇩	⇧	⇧

9.4.2 Decline in agro-biodiversity

Agricultural production has been varying since the year 2000, with attributed reasons of shrinking land-man ratio, depleting vegetation cover, inadequate and unpredictable water resources, high dependence on rainfall and increasing production costs. As per Karnataka Biodiversity Board estimates, there will be extinction of approximately 6% of cereal crops, 85% of legume species and 13% of vegetables species, which is attributed to replacement of local varieties by improved or exotic varieties and species. In view of this, it also suggested that there is need to manage and conserve precious traditional crop varieties, which could become genetic contributors for drought and disease resistant varieties for the future.

9.4.3 Indicators of climate change and their impact in extreme weather

Before going further, let us see how important the climate change is for us. We list here the extreme weather events over north Karnataka, since the start of the 21st Century.

- Successive drought years – 2001 to 2003
- Very heavy rainfall >180 mm in 6 hours on 22nd in semi-arid regions (June 2007)
- Extremely low minimum temperature of 5.6 to 8.2 °C (22nd-25th November 2007)
- Off-season rainfall, cloudiness and high humidity (10th March – 10th April 2008)
- Heavy rainfall for nearly one week due to slow movement of deep depression over peninsular India (28th September – 4th October 2009)
- Prolonged dry spells during monsoon season (2010 and 2011)
- Extremely low minimum temperatures of < 10°C for more than 10 days (2010-11 and 2011-12)
- Failure of rainfall during June 2012, June 2014 and June to August 2015
- Widespread hailstorm during summer of 2014 and 2015
- Severe summer with maximum temperature of more than 40°C for during 2015
- Very long dry period of nine weeks from standard meteorological week (SMW) 25

- Very high maximum temperature during November-December 2015 reaching 33 °C by the end of December 2015 – a hot winter!!
- Mid-season drought period of six weeks from SMW 27 to 32 (15.4 mm) and of four weeks from SMW 34 to 37 during 2018

Some of the events were location specific while others were widespread over entire north Karnataka. Hence, even though we realize climate is what we expect and weather is what we get, these extremes are too conspicuous with their frequency as well as diversity, forcing us to admit advertently or inadvertently to the climate change. Most of the above events have had impact on agriculture, from not taking up sowing, to physiological impact on growth of sorghum during extreme winter, retarded growth and yield of *kharif* crops in years of extended dry spells, adoption of contingency crop plans more frequently due to extreme delayed onset of monsoon, and devastation of not only the produce but also the entire horticultural orchards of grape, lime, pomegranate, papaya etc due to hail storms. So the climate change impacts are already too prevalent.

9.4.4 Impact of dry spells and flash floods

Droughts affect agricultural production in the state to a great extent. Mid season dry spell duration has been on the increase in recent years. In 2009, rainfed *kharif* crops such as jowar, maize, bajra, tur, green *gram*, black *gram*, groundnut, soybean and sunflower in major parts of northern Karnataka were affected by severe stress due to a dry spells lasting about 4-6 weeks between July and August. Worse has been the situation in 2015 and 2018. On the other hand, major parts of northern Karnataka were also affected by widespread heavy rainfall during September 26 to October 4, 2009 in Raichur, Koppal, Ballari, Bagalkot, Vijayapura and Kalaburgi districts. Nineteen lakh hectares of *kharif* crop loss was estimated due to submergence, deposition of silt and deterioration of grain quality, whereas three lakh hectares of sown *rabi* crop was completely washed off in Vijayapura district alone. General changes in cropping pattern have been noticed in many regions in north Karnataka after the low rainfall epoch of 2001-2003. The drought of 2003 impacted livestock as well as on fodder management, as the farmers had to sustain either by selling the cattle or purchasing the fodder at exorbitant rates (Biradar and Sridhar, 2009).

9.4.5 Impact of hailstorms

Hailstorm events during March 2014 were perhaps the most damaging extreme event that north Karnataka has come across in recent years. It damaged harvested rabi crops lying in the field, as well as devastated the grape farms completely, with practically no chance of recovery over large scale resulting in losses to more than three hundred Crores of Rupees.

9.5 Climate change and crop projections

Rise in temperature will provide less opportunity for crop canopies to accumulate dry matter by accelerating the growth rates of the crops, resulting in reduced yield. Projections of Indian Network for Climate Change Assessment (INCCA, MoEF, Govt of India) suggest that in the 2030s, higher temperatures and reduced sunshine may limit biomass production and yield as compared to baseline conditions. The estimated 2.1°C rise in mean temperature and a 4.5% increase in mean precipitation would reduce net agricultural productivity in the state by 1.2%. This sector and regional analysis for the 2030s also provides projections for possible changes in crop production in the western coastal areas of India. Agriculture is perhaps the most vulnerable sector to climate change. The direct effects of climate include variability in temperature, precipitation, and carbon dioxide concentrations while changes in soil moisture content and increased frequency of pest and disease infestations are the indirect affects. As agriculture in Karnataka is mainly rain dependent, the variation in precipitation will require changes in agriculture pattern that may give rise to changes in production to a considerable extent. Studies indicate that any change (increase or decrease) in agricultural production is directly dependent on the scenario to which simulation/ projection has been carried out. MoEF (2004) studied the impact of 2010 climate change scenario on yields of rice, wheat and sorghum and projected that rice and sorghum yield would increase by 1.3-5.8% and 1.0-3.4% respectively in South India. O'Brien *et al.* (2004) reported that the northeastern part of the state is highly vulnerable to a changing climate due to lesser adaptive capacity. North Interior Karnataka was in general placed in the category of high vulnerability. BCCI-K (2011) studied the climate change on crops through Infrocrop simulation model for maize, sorghum, rice, pigeon pea, cotton, potato, soybean and wheat. Based on the simulation studies, the vulnerable districts are presented in Table 9.3. The study found that *kharif* crops may witness greater changes in their productivity, spanning a range from a +9.5% (gain) to -14.4% (loss) from their base level yields. The yield of rainfed rice may undergo changes between +7.2% (gain) to -13.8% (loss) with large regions losing up to -9.6%. Maize and sorghum yields are projected to change from +27.6 % to -19.3 % and +17.2 % to -18.4% respectively from their baseline yield. This is attributed mainly to property of C4 carbon fixation. Any further increase in rainfall (more cloud cover) would result in reduced yields in the *kharif* season. District wise details are given in Table 9.4. Considering the projected 4.5% increase in mean precipitation and rise in mean temperature predicted for the state, BCCI-K (2011) also estimated that Karnataka as a whole could witness a -2.5% decline in net agriculture production with a major reduction in coastal regions. Further, the study observed that state may record a -12.3% loss in food production due to climate change while enhanced CO2 and temperature may cause significantly higher productivity (up to +35%) in certain districts.

Table 9.3: Crops and districts projected as vulnerable to climate change

District	Rice (irrigated)	Rice (aerobic)	Maize	Jowar	Red gran	Cotton	Wheat	Sugarcane
Bagalkot					■			
Belagavi	■	■			■			
Ballari	■	■			■	■		
Bidar					■	■		
Vijayapura	■		■			■		
Dharwad	■		■			■		
Gadag						■	■	
Kalaburgi (+Yadgir)		■			■		■	
Raichur (+Koppal)								
Uttar Kannada		■	■	■	■	■		■

Table 9.4: Projected productivity changes and districts of major crops with respect to baseline

Crops	Projected net yield change in %	< 10 % increase in productivity and identified to increase in growing area	10% to 25% increase in productivity and identified increase in growing area
Rice	–0.3 (irrigated) –0.9 (rainfed)	Bagalkot, Haveri	Raichur Uttara Kannada
Maize	+1.2	Bidar, Gadag, Kalaburgi (+ Yadgir), Haveri	Belagavi
Sorgum	+2.6	Bagalkot, Belagavi, Haveri, Raichur	Ballari, Bidar, Gadag, Kalaburgi (+ Yadgir)
Red *gram*	+1.3		Dharwad, Gadag, Haveri
Cotton	+1.3	Bidar, Haveri	Belagavi, Vijayapura
Ragi	–5.8	Haveri, Gadag	
Sugarcane	+2.6	Belagavi	

9.6 Climate risk management

There are two components in climate risk management in agriculture. One is adoption of resilient technologies in general, and the other is management options within the season. Research, maintenance cum management and promotion of suitable technologies is required in both these options.

9.6.1 Research and Development

- Undertake long-term research into climate change impacts on crop growth and yield, pest and weed growth, water availability, soil moisture and organic matter, rate of evaporation etc.;
- Conduct vulnerability and impact assessments, including assessments of vulnerability under a range of plausible future scenarios

- Development and planting of climate-hardy indigenous cultivars tolerant to drought, thermal extremes, salinity and pests, and cultivars that can utilize water and nitrogen efficiently
- Development or identification of agri-biodiversity sites and support farmers developing indigenous varieties
- Research on carbon sequestration potential of carbon-deficient dry land soils
- Further explore development of resource-efficient dry land techniques

9.6.2 Maintenance and management of resilient procedures and technologies

- Establishment of a seed storage network for climate-hardy cultivars
- Development of tissue culture labs to enhance the germ plasm conservation capacity
- Identification of agro-heritage sites such as the Varada river basin for flood resistant rice, areas under *togaribele* in Gulbarga, millet growing areas in arid and semi-arid regions, horticulture in Bijapur

9.6.3 Technology promotion

- Promotion of dry land agriculture techniques such as low tillage, in-situ soil moisture conservation, raised bed, ridges and furrows, compartment bunding, sand mulching, intercropping systems etc.
- Need based contingency crop planning and management
- Promotion of agro-forestry
- Spreading awareness among farmers of the benefits of multifunctional agro-forestry, domestication of relevant species

9.6.4 Action plan for climate change

Activities are undertaken by Karnataka state government through different stake holders.

- Provision of crop insurance under the National Agricultural Insurance Scheme and (NAIS) and Weather Based Crop Insurance Scheme (WBCIS)
- *BhooChetana* (Land improvement) programme
- Promotion of organic farming

- *Sookshma Neeravari* (Micro irrigation) scheme
- Implementation of the *Rashtriya Krishi Vikas Yojana* (National Agriculture Develop-ment Scheme)
- Implementation of the *Rashtriya Krishi Bima Yojana* (National Agricultural Insurance) programme
- Implementation of the *Mathsya Mahila Swavalambana Yojane (Financial support for fisherwoman)* scheme
- Incentives for fish farming, construction of fish markets and *Mathsyashraya* (housing for fisher folk) programme
- Implementation of the *Suvarna Gramodhaya (Village development)* scheme
- Implementation of the *Suvarna Bhoomi (Land prosperity)* scheme
- Implementation of the *Shramashakti (Skill development)* scheme for skills improvement of artisans and traditional workers
- Implementation of the National Horticulture Mission
- Maintenance and interpretation of historical data on and gradual changes in climate and cropping pattern. This will enhance the understanding on the drivers of change in cropping pattern (environmental, financial or other) and may be used as fundamental principles to formulate policies on change in cropping pattern.

Other policy matters involved are:

- Establishing a State Level Policy body for devising cropping shifts
- Promotion of Dry land farming
- Creation of market for indigenous crops
- Capacity building

9.6.5 Climate Smart Agriculture

9.6.5.1 Research activities for climate resilience

Considerable efforts are made for technology adoption, identifying mitigation strategies, but there is need to identify role of weather conditions in crop yield parameters, or identify crops and crop varieties that can sustain adverse weather situations of temperature, low atmospheric humidity etc. In this connection, a few examples are presented based on experiments at Regional Agricultural Research Station, Vijayapura and Main Agricultural Research Station, Dharwad of University of Agricultural Sciences Dharwad.

9.6.5.2 Weather conditions for different crops and varieties

Groundnut: Out of the seven groundnut genotypes (Dh-3-30, GPBD-4, JL-24, K-134, DSG-1, ICGV-86325 and Kaushal) used in the experiments of AICRP on groundnut, the variety GPBD-4 performed well against adverse conditions of higher maximum temperature and lower relative humidity.

Soybean: From the analysis of four genotypes of Soybean (DSb-1, KHSb-2, PK-1029 and JS-335) experiment conducted by the AICRP on Soybean, it was noticed that yield was influenced by the weather parameters that prevailed during reproductive period and also total crop period influenced the yield. Higher maximum temperature during the crop duration reduced yield of all the genotypes. In case of genotypes PK-1029 and JS-335 higher minimum temperature during reproductive phase was unfavourable, while higher morning humidity was favourable. The yield of genotypes DSb-1 and KHSb-2 was less influenced by weather parameters, and hence considered more versatile with respect to climate.

Sunflower: The important results of sunflower experiment conducted by AICRP on Agrometeorology are given here that, consistently low minimum temperature of less than 17 °C during 7th to 9th week after sowing (WAS), consistently low maximum temperature of less than 30.5 °C in WAS 5- 7, resultant higher diurnal temperature range of more than 12 °C during WAS 6-9 influenced sunflower crop adversely and yield levels were considerably reduced. The growing degree day profile, on the other hand, indicated that the values should be moderate during the first seven weeks for the crop to provide high yields. That is, the either higher or lower values would be adverse during the period. Continuous dry spell WAS 5-9 is the most important weather condition to reduce the sunflower yield to less than 600 kg/ha. This period, coinciding with flower bud initiation and flowering stages, is identified as the most critical period towards reduction in sunflower yield. These inferences can be used as inputs for weather based crop insurance programs.

Pigeon pea: The experiment on pigeon pea by AICRP on Agrometeorology at Vijayapura indicated that decrease in afternoon humidity during flowering stage by 15% and during pod formation and grain filling stage by 17 % , and low rainfall were favorable for reduction in pigeon pea yield from 1300 kg/ha to 450 kg/ha. An example of role of higher temperature range (>14 °C) in flowering stage in reducing the yield of pigeon pea crop is shown in Fig.9. 4.

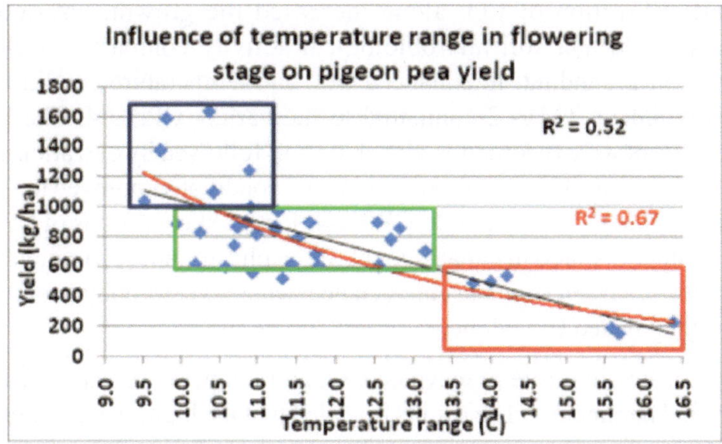

Fig. 9.4: Temperature range in flowering stage and yield of pigeon pea

9.6.5.3 Crop varieties for elevated CO_2 and temperature

Experiments were conducted by Institute for Agricultural Research on Climate Change (IARCC) at UAS Dharwad on crop varieties in controlled conditions of open top chambers having the environments of ambient temperature, elevated carbon dioxide alone, elevated temperature alone by 2 °C and elevated CO_2 and temperature together. Yield, physiological, pest and disease incidence and soil microbial parameters were monitored in these experiments.

Chickpea: The results on chickpea crop in *rabi* season indicated that the varieties JG-14 and BGD-103 recorded highest plant growth, plant physiological, soil microbiological activity and yield compared to rest of the genotypes under elevated temperature condition. Elevated conditions of CO_2 alone resulted in profuse growth, whereas highly stressed situation was noticed when temperature alone was enhanced (Fig. 9.5).

Elevated CO_2 Elevated Temp

Fig. 9.5: Response of chickpea to elevated CO_2 and temperature

Groundut: Elevation of CO_2 alone increased the growth, physiological parameters, yield and soil microbiological activity compared to elevated temperature in groundnut. In enhanced CO_2 and temperature conditions, the yield was higher in TMV- 2 compared to the variety GPBD-4. The highest yield parameters were recorded in elevated CO_2, followed by elevation of both CO_2 and temperature. The altered environmental conditions improved the growth parameters of TMV-2 compared to GPBD-4. Elevated conditions of both CO_2 and temperature resulted in enhancement of the physiological parameters viz., photosynthetic rate and stomatal conductance, whereas chlorophyll content was highest with elevated CO_2 alone.

In the first of its kind, role of elevated CO_2 and temperature was assessed on soil microbial parameters. The results revealed that the highest soil water stable aggregate, organic carbon, inorganic carbon, available P, K and S were recorded with elevated CO_2 whereas soil pH, EC and available nitrogen did not vary significantly. Also, the population of soil bacteria, fungi, actinomycetes, free living nitrogen fixers, fluorescent pseudomonads and P- solubilizers was significantly increased in elevated CO_2 condition, compared to others. Highest nodule number and nodule weight were noticed in elevated CO2 condition, followed by the case of combined enhancement of CO_2 and temperature. The soil enzyme activity viz., dehydrogenase, urease, cellulase, acid phosphatase, alkaline phosphatase, soil biomass carbon and nitrogen were greatly increased by elevated CO_2 condition. The results provide the response of favourable microorganisms to changes in climate change factors. Thus the type of information generated at the UAS Dharwad, will help identify crops suitable to grow, or not suitable to grow (avoid), in specific weather situations /climate, use of suitable microorganisms.

9.6.5.4 Weather based crop management and capacity building of farmers

While it is important to develop and identify technologies for climate resilience, it is of utmost importance to make the farmers aware of the problems and make them adopt preventive measures than for curative measures. Considerable efforts have been made in this connection in north Karnataka.

Climatology helps the Government towards planning and preparedness whereas, in-season weather forecasts and contingency measures guide actual implementation of programs that are remedial or preventive measures. The Gramin Krishi Mausam Seva project (GKMS), AICRP on Agrometeorology (AICRPAM), AICRPAM-NICRA and RKVY funded Climate Change Project of UAS Dharwad have made efforts in making the farmers aware of the problems of climate change, variable weather and its extremes through frequent farmers awareness programs. The awareness is brought into focus also during trainings organized by Krishi Vigyan Kendreas (KVKs) and Department of Agriculture as also during the annual Krishi Melas through lectures and consultancy services. In spite of this, the knowledge on the role of weather based farming is yet to reach the farmers in a big way. In an attempt to address this issue, the University of Agricultural Sciences Dharwad through its AICRPAM Centre at Vijayapura and the RKVY project on Climate Change at Dharwad started providing weather forecasts since 2008 (group messaging) and since 2009 (bulk messaging) to the registered farmers. The main idea under these activities has been to increase the confidence among the farmers on weather forecasts, and instil in them the realization of the need for weather based farming in the coming years, in view of the expected recurrence of various adverse weather events in the climate change eventualities. Telephonic advice also is provided by the Agrometeorologist. These weather forecast messages were followed by messages on crop advisories from KVK (2013) and GKMS (2015). All these have also been incorporated in the University Help Line of UAS Dharwad – 1800 425 1150 since 2016. The effort of the Agrometeorology group in UAS Dharwad gained considerable popularity among farmers, and has become a very important conduit in facing and solving their weather related problems.

Weather forecast based management is very important for reducing the impact of adverse situations. It becomes a preventive management rather than curative, and therefore improves sustenance. Towards this, the seasonal forecasts of India Meteorological Department and SASCOF were utilized in the pre-season crop planning of the Government and of the Agricultural Universities in Karnataka in both the below-normal rainfall years of 2014 and 2015. Based on the analysis performed by the UAS Dharwad, additional information was provided for north Karnataka, according to which, in years of El Nino, the North Interior Karnataka would receive below normal rainfall (*Venkatesh et al, 2015*). Also, the information from ITK in north Karnataka was adopted, which says that if rainfall occurs in *Ashvini Nakshatra* period (April 14-26), the rainfall would be poor during rainy season. All these factors enabled the Government to plan and take up many drought management practices for crops in the drought years of 2014 and 2015. This multi-faceted approach helped in preparing the farmers to face the drought and making them less vulnerable. In order to help the farmers at a higher resolution of a block (Taluk), Microlevel Agromet Advisory Service was started through NICRA project in Belagavi district by Vijayapura Centre of AICRP on Agrometeorology. In this, block level forecasts, ground truth of crop and experienced weather were taken into account for issue of block level advisories and disseminate the same through field information facilitators who would provide the bulletins to the selected farmers.

Vijayapura Centre of AICRPAM and the Institute for Agricultural Research on Climate Change based at UAS Dharwad provided farmer oriented weather forecasts for seven districts under the umbrella of UAS Dharwad. The information was disseminated by SMS through a dedicated control panel to 25000 farmers. The district level advisories are also sent to Karnataka State Disaster Monitoring Centre (KSNDMC) for further transmission of messages to lakhs of their registered farmers. In 2016, the first ever Agromet Mobile App "havaamaana krishi" was launched by AICRPAM Vijayapura Centre. With all these facilities, innovations for dissemination of forecasts and advisories on realtime basis, UAS Dharwad has created a platform for their adoption by more and more farmers. Indeed a multi-dimensional approach is adopted for success and sustenance of the farmers in north Karnataka. Sustained base-work has been done in this connection. When this adoption becomes a regular feature, then our farmers will be the first to be ready for adaptation to climate change eventualities. Invigorating the farmers towards timely adoption of suitable technologies is nothing but climate smart agriculture. Weather forecasts and climate projections are essential feature of this phenomenon. Thus, it is seen that different government agencies have put in efforts to face climate change in different manner – some have done ground level work, whereas some others have made the planning and yet some others implemented at different spatial levels. In spite of the repeated occurrence of adverse weather and climate, farmers in north Karnataka have braced with them successfully, and are in the process of acclimatization to the events.

9.6.5.5 Climate resilience through weather based farming

More importantly, the rainfall climatology including probabilities and variability patterns and weather forecasts should always be the basis on which the crop planning including contingent crop plan and contingent measures to meet the exigencies under aberrant weather conditions in dry land agriculture should be considered to advise the farmer. The possible ways of overcoming any aberrant rainfall situations are: Choice of crops and cropping systems appropriate for the prevailing condition to adopt suitable planting geometry and plant density; adoption of off-season tillage and repeated intercultivation during crop season; application of fertilizer as per recommendation – both in dosage and timing; adoption of suitable *in situ* moisture conservation practices, adoption of weather based integrated pest and disease management practices, adoption of mid season correction like thinning of plant population, removal of alternate rows, ratooning of the crop, stripping of old and non functional leaves, adoption of integrated nutrient management, providing the crop life saving irrigation wherever facilities are available etc. The contingency crop plans should be implemented in tune with the micro-level rainfall characteristics as well as medium and extended range forecasts.

Summary

Karnataka is one of the most vulnerable regions of India to climate change. This chapter has discussed various components of climate change – including climate variability, weather extremes, steps taken by the Government of Karnataka to face the imminent climate challenges, efforts of Karnataka State Natural Disaster Monitoring Centre, University of Agricultural Sciences Dharwad towards vulnerability assessment, mitigation strategies, adverse weather conditions for crops, crop and varietal choices in different climate scenarios, preparedness to face the exigencies climate change. It is noticed that more work at ground level is required to be done towards identifying genetic traits of crops to face the challenges posed by climate change.

References/Suggested Readings

Anandhi, A., 2010. Assessing impact of climate change on season length in Karnataka for IPCC SRES scenario. *Journal of Earth System Science*, 119 (4), 447-460).

Anonymous, 2015. Annual Report (2014) of AICRP on Agrometeorology, Vijayapura Centre.

BCCI-K, 2011. Karnataka Climate Change Action Plan; Final Report.

Biradar Nagaratna and K. Sridhar, 2009. Consequences of 2003 Drought in Karnataka with Particular Reference to Livestock and Fodder. *J Hum Ecol*, 26(2): 123-130.

INCCA, 2010. Climate Change Assessment Report, Ministry of Environment and Forests, Government of India.

The Hindu, Bangalore Edition (June 11, 2010). Disaster response force for State .

Venkatesh H., 2003. *Rainfall Analysis* – As a tool in watershed development in Northern Dry Zone of Karnataka. University of Agricultural Sciences, Dharwad. Tech. Bulletin: 26, ISBN : 81-88367-04-4.

Venkatesh H., Krishna Kumar K, Bellakki M.A., Aski S.G., and S.M. Warad, 2003. Adaptation strategies for farmers in case of adverse impact of climate change on sorghum cultivation in North Karnataka. Proceedings of the Workshop on vulnerability Assessment and Adaptation due to Climate Change on Indian Agriculture, Forestry and Natural Ecosystems. Indian Institute of Science, Bangalore, July 18-19, 2003, pp. 54-67.

Venkatesh H., Kulkarni S.N., Guled M.B., Surakod V.S., and M.S. Shirahatti, 2008. Agroclimatic Atlas of Bijapur and Bagalkot districts. AICRP on Agromteorology Bijapur Centre. Tech. Bulletin. University of Agricultural Sciences, Dharwad. 167 p.

Venkatesh H., Patil B.N., Rao V.U.M., Rajput Rajani B., Naikodi, S.M., Hiremath J.R. and P. Vijayakumar, 2015. El Nino: Its influence on rainfall and crop production in North Interior Karnataka.Technical Bulletin. AICRPAM NICRA, Vijayapura Center and IARCC, UAS Dharwad . 22p.

Venkatesh H., Shivaramu H.S., Rajegowda M.B. and V.U.M. Rao, 2016. Agroclimatic Atlas of Karnataka. AICRPAM, University of Agricultural Sciences, Dharwad – 580 005. 240 p.

10

Climate Change and Agriculture in Andhra Pradesh: Vulnerability, Adaptation and Farmers' Perceptions

C.A. Rama Rao*, B.M.K. Raju and Ch. Srinivasa Rao

ICAR-Central Research Institute for Dry land Agriculture, Santoshnagar Hyderabad – 500059, India

The State of Andhra Pradesh in the present form came into being on June 2, 2014 as a result of the Andhra Pradesh Reorganization Act. The state has 13 districts out of which nine are along the coast of Bay of Bengal and four are land locked in the southern part. A part of Khammam district, which is now in the Telangana state, was also added to the East and West Godavari districts of the State of Andhra Pradesh. The existing data for these two districts do not include this part and this fact is to be borne in mind. The state has a geographical area of 1.602 lakh km^2 with a population of 49.83 million. The state in the current form is more agrarian compared to the undivided state as about 62 per cent of population depends on agriculture as a source of livelihood. Yet, the contribution of agriculture to the state's gross domestic product is only 17 per cent. Though the state is known as 'rice bowl of India', not all is well with agriculture in the state. Some parts, especially the coastal districts, are prone to incidence of cyclones and floods while some other districts, Rayalaseema districts, are chronically drought-prone. The observation of 'crop holiday' when farmers chose not to raise rice in the two Godavari districts in the recent past is a pointer to the problems associated with agriculture in the state. Land degradation associated with indiscriminate use of chemical inputs, sea water ingression, inadequate and undependable rainfall is some of the major problems that cause low productivity of agriculture in the state. Imperfections in the market in terms of inadequate access to markets and market information, inadequacies in market and processing infrastructure are some of the major impediments to realizing the potential returns from crop and animal production. The rising cost of inputs is another detriment to the profitability of agriculture. Aggravating all this problems is the climate change that further brings in the urgency to put in place a variety of measures that help improve productivity and profitability of agriculture. Recognizing all these dimensions, even the strategy document on transforming agriculture in Andhra Pradesh laid emphasis on making agriculture climate smart (GoAP, 2014).

The State consists of a wide range of agro-climatic situations. The Krishna and Godavari zones have high potential in terms of agriculture, the scarce rainfall zone of Rayalaseema presents a challenging situation to making agriculture sustainable. The north coastal zone is unique in the sense that it is relatively well endowed with natural resources and is lagging in the overall socio-economic development.

10.1 Need for building climate resilience and vulnerability analysis

As opposed to climate smart agriculture, the phrase 'climate resilient agriculture' gives relatively more importance to 'adaptation' with a view to protect and enhance productivity in the face changing climate. Climate change is characterized by increasing temperatures (especially night temperature), changes in precipitation/ rainfall pattern, increasing frequency of incidence of extreme weather events such as drought, flood, heat wave, cold wave. Various climate models and the IPCC's Fifth Assessment Report observe that the world is already committed to a certain amount of global warming irrespective of the intensity of mitigation efforts because the benefits of the current mitigation efforts will be realized only after a few decades. What is important is that climate is a spatially variable phenomenon and hence understanding what is likely to happen in future is a useful beginning of planning for measures for making agriculture more resilient to climate change. An appropriately planned assessment of vulnerability to climate change helps identify the regions that are relatively more vulnerable to climate change. Vulnerability of an entity, as defined by the IPCC, is a function of sensitivity, exposure and adaptive capacity. The first two components of vulnerability, sensitivity and exposure determine the potential adverse impact of climate change which can be moderated by adaptation determined by adaptive capacity of the entity. Sensitivity is defined as "the degree to which a system is affected, either adversely or beneficially, by climate-related stimuli". It is determined by demographic and environmental conditions of the region concerned. Exposure is defined as "the nature and degree to which a system is exposed to significant climatic variations". Thus, exposure relates to climate stress upon a particular unit of analysis (Gbetibouo and Ringler 2009). "A more complete measure of exposure to future climate change would require consideration of projected changes in climate in each analysis unit" (Eriyagama *et al.*, 2012). Adaptive capacity is "the ability of a system to adjust to climate change, including climate variability and extremes, to moderate potential damages, to take advantage of opportunities, or to cope with the consequences". It is considered to be "a function of wealth, technology, education, information, skills, infrastructure, access to resources, stability and management capabilities" (McCarthy *et al.*, 2001). Thus, an analysis of vulnerability helps identify, prioritize and target investments and interventions in terms of research, extension, development and policy measures. Table 10.1 presents the relative degree of vulnerability along with the major determinants of vulnerability in case of the thirteen districts of Andhra Pradesh. Sensitivity and adaptive capacity indices were computed using the data on a range of agro-climatic and socio-economic indicators whereas the exposure index was computed using the climate projections made using the PRECIS for the scenario A1B for the period 2021-50. Further details are available in Rama Rao *et al.*, 2013. It is observed from the table that all the four districts of Rayalaseema were found to be more vulnerable to climate change. The three districts of Chittoor, Anantapur and Kurnool were also categorized as districts with 'high/ very high' vulnerability even at the country level analysis (Fig. 10.1). A look at the relative position of the districts in terms of the components and the factors leading to vulnerability gives useful guidance as to where and how the investments and interventions are to be planned.For example, a decrease injuly rainfall is projected for most districts and this decrease is likely to be highest in Chittoor district. Therefore, it is necessary to alter crop calendars or to explore the possibilities of protecting the crops during prolonged drier month, or a combination of both. Research efforts should take into consideration such possible changes in the rainfall patterns. Similarly, low rainfall and high net sown area as a proportion to geographical area were found to be sensitivity-related factors contributing to vulnerability. It is to be noted here that rainfall, which is relatively lower in the Rayalaseema districts, is also projected to decrease underlining the growing importance to the need for water harvesting as well to devise technological and management options that facilitate more equitable and efficient water use. On the other hand, a higher share of the geographical area being cultivated indicates higher dependence on agriculture of the population warranting the attention towards more sustainable land use and also towards enhancing the opportunities for non-farm employment. The most important adaptive capacity factor is related to irrigation either in the form of area having access to irrigation or in the form of groundwater availability. Investments are therefore needed in expanding the irrigation facilities. As indicated earlier, a more efficient use of irrigation water is also equally important in realizing the irrigation potential.

Another important dimension of changing climate is the projected increase in both maximum and minimum temperatures. The projected changes ranged from 1.3°C in East Godavari to 2.0°C in Kurnool in case of maximum temperature and from 1.83°C in East Godavari to 2.17°C in Nellore in case of minimum temperature. Increasing minimum temperatures can adversely impact the productivity of rabi crops. Such changes are likely to affect crop growth and productivity and have to be factored in setting up the research agenda for the future. Thus, better water management and development of crop varieties that match the changing climate are two important parts of any strategy to make Andhra Pradesh agriculture more climate resilient. There are already some promising experiences generated as part of the National Initiative on Climate Resilient Agriculture (NICRA) and elsewhere in this regard.

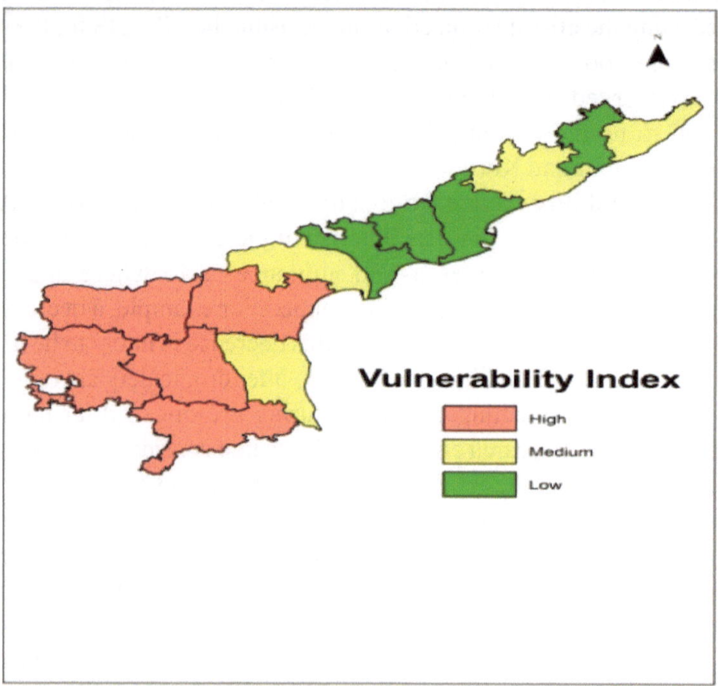

Fig. 10.1: Vulnerability of Agriculture to Climate Change

10.2 Some successful examples

10.2.1 Collectivization of groundwater irrigation

An attempt was made to achieve a more efficient and equitable use of existing groundwater resources by appropriate social and technical interventions in a multi-institutional initiative with support from National Agricultural Innovation Project. This consisted of networking of bore wells into a single system and providing access to irrigation water to those who did not possess any irrigation source. The results indicated that when the existing bore wells were networked and connected to a more efficient irrigation system like sprinkler, there were

Table 10.1: Relative vulnerability and determinants in districts of Andhra Pradesh

District	Rank based on				Most important factor contributing to		
	Vulnerability	Exposure	Sensitivity	Adaptive Capacity	Exposure	Sensitivity	Adaptive Capacity
Chittoor	2	1	7	8	Projected decrease in July rainfall	Low Rainfall	Low groundwater availability
Anantapur	1	2	1	13	Projected decrease in July rainfall	Low Rainfall	Low NIA
Kurnool	3	3	3	12	Projected decrease in July rainfall	Low Rainfall	Low NIA
Kadapa	4	5	2	9	Projected decrease in July rainfall	Low Rainfall	Low groundwater availability
Prakasam	5	8	8	10	Projected decrease in July rainfall	Low Rainfall	Low NIA
Guntur	6	4	4	3	Projected decrease in July rainfall	Low Rainfall	Low groundwater availability
Srikakulam	8	7	10	6	Projected increase in minimum temperature	High NSA availability	Low groundwater
Visakhapatnam	9	11	13	11	Projected increase in minimum temperature	Low Rainfall	Low NIA
Nellore	7	6	9	5	Projected decrease in July rainfall	Low Rainfall	Low groundwater availability
East Godavari	10	9	11	4	Projected increase in minimum temperature	High NSA	Low groundwater availability
Vizianagaram	12	12	12	7	Projected increase in minimum temperature	High NSA availability	Low groundwater
West Godavari	11	10	5	1	Projected decrease in July rainfall	High NSA	Low groundwater availability
Krishna	13	13	6	2	Projected decrease in July rainfall	High NSA	Low groundwater availability

Source: For methodology and measurement of indicators see Rama Rao et al., 2013 and Rama Rao et al., 2016.
NSA- Net Sown Area; NIA- Net Irrigated Area

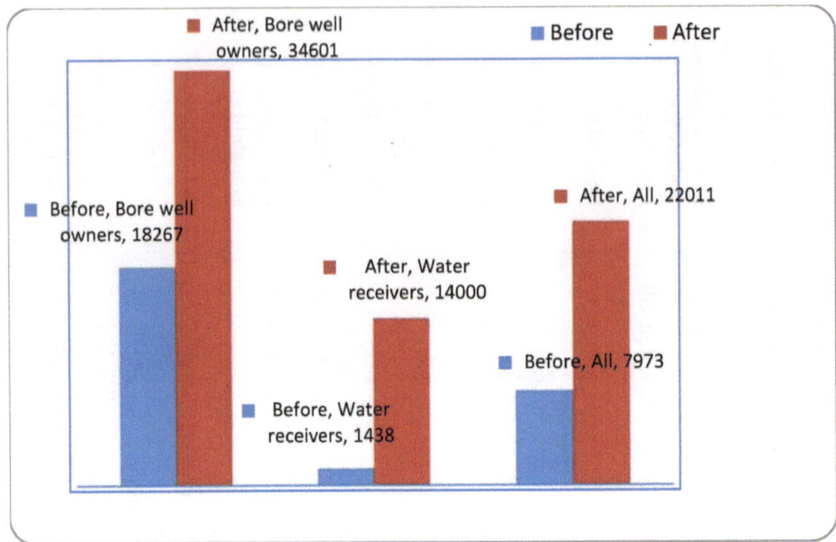

Fig 10.2: Net returns from crop production before and after collectivization of groundwater use (Rs/HH)

significant gains in terms of expansion of cultivated area, increase in cropping intensity and higher crop yields (Fig.10.2). Though both the owners of bore wells and water receivers gained absolutely, the latter gained more in relative terms. The intervention also led to more efficient and equitable use of groundwater. This initiative exemplifies the feasibility and appropriateness of community shared irrigation facilities and highlights the need for well coordinated efforts by different organizations to be effective in achieving the common goals. These efforts were made in Anantapur and Mahabubnagar districts of undivided Andhra Pradesh.

10.2.2 Interventions for yield and income resilience

In an attempt to enhance resilience of crop production to climate variability, various adaptation interventions were evaluated in Anantapur and West Godavari districts as part of NICRA. It was observed that the adaptation intervention in the form of change of crop variety (Indra in place of Swarna) performed better in terms of higher yield and net returns in West Godavari district (Table 10.2). The yield level even during a flood year like 2013-14 is much closer to the normal yield indicating enhanced resilience to flood situation. The yield of this variety was found to be on par with that of Swarna during the normal rainfall year as well indicating that adoption of this (Indra) variety may be a better option so that yield losses are minimized even in the event of occurrence of a flood. Another major finding is that income resilience is betterwhen the livelihoods are more diversified. Even within agriculture, a more diverse cropping pattern may lead to more stable crop income. However, the analysis also hinted at possible lower income with more diversification in 'normal' years. Similarly, various technological adaptation interventions were found to protect yields during a drought year in Anantapur and thus enhance resilience of yield to drought (Table 10.3). The adaptation interventions viz., opening conservation furrows, use of seed drill and soil test based fertilizer application methods were found to enhance yield and income resilience to drought in Anantapur district.

Table 10.2: Effectiveness of an intervention towards adaptation to flood incidence – Introduction of flood tolerant variety: MTU 1061in West Godavari, AP

Variety	Normal year yield (q/ha)	Flood/ heavy rain year yield (q/ha)	Resilience(Stress yield/normal yield
MTU-7029 (Swarna)	21.45 (3.91)	10.15 (3.68)	0.47
MTU-1061 (Indra)	23.47 (3.78)	13.96 (5.10)	0.59

Figures in parentheses are standard deviations.

Table 10.3: Adaptation and resilience to drought in Anatapur

Yield resilience	Yield, q/ha	Yield lost, q/ha	Resilience
Groundnut+Ppea			
Normal	9.94		
No adaptation/FP	2.08	7.86	0.21
Conservation furrow	2.81	7.13	0.28
Seed drill & thresher	4.11	5.83	0.41
STBFA	3.82	6.12	0.38
Income resilience	NR, Rs./ha	NR lost, Rs./ha	Resilience1
Normal	25729		
No adaptation/FP	-11839	37568	-0.46
Conservation furrow	-3462	29191	-0.13
Seed drill & thresher	713	25016	0.03
STBFA	592	25137	0.02

10.2.3 Farmers' perceptions on climate change

A survey of a sample of 45 farmers each in Srikakulam and Anantapur districts of Andhra Pradesh was conducted to understand the perceptions on climate change. It showed that most farmers believed that climate was changing and the change was being manifested in the form of decrease in seasonal rainfall, untimely rains, increasing incidence of drought and increasingly hotter days and nights. Some of these perceptions were in agreement with what is observed through historic data. The increasing temperatures also reflect the projections of various climate models. Farmers were also found to resort to a number of coping ways mostly in the form of consumption adjustments, but increased borrowing was also found to be a coping mechanism leading to indebtedness.

10.2.4 Growth performance in the most vulnerable districts in Andhra Pradesh

Table 10.4 presents the growth performance of major crops during the last decade in the five most vulnerable districts. In Chittor, groundnut is the most important rainfed crop. However, between 2000-01 and 2011-12, area sown decreased from 1.76 to 1.49 lakh ha. The yield of groundnut was found to increase at a rate of 1 per cent annually. The productivity growth in incase of two irrigated crops, rice and sugarcane, were found to be better though the area under the crops also witnessed a decline. Considering the importance of groundnut in ensuring nutritional and livelihood security, it is essential to take necessary measures to enhance the productivity levels in the district. The situation is even worse in case of groundnut in Anantapur as the yield levels were found to decline at an annual rate of 3.56 per cent leading to considerable production decline. Productivity of other important rainfed crops such as pigeonpea, sorghum and castor also recorded significant decline. There was some growth in productivity of crops such as chickpea, sunflower and maize in the district. Cotton, maize and chillies recorded considerable growth in productivity in Kurnool district known for seed production in the state. As in case of other districts, productivity of groundnut decreased in this district also. Growth in productivity of chickpea was not considerable. It is to be noted that area under most crops decreased. But area under chickpea, rice, castor, pigeon pea and maize increased. Increasing area under maize is witnessed in other states as well. Productivity in most of the crops in Kadapa and Prakasam recorded significantly positive growth rates with the exceptions of chickpea, sunflower and pigeonpea in Kadapa and blackgram in Prakasam. The area and productivity of sesamum increased considerably leading to an annual growth rate of 27.26 per cent in its production in Prakasam.

Thus, a detailed analysis of the reasons for slow growth or yield decline in certain crops will help initiate measures for enhancing productivity. It is also equally important that the growth rates in other crops are sustained into future even in the face of changing climate.

Summary

The formation of Andhra Pradesh presents an opportune time to have a critical look at the opportunities for and challenges to the growth required in agriculture as the state is predominantly agrarian. Among various factors that threaten the sustainability of agriculture in general and rainfed agriculture in particular is the changing climate which can act as a major impediment to achieving the required growth in productivity, production and profits. Therefore, a regionally differentiated approach is needed as most of the problems and therefore the

Table 10.4: Changes in the area, production and yield of major crops in the most vulnerable districts of Andhra Pradesh, 2000-01 to 2011-12

District / Crop	Area (lakh ha)		Production (lakh ton)		Yield (kg/ha)		CAGR (%)		
	TE 2002-03	TE 2011-12	TE 2002-03	TE 2011-12	TE 2002-03	TE 2011-12	Area	Production	Yield
Chittoor									
Groundnut	1.76	1.49	1.89	1.50	1074	1008	-0.95	0.09	1.04
Rice	0.71	0.56	1.80	1.75	2548	3125	-1.15	2.31	3.52
Sugarcane	0.40	0.28	32.40	23.83	81838	86009	-2.02	-0.54	1.51
Anantapur									
Groundnut	7.81	7.06	5.17	2.77	662	392	-0.46	-4.11	-3.56
Chickpea	0.45	0.85	0.31	0.73	684	855	6.88	13.63	6.29
Rice	0.58	0.53	1.55	1.47	2690	2762	1.94	2.77	0.75
Pigeonpea	0.31	0.46	0.08	0.07	246	156	3.53	-0.73	-4.29
Sorghum	0.18	0.30	0.24	0.21	1291	687	0.23	-7.02	-7.12
Sunflower	0.47	0.29	0.27	0.20	564	700	-11.34	-5.03	7.03
Maize	0.04	0.20	0.11	0.95	2558	4648	18.27	27.23	7.66
Castor	0.05	0.12	0.02	0.02	326	182	13.04	8.83	-4.31
Kurnool									
Chickpea	1.06	2.19	1.13	2.51	1074	1147	7.89	8.54	0.58
Groundnut	2.12	1.57	2.03	1.07	953	682	-3.06	-7.38	-4.43
Rice	0.83	1.24	2.31	3.97	2791	3215	8.55	11.67	2.87
Sunflower	1.41	0.84	1.11	0.56	788	672	-10.95	-10.89	0.11
Castor	0.10	0.74	0.03	0.29	282	393	29.43	28.53	-0.74
Sorghum	0.97	0.70	1.31	1.36	1356	1928	-5.02	-1.92	3.24
Cotton	0.85	0.62	0.17	0.16	196	260	-1.53	5.12	6.59
Pigeonpea	0.33	0.51	0.14	0.20	415	398	5.46	3.94	-1.47
Maize	0.02	0.22	0.06	1.14	2490	5302	32.65	43.99	8.61
Chillies (dry)	0.13	0.15	0.17	0.35	1273	2370	1.45	7.67	4.94
Blackgram	0.10	0.14	0.05	0.08	468	554	5.89	7.37	1.55
Tobacco	0.11	0.13	0.12	0.22	1098	1695	0.09	6.57	4.60

Contd.

District / Crop	Area (lakh ha)		Production (lakh ton)		Yield (kg/ha)		CAGR (%)		
	TE 2002-03	TE 2011-12	TE 2002-03	TE 2011-12	TE 2002-03	TE 2011-12	Area	Production	Yield
Kadapa									
Groundnut	1.41	1.16	1.07	0.98	762	850	-3.46	5.16	8.90
Chickpea	0.43	0.79	0.40	0.63	946	801	7.97	6.61	-1.30
Sunflower	0.57	0.63	0.42	0.39	733	618	-0.10	-3.71	-3.58
Rice	0.60	0.63	1.48	1.59	2469	2516	3.01	3.28	0.23
Cotton	0.21	0.20	0.04	0.06	205	304	0.03	8.33	8.07
Pigeonpea	0.14	0.14	0.06	0.03	429	244	-4.16	-9.40	-5.88
Sorghum	0.09	0.10	0.07	0.11	810	1049	3.52	15.54	11.77
Prakasam									
Rice	1.20	1.35	3.81	4.78	3180	3551	5.64	8.46	2.66
Chickpea	0.48	0.88	0.83	1.42	1738	1614	5.12	4.59	-0.48
Tobacco	0.59	0.79	0.59	1.31	1000	1654	3.89	8.64	4.57
Pigeonpea	0.71	0.59	0.34	0.23	478	384	-2.01	-1.89	0.14
Cotton	0.36	0.47	0.09	0.20	238	417	4.67	12.96	7.93
Chillies(dry)	0.18	0.22	0.35	0.78	1888	3496	3.24	11.39	7.89
Blackgram	0.44	0.20	0.23	0.09	521	436	-5.59	-6.60	-1.03
Sunflower	0.09	0.20	0.10	0.21	1152	1084	6.61	6.10	-0.40
Sesamum	0.11	0.19	0.02	0.03	153	159	6.37	27.26	19.96
Greengram	0.24	0.12	0.07	0.04	298	328	-13.45	-6.72	6.78
Pearl millet	0.16	0.12	0.22	0.26	1331	2207	-1.97	5.61	7.62

TE- Triennium Ending; CAGR- Compound Annual Growth Rate

solutions are spatially variable. Even the relative importance of a given factor varies spatially. The vulnerability analysis helps identify and target appropriate interventions for making agriculture more resilient to changing climate. It is however important to note that any measure that helps farmers cope with the current climate variability will only place them in a better position to deal with future climate change. Accordingly, what is good at present, especially such practices as integrated nutrient and pest management, better water management, will remain relevant even with climate change as these practices aim at reducing use of purchased inputs and save on cost of cultivation. These practices have to be supported by developing improved crop varieties that thrive better in the changed climate. Also required are adjusting crop calendars, sowing and harvesting time, etc., in view of the changes in rainfall patterns. Appropriately designed insurance products that help smoothen income and consumption patterns during a weather-shock and support technology adoption are also need of the hour. Investments are therefore needed in developing appropriate infrastructure and institutions that help generate and implement the strategies for a more resilient agriculture.

References/Suggested Readings

Eriyagama N, V. Smakhtin, L. Chandrapala and K. Fernando (2012). Impacts of climate change on water resources and agriculture in Sri Lanka : A Review and Preliminary Vulnerability Mapping. IWMI Research Report 135. pp.43

Gbetibouo, G. A and C. Ringler (2009). Mapping South African farming sector vulnerability to climate change and variability: A sub-national assessment.IFPRI Discussion Paper 00885. Washington, DC, USA: International Food Policy Research Institute (IFPRI).

GoAP (2014) Agricultural Transformation in Andhra Pradesh: Equitable, scientific, prosperous and climate smart - A strategy paper on mission for primary sector.

McCarthy, J. J., O. F. Canziani, N. A. Lear y, D. J. Dok ken, and K. S. White (2001). Climate Change (2001). Impacts, Adaptation, and Vulnerability. Cambridge University, Press, 1032 pp.

Rama Rao C A, Raju B M K, Subba Rao A V M, Rao K V, Rao V U M, Kausalya Ramachandran, Venkateswarlu B and Sikka A K (2013). Atlas on Vulnerability of Indian Agriculture to Climate Change. Central Research Institute for Dryland Agriculture, Hyderabad P 116.

Rama Rao C A, Raju B M K, Subba Rao A V M, Rao K V, Rao V U M, Kausalya Ramachandran, Venkateswarlu B, Sikka A K, Srinivasa Rao, M., Maheswari, M. and Srinivasa Rao, H (2016). A District Level Assessment of Vulnerability of Indian Agriculture to Climate Change, *Current Science*, 110(10) 1939-1946

11

Climate Change Vulnerability and Adaptation Strategies for Maharashtra

Kailas Kamaji Dakhore

Agrometeorologist, AICRP on Agrometeorology, Vasantrao Nike Marathwada Krishi Vidyapeeth, Parbhani 431 401, Maharashtra, Inda

Evidence is fast accumulating that the problem of climate change is of a scale not previously encountered by human societies. The scale and scope of climate change challenges to food and water security are unprecedented. World food supply is already facing significant pressures. Even without climate change, ensuring sufficient food for all in the 21st century would be difficult enough. Climate influences food security through many channels. Climate change will increase the severity and frequency of cyclones, flooding and drought. It will intensify rainfall variability, increasing water stress, weeds, pests, and erosion and reducing soil fertility. Above certain temperature ranges, crops develop more rapidly, resulting in lower overall grain production, even moderate increases in temperature can decrease yields in major food crops. Population growth, urbanisation, higher per capita caloric intake, and increasing consumption of meat combine to put more pressure on exiting water recourses. Tension and possibly conflict over water would most likely occur even in the absence of climate change. But climate change greatly exacerbates the problem. There can be little genuine doubt that the world's climate is changing in ways that will seriously affect the life on the earth. There is a widespread acceptance that the troposphere the lower part of the atmosphere that contains most of the world's weather systems is becoming warmer. The vast bulk of the climate analysis supports the view that warming is due to increased emissions of particularly carbon dioxide and methane resulting from human action. There is also a growing understanding of the negative legacy of past greenhouse gas emission. Even if carbon pollution was stopped today the effects of global warming would continue to be felt for at least 50 years into the future. Climate change has significant implications for the food and water security of many of the world's 6.7 billion inhabitants. These implications are especially severe for the 2 billion who are food insecure and/or suffer physically and intellectual limitations due to inadequate nutrition **(ICTSD, 2009).** In relation to India, current climate variability and future climate change at local levels, given the important climatic variations across time and space. While rainfall variability is a major factor causative to instant fluctuations in agricultural production, to overcome for this it is necessitate to move away from this conventional emphasis by also taking into account hail storms, prolonged dry spells, temperature variability and extreme temperatures. Agriculture is the primary source of revenue in the state of Maharashtra. The state has 17.4 million ha of land under cultivation while 84% of the total area under agriculture is rainfed and is directly dependent on the monsoon. A series on Marathwada's battle with three consecutive years of drought is incomplete without understanding the impact of climate change. The rise in the country's annual temperature by 0.7 °C and erratic climatic patterns are sure to affect the region's agriculture which is primarily rainfed with 12 per cent area under irrigation (Dakhore et al, 2017).

11.1 Area, production and productivity trends of major crops

There is ultimate target of increasing the productivity of major field crops in Maharashtra. On the overall region basis area, production and productivity trends are shown in Figs. 11.1 to 11.13 for the major field crops. Rice and maize showed increasing trend in area, production and productivity while in case of pearl millet decreasing trend is noticed in area and production and productivity is showing increasing trend. Groundnut area and production is decreasing and productivity is increasing. This will give an insight to the planners to go further for micro-level analysis regarding sustainability and profitability of a particular crop and cropping system.

11.1.1 Cereals

Paddy, sorghum and pearl millet are the principal cereal crops of Maharashtra. The area under paddy is about 15 lakh ha and there is a marginal increase in its area over the years (Fig. 11.1). The year to year fluctuations in area are mostly due to the drought and scanty rainfall. Drought conditions from 2010 onwards in the state led to drastic reduction of the area under the crop. The paddy production has consistently increased @ 2.21 lakh tons per year. The paddy productivity has risen from 362 kg ha^{-1} to 1538 kg ha^{-1} during the 55 year period from 1960 to 2015. Sorghum is the second largest grown cereal crop due to its commercial value (as cattle feed). It is grown both in *kharif* as well as *rabi* season. The area under *kharif* sorghum has fallen from 31.1 lakh ha to 6.2 lakh ha in last 55 years (Fig.11.2). Productivity level of *kharif* sorghum has increased from 305 kg ha^{-1} in 1972 to 1425 kg ha^{-1} in 2011. However, in the case of *rabi* sorghum, the area has decreased from 37.8 lakh ha in the 1971 to a meager 15.1 lakh ha in 2013 (Fig. 11.3). To an extent a steep increase in the productivity levels of the crop has come handy to commensurate the sharp decline in the area.

Pearl millet is another important cereal crop grown mostly in semi-arid region of Maharashtra. Area under pearl millet crop in the Maharashtra during 1960-2015 witnessed significant decrease in the area. Maharashtra showed slight ups and downs in the pearl millet production but as such an overall increase in a production was observed from 1992 to 2000 and again there was decline in the production 2001 onwards (Fig. 11.4). Maharashtra showed steady increase in the pearl millet productivity during the period of 1960 to 2015. Maharashtra showed slight variation in the wheat area but as such an overall increase in an area under wheat cultivation was observed from 1960 to 2015 (Fig 11.5). The area under wheat improved slightly in 2006 to 2010 and decreased during the years 2011-12 and 2013-14. Production of wheat showed increasing trend from 1960 to 2015. The yield of rabi sorghum improved steadily from 1960 to 2015, the productivity trend was constant through the period.

11.1.2 Oilseeds

Groundnut, soybean and safflower are the major oilseeds crop of Maharashtra. Soybean is the largest area occupying oilseed crop grown in Maharashtra. Area under soybean crop increased from 0.7 lakh ha in 1987 to 36.4 lakh ha in 2014. Its production and productivity have shown a sharp increase over the years (Fig. 11.6). Major shift that occurred in the past decades in favour of soybean crop was due to its early duration, suitability to double cropping, distinct yield advantage, higher price and better market support. The area under soybean has also substituted more of *kharif* sorghum and some cotton acreage. Groundnut is grown in 2.41 lakh ha. The area under this crop was at its peak during 1960-1966 while it averaged 6.2 lakh ha (Fig. 11.7). The production trend of this crop have shown steep decline over the years due to variation in the spread of rainfall. Similarly, area under safflower is also showing a sharp decline but with a marginal decrease in production. This may be due to the introduction of improved cultivars and production technology (Fig. 11.8).

11.1.3 Pulses

The major pulse crops of the state are red gram, black gram and green gram. Area under red gram has picked up in recent years as a good dry land pulse crop. Introduction of high yielding cultivars has raised productivity of the crop from 321 kg ha^{-1} during 1972 to 810 kg ha^{-1} during the period 2007. Red gram area has risen from 5.30 lakh ha during the period 1960-61 to 13.01 lakh ha in 2010. This rise is partly due to introduction of pulses improvement program and partly to market forces (Fig. 11.9). In other years, the inter-annual variability is largely weather driven. Area, production and productivity of black gram have varied over the years. Though there was a marginal rise in all the three parameters during the second part of the 1990s, the crop was replaced by other remunerative crops mainly due to the occurrence of terminal moisture stress (in paddy fallows) and the incidence of YMV disease (Fig.11.10). Area and production under green gram has showed a marginal ascent in recent years but productive.ty levels have been sustained / enhanced due to use of improved package of practices (Fig. 11.11).

11.1.4 Commercial crops

Cotton is the principal commercial crop of the state. It is cultivated on more than 41 lakh ha. Though productivity trends of the crop show large variability and there has been a constant increase in its area and production chiefly due to the introduction of Bt cotton, market forces and partly due to vagaries in the temporal distribution of monsoon rains. In years of early dry spells in the rainy season, more area is sown for cotton owing to its relative drought tolerance. The area earmarked to other crops which are moisture sensitive, are brought under cotton (Fig. 11.12). Apart from the spatial and temporal distribution of rainfall, insect pest damage is another important parameter that affects cotton productivity in Maharashtra. In Tropical zone Maharashtra is the major sugarcane growing state. The sugar industry is instrumental in generating the sizable employment in the rural sector directly and through its ancillary units. It is estimated that about 50 million farmers and their dependents are engaged in the cultivation of sugarcane and about 0.5 million skilled and unskilled workers are engaged in sugar factories and its allied industries. The sugar industry in India has been a focal point for socio-economic development in the rural areas by mobilizing rural resources, generating employment and enhancing farm income. Some of the sugar factories have also diversified into by-products basis industries and have invested and put up distilleries, organic chemical plants, paper, ice board factories and cogeneration plant. A steep increase in sugarcane area and production has come through with a concurrent increase in the area over the years (Fig. 11.13). Very small inter-annual variation in the productivity of the crop may be attributed primarily to the fact that large area under sugarcane crop is irrigated.

11.2 Climate change projections

Climate change projections for Maharashtra were developed after following a process of domain selection and validation of the model outputs with observations. The domain having a higher correlation with observations was then chosen for future projections. The methodology so followed for domain selection and model runs was unique to the state of Maharashtra. Based on this unique domain selection method that sought to represent the regional climate over Maharashtra to a fairly good degree, rainfall and temperature changes have been projected for three time slices- 2030s, 2050s and 2070s.present the

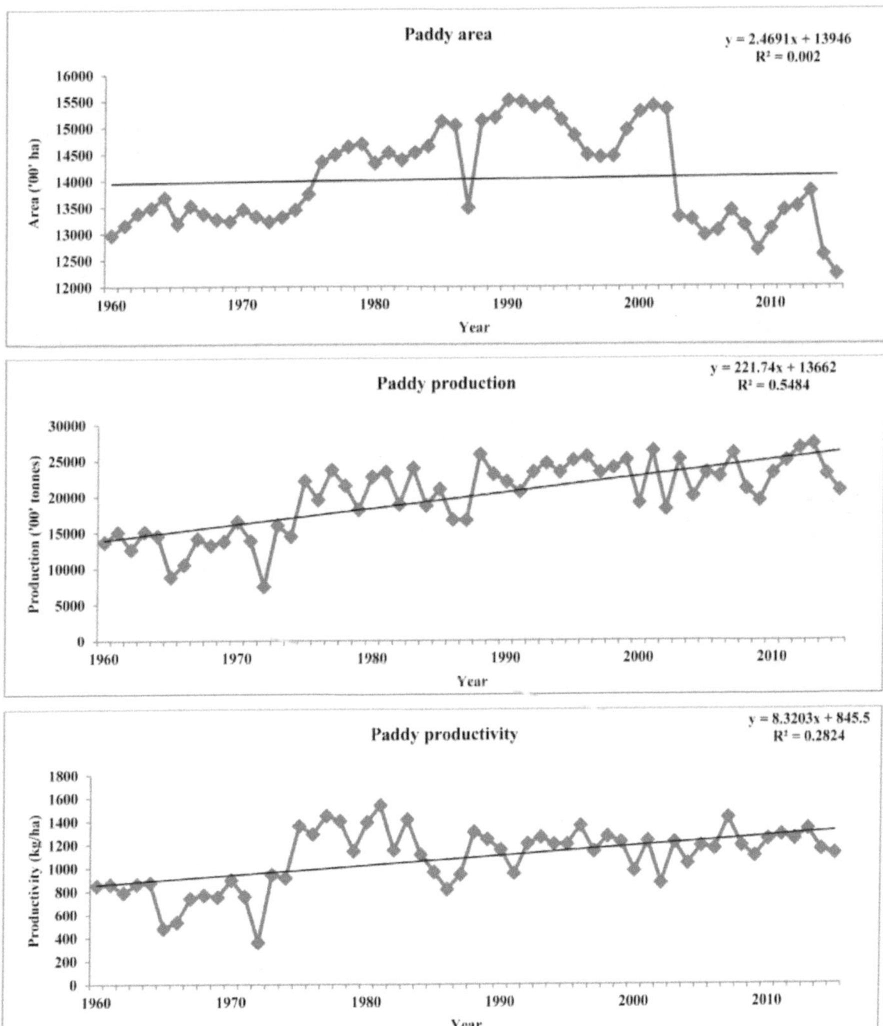

Fig. 11.1: Trends in area, production and productivity of paddy (1960-2015)

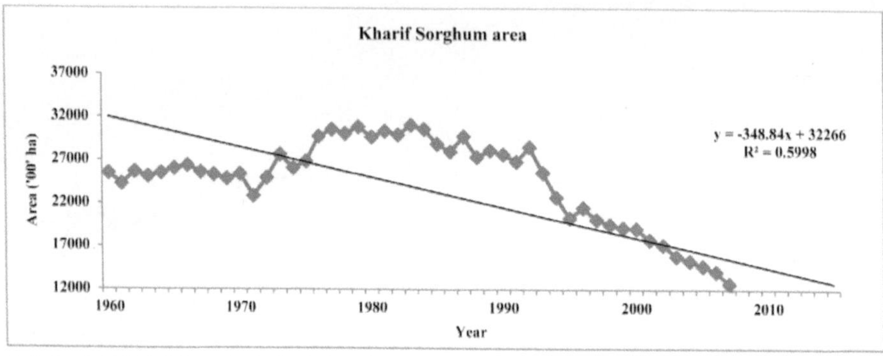

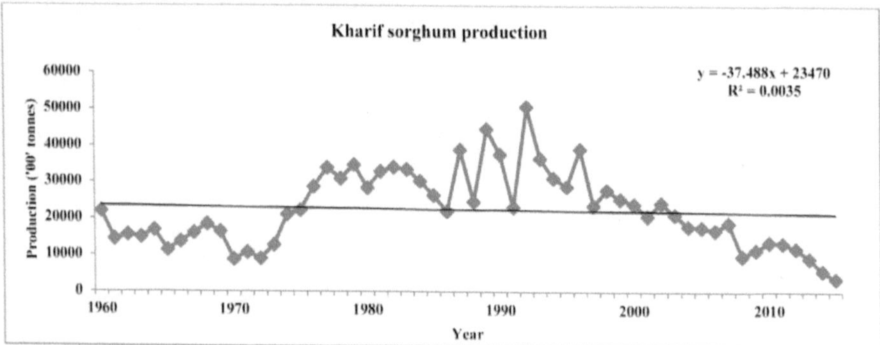

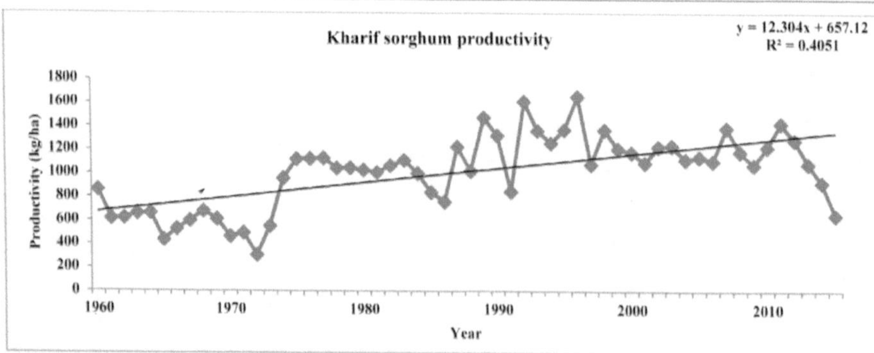

Fig. 11.2: Trends in area, production and productivity of *Kharif* Sorghum (1960-2015)

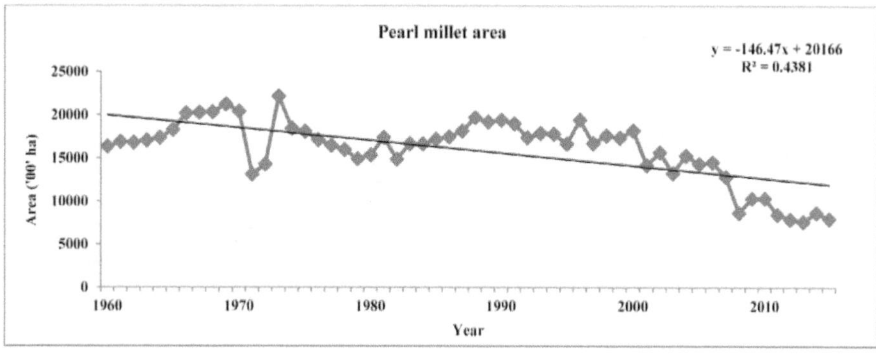

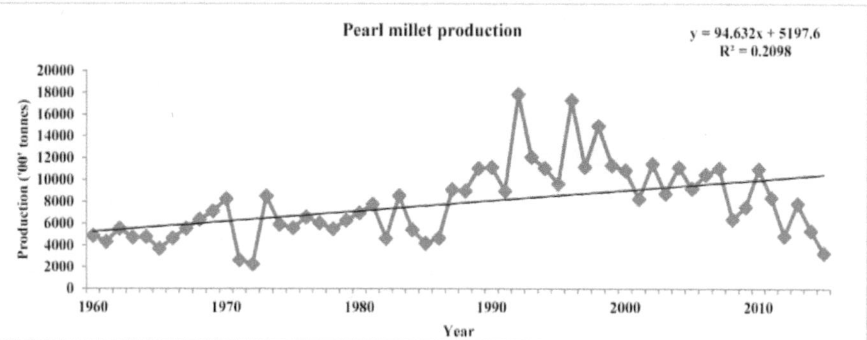

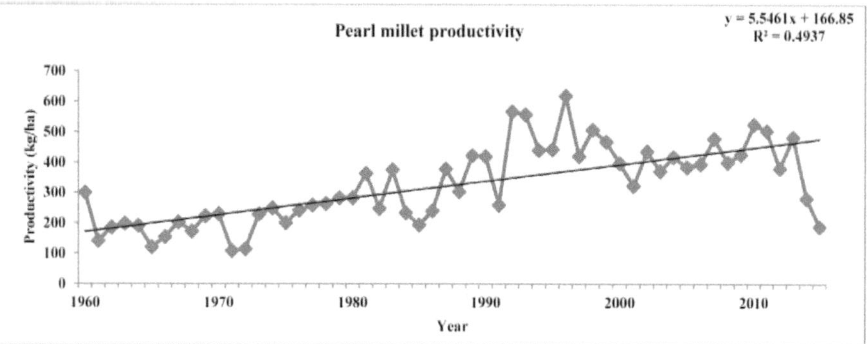

Fig. 11.3: Trends in area, production and productivity of pearl millet (1960-2015)

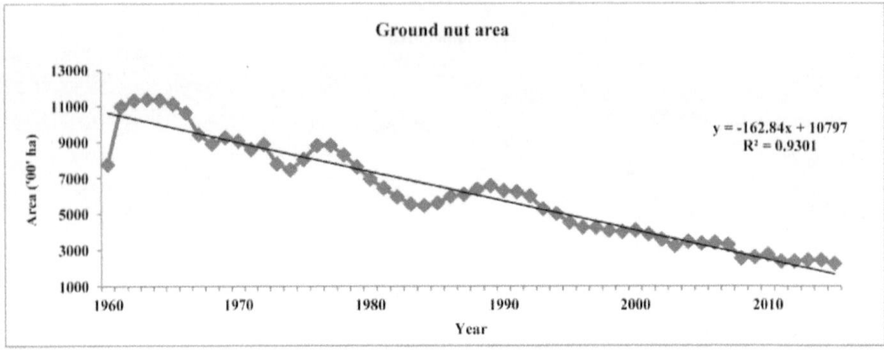

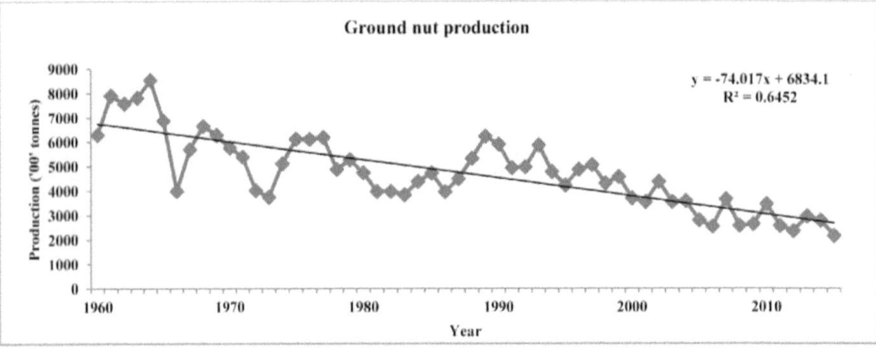

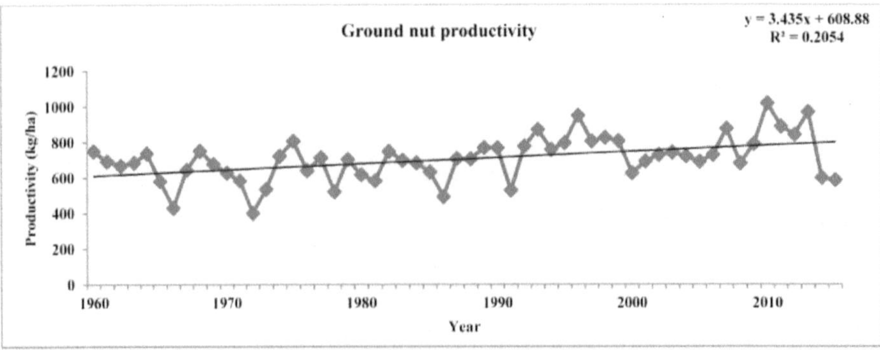

Fig. 11.4: Trends in area, production and productivity of Ground nut (1960-2015)

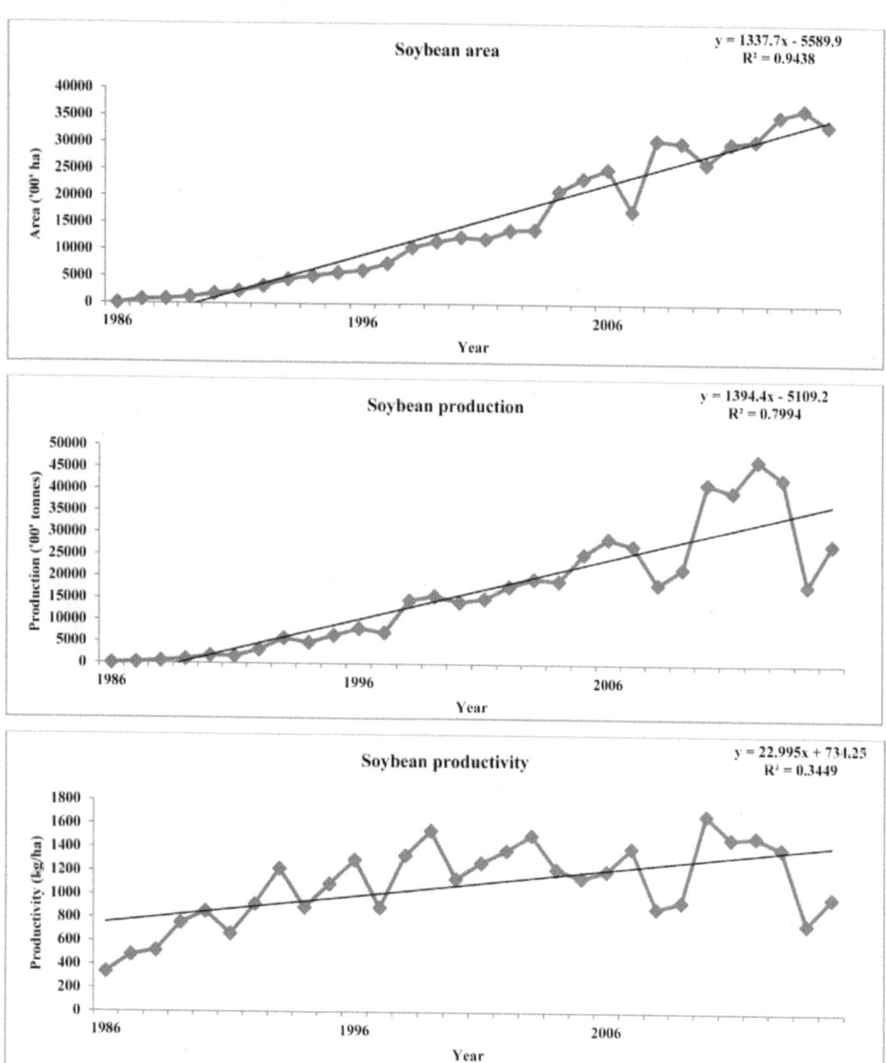

Fig. 11.5: Trends in area, production and productivity of Soybean (1986-2015)

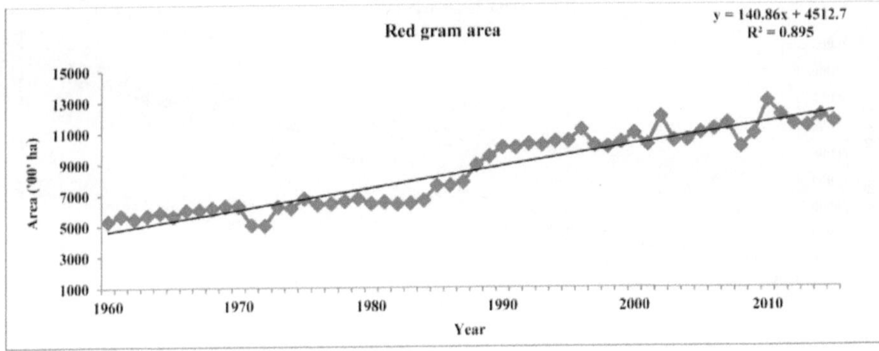

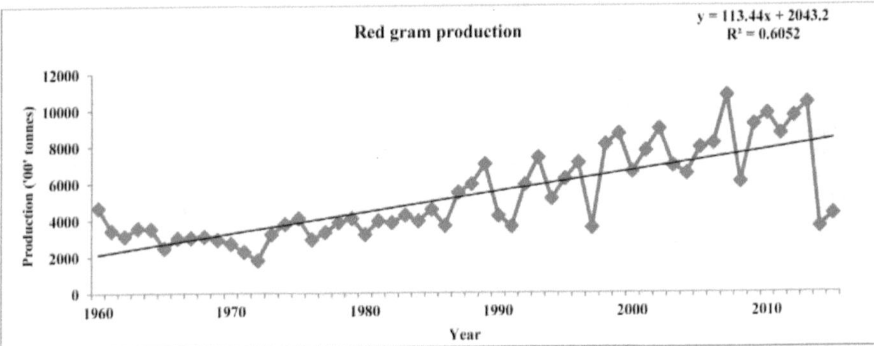

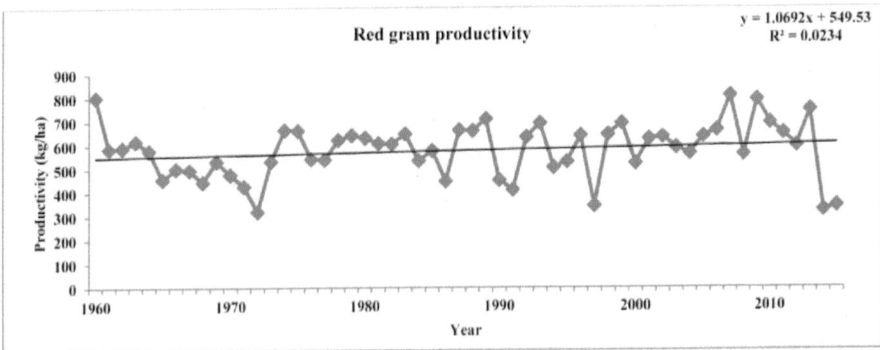

Fig. 11.6: Trends in area, production and productivity of Red gram (1960-2015)

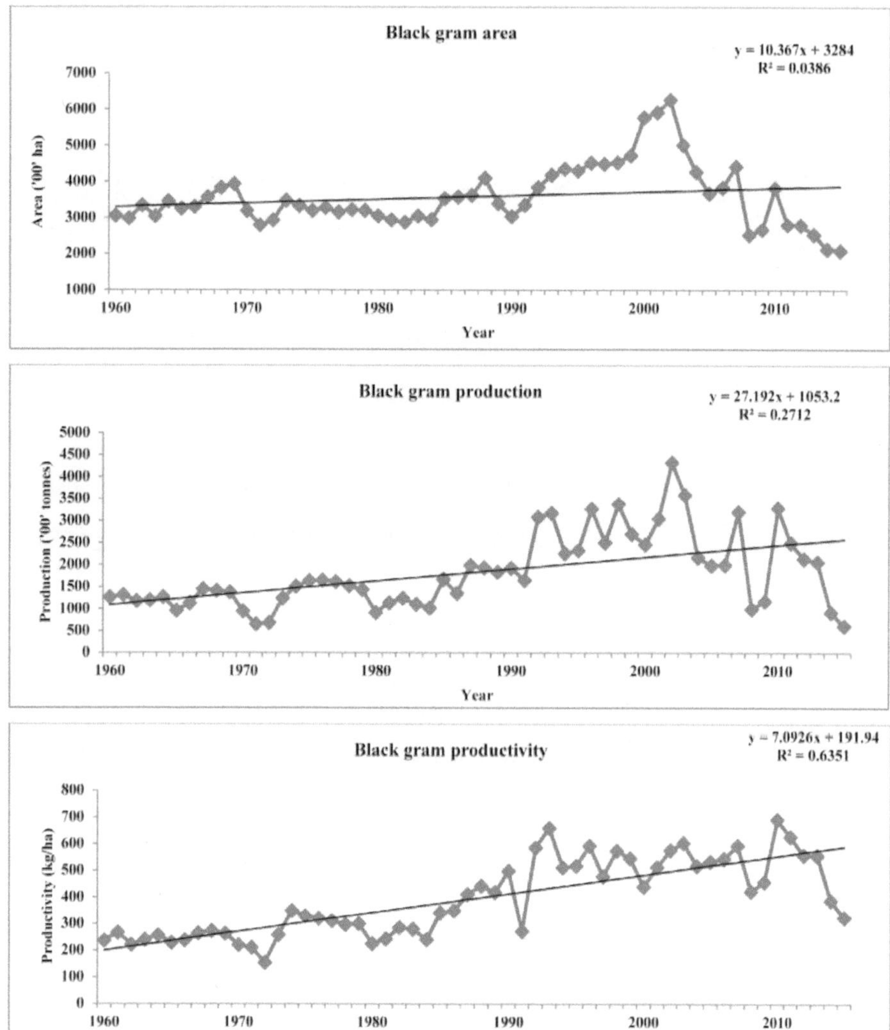

Fig. 11.7: Trends in area, production and productivity of Black Gram (1960-2015)

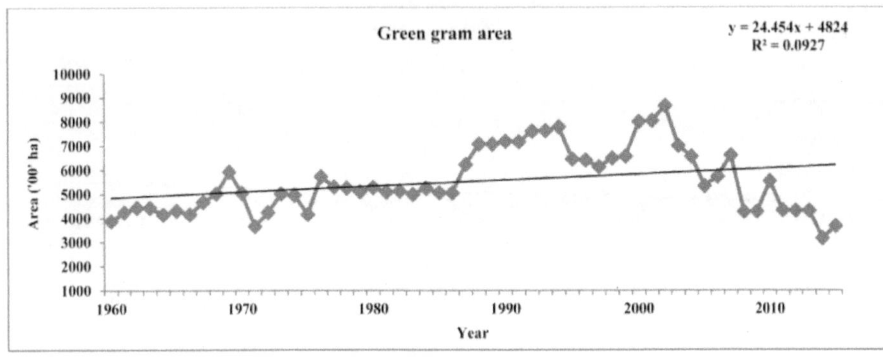

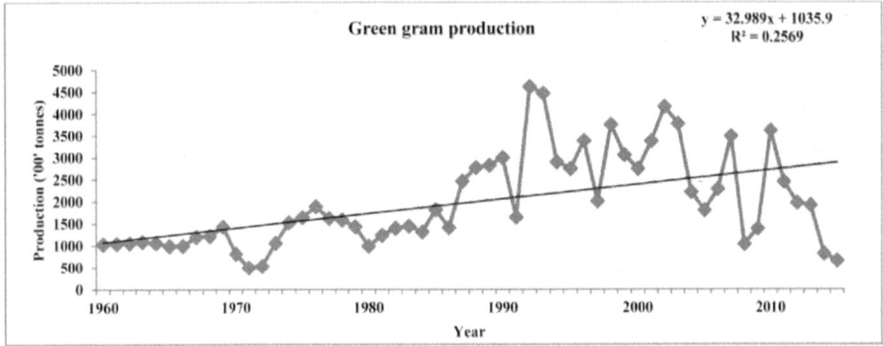

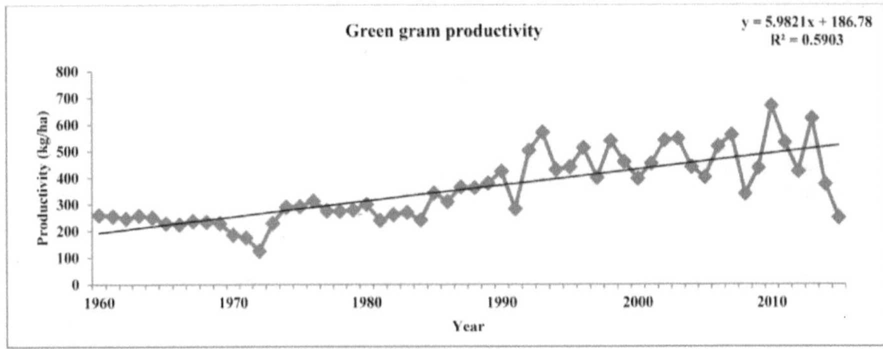

Fig. 11.8: Trends in area, production and productivity of Green Gram (1960-2015)

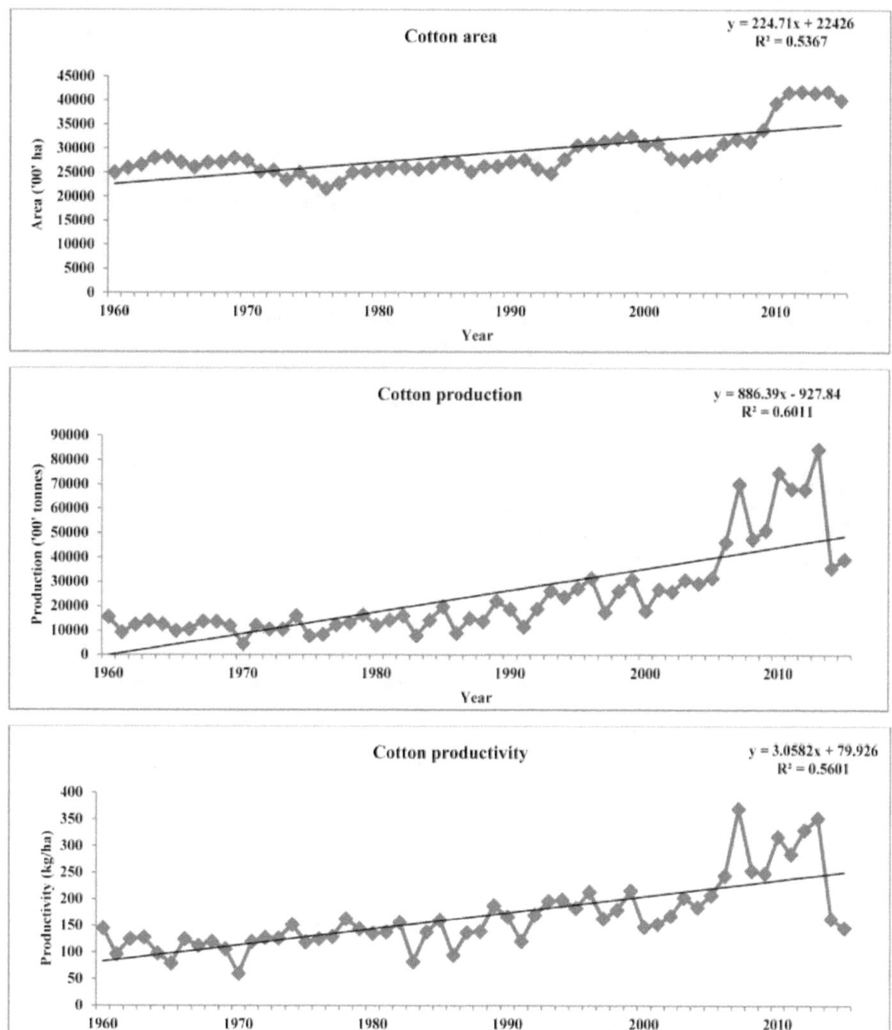

Fig. 11.9: Trends in area, production and productivity of Cotton (1960-2015)

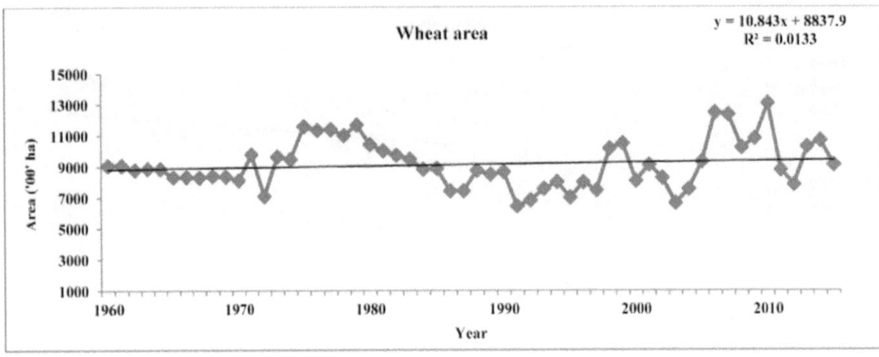

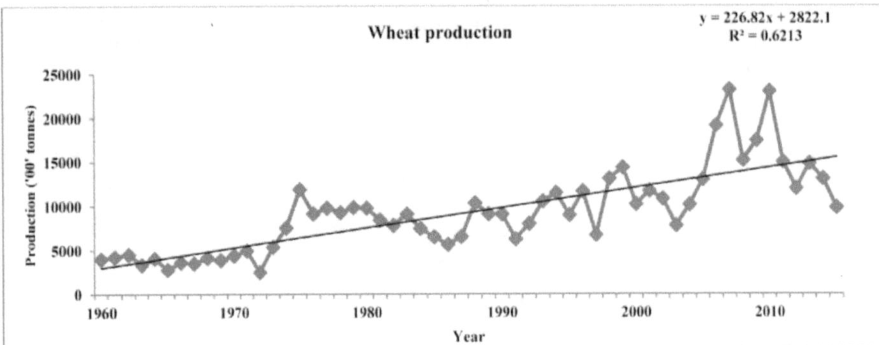

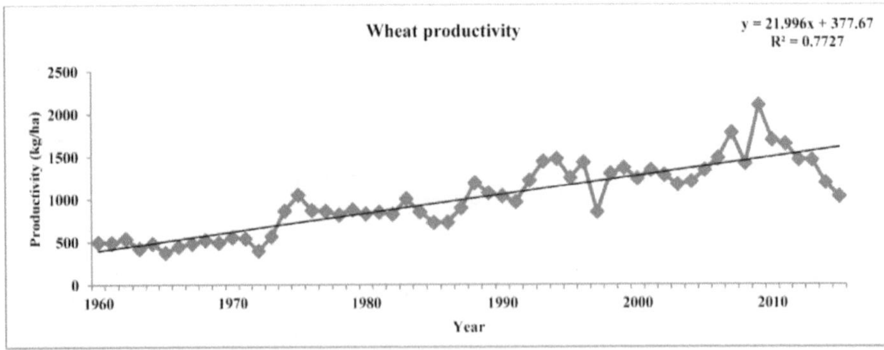

Fig. 11.10: Trends in area, production and productivity of Wheat (1960-2015)

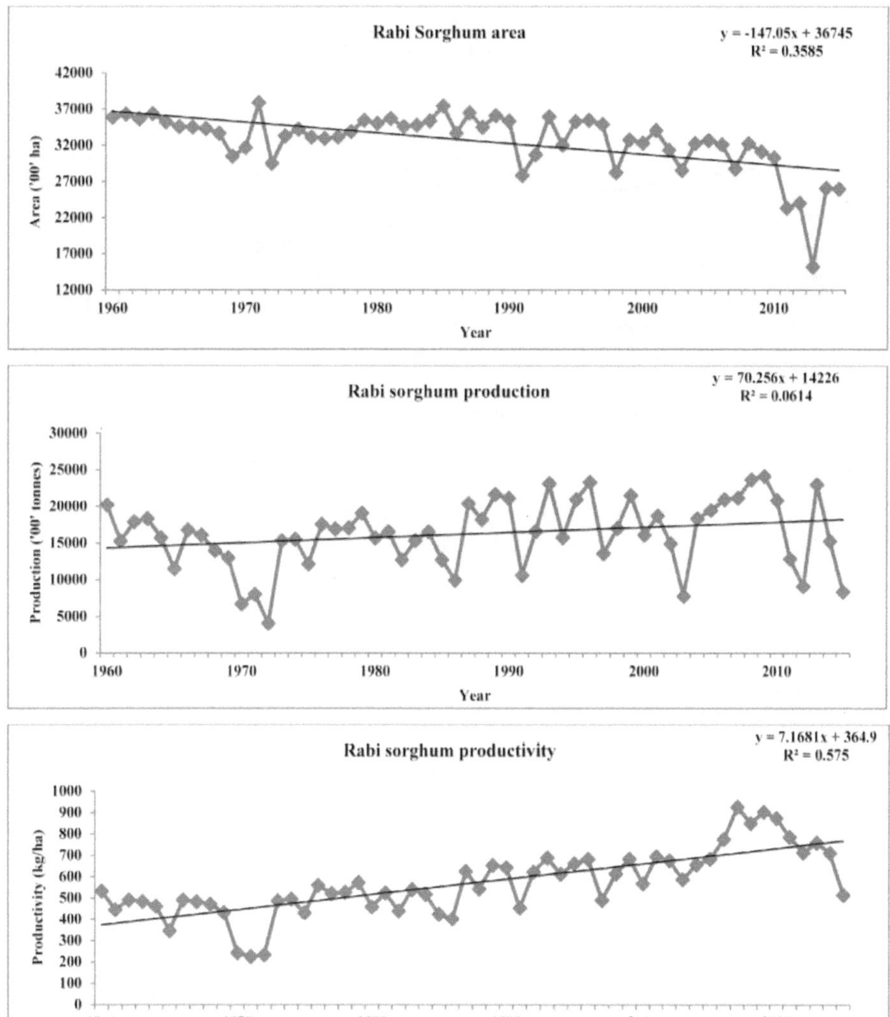

Fig. 11.11: Trends in area, production and productivity of Rabi Sorghum (1960-2015)

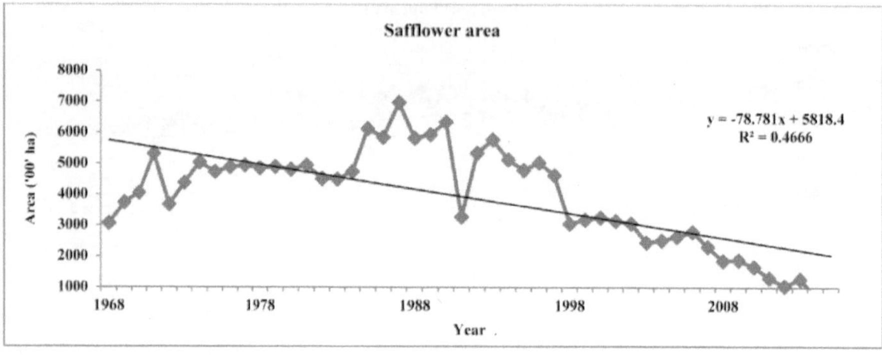

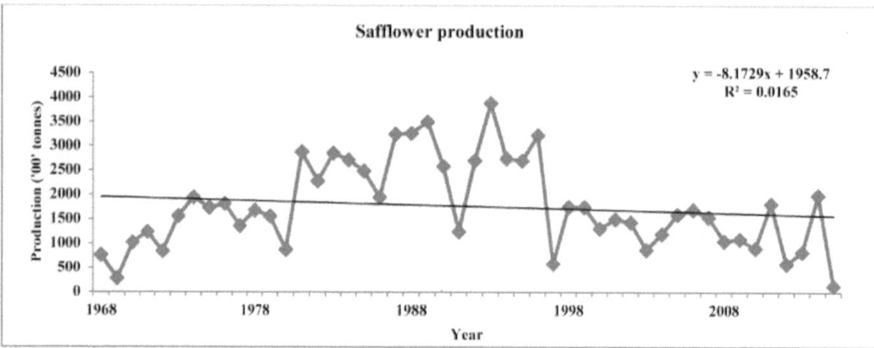

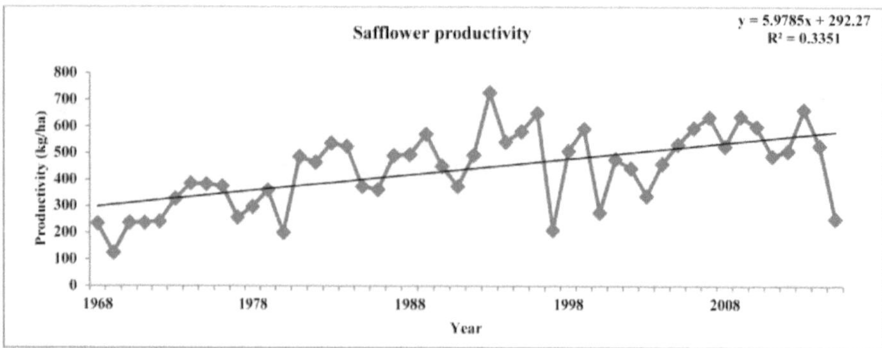

Fig. 11.12: Trends in area, production and productivity of Safflower (1960-2015)

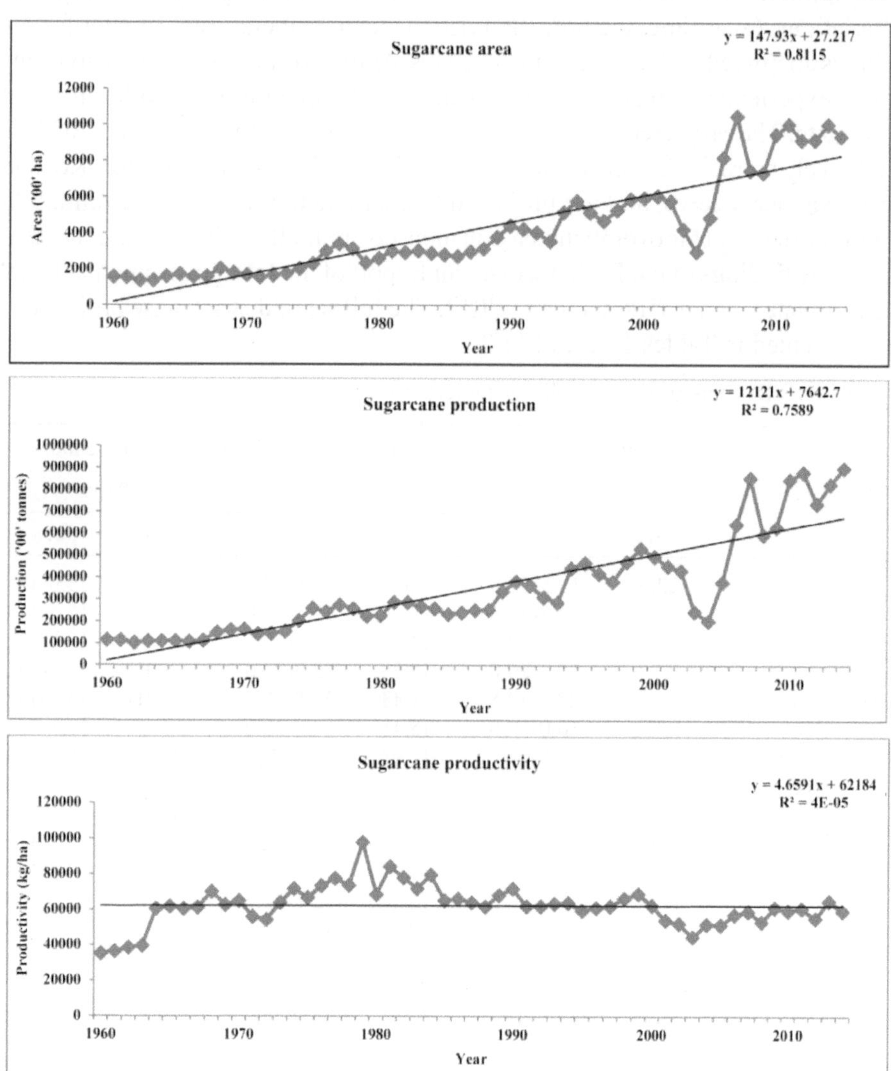

Fig. 11.13: Trends in area, production and productivity of Sugarcane (1960-2015)

projected increases in mean temperature and average monsoon rainfall over Maharashtra, (TERI. 2014). These projections are also summarized by administrative division. The climate modelling results show that temperature and rainfall are projected to increase all over the state though there are regional variations. Over time, the projected rise in mean temperature is greater for the 2070s compared to the 2050s and the 2030s. Amravati and Aurangabad divisions may experience a greater rise in annual mean temperature than other parts of the state. The projected increase in monsoon rainfall by the 2030s and 2050s is relatively more for Amravati and Nashik divisions, though divisions like Konkan and Nagpur receive, and are projected to continue to receive, more rainfall in absolute terms. This overall increase in monsoon rainfall for the state is consistent with the findings of the Fifth Assessment Report of the Intergovernmental Panel on Climate Change (IPCC) IPCC, 2013. The full set of district-wise projections is presented in Tables 11.1 and 11.2.

Table 11.1: District-wise climate change projections

District	Projected increase in monsoon rainfall (% increase from baseline)					
	2030s		2050s		2070s	
	Average	Range	Average	Range	Average	Range
Hingoli	22.9	17.5-27.5	26.3	22.5-30.0	28.16	25.0-32.5
Buldana	26.23	22.5-30.0	28.74	25.5-30.0	21.88	17.5-30.0
Thane	25.16	20.0-30.0	24.71	20.0-30.0	27.32	20.0-32.5
Gondia	17.46	15.0-20.0	18.76	15.0-20.0	25.06	22.5-27.5
Nandurbar	33.83	27.5-37.5	31.43	27.5-37.5	21.01	15.0-30.0
Solapur	15.39	10.0-20.0	15.41	10.0-20.0	25.03	17.5-32.5
Sindhudurg	12.81	7.5-15.0	10.98	7.5-15.0	15.4	10.0-25.0
Kolhapur	15.30	10.0-20.0	15.05	7.5-20.0	22.7	15.0-30.0
Ratnagiri	15.42	7.5-20.0	15.26	7.5-20.0	18.7	10.0-27.5
Sangli	12.97	12.5-20.0	12.04	10.0-22.5	23.6	17.5-32.5
Solapur	14.25	10.0-17.5	13.77	7.5-20.0	22.8	17.5-30.0
Satara	17.90	12.5-22.5	16.25	12.5-22.5	26.1	22.5-35.0
Raigad	20.10	17.5-25.0	19.62	15.0-22.5	20.0	17.5-25.0
Pune	25.76	10.0-32.5	25.03	10.0-30.0	28.7	20.0-37.5
Osmanabad	19.80	12.5-25.0	20.16	12.5-27.5	29.9	22.5-37.5
Latur	17.93	15.0-22.5	21.6	15.0-25.0	32.8	27.5-35.0
Nanded	19.22	12.5-25.0	23.3	17.5-30.0	28.4	17.5-30.0
Beed	22.65	20.0-30.0	26.2	17.5-30.0	34.0	30.0-40.0
Ahmednagar	22.93	15.0-32.5	23.85	15.0-35.0	35.15	30.0-40.0
Nasik	34.78	25.0-40.0	32.92	22.5-40.0	38.6	30.0-52.5
Aurangabad	24.95	20.0-27.5	25.05	20.5-27.5	27.3	20.0-32.5
Jalna	23.40	20.0-25.0	24.1	22.5-27.5	26.98	20.0-32.5
Parbhani	22.64	15.0-25.0	23.5	20.0-27.5	30.1	27.5-32.5
Dhule	36.70	30.0-40.0	33.4	27.5-37.5	24.7	17.5-30.0
Jalgaon	35.29	25.0-40.0	34.9	25.0-37.5	22.6	17.5-27.5

Contd.

Yavatmal	22.35	17.5-27.5	26.8	22.5-32.5	20.15	15.0-27.5
Chandrapur	16.2	12.5-22.5	21.7	17.5-30.0	17.9	15.0-25.0
Gadchiroli	18.15	12.5-22.5	23.4	15.5-30.0	22.6	20.0-30.0
Wasim	23.2	22.5-27.5	28.7	25.0-30.5	22.7	20.0-27.5
Akola	26.2	22.5-30.0	30.0	27.5-32.5	20.6	20.0-25.0
Amrawati	21.7	17.5-27.5	26.2	22.5-30.0	22.4	17.5-25.0
Wardha	18.1	15.0-22.5	20.9	20.0-27.5	18.75	15.0-25.0
Nagpure	17.5	15.0-20.0	21.89	17.5-22.5	25.1	17.5-27.5
Bhandara	17.6	15.0-20.0	20.7	17.5-22.5	22.8	22.5-27.5

Note: The Dhule climate model used for this study is designed to model the climate over Maharashtra state and is not specifically desiJalgaongned for district-scale climate projections. The resolution of the model is 25 km x 25 km. (*Source*: TERI. 2014)

11.3 Climate risk management

As known, rice is a water loving plant widely grown during the *kharif* season. Traditionally, rice is grown under continuous flooded condition and requires 1500-2000 mm water (Kandaih, 1985**).** Though principally a tropical crop, rice requires abundance of water (Huke, 1976). Therefore, this factor has a marked bearing on the growth and yield of rice. The onset of monsoon in Maharashtra state is around 7th to 15th June in the Konkan region including western Maharashtra districts and in eastern part of Vidarbha extends over the entire area by 25th June. The monsoon starts withdrawing from northern parts from mid-September and by 15thOctober it is completely withdrawn from the entire Maharashtra. In the entire Maharashtra, scarcity of water is a critical limitation to adoption of modern technology for increasing productivity of traditional *rainfed* rice growing areas. Most of the farmers construct huge bunds in the rice fields and they impound this water for rice cultivation to avoid the uncertainty factor of monsoonal rainfall. This practice of making huge bunds often becomes adverse for rice crop seedlings as higher bunds submerge the rice seedlings and hence growing of tall, long duration, photo-sensitive varieties which can sustain higher water levels has become a traditional practice of rice cultivation in this area. Farmers in this region grow long duration varieties under broadcast system. These varieties mature by mid-November whereas the south western monsoon withdraws in mid-October. Therefore terminal drought at the reproductive stage of the rice crop is a recurring feature. Under this system, rice seeds are broadcast in a ploughed field immediately after onset of the monsoon around mid-June. *Rabbing* is a practice which consists of burning cow-dung cakes, tree leaves lopping, leaves, grass etc. by making layers on the paddy seed-bed area. *Rabbing* is a sort of partial sterilization of the soil. It improves the physical structure of the soil and increases availability of nutrients in the soil. The practice is, therefore beneficial in raising vigorous seedlings but it involves wastage of valuable organic matter, which can preferably be used in

Table 11.2: Division wise climate change projection

Administrative division	IMD climate normal: annual mean temperature (°C)	Projected increase in annual mean temperature			IMD climate normal: monsoon rainfall (mm)	Projected increase in monsoon rainfall (mm)		
		2030s	2050s	2070s		2030s	2050s	2070s
Amravati	27.21	1.44-1.64	2.2-2.35	3.06-3.46	785.3	17.5-30	22.5-32.5	15-27.5
Aurangabad	26.46	1.44-1.56	2.15-2.3	3.14-3.38	708.8	12.5-27.5	15-30	20-40
Nasik	2.79	1.4-1.68	2-2.4	2.82-3.3	567.5	17.5-40	15-40	15-52.5
Nagpur	27.19	1.18-1.4	1.95-2.2	2.88-3.16	1124.7	12.5-20	12.5-30	15-27.5
Pune	25.22	1.15-1.28	1.65-1.95	2.46-2.74	852.2	10-32.5	10-32.5	12.5-37.5
Konkan	26.99	1.1-1.28	1.5-1.8	2.18-2.6	2578.2	10-30	10-30	10-32.5

Note: The projection for the 2030s is the average of projections for the period 2021-2040. Similarly, the projection for the 2050s in the average of projections for 2041-2060 and that for the 2070s is the average of projections for 2061-2080. **(Source:** TERI. 2014)

compost making. After about 30 to 35 days of onset of monsoon when sufficient water is impounded in the fields, the fields are ploughed in the standing crop. The puddling operation is done mainly to control weeds, to create semi-puddled conditions, to arrest percolation losses, to decrease the initial high population and to slightly adjust the plant population. Some of the constraints in rice productions envisages both soil and atmosphere condition of region, The soil of rice growing area in the region shows a lot of variation in respect of their physico-chemical properties and hence in nutrient availability. Inadequate coverage under high yielding varieties, poor soil fertility, inadequate and poor water management, low fertilizer use and poor fertilizer use efficiency, incidence of pests and diseases, dry spells at critical stages, poor resource base of farmers, scanty and erratic rains affect rice crop growth in the Maharashtra region etc. (Dakhore *et al.*, 2017). There are some areas which are sensitive to water logging conditions and most common problem in pulses is either shortage of rainfall or impeded drainage. Distribution of rainfall is a crucial determinant in determining the final productivity of pulse crops. Pigeon pea and chickpea are the important pulse crops of Maharashtra. Early information about the delayed onset of monsoon helps in the selection of cultivars suitable for reduced LGP. Contingent plans have been developed for the state and emphasis has been laid on growing of pulse crops like pigeon pea, green gram, black gram etc. which can adjust well in the cropping system. This information can also help the concerned parallel departments for the mobilization of inputs like seed, fertilizers etc. Information on the ensuing break monsoon conditions in pulse growing areas can assist in developing agromet advisories on intercultural operations, thinning out and adoption of soil moisture conservation practices. Nutrient deficiencies that develop due to mid-season droughts in pulses can be avoided/ corrected through weather based decision support systems. In case of failure of monsoon in some of the districts, advisories on the revival of southwest monsoon season can help the farmers for sowing of pulses on the receipt of rains late in the season. In such situations advisories on varietal selection and likely pest and disease incidence are crucial.

Oilseeds are sensitive to stagnation of rainwater and impeded drainage which reduces the yield. Delayed monsoon conditions severely affect productivity of oilseed crops particularly groundnut. Seasonal agromet advisories help in the process of selection of appropriate cultivars in groundnut and other oilseed crops. If onset of monsoon is delayed or sufficient rains are received till July, crop alternative to groundnut can be suggested and there are options like niger and sesame in some of the districts. Hence seasonal weather forecast is vital to minimize the losses at farm level. Break monsoon conditions create nutrient deficiency in oilseeds and early information on mid-season dry spell or heavy rains will assist the farmers to undertake measures to mitigate the possible adverse effects on crop. Intercropping of soybean and pigeon pea crops are also coming up recently for providing additional income to farmers. Cotton cultivation is majorly rain dependent and subject to heavy vagaries of monsoon rains. The unpredictability of rainfall and high element of risk leads to low input usage. Cotton is mainly grown on shallow and medium deep soils, which have low available water holding capacity and are highly erosion prone. Most of the rainfall is received during July and August in short and heavy spells, resulting in high runoff and soil loss and 40-80% of the rain water goes unutilized in the absence of proper soil and water conservation measures. Abiotic adversities like cloudy weather, water logging during initial stages, moisture stress at later stages aggravate physiological shedding of buds and bolls. Continuous presence of cotton makes it easy for pests, diseases and other biotic stress agents to survive multiply and cause frequent epidemics. Jassids and bollworms are the major pests of cotton. Unfavourable weather conditions affect the timely pest control in the early stages of the crop. Infection by powdery mildew and bacterial blight diseases cause premature crop cessation and reduction in yield. Resource poor farmers rely on credit based input service system in the cotton zone and the pest management advisory role is mainly by pesticide dealers who are not qualified for the job. Early withdrawal of monsoon in some years creates moisture stress and nutrient deficiency in cotton. Any information on the impending early withdrawal could potentially help the farmers to resort to foliar sprays. During high rainfall events and prolonged wet spells cotton is sensitive to high humidity and boll rot can occur; therefore prophylactic measures should be taken up through timely and precise agromet advisory system. Cane yields are affected to either deficit or excess of moisture and lack of rainfall during July and August results in a substantial reduction in cane yields. Trash mulching @ 3 t ha^{-1} during dry spells improves the cane yields. Like-wise if the crop is subjected to water logging the sucrose content in the cane reduces, thus affecting the recovery of sugar. High winds make the crop to lodge which depresses juice sucrose and cane weight decreases as well. All these field problems can be addressed by proper and timely agromet advisories. During break monsoon conditions, a complex of sucking pest increasingly attack vegetables and the crop undergoes nutrient and moisture stress. Issue of agromet advisories well in advance, would facilitate farmers to resort to intercultural operations to conserve soil moisture, timely plant protection and take up foliar application of nutrients to correct any deficiency. If the monsoon rains are considerably delayed, then advisories on early duration vegetable crops/varieties and suitable planting method will help to sustain the yield at appropriate levels.

11.4 Resilient practice / technology

There is a need for in-situ soil and water conservation and proper drainage technology in deep black soils. Broad bed and furrow (BBF) system involves preparation of a broad bed of 90 cm, furrow of 45 cm and sowing of crop at a row spacing of 30 cm, TERI. 2014. The cost of BBF implement is Rs. 45,000. The BBF technology has many advantages including in-situ conservation of rainwater in furrows, better drainage of excess water and proper aeration in the seedbed and root zone. Similarly, furrow irrigated raised bed (FIRB) planting was promoted for cultivation of different crops in Uttar Pradesh, West Bengal, Punjab, Maharashtra, Karnataka, Rajasthan and Tamilnadu. Ridge and furrow method of vegetable cultivation was promoted in Gunia village of Gumla district and in cotton at Amravati and Aurangabad, Maharashtra. (Table 11.3)

Table 11.3: Performance of soybean in ridge and furrow method of sowing

Particular	Ridge & furrow	Farmers practice	% increase in yield/income
Grain yield (kg/ha)	1937	1152	40.5
Net Return (Rs/ha)	38805	18898	51.3
B:C ratio	3.51	2.41	
Water use efficiency (kg/ha/mm)	2.09	1.24	

(*Source:* TERI. 2014)

11.4.1 Impact

Advantage of BBF planting method:

- Increase in water use efficiency
- Increase in crop productivity (5-10%)
- Less moisture stress during non- rainy days
- Time saving (25-30%) in irrigation
- Requires 20-25% lower seed rate
- Water saving up to 25-30%
- Better weed management
- Reduces crop lodging

11.4.2 Advantages of FIRB planting method

- Management of irrigation water is simpler and more efficient. On an average about 30% less irrigation water was required compared to flat bed method and improved crop yields by more than 20%. FIRB planting saved 30 to 50% wheat seed compared to flat bed planting.

- Better upland crop production is possible under wet spells because of proper drainage.
- Farmers can apply N and irrigation water at grain filling stage to improve protein content without inducing lodging. Reduced lodging can have a significant positive effect on yield as many farmers do not irrigate after heading precisely to avoid lodging.
- Weeds between the beds can be controlled mechanically early in the crop cycle.
- Herbicide dependence is reduced, and hand weeding and hoeing between rows are easier
- Yield of rice transplanted on FIRB was comparable with traditional rice but could save as much as 25-50% in irrigation water.
- Compaction of soil is limited only to the furrows used as tramlines (tractor tracks).

11.4.3 Ridge and furrow method of vegetable cultivation in Gunia village of Gumla district had the following advantages:

- Crop stand improved by 70-75%
- Time saving (25-30%) in irrigation (1.5 hr/ha/irrigation)
- Required 20-25% lower seed rate
- Water saving was up to 25-30%
- Better crop management
- Reduced crop lodging in tomato
- Obtained 10-15% higher yield

11.4.4 Scope for upscaling

BBF / FIRB / Ridge furrow planting of crops was demonstrated in Coochbehar (West Bengal), Kushinagar (U.P.), Baramati, Amravati, Aurangabad, Nandurabar (Maharashtra),Nalgonda (A.P), Kota, (Rajashthan) Datia, Morena, Satna (M.P) Chikballapur (Karnataka) and Villupuram (Tamilnadu) in 412 ha covering over 1000 farmers with an average 10 to 40% increase in yield and benefit cost ratio of 2.2 to 4.7 compared to farmers' practice TER,. 2014. There is scope for upscaling the techonology in soybean,maize and wheat growing areas.

Amongst cereals, Paddy, sorghum and pearl millet are the principal cereal crops of Maharashtra. The area under paddy is about 15 lakh ha and there is a marginal increase in its area over the years. Sorghum is the second largest grown cereal crop due to its commercial value (as cattle feed). It is grown both in *kharif* as well as *rabi* season. Pearl millet is another important cereal crop grown mostly in semi-arid region of Maharashtra. The area under *kharif* sorghum has fallen from 31.1 lakh ha to 6.2 lakh ha in last 55 years. Productivity level of *kharif* sorghum has increased from 305 kg ha in 1972 to 1425 kg ha in 2011. However, in the case of *rabi* sorghum, the area has decreased from 37.8 lakh ha in the 1971 to a meager 15.1 lakh ha in 2013. Pearl millet is another important cereal crop grown mostly in semi-arid region of Maharashtra. Area under pearl millet crop in the Maharashtra during 1960-2015 witnessed significant decrease in the area. (Dakhore *et al*., 2017). The major pulse crops of the state are red gram, black gram and green gram. Area under red gram has picked up in recent years as a good dry land pulse crop. Red gram area has risen from 5.30 lakh ha during the period 1960-61 to 13.01 lakh ha in 2010. This rise is partly due to introduction of pulses improvement program and partly to market forces. In other years, the inter-annual variability is largely weather driven. Area, production and productivity of black gram have varied over the years. Area and production under green gram has showed a marginal ascent in recent years but productivity levels have been sustained / enhanced due to use of improved package of practices. Groundnut, soybean and safflower are the major oilseeds crop of Maharashtra. Soybean is the largest area occupying oilseed crop grown in Maharashtra. Area under soybean crop increased from 0.7 lakh ha in 1987 to 36.4 lakh ha in 2014. Its production and productivity have shown a sharp increase over the years. Major shift that occurred in the past decades in favour of soybean crop was due to its early duration, suitability to double cropping, distinct yield advantage, higher price and better market support. The area under soybean has also substituted more of *kharif* sorghum and some cotton acreage. Groundnut is grown in 2.41 lakh ha. The production trend of this crop have shown steep decline over the years due to variation in the spread of rainfall. Similarly, area under safflower is also showing a sharp decline but with a marginal decrease in production. This may be due to the introduction of improved cultivars and production technology. Cotton is the principal commercial crop of the state. It is cultivated on more than 41 lakh ha. Though productivity trends of the crop show large variability and there has been a constant increase in its area and production chiefly due to the introduction of Bt cotton, market forces and partly due to vagaries in the temporal distribution of monsoon rains. A steep increase in sugarcane area and production has come through with a concurrent increase in the area over the years. Very small inter-annual variation in the productivity of the crop may be attributed primarily to the fact that large area under sugarcane crop is irrigated.

11.5 Agricultural mitigation options

11.5.1 Cropland management

The major sources of land based mitigation from agriculture, following IPCC (2007) are described below:

Improved agronomic practices generate higher inputs of carbon (C) residue, leading to increased soil C storage (Follett *et al.*, 2001). Such practices include using improved crop varieties, extending crop rotations, avoiding use of bare fallow and using cover crops.

Integrated nutrient management can reduce emissions on-site by reducing leaching and volatile losses, improving nitrogen (N) use efficiency through precision farming (Leiva *et al.*, 1997) and improving fertilizer application timing.

Increasing available water in the root zone through water management can enhance biomass production, increase the amount of above-ground and the root biomass returned to the soil, and improve soil organic C concentration. Soil and water conservation measures, such as the construction of soil or stone bunds, drainage measures, and irrigation constitute important aspects of water management.

Tillage management practices with minimal soil disturbance and incorporation of crop residue decrease soil C losses through enhanced decomposition and reduced erosion (Lal, 1997). Systems that retain crop residues tend to increase soil C because these residues are the precursors of soil organic matter.

Agro-forestry systems increase C storage and may also reduce soil C losses stemming from erosion. Options include combining crops with trees for timber, firewood, fodder and other products, and establishing shelter belts and riparian zones/buffer strips with woody species (Oelbermann *et al.*, 2004).

11.5.2 Improved grassland management

Improved productivity through increasing nutrients for plant uptake and reducing the frequency or extent of fires (e.g. improvements in forage quality and quantity, seeding fodder grasses or legumes with higher productivity and deeper roots, reducing fuel load by vegetation management).

Improved grazing management by controlling intensity and timing of grazing (e.g. stocking rate management (Freibauer *et al.*, 2004), rotational grazing, and enclosure of grassland from livestock grazing).

11.5.3 Management of organic soils

Draining organic soils for cultivation leads to high GHG emissions. Avoiding drainage is the best option in terms of reduced GHGs; other practices to minimize emissions include avoiding row crops and tubers, avoiding deep ploughing, and maintaining a shallower water table (Freibauer *et al.*, 2004).

Summary

In the entire Maharashtra, scarcity of water is a critical limitation to adoption of modern technology for increasing productivity of traditional *rainfed* rice growing areas. Use of hydrological models coupled with seasonal rainfall forecast of monsoon has been observed to help in giving advance information on the schedule of water release. By using basin models, the areas likely to be inundated can also be identified. Mapping of the areas prone to frequent floods with the help of historic rain data aids in the design, development and maintenance of appropriate drainage systems in the agricultural regions. In the cotton growing tracts of Maharashtra, advance information with respect to the late arrival of monsoons could help in the selection of cultivars/cropping systems and rescheduling of fertilizer dosages/application and the adoption of soil conservation measures. Information on possible break in monsoons would also help in assessing the likely buildup of abiotic stresses in cotton. High incidence of boll worm (or *spodoptera*) is likely under such situations and agromet advisories could help in the taking up prophylactic measures. Any information on the impending early monsoon withdrawal could potentially help the farmers in rescheduling the fertilizer applications or resorting to foliar nutrient sprays. During high rainfall episodes, and prolonged wet spells are expected, cotton crop is sensitive to high humidity when boll rot occurs; therefore prophylactic measures could be taken up if timely and precise agromet advisory system has been developed and in place. Information on the high probability of break in monsoon conditions in grain legumes growing areas can assist in developing agromet advisories on intercultural operations, thinning out and vigorous adoption of soil moisture conservation practices. In case of prolonged delay in the onset of southwest monsoons, farmers resort to sowing of pulses on the receipt of rains, late in the season. There are some areas which are sensitive to water logging conditions in such situations, advisories on the likely incidence of rainfall are crucial. Seasonal agromet advisories are of great assistance in the process of selection of appropriate cultivars of soybean and other oilseed crops. If onset of monsoon is delayed or sufficient rains have not been received till July, alternate crops to soybean can be suggested, hence seasonal weather forecasts are vital to minimize the losses at farm level. Break monsoon conditions create nutrient deficiency in oilseeds and early information on mid-season dry spell could certainly assist the adoption of timely corrective measures before nutrient deficiencies occur. In the case of sugarcane, its yields are affected by both the deficit or excess of moisture and lack of rainfall during July and August results in a substantial reduction in cane yield. Application of trash mulching @ 3 t ha-1 during dry spells improves the cane yields. Like-wise, the crop when subjected to water logging affects the sucrose content. It is reduced, thus decreasing the recovery of sugar. High winds lodge the crop which depresses juice sucrose and cane weight decreases as well. All these field problems can be addressed by proper and timely availability of agromet advisories.

References/Suggested Readings

Dakhore K. K., Karunakar A., Jadhav J. D., Jagtap D. N., Pawar P. B. and Vijaya Kumar P. 2017. Agroclimatic Atlas of Maharshtra.

Follett, R.F., Kimble J.M., Lal R., The potential of U.S. grazing lands to sequester soil carbon, in The Potential of U.S. Grazing Lands to Sequester Carbon and Mitigate the Greenhouse Effect, R.F. Follett, Kimble J.M., Lal R., Editor. 2001, Lewis Publishers: Boca Raton, Florida. p. 401-430.

Freibauer, A., Rounsevell M., Smith P., Verhagen A., 2004. Carbon sequestration in the agricultural soils of Europe. Geoderma, 2004. 122: p. 1-23.

Huke, R. 1976. Geography and climate of rice. Pages 31-50 in International Rice Research Institute. Climate and rice. Los Baños, Philippines

ICTSD, 2009. ICTSD-IPC, Platform on Climate Change, Agriculture and Trade: Consideration for Policymakers.

IPCC, Agriculture, in Climate Change 2007: Mitigation. 2007, Working Group III Contribution to the Fourth Assessment Report of the Intergovernmental Panel on Climate Change. Cambridge University Press.

IPCC, 2013. Summary for Policymakers. In: Climate Change 2013: The Physical Science Basis. Contribution of Working Group I to the Fifth Assessment Report of the Intergovernmental Panel on Climate Change [Stocker, T.F., D. Qin, G.-K. Plattner, M. Tignor, S.K. Allen, J. Boschung, A. Nauels, Y. Xia, V. Bex and P.M. Midgley (eds.)]. Cambridge University Press, Cambridge, United Kingdom and New York, NY, USA

Kandaih, A. 1985. Modern concepts of water management in rice production. Proceedings of the 16th season of the International Rice Commission, 10-14 June 1985, Los Banos, Laguna, Philippines: Rice progress assessment and orientation in the 1980s. *International Rice Commission Newsletter*, 34(2): 153-162

Lal, R., Residue management, conservation tillage and soil restoration for mitigating greenhouse effect by CO2- enrichment. *Soil and Tillage Research*, 1997. 43:81-107.

Leiva, F.R., Morris, J., and Blackmore, B.S.. Precision farming techniques for sustainable agriculture. in First European Conference on Precision Agriculture. 1997. Warwick.

Oelbermann, M., Voroney R.P, Gordon A.M., 2004. Carbon sequestration in tropical and temperate agroforestry systems: a review with examples from Costa Rica and southern Canada. *Agriculture, Ecosystems and Environment* 104: 359-377.

TERI. 2014. Assessing Climate Change Vulnerability and Adaptation Strategies for Maharashtra: Maharashtra State Adaptation Action Plan on Climate Change (MSAAPC), New Delhi: The Energy and Resources Institute. 302 pp. [Project Report No. 2010GW01].

12

Climate Change and Rice Cultivation over Konkan Region of Maharashtra: A Case Study

D. N Jagtap, M. W Sutar. U. V Mahadkar, S.A Chavan and M. V. Zagade

Dr.Balasaheb Sawant Konkan Krishi Vidyapeeth,Dapoli, Maharashtra- 415712, India

Importance of rice in Konkan area is characterized by heavy rains of 100 mm per week or so over an extended period and contributes 42.9 % of total rice production in the Maharashtra State. Rice is the only suitable crop that can be grown under puddled soil conditions, i.e. with standing water over bunded fields. Tansplanting method of rice is commonly followed by most of the farmers in this region. Puddled rice is said to be of assistance in mitigating the effects of floods. The problem of weeds is minimal in puddled rice culture. Thus, rice serves as a livelihood crop for millions of small, and marginal farmers in Konkan who can afford only low-cost technologies. Even though the crop is robust and versatile, it has faced drawbacks in terms of low yield and reduction in acreage. In Konkan region of Maharshtra, there are two agro climatic zones *viz*., North Konkan coastal zone comprising three districts *viz*., Thane, Raigad and Greater Mumbai and South Konkan Coastal Zone comprising two districts *viz*., Ratnagiri and Sindhudurg. Ratnagiri district receives the highest rainfall of > 3500 mm while Thane district receives the lowest rainfall (2000-2500 mm) in this region. Sindhudurg and Raigad districts receives 3000-3500 and 2500-3000 mm rainfall, respectively. It is generally observed that South Konkan districts record high rice yields in spite of low area while reverse in North Konkan districts. The district wise area and production of rice in Konkan region as the mean for the period of 1997-98 to 2014-15 revealed that, Thane occupies largest average area of 133822 ha under rice, while Raigad has the highest production of rice (308555 tonnes) and Sindhudurga district has the high productivity of 2.75 t ha^{-1}. The trends in area, production and productivity of rice for the period 1997-98 to 2014-15 shows significant decreasing trend in paddy area while in case of production and productivity the trend is increasing. Paddy area decreased from 425000 ha in 1997-98 to only 311300 ha in 2014-15, while productivity increased from 2.44 tonnes ha^{-1} in 1997-98 to 2.88 tonnes ha^{-1} in 2014-15. In this context, it is necessary to understand the agro meteorological constraints for improving the yield and simultaneously reduce the fall in cultivated area.

The main agro climatic constraints are:

- *Uneven Distribution of Rainfall:* High rainfall with highly variable distribution causes uncertain crop yields. Besides its uncertainty, the distribution of rainfall during the crop period if uneven might affect the crop.
- *Bright sunshine hours:* Less number of bright sunshine hours due to continuous cloud cover during the season.
- *Late onset and early cessation of rain:* During delayed monsoon, the sowings of crops are delayed resulting in poor yield. At times, the rains may cease very early in the season exposing the crop to drought during flowering and maturity stages to reduce the crop yield considerably.
- *Prolonged Dry spells during the crop period:* Long breaks in the rainy season is an important feature of Indian monsoon. These intervening dry spells when prolonged during the crop period, reduces crop growth and yield and when these dry spells unduly prolonged the crop ultimately fails.

12.1 Agrometeorology of rice

A long- term experiment was conducted on agrometerological studies in rice to study the effect of weather on phenophase development in rice during 1998-2012 at Agronomy Farm, College of Agriculture, Dapoli, Maharshtra. The experimental field was laid out in strip plot design. Vertical strips having treatments viz. two sowing windows with different number of seedlings viz. 24th MW with one seedling, 24th MW with two seedlings and 26th MW with one seedling, 26th MW with two seedlings were taken. While horizontal strips having different fertilizer doses (kg ha^{-1}) i.e. 100 kg N + 50 kg P_2O_5 + 50 kg K_2O, 150 kg N + 75 kg P_2O5 + 75 kg K_2O and 200 kg N + 100 kg P_2O_5 + 100 kg K_2O were taken. The data presented in Table 12.1 showed that the values of growing degree days and Heliothermal units were different due to variability in duration of phenophases. Among phenological stages, highest accumulated GDD i.e. 718^0 days is noticed in seedling to tillering stage followed by germination to tilllering and flowering to grain development stages. Similar to growing degree days, highest Heliothermal units were accumulated during seedling to tillering stage which was followed by flowering to grain development stage.

Highest grain yield of Sahyadri-2 rice hybrid could be achieved under growing environment-I or 24th SMW sown crop than under growing environment-II (26th SMW) due to comparatively higher rainfall (3378 mm) and minimum temperature (22.9 ^{0}C) besides lower maximum temperature (29.1 ^{0}C) and bright sunshine hours (3.7 hrs). To identify the critical weather parameters and weather sensitive phenological stages influencing yield of rice, correlations between yield over two growing environments and fifteen years of experimentation and phenological stage wise weather parameters (Table 12.2) was worked out. The phenological stages of rice include germination (P1), seedling (P2), tillering (P3), flowering (P4), grain development (P5) and physiological maturity (P6). Out of all the four weather parameters, rainfall was found to have higher correlation with yield than other parameters during flowering and grain development stages. Maximum temperature during tillering stage showed higher correlation with yield than other three weather parameters. Hence, flowering and grain development stages were identified as critical stages for rainfall and tillering stage as critical stage for maximum temperature.

12.2 Climate change and crop projections

A field experiment was conducted at Department of Agronomy, College of Agriculture, Dapoli, Dist. Ratnagiri (M.S.) for generating required experimental data for the calibration of the CERES - Rice model. Daily weather data on weather parameters *viz.*, maximum temperature, minimum temperature, rainfall and sunshine hours were recorded at Department of Agronomy, College of Agriculture, Dapoli. The soil profile data needed for the model was retrieved from the published research work on characterization of Konkan Soils. The initial soil characteristics were determined by collecting samples from the profile extending upto 60 cm depth and analyzing in the laboratory. The calibration of model was based on certain Genetic coefficients. The Genetic coefficients of rice variety Sahyadri were established using the genetic data based system of the model and it was presented in Table 12.1.

Table 12.1: Genetic coefficients of Sahyadri variety computed through CERES -Rice model

Symbol	Description	Genetic coefficients of Rice var. Sahyadri
P1	Time period (expressed as growing degree days [GDD] in ^{0}C above a base temperature of 9 ^{0}C) from seedling emergence during which the rice plant is not responsive to changes in photoperiod. This period is also referred to as the basic vegetative phase of the plant.	620.0
P2O	Critical photoperiod or the longest day length (in hours) at which the development occurs at a maximum rate. At values higher than P2O developmental rate is slowed, hence there is delay due to longer day lengths.	12.5
P2R	Extent to which phasic development leading to panicle initiation is delayed (expressed as GDD in ^{0}C) for each hour increase in photoperiod above P2O.	140.0
P5	Time period in GDD ^{0}C) from beginning of grain filling (3 to 4 days after flowering) to physiological maturity with a base temperature of 9 ^{0}C.	460.0
G1	Potential spikelet number coefficient as estimated from the number of spikelets per g of main culm dry weight (less lead blades and sheaths plus spikes) at anthesis. A typical value is 55.	95.0
G2	Single grain weight (g) under ideal growing conditions, i.e. nonlimiting light, water, nutrients, and absence of pests and diseases.	.0320
G3	Tillering coefficient (scalar value) relative to IR64 cultivar under ideal conditions. A higher tillering cultivar would have coefficient greater than 1.0.	1.00
G4	Temperature tolerance coefficient. Usually 1.0 for varieties grown in normal environments. G4 for japonica type rice growing in a warmer environment would be 1.0 or greater. Likewise, the G4 value for *indica* type rice in very cool environments or season would be less than 1.0.	1.00

The statistics for model runs in DSSAT4.5 versions are presented in Table 12.2.

Table 12.2: Statistics for comparing anthesis dates, maturity date and grain yield predicted by DSSAT/ CERES -Rice model with those observed from field experiments for rice var. Sahyadri

S.N.	Statistic	Mean anthesis day		Maturity day		Grain yield (kg ha^{-1})	
		Observed	Simulated	Observed	Simulated	Observed	Simulated
1.	Mean	84	84	119	118	5487	5158
2.	Std. Dev.	0.748	1.02	0.748	0.980	206.85	189.14
3.	Mean difference	1	1	-1	1	-329	333
3.	Mean (ratio)	1.007		0.992		0.941	
4.	r-square	0.793		0.241		0.294	
5.	RMSE	0.775		1.342		379.94	
6.	d-stat	0.826		0.551		0.528	
7.	Total no. of observations	5		5		5	

The data presented in Table 12.2 indicated that, the root mean square of error (RMSE) of anthesis day, maturity day and grain yield for rice var. Sahyadri was very less i.e. 0.775, 1.342 and 379.94, respectively, The d-stat values of anthesis day, maturity day and grain yield were 0.826, 0.551 and 0.528, respectively. Hence, the genetic coefficients presented in Table 12.2 were found suitable for use in DSSAT/ CERES -rice model for rice var. Sahyadri and it was further tested in studying the impact of climate change scenarios on rice yield. Validation of the DSSAT/ CERES -rice model was done by using observed data for the years 2004-2011.

The data presented in Table 12.3 indicated that, the predicted days to anthesis as well as maturity and grain yield were reasonably close to observed values. The observed and predicted days required for anthesis (averaged over the years) were 83.88 and 84.63, respectively and error percent was also very less i.e. 0.89. During the year 2010, the maximum difference between observed and predicted days for anthesis i.e. 2.4 was observed. In case of days for maturity of rice also the average per cent error was very less i.e. -1.15 per cent. The differences pertaining to observed and predicted grain yield of rice var. Sahyadri were well within the acceptable limits. The average percentage of error for prediction of grain yield was -4.78 per cent. Days taken for anthesis as well as maturity and yield in the years 2020, 2050 and 2080 under four climate scenarios i.e. Representation Concentration Pathways (RCP) 2.6, 4.5, 6.0 and 8.5 were simulated for studying the impact of climate change on phenology and yield of rice.

Table 12.3: Days to anthesis, maturity and grain yield of rice variety Sahyadri as predicted by DSSAT/CERES-Rice model and observed data from the field experiment

Year	Days to anthesis			Days to maturity			Grain yield (kg ha^{-1})		
	Observed	Predicted	error (%)	Observed	Predicted	error (%)	Observed	Predicted	error (%)
2004	85	86	1.2	120	120	0.0	5660	5085	-10.2
2005	84	84	0.0	119	118	-0.8	5784	5399	-6.7
2006	84	85	1.2	118	118	0.0	5445	5130	-5.8
2007	83	84	1.2	119	117	-1.7	5308	5317	0.2
2008	83	83	0.0	120	118	-1.7	5238	4859	-7.2
2009	84	84	0.0	120	117	-2.5	5424	5542	2.2
2010	84	86	2.4	120	118	-1.7	5869	5032	-14.3
2011	84	85	1.2	119	118	-0.8	5148	5329	3.5
Average	83.88	84.63	0.89	119.38	118.00	-1.15	5484.50	5211.63	-4.78

Projection of yield and days required for different phenological stages of rice (Table 12.4) indicated that, in all future climate change scenarios (RCPs), projected yield is likely to be reduced for rainfed rice var. Sahyadri. The range of yield reduction will be from -7.06 to -30.78 across all the years and scenarios. Under all the RCPs (except RCP 2.6), reduction in days taken for anthesis and maturity was higher in the year 2080 compared to year 2020 and 2050. Yield reduction will be comparatively higher in year 2020 under the RCPs 2.6 and 4.6 while it will be very much higher in the year 2080 under the RCPs 6.0 and 8.5. The early maturity and highest yield reduction was observed in year 2080 under scenario RCP 8.5 (18 days early maturity and 30.78 per cent reduction of yield).

Table 12.4: Early maturity and reduction in yield (percentage loss) of rainfed rice var. Sahyadri due to climate change in India.

Sr. No.	Year	RCP	Anthesis (day)	Physiological maturity (day)	Simulated yield (kg/ha)	Per cent increase/ decrease in yield
1	2020	2.6	80 (-5)	112 (-6)	4521	-11.13
2	2050	2.6	78 (-7)	110 (-8)	4728	-7.06
3	2080	2.6	78 (-7)	110 (-8)	4648	-8.63
4	2020	4.5	80 (-5)	112 (-6)	4615	-9.28
5	2050	4.5	78 (-7)	109 (-9)	4627	-9.04
6	2080	4.5	76 (-9)	106 (-12)	4633	-8.92
7	2020	6.0	80 (-5)	112 (-6)	4670	-8.20
8	2050	6.0	78 (-7)	109 (-9)	4648	-8.63
9	2080	6.0	75 (-10)	105 (13)	4377	-13.96
10	2020	8.5	80 (-5)	112 (-6)	4663	-8.33
11	2050	8.5	77 (-8)	107 (-11)	4492	-11.70
12	2080	8.5	71 (-14)	100 (-18)	3521	-30.78
Average 2004 to 2011			85	118	5087	

Figures in the parenthesis indicated decrease in number of days

12.3 Climate Risk Management

12.3.1 Drought

Early season drought (delayed onset) Delay by 2 weeks (3rd week of June)

For Upland medium deep to shallow soils: Prefer early matured variety (Ratnagiri – 73, Ratnagiri – 24, Ratnagiri -1)

For Mid-land medium deep soils: Prefer early matured variety (Ratnagiri – 1, Karjat – 3, Ratnagiri – 24, Ratnagiri – 5, Karjat- 7)

Early season drought (delayed onset) Delay by 4 weeks (1st week of July)

For Upland deep to shallow soils: Prefer very early matured varieties (Ratnagiri – 73, Karjat - 184), Sowing of Sprouted seed

For Mid-land medium deep soils: Early matured variety (Ratnagiri – 73, Ratnagiri – 24, Karjat - 184)

For Low land deep soils: Mid-late duration variety (Ratnagiri – 4, Palghar – 1, Palghar – 2, Karjat - 5)

Normal onset followed by 15 to 20 days dry spell at the time of transplanting

For Upland medium deep to shallow soils

- Increase number of seedlings per hill (5 to 6)
- Increase 25% Nitrogen dose
- For shortage of seedling prepare seedling by mat nursery using short duration variety
- Protective irrigation for nursery
- Adopt closer spacing (15×15 cm), Use water from the outside sources like farm pond, nalas, streams, rivers for puddling operation

For Mid-land medium deep soils and Low land deep soils

- Increase number of seedlings per hill (5 to 6)
- Increase 25% Nitrogen dose
- Protective irrigation , Use water from the outside sources like farm pond, nalas, streams, rivers for puddling operation

Mid season drought (long dry spell, consecutive 2 weeks rainless (<2.5 mm) period) At vegetative stage

For Upland medium deep to shallow soils

- Adopt weed management practices
- Protective irrigation
- Apply spilt dose of Nitrogen after restart of rains.

For Mid-land medium deep soils and Low land deep soils

- Adopt weed management practices
- Protective irrigation
- Maintain the existing water level in the field.
- Apply spilt dose of Nitrogen after restart of rains

Mid season drought (long dry spell) At flowering / fruiting stage

For Upland medium deep to shallow soils: Protective irrigation

For Mid-land medium deep soils and Low land deep soils:

- Protective irrigation
- Maintain the existing water level in the field

Terminal drought (Early withdrawal of monsoon)

For Upland medium deep to shallow soils

- Harvest crop at physiological maturity
- Cowpea, groundnut, water melon, leafy vegetables, raising of seedlings for chilli, brinjal, cabbage, knolknol

For Mid-land medium deep soils

- Protective irrigation
- Harvest crop at physiological maturity
- Horse gram, cowpea, water melons, cucurbitaceous leafy vegerables, raising of seedling for chilli, brinjal, cabbage, knolknol

For Low land deep soils

- Protective irrigation
- Harvest crop at physiological maturity
- Sow field bean, horse gram, cowpea, mustard green gram on residual moisture.

12.3.2 Irrigated Situation

Delayed release of water in canals due to low rainfall

For Mid and low land, Medium deep to deep soils

- Prefer early duration variety (Ratnagiri-73, Ratnagiri-1) or Grow short duration pulses viz. cowpea (Var. Konkan Sadabahar), under control irrigation and tail end area
- Dapog / mat technique of nursery raising. Young seedling transplanting. SRI Technique

Limited release of water in canal due to low rainfall

For Mid and low land, Medium deep to deep soils:

- Prefer early duration variety in low land situation (Ratnagiri- 73, Ratnagiri-1)
- Grow short duration pulses, groundnut, vegetables in midland situation.
- Use SRI Technique of rice cultivation

Non release of water in canals under delayed onset of monsoon in catchment

For Mid and low land, Medium deep to deep soils:

- Field been, horse gram black gram, cowpea, mustard on residual moisture under low land situation
- Minimum tillage and sowing of seed by dibbling

Unusual rains (untimely, unseasonal etc) (for both rainfed and irrigated situations)

If Continuous high rainfall in a short span leading to water logging at maturity stage of rice crop then Drain out water and harvest the crop before lodging. Immediate threshing and drying in shed carry out to reduce the loss. If Outbreak of pests and diseases due to unseasonal rains in rice at seedling stage then Spray Carbendazim 0.1% or Copper oxy chloride 0.25% or control blast disease and if at vegetative stage then spray Carbendazim 0.1% or Copper oxy chloride 0.25% or tricydazole 0.1% to control blast disease.

12.3.3 Floods

Due to flood situations if nurseries washed out then resowing of nurseries / sowing of sprouted seed in puddled field. If flood situation occurs at harvest then Immediately harvesting and threshing should be done and dry produce in shed. If Continuous submergence for more than 2 days occurs during nursery stage and nursery get washed out, then complete resowing of nursery by using mat nursery / sowing of sprouted seed in puddled field. If same situation occurs at vegetative stage then Drain out excess water and apply dose of nitrogen. If situation occurs at harvest then Immediately harvesting and threshing should be done and dry produce in shed.

Summary

Konkan region contributes 42.9 % of total rice production in Maharashtra. Based on area and productivity of rice, *Konkan* region was divided into two productivity zones namely southern rice zone (Ratnagiri and Sindhudurga districts) and northern rice zone (Thane and Raigad districts). Among these two, southern rice zone showed high productivity than northern zone due to high rainfall as well as enriched fertile soils. Long term analysis of area, production and productivity of rice in Konkan region brought out that area under rice had significant decreasing trend due to urbanization, while production and productivity showed increased trend. Highest grain yield of Sahyadri-2 rice hybrid could be achieved under growing environment-I or 24th SMW sown crop than under growing environment-II (26th SMW) due to comparatively higher rainfall (3378 mm) and minimum temperature (22.9 ^{0}C) besides lower maximum temperature (29.1 ^{0}C) and bright sunshine hours (3.7 hrs).

Weather conditions in two major crop stages viz., vegetative and reproductive stages revealed that lower rainfall and higher maximum temperature and sunshine hours during vegetative stage and higher rainfall and lower sunshine hours during reproductive stage were optimum weather conditions for obtaining above average yield and vice versa. Out of all the four weather parameters, rainfall was found to have higher correlation with yield than other parameters during flowering and grain development stages. Maximum temperature during tillering stage showed higher correlation with yield than other three weather parameters. Hence, flowering and grain development stages were identified as critical stages for rainfall and tillering stage was identified as critical stage for maximum temperature. DSSAT/ CERES -rice model for rice was tested to study the impact of Climate change scenario on rice yield. Validation of the DSSAT/ CERES -rice model was done by comparing simulated values with observed data for the years 2004-2011. In all future climate change scenarios (RCPs), projected yield is likely to be reduced for rainfed rice var. Sahyadri. The range of yield reduction was from -7.06 to -30.78 across all the years and scenarios. Under all the RCPs (except RCP 2.6), reduction in days taken for anthesis and maturity was higher in the year 2080 compared to year 2020 and 2050. Yield reduction will be comparatively higher in the year 2020 under the RCPs 2.6 and 4.6 while it will be very much higher in the year 2080 under the RCPs 6.0 and 8.5. The early maturity and highest yield reduction was observed in year 2080 under scenario RCP 8.5 (maturity early by 18 days and reduction of yield by 30.78 per cent).

References/Suggested Readings

Dakhore, K.K. Anil Karunakar, Jadhav, J.D., Jagtap, D.N., Pawar, P.B and Vijayakumar, P. 2017. Agroclimatic Atlas of Maharashtra.234 pp.

Gomez, K.A.; Gomez, A.A. 1984.Statistical Procedures for Agricultural Research.2ed. John Wiley, New York, NY, USA.

Hoogenboom, G., C.H. Porter, V. Shelia, K.J. Boote, U. Singh, J.W. White, L.A. Hunt, R. Ogoshi, J.I. Lizaso, J. Koo, S. Asseng, A. Singels, L.P. Moreno, and J.W. Jones. 2017. Decision Support System for Agrotechnology Transfer (DSSAT) Version 4.7 (https://DSSAT.net). DSSAT Foundation, Gainesville, Florida, USA.

Jones, J.W., G. Hoogenboom, C.H. Porter, K.J. Boote, W.D. Batchelor, L.A. Hunt, P.W. Wilkens, U. Singh, A.J. Gijsman, and J.T. Ritchie. 2003. DSSAT Cropping System Model. *European Journal of Agronomy* 18:235-265.

Rao G.G.S.N., Rao V.U.M., Rao A.V.M.S. and Saikia U.S. 2006. All India Coordinated Research Project on Agrometeorology, Bi-Annual Report 2004-06. Central Research Institute for Dryland Agriculture, Hyderabad. Pp. 170.

Rao G.G.S.N., Rao V.U.M., Rao A.V.M.S. and Saikia U.S. 2007. All India Coordinated Research Project on Agrometeorology, Annual Report 2006-07. Central Research Institute for Dryland Agriculture, Hyderabad.Pp. 162.

Rao G.G.S.N., Vijaya Kumar, P., Rao V.U.M. and Rao A.V.M.S. 2009. All India Coordinated Research Project on Agrometeorology, Annual Report 2009. Central Research Institute for Dryland Agriculture, Hyderabad.Pp. 242.

Rao V.U.M., Bapuji Rao, B., Rajkumar Dhakar, Sarath Chandran, M.A., Vijaya Kumar, P. and Rao A.V.M.S. 2013. All India Coordinated Research Project on Agrometeorology, Annual Report 2012-13. Central Research Institute for Dryland Agriculture, Hyderabad.Pp. 127.

Rao V.U.M., Bapuji Rao, B., Rajkumar Dhakar, Vijaya Kumar, P. and Rao A.V.M.S. 2012. All India Coordinated Research Project on Agrometeorology, Annual Report 2011-12. Central Research Institute for Dryland Agriculture, Hyderabad.Pp. 188.

Rao V.U.M., Bapuji Rao, B., Sarath Chandran, M.A., Vijaya Kumar, P. and Rao A.V.M.S. 2015. All India Coordinated Research Project on Agrometeorology, Annual Report 2014-15. Central Research Institute for Dryland Agriculture, Hyderabad.Pp. 132.

Rao V.U.M., Bapuji Rao, B., Sarath Chandran, M.A., Vijaya Kumar, P. and Rao A.V.M.S. 2014. All India Coordinated Research Project on Agrometeorology, Annual Report 2013-14. Central Research Institute for Dryland Agriculture, Hyderabad.Pp. 116.

Rao V.U.M., Vijaya Kumar, P., Bapuji Rao, B. and Rao A.V.M.S. 2010. All India Coordinated Research Project on Agrometeorology, Annual Report 2010. Central Research Institute for Dryland Agriculture, Hyderabad.Pp. 180.

13

Climate Resilient Agriculture in Chhattisgarh

J.L. Chaudhary

Sr. Scientist, Department of Agrometeorology, IGKV Raipur, India

In order to achieve maximum and sustainable crop production from available farm resources, it is essential to have proper knowledge of the agro-climatic resources of the location/region. Therefore, a thorough understanding of the climatic conditions would help in determining the suitable agricultural management practices for taking advantage of the favorable weather conditions and avoiding or minimizing risks due to adverse weather conditions. better planning of cropping pattern, developing irrigation and drainage plans for an area (Chaudhary *et al.* 2015). Hence, trends of temperature and rainfall are studied in this section with possible impact studies on rice production for Chhattisgarh state. Challenges of rainfed rice production can be understood through trend studies of key weather variables. Analysis of rainfall in the context of rainfed rice production in Chhattisgarh state has been carried out. Analysis has been done at Department of Agrometeorology under All India Co-ordinated Research Project (Raipur) to characterize the key weather parameters statistics.

13.1 Temperature variations and trends

At Ambikapur which is representing Northern Hill Zone, minimum temperature has shown a significant increasing trend during 1954-1983 on annual basis at 1% level while maximum temperature has shown no significant change. At Jagdalpur representing Bastar Plateau ACZ, decreasing pattern in maximum pattern is found for the almost all seasons and almost all periods 1954-1983, 1984-2013 and 1954-2013. Interesting result which has come out for Jagdalpur is that there is almost decreasing pattern in all seasons and also on an annual basis for all time frames. In Raipur, maximum temperature is rising during SWM season in the recent periods and minimum temperature is showing a significant increasing tendency in recent time period 1984-2013 on annual basis, SWM season, NEM season, summer season. At Pendra road, maximum temperature is showing an increasing trend during all time durations and minimum temperature significantly declining pattern only in SWM season (1954-2013). The decreasing trend in minimum temperature during winter season is creating favorable condition for rabi crops particularly in Bastar Plateau zone and there is a potential of increasing cropping intensity with conserved rainwater in monsoonal season. Through long term data analysis, it has been observed that there are variations on different scales in the state. However at Raipur representing Chhattisgarh plains, there are significant increasing trends in minimum temperature at different levels which can have adverse consequences on *rabi* crops.

13.2 Rainfall variations and trends

The knowledge of total rainfall and its distribution throughout the year is very useful for crop planning. The trends of rainfall are studied for different districts. It can be observed that on the database of 1954-2013, there is a significant decreasing trend at different levels in Dhamtari, Jagdalpur, Kanker, Kawardha, Mahasamund, Raigarh and Raipur when annual rainfall is taken into consideration. Over the database of 1984-2013 which is more recent, decreasing trend in annual and SWM rainfall is observed in Ambikapur and significant rising trend in Mahasamund and Kanker. For the southwest monsoon season, there is a significant decreasing pattern in Bilaspur, Kawardha, Raigarh and Rajnandgaon districts during the period 1954-2013. It can be very well observed that there are some districts like Raigarh and Rajnandgaon which are the worst affected and negative departure observed in these districts -20.6 per cent to -15.8 per cent respectively during the period 1991-2013 as compared to 1961-90. Change of south-western rainfall is analyzed and it can be very well observed that in most of the districts, there is negative departure of SW rainfall and it is varying from -0.1 per cent to -20.8 per cent when comparison of these two timeperiods is made. Rainfall qunatity has decreased in recent times by about 20.8 per cent in Raigarh and by about 11.4 per cent in Rajnandgaon. There are some districts where there is positive trend in SWM rainfall and one such district is Korba where increasing and positive trend of 7.6 per cent is being observed in the recent past 1991-2013. Decresaing trend of rainfall is generally going to cause stress in rainfed region of the state as rice based cropping system is pre-dominant.

13.3 Climatic Change Impacts

Crop growth modelling can be used as a tool for climate change impact studies. Here in this case study, calibration and validation of Crop simulation models for different rice varieties has been tried. The potential yield of rice was simulated using CERES-Rice model for 30 years soils (1973-2002) separately for irrigated and rainfed conditions under clay and sandy loam soils and shown in **Table 13.1.** The results indicated that potential yields under irrigated conditions are much higher than rainfed conditions in both soil types. The inter-annual variability in potential yield under irrigated conditions is less when compared to rainfed conditions in both soils. In the same study, impact of climate change in temperature by 1, 2, 3°C over the normal value increased the grain yield under irrigated and to some extent under rainfed conditions. The results revealed that the potential productivity of rice ranged from 8197 to 10550 kg ha^{-1} for irrigated condition, whereas in rainfed condition, the range was 3367 to 8949 kg ha^{-1} for the Kanhar soil and 3955 to 8948 kg ha^{-1} for Matasi (sandy loam) soil. Matasi soil proved better than Kanhar (clay) soil in yield point of view under rainfed condition. On reducing the minimum temperature by 1°C, 2°C and 3°C, the yield increased to 9459 kg ha^{-1}, 9593 kg ha^{-1} and 9655 kg ha^{-1}, whereas, in normal condition the production potential was 9362 kg ha^{-1}. It was found that when the maximum temperature during reproductive stage was increased, the yield was reduced from 9362 kg ha^{-1} in normal condition to 8895 kg ha^{-1}.

Table 13.1: Simulated potential yields of rice under irrigated and rainfed conditions at Raipur (IR-36 variety)

Year	Rainfall (mm)	Potential Yield (q ha^{-1})		
		Irrigated	Rainfed	
			Clay	Sandy loam
1973	1271	92.39	87.99	87.99
1974	798	92.29	33.67	39.55
1975	1209	81.97	73.01	73.04
1976	991	86.11	49.01	57.27
1977	1138	87.35	73.44	74.91
1978	996	88.44	68.82	71.29
1979	623	82.92	42.22	46.85
1980	1586	95.41	72.92	80.56
1981	1132	97.52	84.04	87.63
1982	821	95.99	44.29	50.77
1983	1263	89.96	86.15	86.15
1984	1071	105.5	47.56	54.99
1985	1455	95.87	80.93	83.07
1986	1143	104.84	43.4	45.67
1987	715	96.82	89.49	89.48
1988	669	90.19	71.4	73.58
1989	1219	95.57	73.42	76.93
1990	1097	86.08	82.21	82.21
1991	786	98.83	59.64	62.93
1992	718	103.72	43.6	50.57
1993	1252	99.41	77.92	82.72
1994	1606	99.86	87.71	88.86
1995	1030	97.22	51.07	59.48
1996	1090	97.28	81.21	84.07
1997	1037	87.46	79.63	82.11

Contd.

1998	894	87.21	80.31	80.39
1999	929	86.64	80.64	82.21
2000	669	90.53	50.87	56.01
2001	905	90.06	85.2	84.32
2002	725	97.49	39.74	49.13
	Mean	93.36	67.38	70.82
	SD	6.32	17.78	15.63
	CV	6.77	26.38	22.07

13.4 Seasonal and annual rainfall in El Niño years

Effect of El Niño on seasonal and annual rainfall in different districts of Chhattisgarh state has been studied in the present investigation as per climate change studies. Statistics of rainfall during all El Niño years (weak, moderate, strong El Niño years) as compared to Long Period Average (LPA) of rainfall on the data base of 1951-2012 (Table 13.2) brought out that during southwest monsoon (SWM) season which is the major crop growing season, there is a definite pattern of negative deviation on rainfall quantity, decreasing by about -2 % in Surguja district to -11 % in Korba district. However, no change and slight increase (1%) is noticed in Mahasamund and Kanker districts, respectively. In northeast monsoon (NEM) season, there is decrease in the rainfall form -1 percent (Jashpur) to -23 percent (Dantewada) and positive deviation of 7% in Surguja and 25% in Kawardha. In the case of annual rainfall, it can be seen that during all El Niño years, there is negative departure of annual rainfall in the range of -1 per cent to -10 per cent in different districts of the State. However, there is positive deviation in Kanker district of about 1 per cent.

13.5 Climate change and crop projections

Due to regional climate change, extreme events are arising viz., excess climate anomalies, deficient and flooded rainfall years which have a dramatic impact on the economy as well as on the living conditions of the inhabitants of the affected regions. Therefore, agroclimatic inventory is very vital in crop planning especially in rainfed agriculture. Other georeferenced inventory along with climatic data becomes a strong base not only for planning but also for developing strategies for alleviating the different stresses that come during crop season. The stresses especially climatic stresses vary from drought to floods, cyclones, cold waves, heat waves, wind damage etc.,. Due to some unpredicted extreme events, the farmer suffers loss in the field. Extreme weather in the most obvious sense is the weather that lies outside a locale's normal range of weather intensity. It is therefore by definition, infrequent or rare. Extreme weather is also potentially destructive although not all extreme weather events end in disasters. For some weather events, the idea of what constitutes an extreme could vary from place to place. It often depends on what a region is used to experiencing and what it

Table 13.2: Per cent change in seasonal and annual rainfall (mm) during all El Niño years (weak, moderate and strong) compared to normal rainfall (1951-2012) in Chhattisgarh state

Districts	Southwest monsoon			Northeast monsoon			Annual rainfall		
	Average rainfall	El Niño years	% change	Average rainfall	El Niño years	% change	Average rainfall	El Niño years	% change
Raipur	1051	958	-9	59	49	-17	1169	1073	-8
Durg	1016	914	-10	58	45	-22	1114	1004	-10
Rajnandgaon	1032	942	-9	71	65	-8	1159	1062	-8
Kawardha	876	797	-9	71	89	25	1000	945	-5
Mahasamund	1039	1035	0	53	43	-19	1132	1119	-1
Dhamtari	1005	960	-5	64	53	-18	1113	1069	-4
Kanker	1234	1242	1	69	54	-21	1345	1365	1
Bilaspur	1048	969	-8	66	66	0	1197	1127	-6
Korba	1299	1157	-11	68	59	-18	1396	1333	-4
Janjgir champa	1162	1052	-9	49	42	-12	1265	1151	-9
Raigarh	1208	1111	-8	61	57	-6	1352	1247	-8
ChhattisgarhPlain Zone	1088	1012	-7	63	57	-11	1204	1136	-6
Jashpur	1331	1287	-3	91	90	-1	1535	1478	-4
Koriya	1124	1041	-7	48	39	-18	1212	1115	-8
Surguja	1167	1142	-2	64	69	7	1300	1285	-1
Northern Hills Zone	1207	1157	-4	68	66	-4	1349	1293	-4
Bastar	1243	1234	-1	114	103	-9	1479	1466	-1
Dantewada	1193	1164	-2	108	84	-23	1340	1284	-4
Bastar Plateau Zone	1218	1199	-2	111	94	-16	1410	1375	-3

is prepared for. Study of weather extremes in rainfed agriculture is a vital issue for increasing production and productivity of major crops for livelihood security, employment generation and productivity increase looking into the increasing population pressure. Rainfed agriculture constitutes 80 per cent of global agriculture and plays a vital role in achieving food security. These rainfed areas are vulnerable to poverty, malnutrition, water scarcity, severe land degradation and poor physical and social infrastructure. To eliminate all these severe problems, it is very important to conduct basic and applied research that will contribute to the development of sustainable rainfed farming systems dedicated for every agro-climatic zones separately (Thakur *et al.* 2014). The production of rainfed rice (*Oryza sativa* L.) in drought-prone areas such as eastern India is highly variable and risky due to temporal and spatial variability in rainfall. Rice yields in eastern India only average 1.0-2.4 t/ ha (Tomar, 2002). Extreme weather events induce stress on plant growth and performance. When these extremes exceed the tolerance limits the impact can be severe on plant performance and even its survival capacity often leading to plant damage and mortality. Extreme weather events induce stress on plant growth and performance. Heat waves and cold waves are such climatic aberrations of a short period that can cause severe stress on plant survival and performance. Significant losses have been experienced year after year due to increased occurrence of weather extremes which could be significant part of climate change impact. Both the extremes of maximum and minimum temperatures cause stress to plant productivity (Ramakrishna, 2013). According to Gates (1988), experience of the world is sufficient to convince people that even a temporary change of climate can have profound impact on agricultural production and on the use of energy and water resources. Water stresses are a common feature during growth cycle of crops. Although water in form of precipitation is availble freely and right at the site where it is to be used, yet so tenuous and delicate is the balance between the demand for water by crops and its supply by precipitation that even short term deficit periods often reduce the production significantly (Gupta *et al.* 1990). Interpretation of climatic variables is essential because of these variations (Subramaniam and Raju, 1988).

Chhattisgarh state, situated in eastern India stretches between 80°C 152 to 84°C 242 E longitude and 17°C 462 to 24°C52 N latitude. It covers total geographical area of about 13.5 million hectare. Rice crop is grown in 3.7 million hectares and rainfed rice production has always remained a challenge in this region. Under wide range of farming situations and soil conditions, except upland light soil, rice is widely accepted and grown by the farmers depending upon their socio-economic conditions. During *kharif*, growing of rice is a tradition while, in *rabi*, there are fewer options for the stakeholders to take profitable and/or suitable crops. Under these circumstances, they generally follow rice wheat, rice – mustard and rice – winter vegetables under partially or assured irrigation and rice- fallow, rice – utera (*Lathyrus*, chickpea and linseed) under rainfed situation. The state of Chhattisgarh is endowed with three distinct agro-climatic zones viz., Chhattisgarh Plains, Bastar Plateau and Northern Hills zone.

Agro-climatic Zone	Districts covered
Chhattisgarh Plains Zone	Raipur, Mahasamund, Dhamtari, Gariaband, Balodabazar, Durg, Balod, Rajnandgaon, Kawardha, Bemetara, Mungeli, Bilaspur, Korba, Janjgir-champa, Raigarh and Kanker
Bastar Plateau Zone	Bastar, Kondagaon, Sukma, Bijapur, Naraynapur and Dantewada
Northern Hills Zone	Surguja, Koriya, Jashpur, Surajpur and Balrampur

13.5.1 Impact assessment under different climate change scenarios and identification of adaptation strategies (*Karma mahsuri* variety)

By running DSSAT model, it is observed that in Raipur region the overall predicted rice yield in three different growing environments (11th June, 21th June and 29th June 2011) of sowing shows that second growing environment (21th June 2011) gives higher yield of 5094, 5395, 5463 and 5803 kg per hectare for the year 2011, 2020, 2030 and 2040 respectively (Verma *et al.* 2014). The trend analysis is done by considering current year as a base year value. The trend shows overall increasing yield in second growing environment. Yield increases by 301 kg/ha and 709 kg/ha from the present yield (5094 kg/ha) for the year 2020 and 2040 respectively (Table 13.3 and Fig. 13.1). On the other hand increasing trend of yields for future projections has been observed in 2012 year data in third growing environment (22th June 2012). On running of simulation model it was observed that taking 2012 yield (4565 kg/ha) at reference level, projected changes are by 17.17%, 17.22%, 29.09% and 33.14% in 2020, 2030, 2040 and 2050 respectively and then decreasing by 8.87 % (2080 projection). The calibration model may be operationally used with routinely observed soil, crop and weather parameters. In this region significant deviation has been observed. Thus in central Chhattisgarh, overall positive output in yield with a decreasing trend in later phase is obviously a matter of concern. The planners and farmers should be aware of the fact. Low cost adaptation strategies like finding the suitable growing environment, improved variety with higher water use efficiency etc., must be followed towards yield sustainability.

Table 13.3: Variation of predicted yield in two different years of Raipur station

Raipur Agroclimatic Zone	Growing enviro -nment	Yield (kg/ha)						Yield difference compaired with current year (%)				
		2011	2020	2030	2040	2050	2080	2020	2030	2040	2050	2080
Karma Masuri	I	4849	4668	3936	3369	2097	4668	-3.73	-18.83	-30.52	-56.75	-3.73
	II	5094	5395	5463	5803	5420	5395	5.91	7.24	13.92	6.40	5.91
	III	4000	3912	3768	3719	3795	3912	-2.20	-5.80	-7.03	-5.13	-2.20
		2012	**2020**	**2030**	**2040**	**2050**	**2080**	**2020**	**2030**	**2040**	**2050**	**2080**
Karma Masuri	I*	3486	4609	3829	3216	2105	966	32.21	9.84	-7.75	-39.62	-72.29
	II*	3463	5170	4809	3725	3457	1589	49.29	38.87	7.57	-0.17	-54.11
	III*	4565	5349	5351	5893	6078	4160	17.17	17.22	29.09	33.14	-8.87

*Growing environment created as per nursery sowing dates on 10 June, 15 June and 22 June.

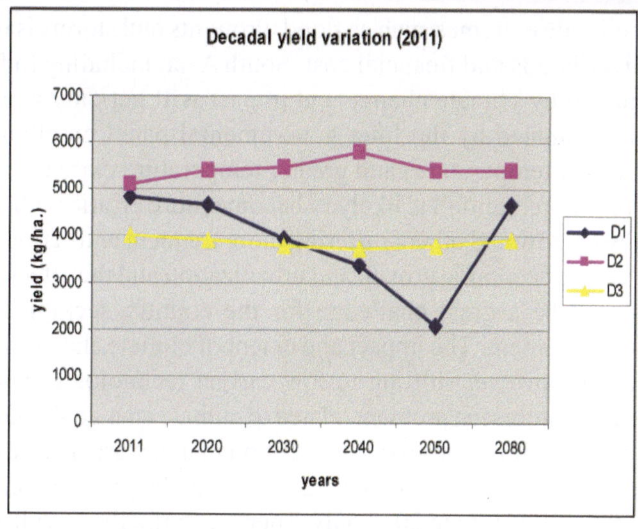

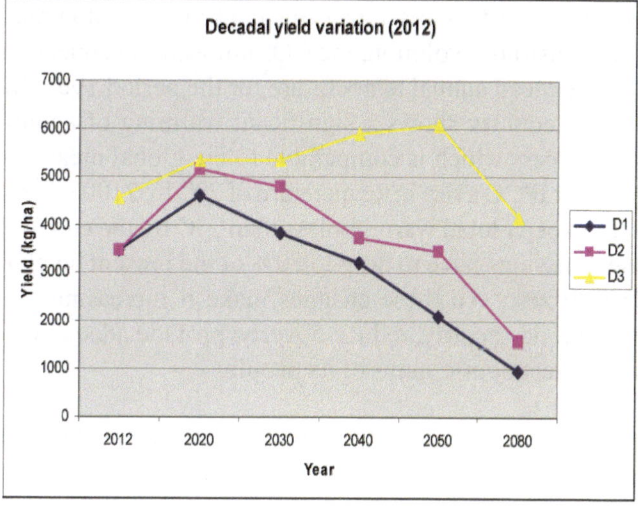

Fig. 13.1: Decadal simulated yield variation of rice: future projections

13.6 Climate risk management

Climate change and global warming have become significant issues in relation to agriculture in recent times. Experience of the world is sufficient to convince people that even a temporary change of climate can have profound impact on agricultural production and on the use of energy and water resources. Climate change poses monumental risks for agricultural systems thereby impacting economic growth and development, poverty reduction, food and livelihood security. The impact of higher temperatures and erratic rainfall are particularly being felt by the poorest and the most vulnerable section of society. Increasing incidence of climate extremes such as flood, droughts and storms is resulting in loss of life, livelihoods and financial cost. South Asia, including India is being seriously affected by climate change and impact will get only worse as time goes on. It is estimated by the Inter governmental panel on climate change (IPCC) that higher temperatures and greater temperature extremes will hit the country. Furthermore, rainfall is likely to become more erratic with some areas more than they do now and others suffering more frequent and serious droughts. Coupled with rapid economic growth and urbanization and the effects of climate change, this will be a great challenge for the country seeking sustainable agricultural development. The impact and extent of climate change in the future is still uncertain, however shifting to low carbon technologies and adaptive strategies to curb unforeseen problem related to climate change should constitute priority policy focus. It is worthwhile to mention that in warm periods of the last 800000 years, CO_2 concentrations peaked at around 300 ppm whereas the same has already passed 400 ppm in 2014 only. There is compelling evidence (IPCC) that the recent spurt in CO_2 emissions is the consequence of human activities. Ever since the industrial revolution, the CO_2 atmospheric concentrations have risen by 40%. The mean annual temperature for the period 1901-88 represents 73 stations of the country shows a significant warming of about 0.4 degree centigrades /100 years which is comparable to the global mean trend of 0.5 ° C/100 years. As per IPCC a rise in temperature of 2°C by 2100 will be dangerous and therefore, to keep global warming less than 2°C by the end of the century, human carbon emissions have to drop to 50% of the present level by 2050 and thereafter drop to zero. All these changes make it increasingly difficult for India to increase food security, reduce poverty, provide adequate housing and improve the socio-economic status of its people.

13.7 Looking into the regional climate change studies in Chhattisgarh state, some policy issues have been documented hereunder:

13.7.1 Policy Issues

1. To develop model for climate resilient agriculture to address scientific questions on attribution and projection of regional climate change.
2. To develop sector specific plans for developing early warning forecasting system for increasing crop productivity.
3. To study the impact of climate change on forests and agricultural biodiversity and develop conservation and management approaches.
4. To study the impact of climate change on farm incomes, livelihood and food security and provide alternatives for mitigation strategies.

5. To identify areas and genotypes suitable for climate resilient agriculture and develop appropriate agro-techniques interventions.
6. Documentations and validation of traditional and indigenous knowledge related to climate resilient agriculture In Chhattisgarh.
7. Promotion and creation of awareness and skilled development of farmers, young educated unemployed youth and woman on climate resilient agriculture.
8. To carry out economic analysis of climate resilient agriculture across the agro- ecological zones of Chhattisgarh.

Even in areas where rainfall is ample, it is unevenly distributed affecting the crop yields due to excess water at one time and due to water stress at the other. Therefore, natural occurrence of rainfall both in time frame and space has to be managed to correspond to agricultural needs. It is obvious that low and fluctuating rainfall during crop growing season is a constraint for improving crop production in rainfed regions. In rainfed agriculture, the importance of rainfall over-rides all other climatic factors which determine the yields. With almost negligible irrigation facilities in rainfed regions, rain water shortages often lead to moisture stress causing substantial reduction in crop yields. This is also true for Chhattisgarh state as the onset of monsoon at the initial stage, break monsoon conditions among the crop growth stages and cessation of rainfall at the terminal stage determine the productivity of rice and other *kharif* crops. Looking into the discussion held, an attempt has been made to study the weather extremes in the state of Chhattisgarh under All India Co-ordinated Research Project on Agrometeorology so that adaptation and mitigation strategies can be documented and applied under field conditions as model testing for its implementation at wider scales. This will also help in adopting and developing risk management strategy.

13.7.2 Policy interventions

(i) Rainwater harvesting

*Farm Ponds

*Percolation tanks

*Dug wells

*Nala Bandhan

*Check dams

(ii) Introduction of Pressure Irrigation:

*Drip irrigation

*Sprinkler

*Pipe irrigation

(iii) Moisture conservation techniques:

*Use of plasticulture

*Contour bunding

*Staggered trenches

*Gabien structures

*Soil mulch/ straw mulches

(iv) Crop/ cropping system management :

*Early duration rice

*Drought resistant varieties

*Crop diversification

*Broadbed furrow system

(v) Farming system approach

*Dependancy on allied enterprises

*Different components of farming system viz, animals, dairy, poultry, fisheries, lac cultivation, honeybee and mushroom etc.

(vi) Use of Farm Mechanization in agriculture

*Use of mechanization for crop establishment, interculture, harvesting etc.

*Custom hiring approach for completing weather sensitive farming operations in time

(vii) Strengthening of real time agro-advisory dissemination

(viii) Technological interventions

*Dry line sowing

*Chemical weed control

*Mechanized crop establishment

*Ridge furrow for dryland crops

*Selection of crops and varieties

(ix) Weather based crop insurance for different crops

(x) Development of assured irrigation sources

*Tubewells/ Dugwells

(xi) Employment generation through entrepreneurship development.

(xii) Capacity building through training programmes at village level and exposure visits.

13.8 Rice varieties against abiotic stress

Study on extreme events and their risk management strategies was conducted at Department of Agrometeorology, Raipur (Chhattisgarh) by using analysis of available meteorological data of Raipur centre. Time trends for various threshold values of maximum and minimum temperature were worked out using weather cock software developed by CRIDA, Hyderabad (Rao *et al.* 2011). Details of climate risk management for different categories are given in Table 13.4

13. 9 Extent of sensitive area and periodicity of drought and dry spells

Whole of Chhattisgarh state is affected due to intermittent dry spells and the *kharif* monsoonal crops mainly rice, maize, arhar get affected during one part or other part of the life cycle of crops. High water requirement crops like rice during *kharif* season get affected severely as compared to other low water requirement crops like Soybean and minor millets. Generally terminal drought and dry spells in rice crop, early season-mid season and late season dry spells in *kharif* season crops i.e. rice, pigeonpea, maize are the factors which cause low yields. In recent years drought occurred in 2002, 2007, 2009, 2011 in the state and dry spells occur almost every year. Adaptation and mitigation strategies recommended are indigenous water harvesting structures and OFRs (On Farm Reservoirs) and local traditional techniques like trenching, indigenous drip method etc. can be applied for mitigation of this extreme event. RCC (Restructured concrete cement) wells and embankment type ponds are some other techniques. Looking into the decreasing rainfall pattern, feasible strategy is to alleviate this limitation is to harvest excess rainwater in a farm pond or OFR (On Farm Reservoir) during the wet season and use conserved water for crop production in both wet (as insurance against drought) and dry seasons by adopting suitable crop and cropping systems (Rathore *et al.* 1996). As far as economics of the technology is concerned, these are low cost technologies and cost can be recovered by increased output and profitability in only few years. Other

Table 13.4: List of varieties of rice crop with abiotic stress tolerance in different agro-climatic zones of Chhattisgarh

(Name of the state : Chhattisgarh state)
Names of varieties recommended for

	Delayed monsoon	Drought tolerance	Flooding/ submergence tolerance	Heat tolerance	Cold tolerance	Source of seed availability
Agroclimate zone I (Chhattisgarh Plains zone)	Vanprabha, Indira Rajeshwari, Indira Sona, Aditya, Danteshwari	Poornima, Tulsi, Kranti, Indira Barani Dhan -1, Samleshwari, Annada, Aditya, Kalinga-3	Jaldubi Baleshwari, Mahamaya	Karma Mahsuri	-N/a-	IGKV seed farm and C.G. Rajya Beej avam Krishi Vikas Nigam
Agroclimate zone II (Northern Hills)	Indira Rajeshwari, Vanprabha, Indira Sona	Poornima, Tulsi, Kranti, Indira Barani-1, Samleshwari, Annada, Aditya,Kalinga-3	Jaldubi Baleshwari, Mahamaya	Karma Mahsuri	-N/a-	IGKV seed farm and C.G. Rajya Beej avam Krishi Vikas Nigam
Agroclimatic zone III (Bastar plateau)	Vanprabha,Annada, Aditya, Samleshwari, Indira Sona	Pradhan, Lalu Chanda, SatkaVandana, Poornima	Jaldubi Baleshwari, Mahamaya		-N/a-	IGKV seed farm and C.G. Rajya Beej avam Krishi Vikas Nigam

traditional low cost technologies like indigenous drip etc. can also be adopted. Adoption and economic benefits of some case studies are that these technologies are recommended and OFRs are established techniques, RCC wells are picking up through watershed projects and already tested in NAIP and NICRA-AICRPDA projects in tribal region of Bastar division. Details of benefitted farmers and economic benefit for NATP farmers are shown in Table 13.2.

Dry spells can be mitigated in different situations through some contingent planning as per points mentioned below:

A. Delayed onset of monsoon (By 2 weeks) under midland inceptisol (matasi-sandy loam/shallow lowland alfisols (dorsa-clay loam or vertisiols (kanhar-clayey)/ Bahra lowland vertisols (kanhar-clayey)

(i) Direct dry seeding in line technique suggested for better crop yield and double cropping

(ii) Line sowing to avoid mortality of germinating seed in case drought follows after scanty rainfall events

(iii) Promote application of post emergence herbicide for timely weed management and avoiding *biasi* operation

B. Early season drought: monsoon delay by 4 weeks (July 3rd week)

(i) 25 % higher seed rate in mung, urad, sesamum , groundnut or Erect variety GG-5/G-20 in groundnut under bharri situation .

(ii) Rice crop can be replaced by groundnut or sesamum/ soybean crop (Indira soy-9, JS93-05, JS-335, JS80-21) or if the farmers have to go for rice crop, Tulsi, Indira barani dhan-1, Annada varieties should be chosen in upland unbunded *bharri* situation.

(iii) In rice crop, varieties like Chandrahasni, IR-64, Mahamaya, Bamleshwari, Karma mahsuri should be chosen.

C. Early season drought: Delay by 6 weeks (August first week)

(i) Instead of Mung, urad crops, horsegram and niger crops should be chosen for sowing under upland unbunded *bharri* situation.

(ii) In groundnut growing regions, urad should be grown replacing groundnut. Varieties suitable for urad crop are PTU4, TU94-2, pant-U31, KU96-3, TAU2. Mung should be grown instead of seasmum.

(iii) In bunded upland *bharri* situation, sowing of sprouted seed (*lai-chaupa*) adopting lehi method of rice cultivation should be adopted. If farmers prefer to replace rice crop at this stage, mixed or intercropping of pigeonpea and mung (4:2) or mixed/ intercropping of sesamum and mung (4:2).

(iv) Rice varieties Indira barani dhan-1, Samleshwari, Danteshwari, MTU-1010, Purnima should be selected in midland/ lowland soils if farmers want to go for rice crop at this stage.

(v) Coriander crop should be chosen instead of lathyrus and safflower.

D. Normal onset of monsoon, mid season-vegetative stage and terminal drought

- Inter tilling for soil mulch
- Mulching with paddy straw or use plastic mulch or other locally available material
- Compartmental bunding, ridge and furrows, tied ridges to conserve rainwater during *kharif* for regular sowing of rabi crops
- Gap filling, sowing of sprouted seed (*lai-chaupa*) adopting lehi method of rice cultivation and sowing of relatively early varieties like IR-64, Chandrahasni Bamleshwari, Karma mahsuri in midland and lowland farming situations.

E. Mid season drought (long dry spell, consecutive 2 weeks rainless (<2.5 mm period) at vegetative stage in all soil types for rice and any other crop grown viz. mung, urad etc.

- Weeding and protection against insect and pests
- Avoid top dressing of urea
- Supplemental irrigation from water harvesting structures using micro-irrigation i.e. drip and sprinklers
- Increase 25 percent seed rate of rabi crop and sowing of rabi crops adopting zero tillage technique

F. Terminal drought (Early withdrawal of monsoon)

Harvest mature plants, thin out plant population, life saving irrigation if available as crop management practices and mulching and inter-tilling as agronomic measure.

G. Drought - Irrigated situation or delayed release of water in canals due to low rainfall

- Direct seeding of rice preferably dry seeding in line
- In case of failure of crop or poor crop stand then Sowing of sprouted seed (*lai-chaupa*) adopting lehi method of rice cultivation

- Avoid transplanting of rice
- Weed control by herbicide

13.10 Flood and submergence, waterlogging and wet spells

Mainly districts of Bijapur, Konta, Sukma of Godavari plains are affected by floods though temporary waterlogging phenomenon in lowlying areas is a routine feature in all the districts. Jowar, maize crops are getting affected in areas of streambanks during July-Aug when water table is high and situation is aggravated when overflowing occurs. Latest year of flooding was in year 2010. adaptation and mitigation strategies against flooding recommended is indigenous broadbed furrow maker through tractor as crop management technique, secondary reservoir techniques. SRI marker for grid making, tractor drawn raised bed planter, tractor drawn ridger, SRI technique for drainage Mechanization can help a lot in mitigating wet spells or flooding situations through low cost technology. All these techniques are recommended techniques. In horticultural and vegetable crops, it is recommended to go for surface drainage in tomato/ brinjal/ coriander/ garlic/ onion crops.

13.11 Hailstorms

These kinds of events are occasionally observed. Such type of hailstorm did not occur earlier at Raipur but now some places are experiencing hailstones particularly in the month of March. Latest such instance in Raipur was in 2006 and 2014. The diameter of hailstones was as much as 3-4 cm. The intensity has been observed very high as the fields and lawns were covered with 2-3 cm layer of ice within 10-15 minutes. Severe crop damage was reported in chickpea, wheat, mustard, lentil and linseed crops which were in maturity stage. During this period 14.0 mm rainfall was received at Agromet observatory, IGKV, Raipur. The hailstorm has been observed in 4-5 km radius of Raipur city while the weather remains clear in other parts of the state. Hailstorm occurred on 07.03.06 in between 2.05 to 2.20 PM Losses in wheat crop were estimated in the AICRP expt. Losses to the extent of 30-40 per cent was reported in mustard, gram, linseed and tall varieties of wheat. Vegetables like cabbage, cauliflower, onion and leafy vegetables also get affected. Villages Jarod, Mahub-hata, Nawagaon, Chanderi, Ghatoli, Dongaria, Khapradih, Khokhli , Diggi, Suhaila (District Baloda-bazar were affected by hailstones in 2014. Gram, wheat, mustard and fieldpea mainly affected as the crops were in physiological maturity stage and affected by grain shattering, fungal infection and sprouting. Wheat crop which is in soft dough stage have shown fungal infection, affected severely due to lodging effect in these villages. Vegetables like chillies, cabbage, capsicum and tomato are affected due to fruit dropping and rotten in the field. Horticultural and vegetable crops viz., Papaya, onion, okra, banana also affected in these villages. Animal fodder mainly paddy straw is damaged. Harvested produce which is kept in the field is damaged significantly. Summer fruits like watermelon, muskmelon, cucumber and cowpea are also damaged.

Agromet advisory issued for the framers was that apply booster dose of fertilizer in standing crops where lodging and uprooting has been observed. Farmers were advised to postpone the harvesting operation. It was also advised to the farmers for pre-poning the plucking of vegetables.

13.2 Cold wave

Surguja district in Chhattisgarh and sometimes some districts in Chhattisgarh plains during winter season get affected by cold wave. Rabi crops like chickpea, linseed, lathyrus get affected to different degrees. Cold wave effect is mainly due to metabolic action and physiological mechanism. Extent of damage reported is not assessed and very less effect and that too on few crops. Adaptation and mitigation strategies recommended/ indigenous is heaters, sprinkler irrigation to rabi crops and providing light irrigation to the rabi crops grown including vegetables.

13.13 Duststorms

It is very occassional feature and there is general crop damage as per severity. Not assessed and very less effect and that too on few crops.

13.14 Heatwave

It is routine feature in districts of Chhattisgarh plains and there is general crop damage as per severity. Horticultural crops and vegetable crops are more seriously damaged. Poultry and livestock get affected due to shortage of drinking water. Moderate effect on crops but poultry and livestock get affected extensively. To mitigate this stress, adaptation and mitigation strategies are recommended is to go for heat tolerant varieties for vegetables and fruit crops, sprinkler or indigenous shade structures, mulching and latest developing green house/ net house.

Summary

There are evidences to suggest that there is a considerable regional climate change including rainfall departures on negative side impacting rural livelihood system and agricultural production –the source of food, income and employment security for majority of its people. Chhattisgarh state has among the lowest income, agricultural productivity and poverty concentration. Thus, there is a direct link between climate change and food and economic security of the agrarian people of the state. The most vulnerable groups of the state live in drought and flood prone areas, forests and rainfed farming situations. Even small unforeseen changes in the climate could cause irreversible losses and may force these people into destitution. The proposed policies should aim to detail how climate change is likely to impact particular districts and sectors within the state and generate some credible estimates of the future scenarios in addition to suggesting mitigating and adaptive strategies for sustainable agricultural development, increasing production and productivity, food and livelihood security. Regarding temperature at Jagdalpur station representing Bastar district, minimum temperature has gone down in winter season and this is creating favourable crop conditions for rabi crops and this should be made use of by the policy makers and researchers, extension workers. At Ambikapur, representing northern region of the state, minimum temperature has shown significant decreasing pattern at different levels during NEM, SWM, winter seasons and also on annual basis during 1991-2013.

Policy interventions like Rainwater harvesting (Farm Ponds, Percolation tanks, Dug wells, Nala Bandhan and check dams) are being promoted trough MNREGA scheme. Moreover there is stress on pressure Irrigation (drip irrigation, Sprinkler and Pipe irrigation) which should be made use of. Moisture conservation techniques need to be stressed and in this plasticulture, contour bunding, staggered trenches, gabien structures and soil mulch/ straw mulches should be made use of. Through research in Indira Gandhi Agricultural University and field experiences, crop/ cropping system management like early duration rice varieties, drought resistant varieties need to be chosen looking into the decreasing rainfall quantum and inter-mittent dry spells being experienced among crop growing season. Crop diversification is another technique which need to be followed as replacing high water requirement rice crop by oilseed and pulses is need of the hour to save huge quantities of water and spread acreage under other crops. Broadbed furrow system has been found effective in Soybean based copping system. Moreover, nowadays time has come of IFS techniques in which allied enterprises and different components of farming system viz, animals, dairy, poultry, fisheries, lac cultivation, honeybee and mushroom etc. can be included in the crop mainstream. Use of farm Mechanization in agriculture for crop establishment, interculture, harvesting etc. should be stressed and in this custom hiring approach for completing weather sensitive farming operations in time has become quite popular. Rural youths can be involved to meet these rural needs. Strengthening of real time agro-advisory dissemination has become a challenging task in changing climate scenario and ICT techniques can help a lot in achieving this target. There are technological interventions like dry line sowing, suitable chemical weed control, mechanized crop establishment, ridge furrow for dryland crops and selection of appropriate crops and varieties which can help in climate change mitigation techniques. Weather based crop insurance for different crops is coming up nowadays as national and state polices and therefore it needs to be applied for field crops and horticultural crops. This will largely compensate the farmers in case extreme events and recurring drought and dry spells and occasional wet spells cause losses for vegetable, horticultural and field crops. Further, there is a need that farmers are going for development of assured irrigation sources in their fields though tubewells/ dugwells through state govt. aided schemes so that life saving irrigation is available. In addition to it, employment generation through entrepreneurship development will largely help in mitigating the adverse effects as this will largely help in increasing the profitability of field crops and allied enterprises. To add to all this, capacity building of farmers through training programmes at village level and exposure visits of farmers to profitable and functional models will largely help the farmers in mitigating the adverse effects of climate change.

References /Suggested Readings

Chaudhary, J.L., Neha Sinha, S.R.Patel, Rajesh Khavse and N. Manikandan. 2015. Analysis of rainfall for strategic crop planning under rainfed rice based cropping system in Chhattisgarh. *Ann. Agril. Res. New Series* 36 (2): 184-190

Das, S., 2013. Thunderstorms, hailstorms and cloudbursts. Agrometeorological aspects of extreme weather events, Central Research Institute for Dryland Agriculture, Hyderabad, 113 pp.

Ghadekar, S. R. and K. K Thakare 1991. Some studies on rainfall climatology of the Nagpur region. *Mausam* 42 (1): 57-64.

Nayak, M. A., and Ghosh, S. 2013. Prediction of extreme rainfall event using weather pattern recognition and support vector machine classifier. *Theoretical and Applied Climatology.* 114(3/4):583-603.

Pali, G.P., Vikas Pali, G.K. Sharma, J.L. Chaudhary, R.S. rajput and R.M. Savu. 2012. Water harvesting for drought alleviation: lesson from Chhattisgarh. *Jornal of Agricultural Issues,* 17 (1 &2): 7-13.

Rao, V.U.M., A.V.M.S. Rao, G.G.S.N. Rao, T. Satyanaryana, N. Manikandan and B. Venkateswarlu, I. Ramamohan. 2011. Weather Cock software. Central Research Institute for Dryland Agriculture, Hyderabad.

Ramakrishana,Y. S., 2013. Heat and cold waves their effect on agriculture and allied sectors. *Agrometeorological aspects of extreme weather events, Central Research Institute for Dryland Agriculture, Hyderabad,* 137 pp.

Rathore, A.L., Pal, A.R., Sahu, R.K. and Chaudhary J.L. 1996. On-farm rainwater and crop management for improving productivity of rainfed areas. *Agricultural Water Management* 31: 253-267

Subramaniam, A.R. and P.A.N. Raju. 1988. Rainfall variability and crop production in south coastal Andhra. *Fertilizer News,* 33 (4): 39-49.

Thakur, D.S., G.K. Sharma, R.K. Naik, D. Kahlkho, J.L. Chaudhary and S.K Patil. 2014. Evaluation of rice establishment methods with nutrient and weed management options under rainfed farming situations of astar Plateau. *Journal of Soil and Water Conservation* 13 (1): 36-41.

Tomar, V.S. 2002. The beushening system of rice crop establishment in eassternm India. In: Pandey, S., Mortimer, M., Wade, L., Tuong, T.P., Lopez, K., Hardy B. (Eds.), Direct seeding: Research strategies and opportunities. International Rice Research Institute Los Banos, Philippines, pp. 53-73.

Verma, Praveen K. S. Bhelawe, J.L. Chaudhary and S.R. Patel. 2014. Estimating productivity of Rice (Oryza sativa) in Chhattisgarh Plain zone for the future climate conditions using DSSAT v. 4.5. rice model, *Int. J. Agricult. Stat. Sci.,* 10(1): 217-220, ISSN : 0973-1903- (NAAS score 5.13).

14

Climate Change and Agriculture in Gujarat: Retrospective and Prospective View

M M Lunagaria and Koyel Sur

Department of Agricultural Meteorology, BACA, AAU, Anand, Gujarat, India

The Inter-governmental Panel on Climate Change has confirmed that change in climate system is gradually leaving a strong impact. Observations undoubtedly show that there is increasein global temperatures, wide spread melting of snow and ice, and rising global mean sea level. Global climate has shown warming of 0.89 [0.69 to 1.08] °C over the period 1901–2012 which is mainly attributed to anthropogenic activities. Scientific findings indicate that warming is more pronounced than expected and the impact would be particularly severe in the tropical areas, which mainly consist of developing countries, including India (Shukla, 2016). Thus, climate change is a great challenge which is continuously being faced by the sustainable economic growth of the India. Climate sensitive sectors like agriculture, forestry, hydrology oriented sectors faces major threat due to projected changes in climate. India Meteorological Department (IMD) in 2012 reportedincreasing temperature trends during last 112 years and increase in heavy rainfall events and decrease in low and medium rainfall events over India. Changes inrainfall and temperatures have also been reported at various place by Dash *et al.*,2009, Arora *et al.*, 2005, De *et al.*, 2005, Guhathakurta and Rajeevan., 2008, MoEF 2010, Jones and Briffa.,1992, Kothawale*et al.*, 2010, Tyagi and Goswami., 2009 and others. Agriculture is likely to be hit by climate change and the Indian economy mostly agrarian based will get worst impact.Several international and national level interventions have taken initiatives foranalyzing the behavioral pattern of climate change.However, it was felt that there is a need for the regional level analysisparticularly, within the state in order tofindtrends that depicts immediate attention at root level. Keeping this in view, many researchers carried out studies at regional level. Systematic collection of literature with reference to climate change in regional level over most of the states of India is being done. These assessments will help in finding relationship between climate change and the vulnerability to human life in long term.

Gujarat is the western-most state of India, with coastline of 1,600 km on Arabian Sea it displays considerable heterogeneity in terms of agro-climatic conditions. Gujarat covers an area of 19.6 million hectares it is the fourth largest state of India, administratively, it is divided into 33 districts. As per Census 2011, about 3.47 cores people of the state live in rural areas forming about 57.4 percent of its total population (GOI, 2011). Gujarat's agricultural and allied sector has outshined its peer states in the last decade by clocking an average annual growth of 11 per cent, compared to an all-India average of 3 per cent between 2001-02 and 2011-12 therefore automatically it forms a source of interest for all researchers.Traditionally Gujarat is broadly divided into three major Agro-Climatic Zones (i) the main land plains extend from the Rann of Kutch and the Aravalli hills in the north to Damanganga in theSouth, (ii) the hilly peninsular region of Saurastra and the rocky areas of Kutch and (III) the north-eastern hill tract. However it is divided into 8 sub agro-climatic zones (Fig. 14.1). The agro climatic zones of Gujarat based on climate and soil types are furnished in Table 14.1.

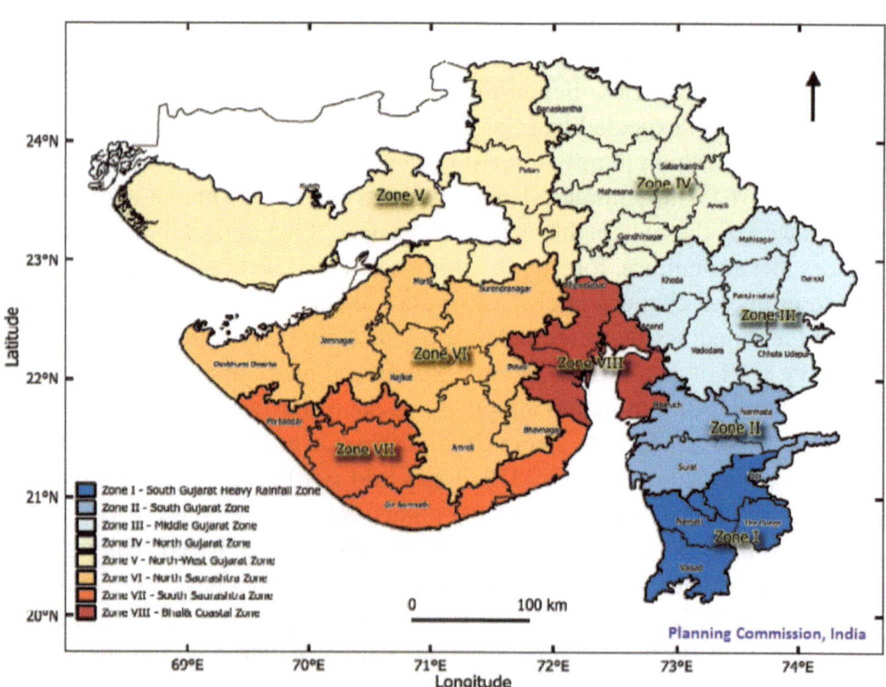

Fig. 14.1: Agroclimatic zones of Gujarat

Table14.1: Soil and rainfall of Agroclimatic zones of Gujarat

Agro climatic zone	Soil	Rainfall (mm)
South Gujarat (Heavy Rain Area)	Deep black with few patches of coastal alluvial, laterite and medium black	1500 and more
South Gujarat	Deep black clayey	1000-1500
Middle Gujarat	Deep black, medium black to loamy sand	800-1000
North Gujarat	Sandy loam to sandy	625-875
Bhal & Coastal Area	Medium black, poorly drained and saline	625-1000
South Saurashtra	Shallow medium black calcareous	625-750
North Saurashtra	Shallow medium black	400-700
North West Zone	Sandy and saline	250-500

Gujarat mostly falls under mega thermic category with mean soil temperature exceeding 28°C. Minimum air temperature remains over 10°C and maximum temperature in May goes around 40°C in north and north-west Gujarat however coastal areas of Saurashtra and south Gujarat has moderate type of climate. Average rainfall received is around 845 mm, spatially it ranges from about 250 mm in north-west to 2,000 mm in south-east. Kutch, Saurashtra and North Gujarat regions are more prone to drought due to poor rainfall received. This reduces yield of important cash crop as well as food crops, leading to the socio economic prospect of the State.

Trend analysis of climate change pattern overany region has received a great response of the scientific society recently (Tabari and Talaee, 2011). Gujarat state as a whole shows warming trends with decrease in diurnal temperature range and increase in precipitation (Rathore *et al.*, 2013). Though, location specific climatic variables' trends at different stations of Gujarat reveal ambiguity in change pattern (Lunagaria *et al.*, 2012). Most of the studies do not show uniform change in pattern at all parts of the state since changes cannot be seen only by shift of mean weather variables but also in frequency and trends of climate extremes. Analysis of temperature and rainfall over the Gujarat using the IMD gridded data for 30 years clearly indicated an increase by 0.11°C during the past 40 years (1969-2005) (Ray *et al*, 2009).Station wise trend showed increase in temperature over Saurashtra region, apparent clearly than the rest of the state. Cold wave conditions have decreased considerably over the past decade indicating an increase in night temperature. It is observed that the heat wave conditions have increased in southern part of the state. Table 14.2 and Table 14.3 give decadal frequency of heat wave and cold wave conditions in selected stations of Gujarat.

Table 14.2: Decadal Frequency of Heat Wave Conditions in Selected Stations of Gujarat

Period Station	1969-1978	1979-1988	1989-1998	1999-2008
Ahmedabad	20	4	17	13
Baroda	16	12	9	28
Deesa	20	5	10	29
Bhuj	31	27	42	16
Rajkot	9	1	14	22
Veraval	53	52	34	72
Naliya	17	35	30	52
Okha	5	1	2	1
Surat	54	43	25	32

(*Source:* Ray *et al.*, 2009)

Table 14.3: Decadal Frequency of Cold Wave Conditions in Selected Stations of Gujarat

Period Station	1969-1978	1979-1988	1989-1998	1999-2008
Ahmedabad	52	72	35	46
Baroda	67	47	22	6
Deesa	60	32	15	47
Bhuj	146	111	144	10
Rajkot	95	33	10	21
Veraval	54	43	21	4
Naliya	107	33	19	27
Okha	4	3	4	2
Surat	48	25	0	62
Bhavnagar	116	50	8	5

(*Source:* Ray *et al.*, 2009)

Heat waves not only influence rise in temperature during summer but effect monsoon season of the state also. Strong correlations have been noticed with heat wave and Multivariate Drought Index by the researchers because average rainfall amount is very low compared to the average potential evapo-transpiration rate resulting in moisture deficit (Bandyopadhyay *et al.*, 2015). Temporal and spatial analysis of rainfall variability also shows that floods and drought occurs side by side over the state (Pandey *et al.*, 1999). Mean seasonal rainfall analysis shows that rainfall has increased over Saurashtra and South Gujarat region (along west coast) and has remained almost same over North Gujarat region and adjoining Kutch region. Dangs witness highest rainfall (1,792 mm) and lowest rainfall variability (29%) were as Kutch has less rainfall (<350mm) and highest annual variability (57%) and on monthly basis coefficient of variability is still higher in Kutch even in monsoon months (Pandey and Patel, 2011).

Table 14.4 Showed decadal frequency of heavy rainfall events in selected stations of Gujarat. According to this table rainfall pattern shows almost a uniform trend over the decades, variation observed over the years is very less.

Table 14.4: Decadal Frequency of Heavy Rainfall Events in selected stations of Gujarat

PeriodStation	1969-1978	1979-1988	1989-1998	1999-2008
Ahmedabad	42	39	50	50
Baroda	21	14	17	34
Deesa	13	7	21	25
Bhuj	5	10	12	10
Rajkot	13	18	15	21
Veraval	21	27	26	33
Naliya	10	13	10	11
Okha	16	25	19	22
Surat	92	88	39	101
Bhavnagar	13	10	13	17

(Source: Ray *et al.*, 2009)

The joint CCl/CLIVAR/JCOMM Expert Team on Climate Change Detection and Indices (ETCCDI) recommended 27 indices based on extremes of temperature and precipitation (Alexander *et al.*, 2006). The extreme indices were computed for Gujarat using daily records of 32 geographically well distributed stations by Lunagaria *et al.*, 2015. There was spatially mixed pattern in most cases temperature and rainfall extreme indices over Gujarat. However, Fig.14. 2 clearly shows that, increase in numbers of hot nights (TR25: Warm nights), decrease in cold days (FD10: Cold days), increase in minimum temperature (TNn: Minimum and minimum of temperature),decrease in frequency cool nights (TN10p: Cold nights) and increase in frequency of warm nights (TN90p: Warm nights) depicts uniform pattern, indicating signature of warming in Gujarat. Further rainfall indices also showed no uniformity for any negative or positive trends over Gujarat. Total annual rainfall (PRCPTOT: Annual total rainfall) and extremely wet days (R99p: Extreme wet days) were found to increase at more numbers of stations and (DTR: Diurnal temperature) Diurnal temperature change showed a mixed pattern over Gujarat. As the rainfall is the parameter having very high variability, very few stations showed statistically significant trends.

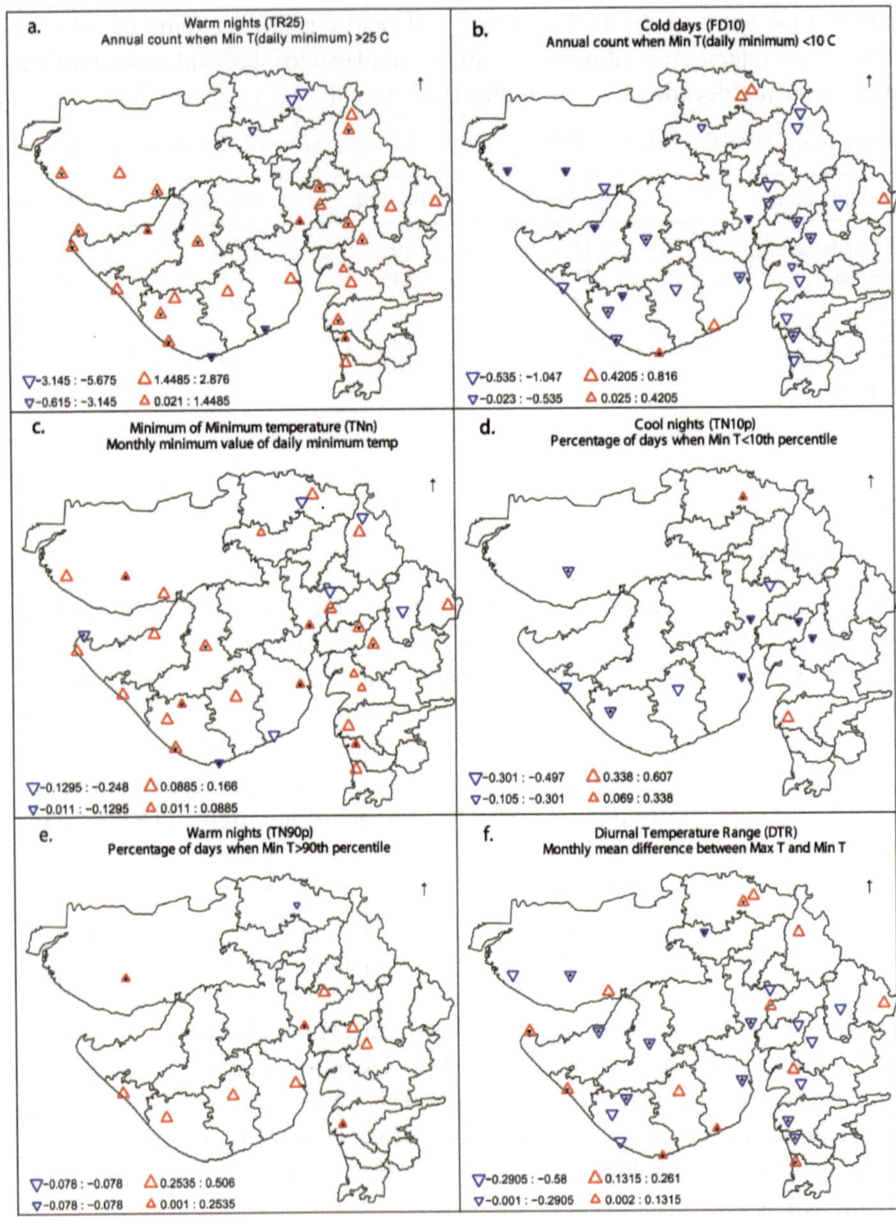
a.
Warm nights (TR25)
Annual count when Min T(daily minimum) >25 C
-3.145 : -5.675
1.4485 : 2.876
-0.615 : -3.145
0.021 : 1.4485
b.
Cold days (FD10)
Annual count when Min T(daily minimum) <10 C
-0.535 : -1.047
0.4205 : 0.816
-0.023 : -0.535
0.025 : 0.4205
c.
Minimum of Minimum temperature (TNn)
Monthly minimum value of daily minimum temp
-0.1295 : -0.248
0.0885 : 0.166
-0.011 : -0.1295
0.011 : 0.0885
d.
Cool nights (TN10p)
Percentage of days when Min T<10th percentile
-0.301 : -0.497
0.338 : 0.607
-0.105 : -0.301
0.069 : 0.338
e.
Warm nights (TN90p)
Percentage of days when Min T>90th percentile
-0.078 : -0.078
0.2535 : 0.506
-0.078 : -0.078
0.001 : 0.2535
f.
Diurnal Temperature Range (DTR)
Monthly mean difference between Max T and Min T
-0.2905 : -0.58
0.1315 : 0.261
-0.001 : -0.2905
0.002 : 0.1315

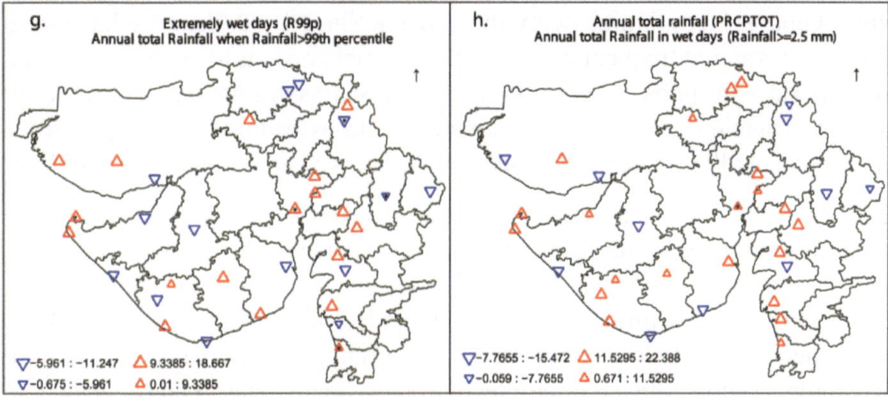

Fig. 14.2: Spatial mapping of climatic extreme indices (trends) over Gujarat. [Triangle indicates positive trend/index value, Inverted triangle indicates negative trend/index value, Asterisk (*) in marker indicates statistically significant value at 0.05 level.] (Source Lunagaria, *et al* in 2015)

14.1 Impacts of Climate Change

The National Action Plan on Climate Change (NAPCC) in India recognizes that climate change would increasing scarcity of water, reduce in yields of biomass, and increase risks of human beings with maximum possibility of decline in theavailability of food grains(Government of India, 2008).In context of Gujarat, major factors which makes it vulnerable towards climate change isits 1600km long coastline which is thelongest coastline among all Indian states. Gujarat has rich biodiversity but it is prone to several hazards, including extreme cyclones, salinity intrusion i.e gradual seeping of salt into the soil and into freshwater aquifers, flooding, and drought (Ahmed & Fajber, 2009).Thus impacts of climate change are invariably seen on agriculture, aquatic ecosystem, energy, and economics.

Gujarat falls in the area of tropical cyclones and is highly vulnerable since it is often effected by its associated hazards such as floods, storm surges etc. Over 120 cyclones originating in the Arabian Sea has passed through Gujarat over a period of 100 years. Eighteen districts of Gujarat namely Junagadh, Kachchh, Bhavnagar, Jamnagar, Porbandar, Amreli, Ahmedabad, Anand, Surat, Navsari,Valsad, Bharuch, Rajkot, Vadodara, Surendranagar, Kheda, Daman and Diu lies within 100 km of coastline are very vulnerable. Cyclones in coastal areas severely affect agriculture directly by high speed wind, torrential rain and extensive flooding. High tide may bring in saline water and sand mass making the fields unsuitable for agriculture. The indirect effects include infection and disease of farm animals, fish and crop plants. Agricultural marketing and trade is adversely affected due to lean season of animal, fish and crop production. Consequently Gujarat is also chronic drought prone state of India, Kutch and

North Gujarat faces famine every three years. Since 1900, the state has faced scarcity of water at frequent intervals. Substantial portions of the State are arid to semiarid with parts of North Gujarat and Saurashtra having no source of alternate irrigation, groundwater exploitation which is leading to increased threats of droughts. Falling water tables have added stress on crops and water supplies.

Climate variability is heterogeneous in nature, cyclones are likely to increase in strength and intensity, though perhaps not in frequency, this consequence may lead to more damage to crops, buildings, and other infrastructure. Thus Climate Change has a great thrust over Land degradation and desertification. This phenomenon leads to loss of potential productivity of soil, as a result of natural or anthropic factors which is often altered from its original state. Land degradation Atlas (SAC, ISRO, 2016) marks that Gujarat is undergoing third highest area under Land Degradation with respect to other states of India, 2011-2012 records 52.29% area under degradation whereas 2003-2005 records 51.35 % of degradation. Numerous micro and macro level research shows many Gujarat is affected by land degradation in several ways in recent times. In Kheda district of Gujarat due to salinity problem, productivity loss estimated was about 25% in the Potato and around 70% in the case of paddy and wheat. Productivity losses range from 10% in the low salinity to 85% in the case of high salinity for paddy (Joshi and Agnihotri, 1984). In Sultanpur district area of paddy and wheat yield went down by more 15% and 56% respectively on salt affected area (Joshi and Jha, 1992).

Mangrove cover changes have globally been raised as the apprehensions influenced by the coastal climate changes(Srivastava *et al.*, 2015). Mangroves serve as carbon sinks as well as they also protect soil from climatic variability. The coast of Gujarat is of strategic economic significance, serving as the natural gateway to India. Several of the State's richfishing grounds, especially in the Gulf of Kutch, are directly or indirectly dependent on mangrovesfor their sustained yield. Degradation of mangroves poses a serious threat to this resource and to thedependent fishing community. It is also leading to increased soil erosion in coastal villages, the damage caused bycyclones is also increased due to absence of mangroves, although severe cyclones often causes destruction of mangroves. The natural barrier to the salt-laden winds is also lost due to destructionof mangroves leading to increased spread of soil salinity which adversely affects agriculturalproduction. This leads to declining livelihood opportunities among agricultural laborers andmarginal workers of villages (GEC, 2012).

14.2 Climate projections and agriculture

India and other countries of SAARC region are most vulnerable to climate change that is affecting agriculture seriously (SAEO, 2009).Regional analysis of agricultural crop is therefore necessary to analyze the impact of climate change. Gujarat's agricultural sector's contribution in State Gross Domestic Productivity has increased over the years in from 24,802.6 cores in 2005 to 30,917.5 cores in 2010 (CMIE, 2012)therefore studying cropping pattern with respect to climate change became a point of interest for the people. Cropping is determined by a number of parameters like soil type, temperature, rainfall, socio-economic factors and technological developments. Thus impact of climatic parameters and socio economic factor over the years can show some defined trend. The regional climate model PRECIS (Providing Regional Climates for Impacts Studies) developed at Hadley Centre, United Kingdom can generate climate change scenarios (Jones *et al.*, 2003) for analyzing agricultural pattern. The weather parameters generated by this model are widely used for various climate change studies. The model has been calibrated and validated for different regions on different spatial scale by several researchers (Rupa Kumar *et al.*, 2006; Yinlong *et al.*, 2006; Islam, 2009; Lunagaria *et al.*, 2011). PRECIS was also used for studying different climatic environments of Gujarat. Lunagaria *et al.*, 2012 reported that there existed large uncertainty in simulation of different climatic parameters of Gujarat. Detailed study on impact of projected climate on various crops viz., wheat, maize, paddy, pearl millet and groundnut were studied using InfoCrop and DSSAT models and projected PRECIS induced climatic data of Anand and Junagadh. Former was selected for all the crops except Groundnut. The models were run for both the periods i.e., 1961-90 and projected 2071-2100. The differences in yield were presented in Fig. 14.3. It may be seen that the highest yield reduction (-61%) was noted under wheat crop followed by kharif maize (47%), kharif paddy (32%) groundnut (24%) and kharif pearl millet (14%), rabi maize (10%) and summer pearl millet (8%). There was not much differences amongst the cultivars. Reduction in yield is mainly attributed to increase in temperature during projected period. Table 14.5 shows analysis of selected stations of Gujarat 1961-1990 and Table 14.6 shows percent variation of climate at various locations in relation to the baseline.

Table 14.5: Analysis of selected stations of Gujarat 1961-1990

Weather Parameter	Period/ Season	Bhavnagar	Bhuj	Kesod	Rajkot	Junagadh	Vadodara	Anand
Maximum	Winter	-0.006	0.031	-0.002	-0.012	-0.002	0.036	0.03
Temperature	Summer	0.004	0.033	-0.012	-0.0001	-0.012	0.033	0.017
	Monsoon	0.001	0.044	0.022	0.015	0.022	0.018	0.016
	Post-Monsoon	-0.003	0.052	0.001	0.012	0.001	0.039	0.039
	Annual	-0.001	0.039	0.021	0.005	0.021	0.031	0.027
Minimum	Winter	0.007	-0.032	0.009	0.082	0.009	0.021	0.017
Temperature	Summer	0.018	-0.02	0.007	0.045	0.007	0.04	0.027
	Monsoon	0.016	0.014	0.021	0.037	0.021	0.017	0.017
	Post-Monsoon	-0.012	-0.019	0.021	0.092	0.021	0.026	0.025
	Annual	0.007	-0.016	-0.467	0.064	-0.467	0.022	0.024
Rainfall	Winter	0.075	-0.032	0.098	-3.28	0.098	0	0
	Summer	-0.18	-0.045	-0.019	-0.158	-0.019	0	0
	Monsoon	-0.296	-1.508	-1.414	0.548	-1.414	10.98	16.51
	Post-Monsoon	2.344	-0.216	-0.237	0.443	-0.237	0	0
	Annual	0.362	-0.449	0.002	-0.613	0.002	10.12	13.25

(*Source:* Patel *et al*., 2015)

Earlier to this study Pandey *et al.*2007 have reported that increase in temperature by 1 to 3°C caused reduction in wheat yield by 8 to 31 per cent. Similarly, Patel *et al.* 2012 reported 5 per cent reduction in maize yield with increase in temperature by 3°C. This clearly suggested that under the threat of projected climate change the wheat crop could be severely affected followed maize, paddy and groundnut during kharif season.Rabi maize and summer pearl millet would be least affected.

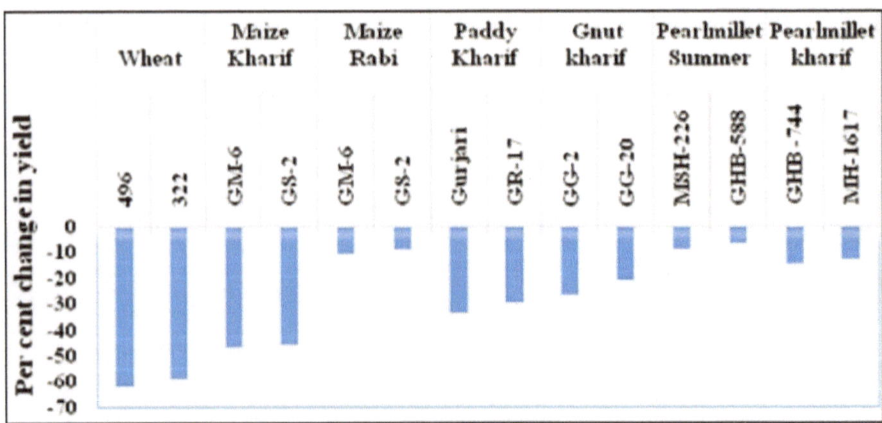

Fig. 14.3: Percentage change in yield of various crops of Gujarat under projected climate. (*Source:* Patel *et al*., 2015)

Table 14.6: Shows percent variation of climate at various locations in relation to the baseline

Station	Rainfall			Maximum Temperature			Minimum temperature		
	Base Line (1960-90)	Projected (2071-2100)	Percentage change	Base Line (1960-90)	Projected (2071-2100)	Difference	Base Line (1960-90)	Projected (2071-2100)	Difference
Anand	919	1259	37	29.8	37.5	7.7	19.1	24.3	5.2
Vadodara	980	1127	15	29.7	33.3	3.6	19.3	23.5	4.2
Junagadh	836	1513	81	33.3	37.3	4	19.9	24.1	4.2
Bhavnagar	627	953	52	33.9	38	4.1	21.3	25.1	3.8
Bhuj	389	782	101	34.7	37.5	2.8	19.4	23.3	3.9
Rajkot	660	1181	79	33.7	37.7	4	21.1	24.5	
Kesod	836	1456	74	33.3	37	3.7	19.9	24	

(*Source:* Patel *et al.*, 2015)

14.3 Climate change risk management

Agriculture is directly as well as indirectly interconnected with climatic phenomenon. Climatic information at all-time scales and at proper precision level is crucial to layout advance risk management plans to improve sustainable production. Agriculture in Gujarat has witnessed significant growth in recent decades. Nonetheless, the sector reports year to year fluctuation in agricultural production and particularly in the production of food grains indicating that this sector has an inherent vulnerability which does not allow it to with stand with climatic shocks and threats (Shukla, 2016). Therefore, it is important to adapt planning measurements.

Field application of climate risk management involves a series of analyses of the historical datasets. Climatic records specifically oriented to crop production are analyzed for understanding the trends. Historical data helps in the risk assessment for present orplanned cropping systems and quantification of risks related to actual dates when rainfallfirst meets crop season onset requirements. Thus historical risk patterns for crops and cropping systems,rainy season characteristics, onset, end and length of growing period, intraseasonal rainfallvariations including wet and dry spells, risks for crops/varieties related to rainfall amountand duration.Climate risk assessment should be followed by formulation of management decisionsthat avoid risk to the greatest extent possible.

Cropping pattern trend and probabilities can be made more informative with specific detailson adaptive management strategies. The management strategies should be optimized forearly and late onset of rainfall, good, average and bad seasons. Modern computer technology coupled with decision support systems can facilitate the comparison of management strategies by simulatingthe crop yield and gross margins. Economic optimization techniques are the tools to befind out for the optimal management practices to reduce the impact of possible climatic change. The key activities of this component are linking past performance of risk factors to crop performances optimizing management practices, considering the farmers' objectives and adjusting cropping practices to fit selected crops and targeted rainfall. Analysis of historical rainfall, data from real-time observation and modern climate monitoring and seasonal climatic impact provide guidelines for selection of suitable management options.

Farmers not well educated and usually use poor means of cultivation. Farmers therefore need to be well advised and educated in farming based activities. They should be well informed on factors like rainfall probability analysis of a place so that optimal cropping systems and management practices related to seasonal climate can be utilized for cropping there farms. Advice to the farmers needs to be simple and targeted on key management factors, such asadapted species or matching cultivar maturity, meeting the requirements of expected durationof the rainy season. These types of agro-advisories need to be prepared byAgrometeorological services and other intermediary organizations dealing with Climate change risk management.

Weather forecasting and early warning systems are very useful elements for decision-making process in climate change and risk management (Challinor, 2009). Intraseasonal as well as interannual climate variability impacts the agriculture sector thus accuracy in prediction can improve the yield. Donald *et al.* (2004) emphasized that intraseasonal forecasts are considered essential for making appropriate decisions reduce vulnerability for small farm holders. There are several indices and methods used to explain the intraseasonal climate variability relevant to agricultural applications this methods should be focused. Advanced information in the form of seasonal climate forecasts has the potential to improve farmer's decision-making, leading to reduced risks and increased opportunities. Proper climate information has the potential to improve livelihoods, enable farmers to adopt improved technology, intensify production and enhance soil fertility, and farmers are capable of investing in more profitable enterprises when conditions are favorable (Hansen *et al.,* 2007). Early warning system developments have the capability to improve the economic return for both smallholder farmers and rural companies, such as grain traders, sugar mills and cotton gins (Meinke *et al*., 2006). Although significant developments have occurred in prediction and dissemination of weather and climate forecasts in recent decades, this information has not yet been utilized adequately and not integrated effectively into agricultural development. The demand for climate information is diverse at different spatial and time-scale thus meaning full understanding of climatic data sets is required which needs expertise hand. Diverse cropping systems and decision cycles associated with crop and livestock activities in small farming systems pose difficulties in customizing information products to match the local needs and activities. The crop monitoring and yield forecasting capabilities are weak and need strengthening at the national level with more emphasis on collection of data such as meteorological, agro-meteorological, soil, remote sensing and agricultural statistics. Identification of the future areas of concern and likely hotspots of vulnerabilities and sensitivities are critical to reduce the risks.

Summary

Climate change is a process which walks in slowly from the root level, its impact is slowly realized in the course of time. Even the most stringent mitigation efforts for controlling or neglecting impacts of climate change over the next few decades won't let the impact completely be buried down. Indeed, we are beginning to see these impacts now which will make our adaptation essential, particularly in addressing near-term impacts. However, unmitigated climate change would, in the long term, be likely to exceed our capacity to adapt. It is essential, then, to develop a portfolio or mix of strategies that includes mitigation, adaptation, tech-nological development (to enhance both adaptation and mitigation) and research (on climate science, impacts, adaptation and mitigation). But analysis of the benefits of various mixes of strategy is severely restricted at present by uncertainty information especially, by lack of understanding of how these impacts will vary under different socio-economic development pathways. It is important that these gaps in our knowledge are filled quickly and thereby local climate changing behavior should be considered in planning and research in different aspects of agriculture. Research approaches to explore the mitigation and adaptation strategies for sustainable agriculture involves uncertainty contributed by scattered meteorological observation network on vast geographic span, climate projection under different scenarios and crop simulation models. As location specific warming and precipitation pattern changes are yet to determine, formulation of crop improvement programmes are based on warming assumption at local scale. Climate change awareness amongst farming community is the urgent need of the time to initiate the mitigation and adaptation efforts in agriculture.

References/Suggested Readings

Ahmed, S., and Fajber, E. (2009). Engendering adaptation to climate variability in Gujarat, India. *Gender and Development*, 17(1), 33-50.

Alexander, L. V., Zhang, X., Peterson, T. C., Caesar, J., Gleason, B., Klein Tank, A. M. G., & Tagipour, A. (2006). Global observed changes in daily climate extremes of temperature and precipitation. *Journal of Geophysical Research: Atmospheres*, *111*(D5).

Arora, M., Goel, N. K., and Singh, P. (2005). Evaluation of temperature trends over India/ Evaluation de tendances de tempèrature en Inde. Hydrological sciences journal, 50(1).

Bandyopadhyay, N., Bhuiyan, C., and Saha, A. K. (2015). Temperature extremes, moisture deficiency and their impacts on dryland agriculture in Gujarat, India. Drought: Research and Science-Policy Interfacing, 119.

Challinor, A. 2009. Towards the development of adaptation options using climate and crop yield forecasting at seasonal to multi-decadal timescales. *Environ. Sci. Policy*, 12: 453–465.

Collins, M., Knutti, R., Arblaster, J., Dufresne, J. L., Fichefet, T., Friedlingstein, P. and Shongwe, M. (2013). Long-term climate change: projections, commitments and irreversibility.

Dash, S. K., Kulkarni, M. A., Mohanty, U. C., and Prasad, K. (2009). Changes in the characteristics of rain events in India. Journal of Geophysical Research: Atmospheres, 114(D10).

De, U. S., Dube, R. K., and Rao, G. P. (2005). Extreme weather events over India in the last 100 years. *J. Ind. Geophys. Union*, 9(3): 173-187.

Desertification and Land Degradation Atlas of India (Based on IRS AWiFS data of 2011-2013 and 2003-2005), Space Application Centre, ISRO, Ahmedabad, India, 219 pages.

Donald, A., Meinke, H., Power, B., Wheeler, M. and Ribbe, J. 2004. Forecasting with the Madden-Julian Oscillation and the applications for risk management, new directions for a diverse planet. In T. Fischer, N. Turner, J. Angus, L. McIntyre, M. Robertson, A. Borrell and D. Lloyd eds. Proc. 4th Int Crop Science Congress. Gosford, Australia, The Regional Institute (available at www. cropscience.org.au/icsc2004/).

GOI (2008), National Action Plan on Climate Change, Prime Minister's Council on Climate Change, Government of India

GOI (2011), Provisional Population Totals, Office of the Registrar General and Census Commissioner, Government of India, New Delhi

Guhathakurta, P., and Rajeevan, M. (2008). Trends in the rainfall pattern over India. *International Journal of Climatology*, 28(11), 1453-1469.

Gujarat Ecology Commission. (2012). Mangrove Restoration Through Public-private Partnership, Gujarat. Gujarat Ecology Commission

Hansen, J.W, Baethgen, W., Osgood, D., Ceccato, P. and Ngugi, R.K. 2007. Innovations in climate risk management: protecting and building rural livelihoods in a variable and changing climate. J. Semi-Arid Trop. Agric. Res., 4(1).

Islam, N.(2009). Rainfall and Temperature Scenario for Bangladesh. *The Open Atmos. Sci. J.*, 3: 93-103.

Jones, P. D., and Briffa, K. R. (1992). Global surface air temperature variations during the twentieth century: Part 1, spatial, temporal and seasonal details. *The Holocene,* 2(2), 165-179.

Jones, R., Hassell, D., Hudson, D., Wilson, S., Jenkins G. and Mitchell, J. (2003). Workbook on generating high resolution climate change scenarios using PRECIS. Hadley Centre for Climate Prediction and Research, Met Office, Bracknell, UK.

Joshi, P. K., and Agnihotri, A. K. (1984). An assessment of the adverse effect of canal irrigation in India. Indian Journal of Agricultural Economics.

Joshi, P. K., and Jha, D. (1992). An economic enquiry into the impact of soil salinity and waterlogging. *Indian Journal of Agricultural Economics*, 47(2), 195.

Kothawale, D. R., Munot, A. A., and Kumar, K. K. (2010). Surface air temperature variability over India during 1901–2007, and its association with ENSO. *Climate Research*, 42(2), 89-104.

Kumar, K. (2009). Impact of climate change on India's monsoon climate and development of high resolution climate change scenarios for India. MoEF, New Delhi on October, 14, 2009.

Lunagaria, M. M., H. P. Dabhi, and Vyas Pandey. "Trends in the temperature and rainfall extremes during recent past in Gujarat." *Journal of Agrometeorology* 17.1 (2015): 118.

Lunagaria, M. M., Pandey, V., & Patel, H. R. (2012). Climatic trends in Gujarat and its likely impact on different crops. *J. Agrometeorol, 14*(1), 41-44.

Lunagaria, M. M., Patel, H. R., Shah, A. V., Yadav, S. B., Karande, B. I., and Pandey, V. (2011). Validation of PRECIS baseline (1961-1990) simulation for middle Gujarat agroclimatic zone. *Journal of Agrometeorology*, 13(2), 92-96.

Meinke, H., Nelson, R., Kokic, P., Stone, R., Selvaraju, R. and Baethgen, W. 2006. Actionable climate knowledge: from analysis to synthesis. *Clim. Res.*, 33: 101–110.

Pandey V, Patel HR, Patel VJ (2007) Impact assessment of climate change on wheat yield in Gujarat using CERES-wheat model. *J Agrometeorology* 9(2):149-157

Pandey V, Shekh AM, Parmar RS (1999). Occurrence of drought and floods over Gujarat. *Journal Agrometeorology* 1(2): 177-181

Pandey, V., and Patel, H. R. (2011). Climate Change and Its Impact on Wheat and Maize Yield in Gujarat. In Challenges and Opportunities in Agrometeorology (pp. 321-333). Springer, Berlin, Heidelberg.

Patel, H. R., Lunagaria, M. M., Karande, B. I., Pandey, V., Yadav, S. B., Shah, A. V. & Nareshkumar, S. (2012). Impact of projected climate change on wheat and maize in middle Gujarat agro-climatic zone. *Journal of Agrometeorology*, *14*(2):134-137.

Patel, H. R., Lunagaria, M. M., Karande, B. I., Yadav, S. B., Shah, A. V., Sood, V. K., and Pandey, V. (2015). Climate change and its impact on major crops in Gujarat. Journal of Agrometeorology, 17(2), 190.

Rathore, L.S., Attri, S. D., Jaswal, A. K. (2013). State level climate change trends in India, Meteorological Monograph No. ESSO/IMD/EMRC/02/2013, IMD, New Delhi.

Ray, K., Mohanty, M., and Chincholikar, J. R. (2009). Climate variability over Gujarat, India. ISPRS Archives, 38(8), W3.

Rupa Kumar, K. , Sahai, A. K. , Krishna Kumar, K. , Patwardhan, S.K. , Mishra, P.K., Revadekar, J.V. , Kamala, K. and Pant, G. B.(2006). High-resolution climate change scenarios for India for the 21st century. *Current Sci.*, 90(3):334-345.

Shukla, S. (2016). Measuring Vulnerability of Agricultural Sector towards Climate Change in Gujarat. Transactions, 38(1), 69.

South Asian Economic Outlook-2009. SAARC Secretariat, Kathmandu, 2009, p.15.

Srivastava, P. K., Mehta, A., Gupta, M., Singh, S. K., and Islam, T. (2015). Assessing impact of climate change on Mundra mangrove forest ecosystem, Gulf of Kutch, western coast of India: a synergistic evaluation using remote sensing. *Theoretical and Applied Climatology*, 120(3-4), 685-700.

Stocker, T. F., Qin, D., Plattner, G. K., Tignor, M. M. B., Allen, S. K., Boschung, J.& Midgley, P. M. (2013). IPCC, 2013: Climate Change 2013: The Physical Science Basis. Working Group I Contribution to the Fifth Assessment Report of the Intergovernmental Panel on Climate Change.

Tabari, H., & Talaee, P. H. (2011). Analysis of trends in temperature data in arid and semi-arid regions of Iran. *Global and Planetary Change*, 79(1-2), 1-10.

Tyagi, A., and Goswami, B. N. (2009). Assessment of Climate change and adaptation in India. Climate Sense, 68-70.

Yinlong , X., Xiaoying, H., Yong, Z., Wantao, L. and Erda, L.(2006). Statistical analyses of climate change scenarios over china in the 21st century. Adv. Clim.

15

Haryana Towards Climate Resilient Agriculture

Chander Shekhar, Anil Kumar and Anurag

Department of Agricultural Meteorology
CCS Haryana Agricultural University, Hisar, Haryana, India

India, as a developing country has reasons to be concerned about the adverse impact of climate change on its economy. A large part of its population depends on climate sensitive sectors for livelihoods which makes it highly vulnerable to climate change. Climate change can have serious impact on its crops, forests, coastal regions, etc. which can in turn affect the achievement of its important national development goals. The issue of climate change cannot however be taken up without linking it to developmental needs such as poverty, health, energy access and education. Traditionally climate change experts have focused on mitigation measures, adaptation measures have also been acknowledged of late as effective and equitable means to deal with climate change impacts. Most of the mitigation measures are high in terms of technology and capital. Therefore, while developed economies choose to mitigate climate change by making heavy investments, developing economies choose to adapt (Monika and Hooda, 2018). There is increasing recognition that the world's current progress on reducing greenhouse gas emissions is not occurring rapidly enough to avoid impacts from climate change in the coming century. Because of this, the world is "committed" to a certain level of global warming and therefore, subject to a degree of impacts that will require adaptive responses by nations and communities (Bizikova *et al.*, 2008). IPCC have specific observations on India in relation to climate change and its impact in its AR 5 Report. Several workers analyzed monthly, seasonal, annual temperature (1901-2007) and rainfall (1873-2011) variability for entire India over two different spatial scales of 30 sub-divisions and 7 regions of India. The net change in rainfall and mean temperature indicates decreasing rainfall and increasing temperature in most of the months and seasons. In India, an increase in temperature over the country and the subcontinent was observed. As discussed in Intergovernmental Panel on Climate Change (IPCC) 5th Report (IPCC, 2014), the global ocean will continue to warm during the 21st century, with the strongest warming projected for the surface in tropical and Northern Hemisphere sub-tropical regions. In addition, changes in precipitation will not be uniform (Pachauri and Meyer, 2014). In the light of these projections, it can be assumed that climatic pattern of Haryana can also undergo considerable changes (Anurag *et al*., 2017). Increasing trend of annual rainfall in Haryana was also reported by India Meteorological Department (Rathore *et al.*, 2013).

The State of Haryana is situated in the northern part of India and is bounded by Punjab and Himachal Pradesh to the north, and by Rajasthan to the west and south. The river Yamuna defines its eastern border with Uttarakhand and Uttar Pradesh. Haryana also surrounds Delhi on three sides, forming the northern, western and southern borders of Delhi. Haryana is a landlocked state in northern India with total geographical area of the state is 44,212 km^2, which is 1.4 % of the geographical area of the country. It is also a leading agricultural state of the country. Animal husbandry has been taken up as an integral component of diversified agriculture. There are no perennial rivers in Haryana, Ghaggar is the only seasonal river, which flows through the northern fringes of the state. The availability of water *vis-a-vis* demand is less, which has created an imbalance and sometimes leads to conflict. In view of the scarce availability of surface irrigation water, development of canal network assumed vital importance for the State.

15.1 Climate change impacts over Haryana

Haryana is mainly an agrarian state where 83% geographical area is under agriculture as per Statistical Abstract of Haryana, 2014. There are two agro climatic zones in Haryana. The western agroclimatic zone has semi-arid climatic conditions and no perennial river flowing in the region. Intensive cropping pattern of agriculture prevails in the state and in-depth analysis of rainfall viz., monthly, seasonal etc. is quiet important as it affect various phases of crop development. The rainfall data of 45 years (1970-2014) showed an increasing trend of rainfall @ 2.3 mm $year^{-1}$. Decreasing trend was observed in January, July, August and November months whereas February, April, May, June, September and October showed an increasing trend. Mann-Kendall test (Z value) showed that June and September had statistically significant increasing trends of rainfall @ 0.7 and 1.1 mm $year^{-1}$ respectively. Seasonal statistics showed significant increasing trend for pre-monsoon period (@ 1.0 mm $year^{-1}$). A study was conducted by Mamta *et al.* (2018) in order to analyze the trends and variability in the extreme ends of the daily minimum and maximum temperatures and rainfall distribution over Haryana state and to strategize the resilient resource management for climate change effect mitigation in agriculture. To quantify the impacts of climate change in terms of extreme rainfall and temperature related indices defined by the Expert Team for Climate Change Detection Monitoring and Indices (ETCCDI, 2013) was used. During the study two rainfall extremes indices i.e. Max 1-day precipitation amount and Max 5-day consecutive precipitation amount (RX1day and RX5day, respectively) and nine temperature extreme indices including maximum and minimum temperature behaviour, trends of cool days and nights, warm days and nights and diurnal temperature range were analysed using the long-term meteorological data of 30 years period (1985-2014) for different representative locations in Haryana state.

The temporal distribution of rainfall indices indicates increase of rainfall distribution, and similarly the temporal distribution of temperature indices on warm nights and tropical nights showed increasing trend @ 0.18 days and 0.41 days, respectively. The increase of minimum temperature in the state led to decrease in the amplitude of diurnal temperature range (DTR) @ 0.04 °C. The indices related to warm days and nights, hottest nights and summer days recorded marginal or no change in the state. The spatial distribution of rainfall indices on annual scale as observed in the state suggested that the northern parts in the state (Panchkula and Yamunanagar) had highest rainfall occurrence and western parts (Sirsa, and parts of Fatehabad, Hisar and Bhiwani) had the least rainfall occurrence. Spatial distribution of temperature indices on annual scale in the state suggested that the northern parts in the state had most occurrence of cold related extremes events, whereas, the western parts had the most hot events occurrence (Plate 15.1). The hot extremes events are in the increasing order on annual basis. Trend and slope of night extremes appear to have been higher than those in day extremes resulting in the decrease of day-night temperature variation which is also indicated by diurnal temperature range. Furthermore, the seasonal response is seen only in winter and monsoon seasons where the DTR has decreasing trend due to narrowing of difference of day and night temperatures with increase of minimum temperature extremes. A decreasing trend in DTR indicated that winters have warmed throughout the state. The DTR decrease was significant during the January month and the warming trend is higher in the western part of Haryana. Overall, the occurrence of extremes events particularly the hike in frequency of night temperature extreme indices in the state is a cause for concern. These indices vary significantly with region and season. In addition, the temporal distribution and the trends of extremes both temperature and rainfall events suggest of an enhanced wetting and warming conditions over the entire state. In general, the rainfall related indices demonstrate trends consistent with increasing rainfall over time and the temperature indices trends are largely consistent with warming. In recent decades the widespread significant changes in temperature trends consistent with warming, especially for daily minimum temperature are observed. The results also show that there has been no substantial change with respect to maximum

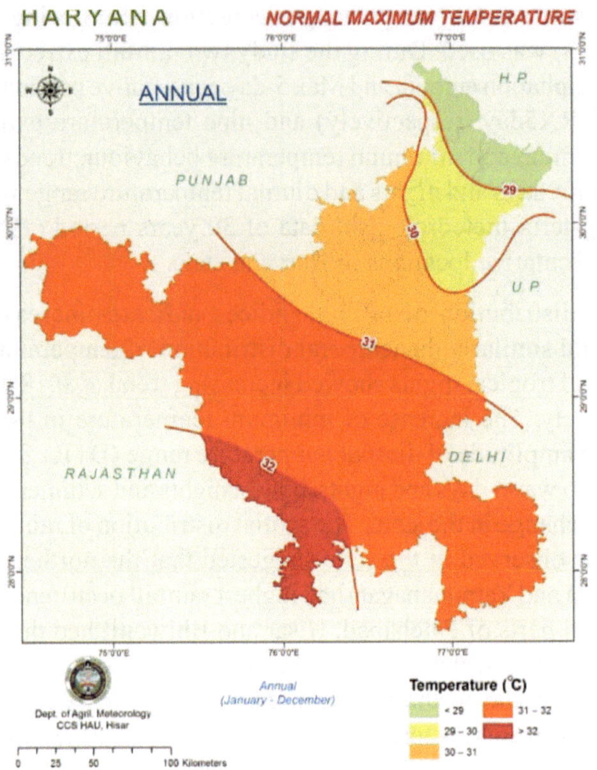

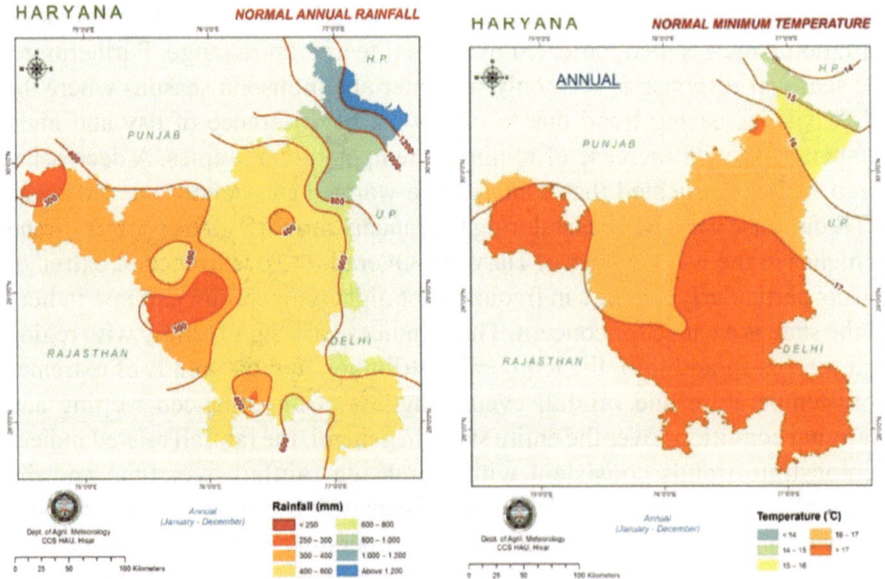

Plate 15.1: Normal maximum and minimum temperatures and rainfall over

temperature values. The temporal trend analyses also revealed that numbers of frost and fog days are on increase, whereas, the number of days with occurrence of dew are decreasing (Plate 15.2). Studies were also conducted to assess of the impact of projected climate change on water and forests in Haryana (SAPCC, 2011). The impact of climate change on water resources may result in 46% (33 mm) increase in runoff to the stream flow and 7% (5 mm) increase in base flow, and negligible decrease to the ground water recharge is projected. Evapotranspiration is projected to increase by 39% (28 mm). The dynamic vegetation model outputs showed that during the short term period of 2030s, out of the 14 forested grids in Haryana, 4 (28.6%) will be impacted by climate change. The number of currently forested grids projected to be impacted by climate change remains the same towards 2080s. The environmental conditions that induce heat stress on dairy animals may exacerbate and may lead to decline in milk productivity and may impair reproductive functions and efficiency of livestock species.

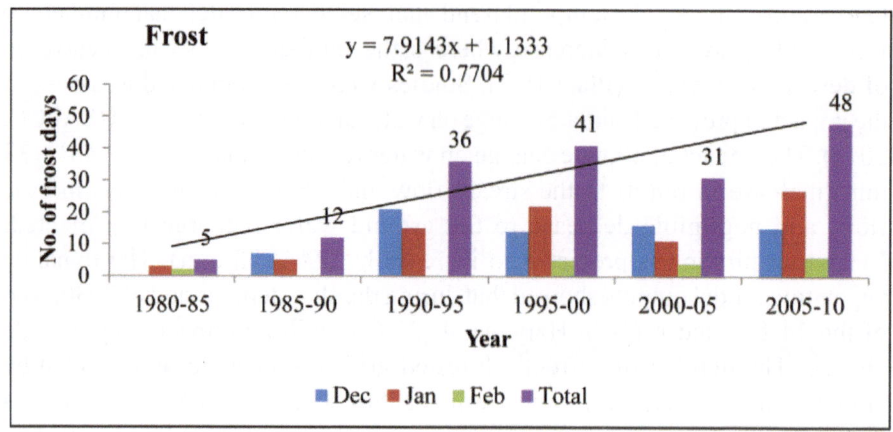

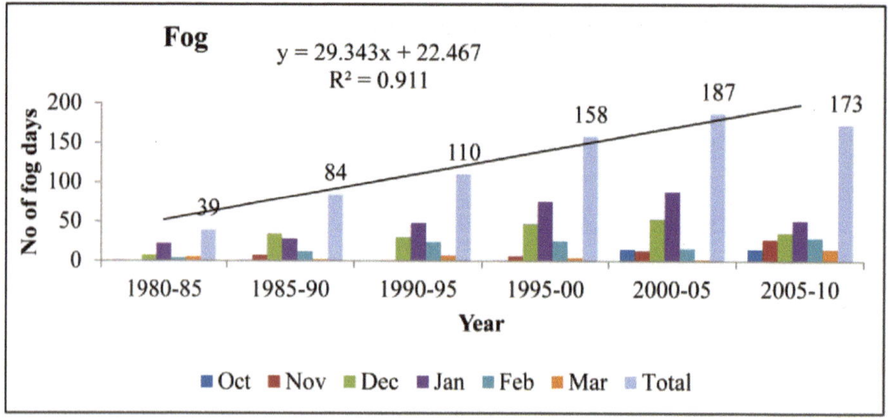

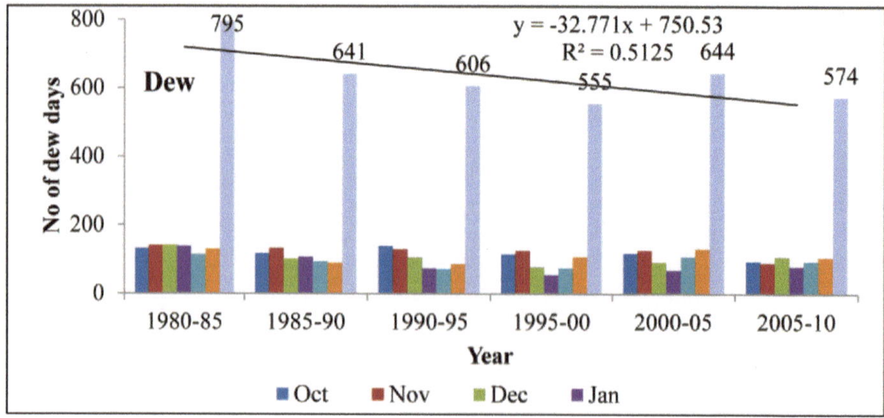

Plate 15.2: Monthly trends of frost, fog and dew days over past decades

15.2 Climate Change Projections for Haryana

The recent report from the IPCC highlights that as one of the most densely populated regions in the world, South Asia is among the most vulnerable to climate change and climate variability. In the absence of adaptation and mitigation, climate change can have major consequences on food security and development in the decades to come. Chaturvedi *et al.* (2012) analysed projections for the two main variables: surface temperature (henceforth referred to as temperature) and precipitation for India with multi-model ensemble mean of 18 GCMs (CMIP5).

Temperature change projections under different climate scenarios (RCPs)

- In the short-term (2030s), annual temperature is likely to rise by 1.5 to 2.5°C
- In the medium-term (2060s), this rise is likely to be between 2 to 4°C
- In the long-term (2080s), it is projected to rise between 2 to 6.5°C

Precipitation change projections under different climate scenarios (RCPs)

- Projections shows increase in annual precipitation in Haryana
- This increase will be approximately up to 10% in the short-term,
- 5 to 15% in the medium-term, and 5-30% in the long-term

15.3 Climate Change and Agriculture

In present day context, agriculture is most vulnerable to weather and climate changes because of its seasonality and narrow range of weather conditions influencing crop and livestock production. People across the globe witnessed above normal temperatures and more rapid warming that occurred during the last half of the 20th century. Climate change presents a profound challenge to food security *vis-a-vis* livelihood and all around development. Since climatic factors serve as direct inputs to agriculture, any change in climatic factors is bound to have significant impact on crop growth, yields and production. Studies have shown significant effect of change in climatic factors on the average crop yield. India is likely to witness one of the highest agricultural productivity losses in the world as a consequence of climate change pattern observed and projected. Climate change projections made up to 2100 for India indicate an overall increase in temperature by 2-4 °C with no substantial change in rainfall quantity as use of fossil fuels increased rapidly in one hand, and on the other hand, forests, the natural buffering system for climate change, are being destroyed indiscriminately for want of fuel, fodder, timbers and urbanization. The magnitude of impact of climate change on agriculture in India varies greatly by region; it is still believed to impact agricultural productivity and shifting crop patterns gradually each year. Climate change can affect crop yields (both positively and negatively), as well as the types of crops that can be grown in certain areas, by impacting agricultural inputs like water for irrigation, amounts of solar radiation that affect plant growth, as well as the prevalence of pests.

Agriculture is highly vulnerable to climate change because of its wide exposure through changes in temperature, precipitation, pest attack and diseases. The loss of agricultural biodiversity is a serious concern. The 5th Assessment Report (AR5) of the IPCC shows that though climate change will have both negative and positive impacts on crop yield initially, the negative impact will overshadow the positive impact by the end of this century.

15.3.1 How climate change impacts major crops and livestock

Wheat: It is the major rabi (winter) crop of the state. Wheat is highly vulnerable to heat stress and requires optimal temperatures at different stages of its growth. It grows best where the temperature is between 20-25°C. Rising temperatures could have significant impacts on wheat yields and productivity across the plains. Studies project that climate change will reduce wheat yield in India in the range of 6 to 23% by 2050 and 15 to 25% by 2080s.

Rice: Conventional tillage-based, irrigated rice consumes more than 50% of water used for irrigation in Asia. Rice requires 3-4 times more water per unit grain produced than crops such as wheat and maize. In a water-scarce region, a move towards adaptation practices that conserve water while increasing yields is critical. In Haryana, rice is sown in the summer months which consumes 150-200 cm water in each crop cycle and is leading to depletion of ground water tables. The government is also encouraging farmers to replace water-guzzling rice varieties and diversify their crops to include maize, pulses, and short-duration rice varieties.

Livestock: Climate change has direct and indirect effect on animal production, reproduction and emergence of many animal diseases. It has been estimated that the annual loss in total milk production due to heat stress on livestock is around 2% in the country's total milk production.

15.3.2 Likely Impacts of Climate Change on Agriculture (SAPCC, 2011)

- Short periods of exposure of wheat crops to temperatures of 28 °C to 32 °C result in significant decrease in yield by 20% or more. North-Western parts of India where sowing of wheat is often delayed due to late harvest of rice and cotton in wheat-rice-cotton cropping system and range of 28 °C to 32 °C temperatures are commonly encountered in the later part of the growing season. As a result, significant losses in yield occur due to heat stress during grain filling period. In view of global warming and increased total demand for wheat in India during next 12 years, development of heat-tolerant cultivars is of major concern in wheat breeding programs.

- In traditional way of irrigated rice production, rice consumes more than 50% of the water used for irrigation in Asia. Rice requires 2-3 times more water per unit grain produced than crops such as wheat and maize. Traditional rice irrigation faces threat with projected less water availability in future.
- Indian mustard cultivation is adversely affected by drought and salinity stress causing production losses as seedlings and mature plants show stress related symptoms including delayed germination, poor growth and poor flowering. With projected increase in drought frequency, erratic rainfall in future mustard yield may be severely affected.
- Unpredictable moisture deficits during crop growth are a major constraint to productivity, adaptation and stability of chick pea performance throughout the world. Due to this single factor the annual yield losses are globally very high, ranging from 30-40 %. Deficit soil moisture conditions are projected to increase in future.
- The disease is favoured by average temperature (18-25 °C) and high humidity (90-95% RH) along with wind currents

Sinha and Swaminthan (1991) analyzed that a 2 °C increase in mean air temperature could decrease rice yield by about 0.75 ton/hectare in the high yield areas and by about 0.06 ton/hectare in the low yield coastal regions. Further, a 0.5 °C increase in winter temperature would reduce wheat crop duration by seven days and reduce yield by 0.45 ton/hectare. An increase in winter temperature of 0.5 °C would thereby translate into a 10% reduction in wheat production in the high yield states of Punjab, Haryana and Uttar Pradesh.

15.3.3 Likely Impacts of Climate Change on Livestock

Climate change has both direct and indirect effect on animal production, reproduction along with emergence and re-emergence of many animal diseases. It has been estimated that the annual loss in total milk production due to heat stress on livestock is around 2% in the country, amounting to over 2.661 crore.

15.4 Impact on production systems

The environmental conditions that induce heat stress on dairy animals can be calculated using temperature humidity index (THI). The heat stress begins to occur in dairy cattle wherein the THI is more than 72(<72 no stress, 72-79 mild, 80-89 moderate, 90-98 severe and above 98 danger). The decreases in milk production can range from 10 to 25%. It has been estimated that with a temperature rise of 1.0 or 1.2 °C with minor change in precipitation during May to August, milk productivity is likely to be marginally affected and during other months productivity will remain relatively unaffected. The negative impact of temperature rise on total milk production for India has been estimated about 1.6 million tonnes in 2020 and more than 15 million tonnes in 2050. An average adult cow or buffalo producing 10-15 lit milk per day requires about 40- 45 lit/ day as drinking water on hot days and about 40- 60 lit for other related work thus requiring a minimum of 100 lit/ day/ animal. An organized animal farm following standard management practices and disposal of animal wastes requires additional water about 50-100 liter/day/animal. Any deficiency in water availability will certainly lead to decline in milk productivity.

15.4.1 Impact on animal growth and reproduction

Heat stress due to temperature or temperature-humidity impairs reproductive functions and efficiency of almost all livestock species. Various studies have shown that heat stress challenges the reproductive performance of cattle and buffaloes (Upadhyay *et al*., 2009). Rising temperatures negatively impact growth and time to attain puberty of livestock species, it is likely to slow down from a growth rate of 500g/day or more to 300-400 g/day of growing cattle. Crossbreds are more sensitive to rise in THI than indigenous breeds. Analysis of the potential direct effects of climate change and global warming on *Murrah* buffaloes indicated that a temperature rise of more than 2°C over existing temperatures in 2050s will cause higher incidence of silent estrus, short estrus and decline in reproduction efficiency of buffaloes. Such impacts are also expected on indigenous cattle.

15.4.2 Impact on physiological responses and functions

The sensitivity of livestock to increasing ambient temperatures under open ambient conditions and in climatic chamber have been evaluated by exposing Zebu, crossbred cattle and Murrah buffaloes to warm/hot ambient temperature (26-40 °C) and low/cool temperatures (6-16 °C) at NDRI Karnal. Body heat storage increased beyond their capacity to tolerate heat particularly on days, when THI exceeded 80 during summer and hot-humid conditions. The study also revealed that Zebu animals under hot dry/hot humid conditions have better heat tolerance than crossbreds or buffaloes. The sensitivity of buffaloes to temperature rise above 35 °C was observed to be higher than either Zebu or crossbreds. The physiological responses, such as respiratory frequency, heart rate and energy expenditure doubled or trebled for an increase of 1.0 °C in temperature.

15.4.2.1 Affect on feed and fodder

Water scarcity not only affects livestock drinking water resources, but also it has a direct bearing on livestock feed production systems and pasture yield. Rising temperatures also have an additional impact on the digestibility of plant matter. Raised temperatures increase the lignifications of plant tissues and thus reduce the digestibility and the rates of degradation of plant species. This not only affects the health of an animal but also results in the reduction in livestock production which in turn has an effect on food security and incomes of small livestock keepers. Studies have shown that dry matter intake decreases in animals subjected to high temperatures. This decrease in dry matter intake can be either short term or long term depending on the length and *duration of heat stress.*

15.4.2.2 Feed and Fodder Availability

In Haryana there is a shortage of green fodder availability mainly in southern parts of the State. The State is surplus in dry fodder and rightly known as dairy State. The estimated area under fodder production is to the extent of 6-8%. The cattle feed is regulated through cattle feed order-99 for insuring qualitative feed to the livestock.

15.4.2.3 Animal diseases and livestock health

Climate change is likely to cause a rise in animal diseases that are spread by insects and vectors mainly due to temperature and humidity rise that favour their spread and growth. Climatic conditions favourable for the growth of causative organisms during most part of the year due to temperature rise will facilitate spread of diseases in other seasons and also increase area of spread. Higher temperatures and changing rainfall patterns can enhance the spread of existing vector borne diseases (Bhattacharya *et al.*, 2006) and macro parasites, accompanied by the emergence and circulation of new livestock diseases. Climate change will modify the dispersal, reproduction, maturation and survival rate of vector species and consequently alter viral and bacterial disease transmission. In some areas, climate change is likely to generate new transmission models. Temperature and humidity variations could also have a significant increase in helminth infections, protozoan diseases such as Trypanosomiasis and Babesiasis. Some of the viral diseases may also reappear affecting both small ruminant population as well. Frequency and incidence of mastitis and foot diseases affecting crossbred cows and other high producing animals may increase due to increase in number of stressful days.

15.5 Impact on poultry

As the ambient temperature reached 34 °C the mortality due to heat stress was significantly high in heave meat type chickens (8.4%) as compared to light layer type (0.84%) and native type (0.32%) chickens. Feed consumption decreased from 108.3 g/bird/day at 31.6 °C to 68.9 g/bird/day at 37.9 °C. Egg production also decreased both in broiler (by 4.5%) and layer (6.4%) breeders as compared to their standard egg production. The body temperature increased from 41 °C to 45 °C as the shed temperature rose from 28 °C to 42 °C and the critical body temperature at which the birds succumbed to death was 45 °C, which was observed at the shed temperature of 42 °C. Naked neck birds performed significantly better than the normal birds with respect to thermo tolerance, growth, feed efficiency and immunity at high temperatures.

15.6 Different approaches towards climate resilient agriculture

15.6.1 Weather based farm advisories for risks minimization

Climatic risk is major challenge of today's agriculture in the country and globe also. Agromet advisories are one such means to pass on climate risk management information to mitigate the losses incurred by farmers. Unfavorable weather condition during growing period like drought, extreme weather events (heat waves, cold waves, unseasonal rains, hailstorm, high wind speed, terminal heat stress etc.) are major concern to the state or nation. The advanced prediction of these weather events, and dissemination of contingency crop planning measures on real time basis using modern information technology would help the farmers in reducing the crop losses under aberrant weather situations.

15.6.2 Agrometeorological applications of remote sensing and GIS

Agricultural is always influenced by climatic and meteorological phenomena. The meteorological data are collected through a spatial network of point station observation (observatories). These stations are neither evenly distributed nor in accordance with the topographic variabilities. Moreover, conventional agro-meteorological techniques have severe limitations to use their data for real time agricultural monitoring and yield forecasting. The development of satellite meteorology has allowed obtaining frequent and accurate measurements of a number of basic agro-meteorological parameters (e.g. surface albedo, surface temperature, evapotranspiration, solar radiation, rainfall etc.). The satellite estimated agro-meteorological parameters have several advantages compared to conventional measurements of agrometeorological data in ground meteorological network. The spatial scale ranging from 50 m to 5 km (depending on the satellite resolution) is more precise than ground climatic data (typically near 100 km for synoptic stations and 10-50 km for local less regular stations). Besides spatial resolution, satellite observation can be accessed regularly over a short period (e.g. every half an hour for METEOSAT) means, high temporal resolution data.

15.6.3 Water management for climate resilient agriculture

More intense precipitation at irregular intervals (resulting into droughts, floods and temporary water logging), reduction in total precipitation in arid and semi-arid regions (subjecting these regions to water scarcity), global warming (causing higher rates of evaporation) and melting glaciers (consequential rise in sea level leading to sea water intrusion) due to climate change as well as population growth (rising demand for food production), intensification of agriculture and urbanization (overexploitation of available water resources), reduced allocation of water to agriculture (increased competition for water among different sectors), intensification of different developmental activities (water pollution due to effluent disposal and chemical leaching) and poor efficiency of water use, are some of the threatening signs for the fresh water resources. Availability of sufficient quantity of water with reasonable quality for irrigation has always been a major factor for crop production. Reduced availability of fresh water *vis-a-vis* increased demand, may pose threat to food security, which would further magnify with climate change. Therefore, adaptation in the current water management practices or their replacement with suitable alternative practices would be required to ensure desired crop production under changing climatic conditions and limited freshwater availability.

Therefore, integrated water management practices would be required to ensure risk free crop production under changing climatic conditions. Different components of proposed integrated water management for crops production shall include water storage facility, reliable irrigation facility and suitable drainage facility. Considerable efforts are required to develop proper water accounting and water allocation mechanisms among different stakeholders for enhanced water productivity. In agricultural sector, production thinking of the past (maximizing kg of grains per unit area) has to be replaced with the concept of specific water use (amount of water required per unit of food produced) for water allocation prioritization.

15.6.4 Climate resilient soil nutrient management

Soil organic matter quality and quantity show a considerable site variation caused by levels of organic matter input, related biological activity, overall nutrients status of the soil and climate. Long-term experiments provide insights into the consequences of land management strategies that can be used to evaluate the effects of agricultural practices on soil. Significant changes in organic C contents of soils due to alternative managements or climate change can only be observed after long time periods. Experimental observations in various agro-ecosystems across the globe have proved beyond doubt that the benefit of C fertilization is observed only when soil has enough available nutrients (particularly N and P) to sustain the enhanced plant growth under elevated CO_2. Several studies projected the increase or decrease in yield of cereals, legumes and oil seeds depending on the net effect of an increase in CO_2, and temperature. A drastic decrease in the concentrations of protein, P, Ca, S, Mg, Fe, Zn, Mn and Cu was observed in major food crops (wheat, rice, barley, soybean and potato) when grown under the combined elevation of atmospheric CO_2 and temperature. However, the reduction in protein content can be arrested by using higher doses of fertilizer N. About 50% of crop residues produced in India are burnt in open fields. Burning of residues not only results in loss of organic matter and nutrients but also causes atmospheric pollution due to emission of GHGs. Cropping system is one of the important components of sustainable farming system as it serves as a component of integrated plant nutrient system, for sustaining the productivity of the system through efficient nutrient cycling. Therefore, balanced fertilization must be based on the concept of the cropping system rather than a crop, so that productivity of a system as a whole is sustained.

15.7 Resource conservation for resilience in agriculture

In India, the green revolution is one of the most remarkable success stories. This revolution in agricultural production in general, and cereals in particular, has been possible because of development of input responsive high yielding varieties, increased use of inputs and over exploitation of natural resources. In this process, there has been remarkable degradation of natural resource base which is posing a serious threat to sustainability of our food production system. The situations will be further aggravated by changing climatic conditions. These climatic changes may pose a serious threat to productivity and sustainability of agriculture. In addition to this, with the increasing cost of inputs and energy and stagnation in productivity of cereal based production system i.e. rice-wheat in north western parts of the country, the income of the farmers is shrinking. Existing agricultural practices of cereal based production system require a large amount of labour, water, time and energy.

In view of above scenario, cost effective and energy efficient resource conservation technologies (RCTs) for shaping future agriculture with the goal of increased productivity have become a critical component to growth in agriculture in India. At the same time, efficient use of natural resources ensuring environmental benefits and improving the livelihoods of farmers is a real challenge in view if changing climatic conditions. The resource conservation technologies (RCTs) which include the practices that increase resource use efficiency and offers benefits like reduced production costs; water, fuel and labour saving; crop establishment in time leading to increased yields. These technologies have been extensively adopted by the farmers in the rice-wheat belt of Haryana state because of several advantages. In present scenario of degradation of natural resources, the resource conservation technologies can play a major role to improve the use-efficiency of natural resources. These technologies may prove very useful to achieve the sustainability of agriculture by minimising green house gases and mitigating the climate change impact on farming. In water scarce areas, rainwater conservation, harvesting and its recycling/recharging of ground water needs to be emphasized on large scale. The climate resilient farming on sustainable basis requires greater emphasis on promotion and adoption of resource-conserving technologies in participatory mode with all the stack holders and possible support from public and private organizations.

15.8 Protected cultivation technology for climate resilient agriculture

Can we manipulate the climatic conditions or can we provide protection to the crops against climatic fluctuations and various other related stresses. Yes, protected cultivation technology has the answer to this but, it‘s a tricky technology highly depending upon intelligent implementation of protected structures for high value short duration crops such as vegetable cultivation by having a knowhow on "What, When, Where and Why" to implement. Every protected structure has its own limitations and advantages, but the basic benefit is its extra protective shelter restricting or minimizing the exposure of the crops to various adverse factors, which are high in open conditions. Protected cultivation of crops can used as tool for combating climate change and offers several advantages to produce vegetables and flowers of high quality and yields, thus using the land and other resources more efficiently. This becomes relevant to growers who have small land holding, say less than one hectare. They would be interested in a technology, which helps them to produce more crops each year from their land, particularly during off-season when prices are higher. However, growing vegetables under protected conditions requires comparatively high input cost and good management practices, which have direct bearing on the economic viability of the production system. Even if the protective structures are cost effective, proper planning, management and attention to details are needed to achieve maximum benefits. Growing vegetables and flowers under location-specific suitable structures may accrue substantial benefits even to small growers.

15.9 Decision Support Systems helpful towards climate resilient agriculture

For easy understanding of a system scientists have quantified each individual process, transform them into some logical mathematical expressions. These processes are systematically arranged and linked to each other as well as to the external factors, to mimic a complete real system. This has led to the beginning of crop growth simulation models. Simulation is mimicking of a real world system in a virtual environment. This is an improved tool to understand the functioning of a plant in the computer with the same precision as if it has been grown actually in the field. There is a host of crop growth simulation models developed for different crops including rice, maize, wheat, legumes, oilseeds, vegetables etc. A number of such individual models are embedded in a single platform for easy imposition/testing of various management and environmental conditions on the growing environment, and it facilitates the process of decision making by the user groups. Examples of commonly used such platforms are DSSAT, WoFost, InfoCrop, AquaCrop etc. These are also called "Decision support systems (DSS)". Agro-climate tools as decision support systems are contributing to climate-smart agriculture by closing the divide between information and effective action. By acting as DSS, they are fostering new ways of thinking by promoting dialogue between groups that may not have previously interacted. By enabling more effective management of current climate risk, they can help to bridge the gaps between climate change and development agendas. Adapting or creating appropriate agro-climate tools is vital to unlocking the potential of climate smart and resilient agriculture in developing countries.

Haryana being a progressive agriculture state adopted various climate smart practices to sustain its agricultural production in changing climate. Following are the few key climate-smart agriculture practice/technologies adopted:

Direct seeded rice: Traditional rice cultivation involves sprouting rice in a nursery and then transplanting the seedlings into an intensively tilled field with standing water. With direct seeded rice, the rice seeds are sown directly in dry seedbed just like any other upland crop. This eliminates the laborious process of manually transplanting seedlings, significantly reduces the crop's water requirements and improves the soil's physical conditions.

Alternate wetting and drying in rice: In alternate wetting and drying, rice fields are alternately flooded and drained. The use of a monitoring instrument like a tensiometer can help farmers decide when to irrigate their fields. Alternate wet and drying reduces methane emissions by an average of 48% compared to continuous flooding. Combining this with precision fertilizer tools can further reduce greenhouse gas emissions.

ICT services to access weather and agro advisories: The weather based agro advisory services to the farmers' of Haryana through various projects such as All India Coordinated Research Project on Agrometeorology (AICRPAM), National Innovations on Climate Resilient Agriculre (NICRA), Gramin Krishi Mausam Sewa (GKMS) etc. are provided. Farmers get voice and text messages that inform them of weather forecasts, new seed varieties, climate smart farming practices and tips on conservation agriculture.

Zero-tillage: Zero-till or no-till farming is a way of growing crops without disturbing the soil through tillage. It increases the amount of water that infiltrates into the soil and increases organic matter retention and the cycling of nutrients in the soil. Zero-tillage improves soil properties, making it more resilient. A CCAFS-CIMMYT study comparing zero-tillage and conventional tillage in Haryana showed that zero-tillage provided both economic and climate gains. The study shows that shifting from conventional tillage to zero-tillage based wheat production reduces greenhouse gas emissions by 1.5 Mg CO_2-eq per hectare per season (Aryal *et al.*, 2014). Zero-tillage also helps in buffering the terminal heat effects which is one of the key climate change related constraints in wheat production in Haryana.

Laser land levelling: A laser-leveller is a tractor-towed, laser-controlled device that achieves an exceptionally flat surface. Levelling the field ensures equitable reach and distribution of water and increases crop productivity. It also increases energy efficiency as less water means less need to run electric pumps. Most of the farmers in Haryana rent the equipment on an hourly basis. Farmers pay between[1] 600-700 per hour to use the machinery. The practice of laser-land levelling has increased exponentially in the state. In Haryana, at the current adoption scale, the estimated amount of irrigation water saving is 933 million cubic meters per year. The estimated greenhouse gas mitigation is 163,600 MT of CO_2-eq per year (CIMMYT-CCAFS, 2014).

Residue management/mulching: Crop residue mulching is a system of maintaining a protective cover of vegetative residues and stubble on the soil surface. It adds to soil organic matter, which improves the quality of the seedbed and increases the water infiltration and retention capacity of the soil. Rice crop residue burning is one of the major issues in Haryana. Using innovative planting machinery like Turbo Happy Seeder, crops can be directly drilled without tillage while residue on surface acts as mulch.

Crop diversification: The aim of crop diversification is to increase farmers' crop portfolio so they are not dependent on a single crop for an income. This also diversifies a farmer's climate risk and contributes to increased household food security. Also, the innovative crop portfolio and system optimisation based diversification options help in sustainable intensification. Introducing legumes in the crop rotation cycle helps fix Nitrogen in the soil.

Agroforestry: Planting trees with crops or vegetables helps sequester carbon in the soil and prevent soil erosion. Trees provide shade to crops and are a source of timber, fruit, fodder and fuel-wood for farmers. They enhance biodiversity by providing habitats to varied species of birds, insects and animals and contribute to a healthy ecological landscape.

Precision nutrient management: Tools such as Nutrient Expert Decision Support Tool, Leaf Colour Chart and Green Seeker sensors are used by farmers to determine optimum fertilizer dosage for crops and assess crop vitality. Overuse of fertilizers increase production costs, damage the soil, contaminate groundwater and add to greenhouse gas emissions.

Nutrient Expert-decision support tool helps farmers decide location specific use of correct fertilizers in the hands of individual farmers. This site specific nutrient management tool adds value to soil testing and guide farmers for precision prescriptions even in absence of farmer access to soil testing. The Nutrient Expert is interactive software and is available for free use on websites.

Green Seeker is a handheld, easy to use crop sensor that is easily calibrated locally. When held above the crop canopy, it calculates the Normalised Difference Vegetation Index (NDVI) which suggests the crop health and nitrogen requirement for a particular plot/field. Using Green Seeker, farmers optimise the fertilizer N use increase crop yield and profits as well as reduce environmental foot prints.

Leaf color chart is a visual chart used for measuring the greenness of the leaves to quantify the nitrogen to be applied to rice fields to get a maximum productivity. It is also suitable for maize and wheat and provides farmers with a good diagnostic tool for detecting nitrogen deficiency.

Summary

Initial testimonials from farmers and policymakers indicate promising opportunities to scale out the climate resilient agriculture practices and to mainstream climate-smart agriculture into policies and development plans. By documenting the evidence on the synergies and trade-offs of a portfolio of climate resilient agriculture interventions for a specific region, it is hoped that climate-smart agriculture can prepare farmers to respond to the uncertainty that comes with climate change and its impacts on food security and livelihoods. Climate resilient agriculture practices are already prioritised and included in the agricultural plans of the state for scaling-out these interventions with the aim of increasing agricultural productivity and farmers' incomes, while safeguarding the natural resource base from exploitation and protecting biodiversity.

References /Suggested Readings

Anurag, Anil Kumar, Diwan Singh, Raj Singh, Surender Singh and Chander Shekhar. (2017). Evaluating Rainfall Trends at Hisar (Haryana) in the Semi-arid Zone of North India. *Annals of Arid Zone,* 56 (3&4): 83-87.

Aryal, J.P., Sapkota, T.B., Jat, M.L. and Bishnoi, D.K. (2014). On-farm economic and environmental impact of zero-tillage wheat: a case of north-west India. *Experimental Agriculture*, 51(1): 1-16.

Bhattacharya, S., Sharma, C., Dhiman, R.C. and Mitra, A.P. (2006). Climate change and malaria in India. *Current Science*, 90: 369-375.

Bizikova, L., Neale, T. and Burton, I. (2008). Canadian communities' guidebook for adaptation to climate change. Including an approach to generate mitigation co-benefits in the context of sustainable development. 1st Edition. Environment Canada and University of British Columbia, Vancouver.

Chaturvedi, R. K., Joshi, J., Jayaraman, M., Bala, G. and Ravindranath, N. H. (2012). Multi-model climate change projections for India under representative concentration pathways. *Current Science,* 103 (7): 791-802.

Climate-Smart Villages in Haryana, India. CGIAR Research Program on Climate Change, Agriculture and Food Security (CCAFS). International Maize and Wheat Improvement Center (CIMMYT).

ETCCDI. (2013). Expert Team (ET) on Climate Change Detection and Indices (ETCCDI). http://etccdi.pacificclimate.org/.

http://www.icar.org.in/files/reports/icar-dare-annual-reports/2010-11/climate-change-AR-2010-11.pdf

IPCC. (2014). Climate Change: Impacts, Adaptation, and Vulnerability. Part A: Global and Sectoral Aspects. Contribution of Working Group II to the Fifth Assessment Report of the Intergovernmental Panel on Climate Change, Summary for policymakers. Cambridge University Press, Cambridge, United Kingdom and New York, NY, USA.

Monika and Hooda, R. (2018). Adaptive and mitigative strategies on issue of climate change in India. *International Journal of Advanced Research and Development*, 3(1): 537-541.

Pachauri, R.K. and Meyer, L. (2014). Climate change 2014 (Synthesis report) Summary for policymakers, IPCC. http://www.climate-service-center.de/imperia/md/video/csc/syr_ar5_spmcorr1.pdf.

Rathore, L.S., Attri, S.D. and Jaswal, A.K. (2013). State level climate change trends in India. Indian Meteorological Department (Govt. of India), New Delhi, India.

SAPCC. (2011). Draft Report: Haryana State Action Plan on Climate Change. Government of Haryana.

Sinha, S. K. and Swaminathan, M. S. (1991). Deforestation, climate change and sustainable nutrition security: A case study of India. *Climatic Change*, 19 (1-2): 201-209.

Upadhyay, R.C., Ashutosh, Raina, V.S. and Singh, S.V. (2009). Impact of climate change on reproductive functions of cattle and buffaloes. In: Global Climate Change and Indian Agriculture, P.K. Aggarwal (Editor), ICAR New Delhi: pp 107-110.

16

Climate Resilient Agriculture – Punjab Scenario

Prabhjyot-Kaur and Harpreet Singh

Department of Climate Change and Agricultural Meteorology
Punjab Agricultural University, Ludhiana – 141 004, Punjab, India

Climate change is one of the most important global environmental challenges facing humanity and as per the instrumental records, the earth's climate system has demonstrably changed on both global and regional scales since the preindustrial era. Further evidence shows that most of the warming (of 0.1°C per decade) observed over the last 50 years, is attributable to human activities. The levels of greenhouse gases namely carbon dioxide (CO_2), methane (CH_4) and nitrous oxide (N_2O) have increased markedly as a result of human activities. The CO_2 concentration in atmosphere is now more than 400 ppm and the decadal increase is from 0.86 ppm per year to 1.99 ppm per year (Table 16.1) which is the highest ever. The Intergovernmental Panel on Climate Change (IPCC, 2007) projects that the global mean temperature may increase between 1.4 and 5.8 °C by 2100. This warming of the earth-atmosphere system is expected to have severe impacts on the global hydrological system, ecosystems, sea level, crop production and related processes. Agriculture sector is the most vulnerable entity in light of the global warming and its impacts. Global mean temperatures of today are already about three quarters of a degree warmer than that of pre-industrial times, and even if carbon emissions were completely stopped today the warming trend would continue for many decades. Moreover, the United Nations (2015) report has projected the world population to increase from current 7.3 billion to 8.5 billion (16%) by 2030, 9.7 billion (33%) in 2050 and 11.2 billion (53%) in 2100. With more population to feed the agriculture of tomorrow will be at cross roads if it is not sustainable as well as productive. Since the beginning of the early civilizations, agriculture and its adaptation to climate change or anything else is an ongoing process. On a long term basis, the agricultural crops will have to adapt to at least a warming of about 2°C. These low cost adaptations in agricultural crops (such as changes in sowing periods / dates, cultivars, fertilizer and irrigation use and crop management practices) can reduce the impact on crop yields by often more than half, relative to the cases of no-adaptation.

Table 16.1: Recent changes in carbon dioxide level during pat five decades

Decade	Average year–over-year increase (ppm/yr)
March 1963 to March 1972	0.86
March 1973 to March 1982	1.44
March 1983 to March 1992	1.55
March 1993 to March 2002	1.55
March 2003 to March 2012	1.99

Punjab is a booster State for the agricultural economy of India particularly for the two major food crops, i.e., wheat and rice. Several studies have reported an increase in temperature, particularly the night time temperatures, and rainfall pattern over the past four decades in Punjab state. The backbone of the Punjab agriculture, particularly the wheat crop is highly susceptible to terminal heat stress. Aggarwal (2008) has reported that an increase in temperature by 1°C is projected to reduce wheat production in India by 4 to 5 Mt, even after taking CO_2 fertilization into account. Hence, agriculture in Punjab will also have to undergo many such low cost adaptive options.

16.1 Impact of climate change

Climate change is impacted as well as impacts the agriculture (Kang and Banga, 2013). Climate change plays a very important role in food security and economy of a nation, especially in a developing country like India. In reality the climate change affects all four dimensions of food security, i.e., food availability, food accessibility, food utilization and food systems stability (Killman 2008) and as a result all agricultural commodities are sensitive to climate change / variability (Kumar *et al.*, 2011). Increase in atmospheric CO_2 promotes growth and productivity of plants with C_3 photosynthetic pathway but the increase in temperature, on the other hand, can reduce crop duration, increase crop respiration rates, affect the survival and distribution of pest populations, and may hasten nutrient mineralization in soils, decrease fertilizer use efficiency and increase evapo-transpiration.

16.2 Effect of elevated concentrations of CO_2 on crop growth

CO_2 is vital for photosynthesis and hence for plant growth (Kimbal *et al,* 2002). Globally several simulation studies have been conducted to predict the likely effects of elevated CO_2 on crop yield. The increasing concentrations of CO_2 favored the plant physiological processes thereby increasing the biomass accumulation, leaf area duration, and net assimilation rate. The grain yield was increased by nearly 32 % when the CO_2 concentration was increased from 330 to 660 ìmol CO_2/mol air (Baker *et al*, 1990). In a simulation study conducted by Lal *et al* (1998), the results revealed that in northwestern India at elevated (doubled) CO_2 concentrations the yields of rice and wheat increased by 15% and 28%, respectively. Pandey *et al* (2007) reported that with the gradual increase in CO_2 concentrations from 440 to 660 ppm wheat yields increased from 21 to 68% under optimal conditions, whereas under suboptimal conditions, similar responses were observed with slightly lower magnitudes (19 to 57%). De Costa et al (2006) studied the growth and yield attributes of rice under elevated levels of CO_2 in field (Table 16.2). They found that though the percentage of filled grains were significantly higher under the elevated CO_2 concentration but the grain weight and harvest index did not differ significantly between the elevated and ambient CO_2 treatments. Grain yield under elevated CO_2 was 24% greater than that of the ambient CO_2 concentration.

Table 16.2: Yield and yield attributes of rice grown under elevated (570 ppm) and ambient (370ppm) carbon dioxide in open top chambers and in open field conditions (De Costa *et al,* 2006)

Parameters	Treatments		
	Elevated CO_2	Open field	Ambient CO_2
Grains (no m^{-2})	47007 (11)	41793	42336
Filled grains (%)	82.9 (9)	72.1	76.0
Grain weight (mg)	24.9 (2)	24.0	24.5
Grain yield (g m^{-2})	871 (24)	723	783
Harvest index	0.47	0.41	0.45

* The values in parentheses represent percentage increase over ambient CO_2 concentration

16.3 Effect of increasing temperature on crop growth

The increase in temperature increase crop-respiration rates; reduce crop duration, number of grains, and crop yield; inhibit sucrose assimilation in grains, affect survival and distribution of pest population, hasten nutrient mineralization in soils, decrease fertilizer-use efficiency, and increase evaporation. Prabhjyot-Kaur and Hundal (2006) reported that an increase in temperature by 2°C resulted in 3-10% decrease in grain/seed yields of *kharif* season crops (rice, groundnut, and soybean) and 29% decrease in grain yield of *rabi* season crops (wheat and gram). Peng *et al* (2004) analyzed the relationship between yield of rice and minimum temperature across the range of 22.1-23.7 °C using a quadratic equation and reported that rice yield declined by 10% with a rise in each 1 °C minimum temperature and yield declined by 15% with each 1 °C rise in mean temperature. Prabhjyot-Kaur and Hundal (2010) analyzed the effect of within-season increase in temperature on wheat using the CERES-Wheat model (Table 16.3). The increase in temperature during February affected the yield of early, normal and late own wheat crop, however the effect varied according to the stage of the crop. The reduction in yield was mainly because of a reduction in grain-filling period caused by an increase in temperature. Aggarwal and Kalra

Table 16.3: Effect within season temperature increase from normal on grain yield (% deviation from normal) of wheat using CERES-Wheat model (Prabhjyot-Kaur and Hundal, 2010)

Time period	Date of sowing	Temperature increase from normal (°C)					
		+1.0	+2.0	+3.0	+4.0	+5.0	+6.0
1st fortnight of February	Early sown (28th October)	-3.4	-3.7	-7.6	-11.5	-13.0	-17.2
	Normal sown (8th November)	+1.7	-1.6	-1.8	-3.9	-7.7	-7.3
	Normal sown (15th November)	-0.5	-2.7	-1.5	-2.0	-1.3	-1.9
	Normal sown (25th November)	+0.5	+2.4	+2.8	+4.7	+4.9	+7.0
	Late sown (2nd December)	+0.7	+0.6	+0.6	+3.4	+3.6	+3.7
2nd fortnight of February	Early sown (28th October)	-2.4	-2.8	-5.2	-8.1	-10.9	-13.8
	Normal sown (8th November)	-0.4	-4.1	-5.1	-9.9	-14.2	-16.4
	Normal sown (15th November)	-2.0	-5.8	-6.0	-8.7	-9.7	-14.2
	Normal sown (25th November)	+2.5	+1.1	+3.4	-0.6	-2.6	-3.3
	Late sown (2nd December)	-0.5	-0.4	-1.7	-2.3	-3.1	-3.6
1st fortnight of March	Early sown (28th October)	-2.3	-4.6	-6.8	-13.8	-8.2	-10.4
	Normal sown (8th November)	-2.7	-3.3	-6.0	-9.5	-9.5	-13.0
	Normal sown (15th November)	-4.8	-9.3	-10.1	-14.2	-16.0	-20.8
	Normal sown (25th November)	-0.5	-5.4	-6.7	-3.3	-16.0	-19.4
	Late sown (2nd December)	-2.3	-1.6	-6.8	-7.6	-12.5	-17.7
2nd fortnight of March	Normal sown (8th November)	+1.1	-1.5	-0.5	-0.1	-1.9	-1.5
	Normal sown (15th November)	-2.5	-1.6	-4.3	-6.9	-5.9	-8.1
	Normal sown (25th November)	-0.1	-4.7	-5.6	-9.2	-10.1	-11.2
	Late sown (2nd December)	-5.5	-6.6	-12.3	-14.5	-19.1	-21.4

(1994) also reported that the reduction in wheat yield with an increase in maximum temperature was mainly attributable to a reduction in duration of anthesis and in grain filling with a rise in ambient temperature, and vice-versa.

16.4 Interactive effect of changing climatic factors on crop growth

Temperature, solar radiation, and water directly affect the physiological processes involved in grain development and indirectly affect the grain yield by influencing the incidence of disease and insects. Rice grain yield was positively correlated with mean solar radiation and negatively correlated with daily mean temperature during the reproductive stage (Yoshida and Parao, 1976). Relatively low temperature and high solar radiation during the reproductive stage had a positive effect on the number of spikelet and hence increased grain yield. Solar radiation had a positive influence on grain filling during the ripening period. Abrol *et al* (1991) reported that in India the adverse effects of 1-2 °C rise in temperature could be absorbed with a 5-10% increase in precipitation. A grain yield increase of 20-30% might be possible on about 70% area in the rice-wheat cropping system in India. In northern India, warming could offset some losses in yield by early pod setting in winter grain legumes, such as chickpea and lentil. Hundal and Prabhjyot-Kaur (1996) reported that in rice crop, the that adverse effect caused by an increase in temperature of up to 0.5 °C could be compensated for by concentrations of CO_2 above 400 ppm. Further increase in temperature of up to 1.0 °C did not decrease maximum LAI, biomass, and grain yield when CO_2 was nearly doubled (600 ppm) relative to the normal temperature. But in scenarios with nearly doubled CO_2 concentrations of 600 ppm, temperature increases of more than 1.0 °C above normal reduced the maximum LAI, biomass, and grain yield of rice. Pandey *et al* (2007) employed the CERES-wheat model and found a gradual decrease in yield of wheat from 3546 to 2646 kg ha^{-1} (8 to 31% less than the base yield) under optimal moisture conditions with an increase in maximum temperature from 1 to 3 °C (Table 16.4). Similarly, under suboptimal conditions, wheat yield were reduced from 2841 to 2398 kg ha^{-1} (9 to 23% less than the base yield).

Table 16.4: Change in simulated wheat yield due to varying temperature under optimal and suboptimal moisture conditions (Pandey et al, 2007)

Change in maximum temperature relative to base temperature (°C)	Simulated grain yield (kg ha^{-1})		Change in simulated grain yield relative to base yield (%)	
	Suboptimal	Optimal	Suboptimal	Optimal
+3	2398	2646	-23	-31
+2	2668	3091	-14	-19
+1	2841	3546	-9	-8
-1	3190	4206	3	10
-2	3358	4485	8	17
-3	3641	4817	17	26

16.5 Effect of climate change on quality of produce

Under elevated CO_2 levels the amylose content of the rice grain (major determinant of cooking quality) is increased (Conroy *et a,l* 1994). Also, the cooked rice grain from plants grown in high-CO_2 environments was firmer, though, the concentrations of iron and zinc (important for human nutrition) were lower (Seneweera and Conroy, 1997). Moreover, the protein content of the grains decreased with combined increases in temperature and CO_2 concentration. Studies have shown that higher CO_2 concentrations lead to reduced plant uptake of nitrogen (N) and trace elements, such as zinc, resulting in crops with lower nutritional value (Taub and Wang, 2008). This would primarily impact people in poorer countries, who are less able to compensate by eating more food and have less varied diets. A study by Mulchi *et al* (1992) showed that the protein content of soybean grain decreased with increases in CO_2 concentration; however, because of increased grain yield, the total quantity of nutrients accumulated in grain per hectare still increased with high CO_2 concentrations. In the same study, increases in CO_2 concentrations from 360 to 510 ppm increased grain oil from 20.4 to 22.3% and decreased grain protein content. The N content of plants is likely to decrease with elevated levels of CO_2, implying reduced protein levels. A study conducted at IRRI on rice (cv. IR 72) grown under three different CO_2 concentrations (ambient, ambient +200, and ambient +300 $\mu L\ L^{-1}$ CO_2) and two different air temperatures (ambient and ambient +4 °C) using open-top field chambers revealed that increasing both CO_2 and air temperatures reduced grain quality (e.g., protein content). The combined effects of CO_2 and temperature suggested that, in warmer regions (i.e., >34 °C) where rice is grown, quantitative and qualitative changes in rice supply would be possible if both CO_2 and air temperatures continued to increase (Ziska *et al* 1997).

16.6 Projected climate change and its effect on crop production

The IPCC reports and a few other global studies indicated a probability of 10-40% loss in crop production in India with increase in temperature by 2080-2100 (Fischer *et al.*, 2002; Majumdar, 2008). Studies conducted in India (Lal *et al.*, 1998; Aggarwal, 2003) have confirmed similar declining trends in agricultural productivity due to climate change. KaviKumar (2010) has reported that the temperature increase is likely to be much higher during *rabi* season than in rainy *(kharif)* season. Precipitation is likely to increase almost in all the months except during December-February when it is likely to decrease. Furthermore, climate change projections made up to 2100 for India indicate an overall increase in temperature by 2-4°C (Table 16.5) with no substantial change in precipitation. Kaur and Prabhjyot-Kaur (2016) used the PRECIS model to analyze the projected changes in temperature and rainfall in Punjab. The results revealed a linear increase in temperature and rainfall under the A1B scenario for the mid century. At the end of the century, the rise in temperature in B2 scenario will continue but at slower rate than the A1B and A2 scenarios (Figs.16.1-16.3). There would be deficit of rainfall by 78 and 30 % during the winter season under the A2 and B2 scenarios of the end century. The projections by the mid and end of 21st century showed an increase in the incidence of extreme of high (March to June) and low (December and January) temperature events thereby indicating increase in heat wave and frost conditions in crops.

Table 16.5: Projected changes in climate in India (2070-2099) (Kavikumar, 2010)

Region	January-March	April-June	July - September	October - December
Change in temperature (°C)				
Northeast	4.95	4.11	2.88	4.05
Northwest	4.53	4.25	2.96	4.16
Southeast	4.16	3.21	2.53	3.29
Southwest	3.74	3.07	2.52	3.04
Change in precipitation (%)				
Northeast	-9.3	20.3	21.0	7.5
Northwest	7.2	7.1	27.2	57.0
Southeast	-32.9	29.7	10.9	0.7
Southwest	22.3	32.3	8.8	8.5

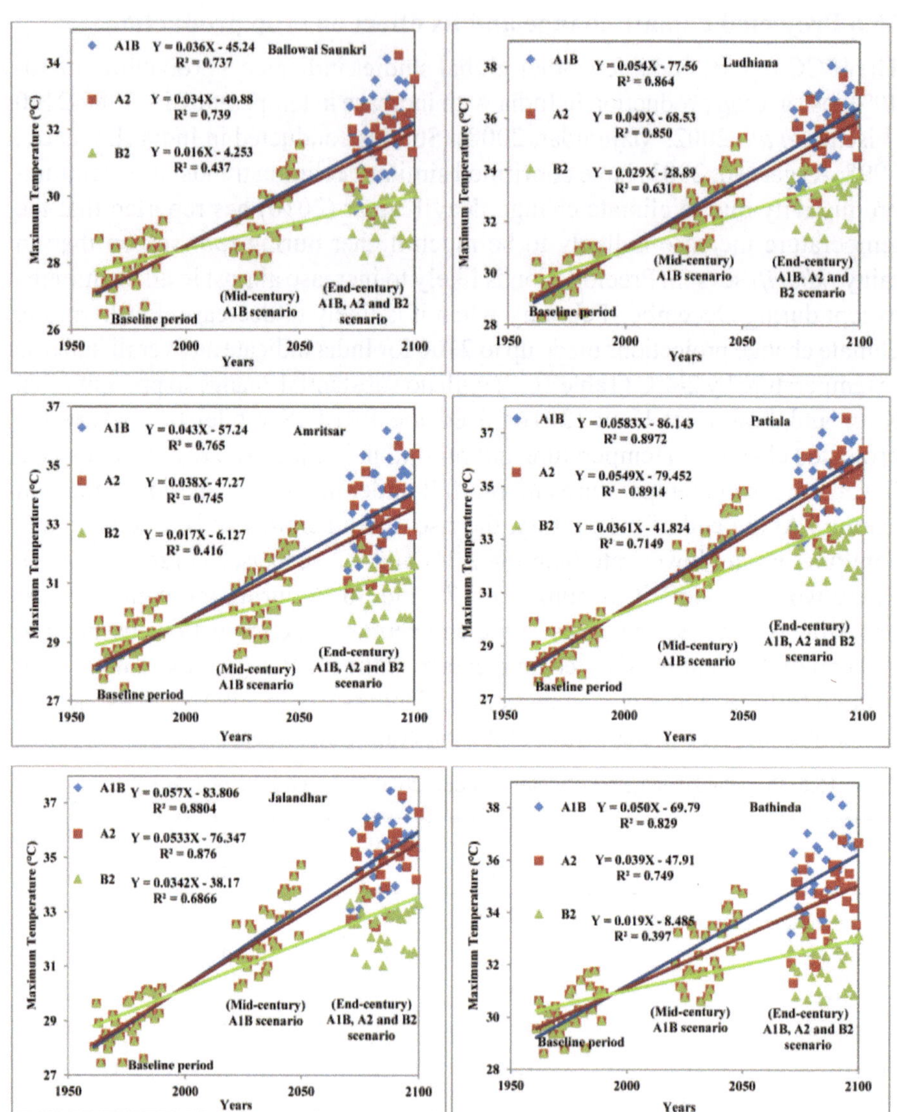

Fig. 16.1: PRECIS-model simulated changes in maximum temperature under different scenarios of climate change in Punjab (Kaur and Prabhjyot-Kaur, 2016)

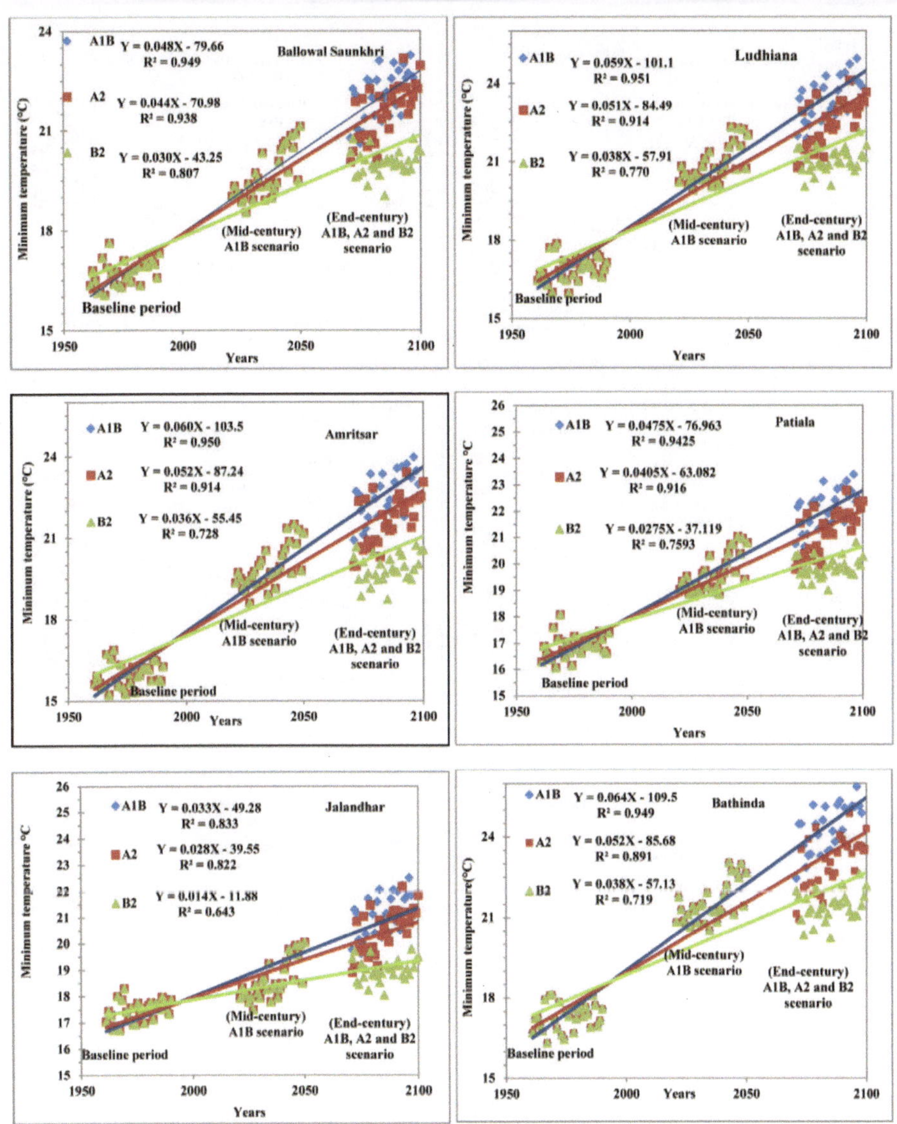

Fig. 16.2: PRECIS-model simulated changes in minimum temperature under different scenarios of climate change in Punjab state (Kaur and Prabhjyot-Kaur, 2016)

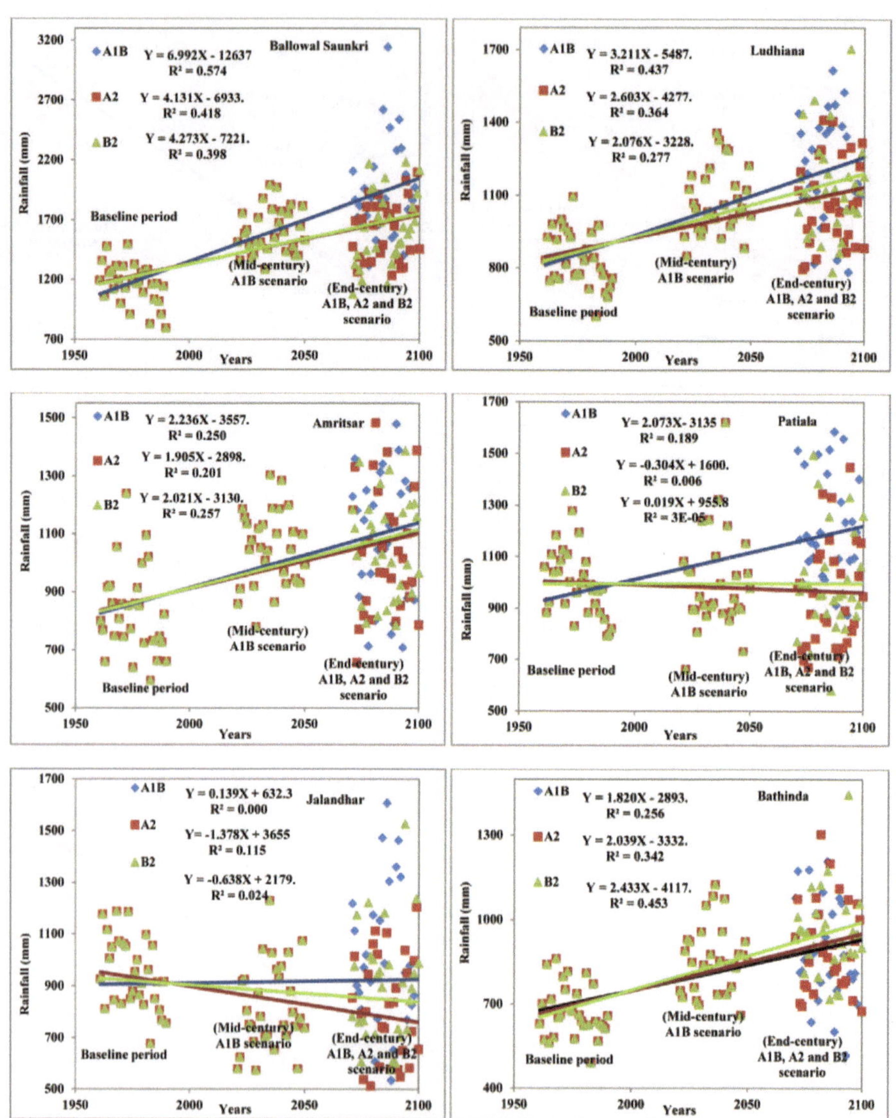

Fig. 16.3: PRECIS-model simulated changes in rainfall under different scenarios of climate change in Punjab (Kaur and Prabhjyot-Kaur, 2016)

16.7 Strategies to mitigate the impact of climate change

India will need to produce 120 million tons (mt) of wheat, 25 mt of pulses, and 100 mt of oil seeds (50 mt for biofuel purposes) by 2025 (Kumar, 2006). Dar and Gowda (2013) suggested that improved crop, soil, and water management practices and stress-tolerant varieties should overcome detrimental impacts of climate change, which in turn would lead to improved food security, livelihoods, and environmental security. They further pointed out that negative effects of climate change on food security can be counteracted by broad-based economic growth, particularly improved agricultural productivity, and robust international trade in agricultural products that can offset regional shortages, and agricultural productivity investments.

16.7.1 Crop management based adaptations

Improved agronomic practices, like altering planting dates, helped in minimizing the effect of high temperature responsible for yield instability in crops (Krishnan *et al*, 2007). In Punjab the major crops include wheat, rice, maize, cotton, rapeseed-mustard etc. So for adapting the agriculture in the state to the brunt of climate change some of the techniques include:

16.7.1.1 Adjustment in management practices of crop

Adjustment in sowing dates is simple, yet powerful tool for adapting to the effects of potential global warming. Mahajan *et al* (2009) reported that the manipulation of sowing periods is effective in reducing yield instability by keeping flowering from coinciding with hottest growing season. Recently the problem of "terminal heat stress in wheat" has become a big issue in de-stabilizing the wheat yield in north India (Gupta *et al,* 2010). During the winter of 2009-10, despite favorable weather conditions an abrupt rise in night temperature during the grain-filling stage in wheat adversely affected wheat productivity in the Indo-Gangetic Plains (IGP) and other northern states of India. The incidence of terminal heat stress on wheat in Punjab during 2009-10 led to a yield decrease of 5.8% compared with the previous year. However, yield losses were as high as 20% in a few districts of Punjab and other transects of the IGP. The magnitude of losses varied depending on planting time, cultivars, and other management practices. In wheat crop the application of foliar application of urea and light irrigation during the midway of the heat stress period were able to reduce the severity of stress (Singh and Prabhjyot-Kaur, 2018). The light irrigation helps to reduce the soil and canopy temperature and the incorporation of chopped paddy straw is a very viable strategy as it adds to the organic matter content which helps the wheat plant to tolerate fluctuation of temperature in the month of February (Fig. 16.4).

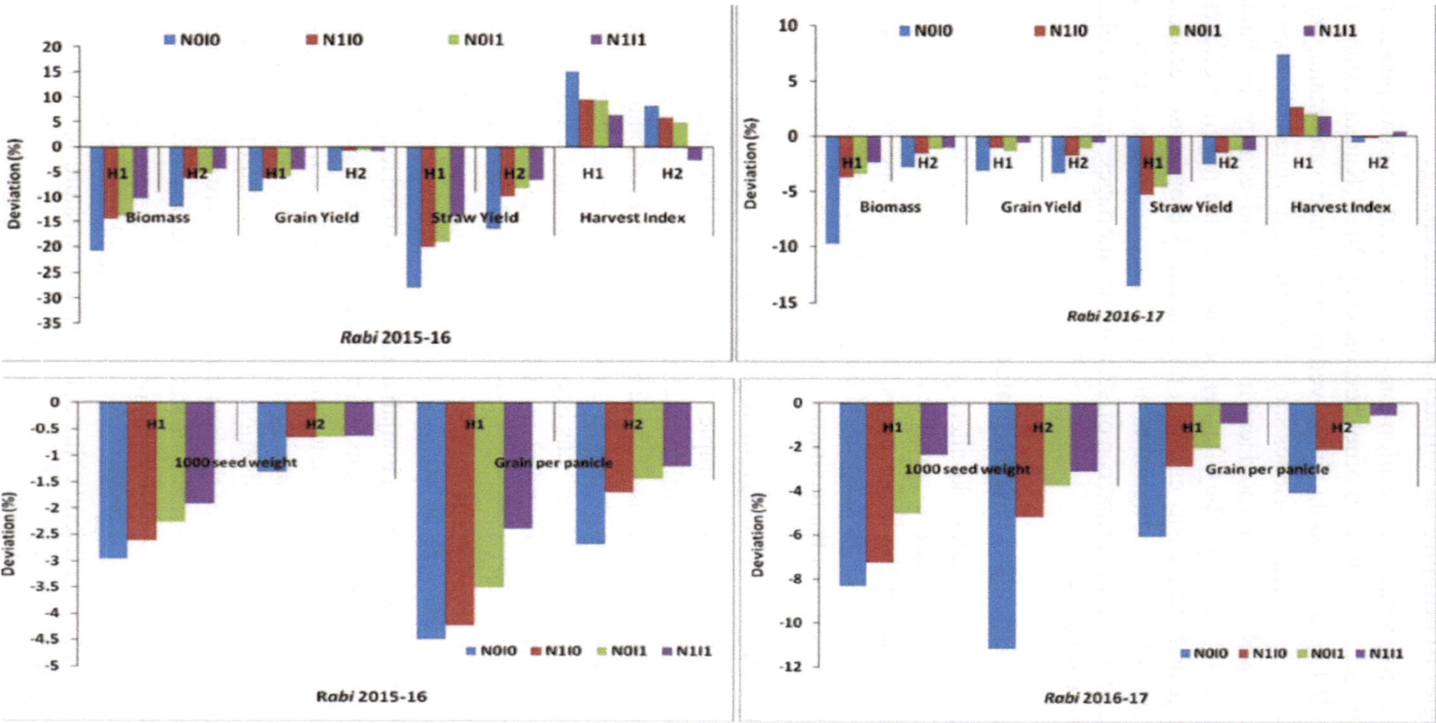

Fig. 16.4: Deviation in yield and yield attributes of wheat cultivar WH 1105 from control under terminal heat stress treatments and their adaptive strategies, *Rabi* 2015-16 and 2016-17 (Singh and Prabhjyot-Kaur, 2018)

16.7.1.2. Effective water management practices in crops

Bed-planting of crops is an effective crop-water management strategy in wake of climate change as it increases water-use efficiency, reduce water logging, better access for inter-row cultivation, weed control, better stand establishment, less crop lodging, and reduces seeding rates (Bhardwaj *et al* 2009 and Chauhan *et al,* 2012). The technique of zero tillage in wheat cultivation under irrigated conditions has successfully reduced the demand for water and other resources (e.g., diesel, herbicides) and zero-till systems are now considered a viable option to combat climate change. Intercropping is another well adapted practice in the light of climate change. If one crop fails because of flooding or drought, the second crop provides a minimum assured returns for livelihood security of the farmers Mittal and Singh (1989). Several workers have reported of reduced irrigation amounts or time, with similar or higher yields for wheat on raised beds compared with conventional tilled wheat (Dhillon *et al,* 2000; Bhardwaj *et al,* 2009). These workers have listed several benefits of bed planting particularly in corn-wheat system regarding its role in arresting soil structural degradation and saving of irrigation water up to 30-40% in wheat and rice and reduced cost incurred in seed saving and labour. They have highlighted the benefits of bed planting particularly in rice-wheat cropping system regarding its role in arresting soil structural degradation and saving of irrigation water up to 30-40% and reduced cost incurred in seed saving and labour.

16.7.1.3 Direct seeded cultivation (DSR) of rice

It has been found that the conventional puddle transplanted rice consumes 120 to 180 cm ha of water (Chauhan *et al,* 2012). Several research experiments reveal the possibility of cultivating aerobic rice, which requires dispersible low irrigation water (Sudhir-Yadav *et al,* 2010; Mahajan *et al,* 2011). The alternative establishment techniques of rice, such as DSR (both wet and dry seeded) have shown promise in different production environments of South Asia and are already practiced in many medium deep and deep water rice ecologies of India and Bangladesh, on Western Himalayan region and the Ghats along the East and West Coast. The area under DSR in India, Pakistan and Bangladesh is reported to be 14.2 million ha, which is about 25% of the total rice area (Pandey and Velasco, 2005). In some trials of irrigation, water saving of 12 to 60% for direct sowing rice (DSR) and transplanting rice on beds, with similar or lower yields have been reported (Balasubramanian *et al,* 2003). Other studies (Bhushan *et al,* 2007 and Mahajan *et al,* 2011) also reported that in DSR the amount of irrigation water was 25-28 % lesser than in puddle transplanted rice. Further they reported that application of N at sowing time in dry DSR can be skipped because the N may not be immediately used by the emerging seedling and application at anthesis stage may further boost the productivity of dry DSR. They advocated that application of N in four splits resulted in higher grain yield of dry DSR than its application in three splits.

16.7.1.4 Precision laser land-leveling in crop

Water table is fast declining in Punjab state and the viable measures need to focus on arresting this decline in water table. In this context the laser land-leveling is one such option that helps in saving of irrigation water, improved input-use efficiency and achieving a better crop stand (Jat *et al.,* 2011). The use of precision land leveling has been reported to save up to 50% of irrigation water (Rickman, 2002), improve crop productivity up to 30%, and reduce the labour requirement for weeding from 21 to 5 days ha^{-1} in rice. Currently this technique is being practiced over 1.5 million ha in South Asia (Jat *et al.*, 2011).

16.7.1.5 Diversification of crops

The diversification of crops is helpful in providing higher protection against risk associated with climate change in addition to assured net returns to the farmers. Risk reduction through crop diversification related to climatic and biotic vagaries particularly in fragile ecosystems and commodity fluctuations by expanding locally adapted or introducing novel varieties and related production systems will contribute to improved food security and income generation for resource-poor farmers and protect the environment (Behera *et al.,* 2007). Rice-wheat cropping system is the most important cropping system for food security in South Asia but the area under the system is static and the productivity and sustainability of the system are threatened because of the inefficiency of current production practices, shortage of resources such as water and labour, and socio-economic changes (Ladha *et al.,* 2003). There is sufficient scope to replace rice-wheat cropping system with other cropping systems without any decline in economic yield rather it improved substantially (Gangwar and Singh, 2011). Hira (2009) suggested that for reducing the area under rice in Punjab by about 1 million ha and cultivating BT (*Bacillus thuringiensis*) cotton, *kharif* maize, soybean and groundnut, which require lesser number of irrigations could be a boon for agricultural production in the state. In cotton, broad beds with spacing of 135 cm and planting cotton in furrows in paired rows increased yield by 44% and saved 40% of irrigation water compared with row spacing of 67.5 cm in flat bed system. Some areas under wheat can also be replaced by oil-seed (rape seed mustard) crops, which require only 2-3 number of irrigations. Thus there is feasibility to shift the part of rice area located in higher landscapes to maize, soybean or other suitable crops. Gangawar and Singh (2011) have reported that replacement of area under rice-wheat cropping system (2.6 million ha) by these alternative crop sequences will amount to the saving of about 135850 ha-m irrigation water with additional net returns of Rs. 4650 million to the farmers and saving of 162 million electrical unit amounting Rs. 378 million. This diversification will also help in reducing methane emissions from anaerobic rice cultivation.

16.7.1.6 Breeding of cultivars with climate resilient and adaptive traits

With climate change, the IGP could continue to be the major contributor of foodgrains despite scarcity of irrigation water, provided new cultivars are grown in judiciously selected cultivation schedules. Crop-breeding programs with the objective of developing climate-resilient (temperature- and drought-tolerant), high yielding cultivars of the identified crops should be given high priority, so that the desired kinds of cultivars are available in the wake of climate change. A combination of conventional and molecular-marker assisted, mutational, and transgenic-breeding approaches will be required to develop the desired kinds of crop cultivars. Crop-based coordinated programs need to be initiated as early as possible to develop climate-resilient cultivars. Recently, the Indian Agricultural Research Institute (IARI) has released an early-maturing basmati (fragrant) rice and a wheat cultivar suitable for late planting (Swaminathan and Kesavan, 2012). At Punjab Agricultural University (Ludhiana) the major focus in the crop breeding programs is to develop short duration cultivars of rice (cv PR 126). It appears that the desired traits of cultivars can also be selected in some of the ongoing breeding programs and the climate-change issues need to be converted from a 'problem' into an 'opportunity'. Germplasm of wild relatives and local land races could be used for developing climate-resilient crop cultivars. There is a need to revisit germplasm that has tolerance to heat and cold stresses but has not been utilized in the past because of low yield potential. Recently Tao *et al.* (2008) identified rice hybrid Guodao 6 as heat tolerant. Genetic enhancement of heat-tolerant genotypes, especially in pulses, by identifying and validating markers for high-temperature tolerance with high yield potential is one of the key technological advances that can prove to be a significant strategy for adapting to climate change. Improving adoption and dissemination of short-duration crop cultivars can enhance the ability of farmers to cope with variable climatic conditions. Similarly, for developing drought-tolerant strains of rice and other crops, *Prosopis juliflora* L. is being used as a donor of a gene for drought tolerance (Swaminathan and Kesavan, 2012). With the change in climatic scenario, maize could emerge as an alternative crop in the rice-wheat cropping system, replacing rice. As the temperate germplasm is highly productive, so it may be used for introgression of desirable genes into promising tropical/ subtropical backgrounds for the development of inbreds suitable for different agro-climatic conditions. A holistic approach towards the management of water, land, crop, and other natural resources, coupled with drought-tolerant cultivars, can go a long way towards alleviating the adverse impacts of drought (Sharma *et al*, 2010).

Summary

The climatic changes as a result of increasing concentrations of greenhouse gases and the resultant rise in global temperature and the uncertainties in rainfall have significant direct and indirect consequences for crop production and hence food security (Sinha and Swaminathan, 1991). The combined effects of climate change may have implications for dryland as well as irrigated crop yields. However this effect on crop production is expected to vary by crop and location as well as by the magnitude of warming and the direction and magnitude of precipitation change. Projections by the IPCC and few other global studies are that unless we adapt, there is a probability of 10-40% loss in crop production in India by 2080-2100 as a result of global warming (Parry *et al.*, 2004; IPCC, 2007) despite the beneficial aspects of increased CO_2 concentrations (Kumar *et al*, 2011). In Punjab state the PRECIS model under the A1B, A2 and B2 scenario predict more increase in minimum temperature than maximum temperature during the mid and end 21st century (Kaur and Prabhjyot-Kaur, 2016). This implies that increase in nighttime temperature will be more than daytime temperature, thereby the diurnal range of temperature will decrease. This will directly hamper the net photosynthesis rate as the respiratory losses will be more due to higher nighttime temperatures. Coupled with the decrease in diurnal range of temperature the extremes of the rainfall with 30 % more or less rainfall than the normal in all the agroclimatic regions have also been predicted to increase in the future. These simulated changes pose a direct threat to the agricultural production in the State. Some of the climate risk associated problems in Punjab include within season incidence of temperature extremes, receding water table, changes in pest scenarios etc. So there is a need for up-scaling of modern climate resilient / adaptive technologies such as conservation and smart agriculture, judicious utilization of available water for agriculture through micro-irrigation and water saving technologies, developing multiple stress tolerant crops through biotechnological tools etc at regional level. Reliable early warning system of environmental changes and their spatial and temporal magnitude will be of quite help and policies to support the diffusion of this information and to help interpret these forecasts in terms of their agronomic and economic implications are required to help farmers in a big way.

References/Suggested Readings

Abrol Y P, Bagga N, Chakravarty N V K and Wattal P N (1991). Impact of rise in temperature on the productivity of wheat in India. Abrol Y P, Gnanam A, Govindjee, Ort D R, Teramura A H and Wattal P N (ed) Proc Indo-US Workshop on Impact of Global Climate Change on Photosynthesis and Plant Productivity. pp 787-98. Oxford and IBH Publishing Co. Pvt Ltd.

Aggarwal P K (2003). Impact of climate change on Indian Agriculture. *J Plant Biol* 30:189-198.

Aggarwal P K (2008). Global climate change and Indian agriculture: impacts adaptation and mitigation. *Indian J Agric Sci,* 78: 911"919.

Aggarwal P K and Kalra N (1994). Analyzing the limitations set by climatic facters, genotype and water and nitrogen availability on productivity of wheat II. Climatically potential yields and management strategies. *Field Crops Res,* 38: 93-103.

Baker J T, Allen L H, Boote K J (1990). Growth and yield responses of rice to carbon dioxide concentration. *J Agric Sci,* 115: 313–320.

Balasubramanian V,Ladha J K, Gupta R K, Naresh R K, Mehla R S, Singh B and Singh Y (2003). Technology options for rice in the rice-wheat system in south Asia. In: Improving the Productivity and Sustainability of Rice-Wheat Systems: Is-sues and Impacts. ASA Special publication 65.

Behera U K, Sharma A R and Mahapatra I C (2007). Crop diversification for efficient resource management in India: Problems, prospects and policy. *J Sustainable Agric,* 30: 97-127.

Bhardwaj V, Yadav V and Chauhan B S (2009). Effect of nitrogen application timings and varieties on growth and yield of wheat grown on raised beds. *Arch Agron Soil Sci,* 56: 211-222.

Bhushan L, Ladha J K, Gupta R K, Singh S, Tirol P A, Saharawat Y S, Gathala M, and Pathak H (2007). Saving of water and labor in a rice–wheat system with no-tillage and direct seeding technologies. *Agron J,* 99:1288–1296.

Chauhan B S (2012). Weed ecology and weed management strategies for dry-seeded rice in Asia. Weed Technol. 26, 1e13. Rice-Wheat Sustainability in Indian Subcontinent 357. *Adv Agron,* 77: 293-368.

Conroy J P, Seneweera S, Basra A, Rogers G and Nissen W B (1994). Influence of rising atmospheric CO_2 concentrations and temperature on growth, yield, and grain quality of cereal crops. *Aust J Plant Physiol,* 21: 741–758.

Dar W D and Gowda C L L (2013). Declining Agricultural Productivity and Global Food Security. *J Crop Imprt*, 27: 242-254. ISSN 1542-7536

De Costa, Weerakoon W A J, Herath W M W, Amaratunga H M L K, Abeywardena R M I (2006). Physiology of yield determination of rice under elevated carbon dioxide at high temperatures in a subhumid tropical climate. *Field Crop Res,* 96: 336–347.

Dhillon S S, Hobbs P R and Samra J S (2000) Investigations on bed planting system as an alternative tillage and crop establishment practice for improving wheat yields sustainability. In: Proc. 15th Conf. of the *Intl. Soil Tillage Res*, Org. Fort Worth, Texas, USA.

Fischer G, Mahendra Shah andVelthuizen H V (2002) 'Climate Change and Agricultural Vulnerability. A special report prepared by the International Institute for Applied Systems Analysis as a contribution to the World Summit on Sustainable Development, Johannesburg 2002.

Gangwar B and Singh A K (2011). Efficient alternate cropping system. Project Directorate for Farming Systems Research, Indian Council of Agricultural Research, Modipuram, Meerut-250 110 (UP), India, pp 300–339.

Gupta R, Gopal R, Jat M L, Jat R K, Sidhu H S, Minhas P S, Malik R K, Gupta R, Gopal R, Jat M L, Jat R K, Sidhu H S, Minhas P S and Malik R K (2010). Wheat productivity in Indo-Gangetic plains of India during 2010: terminal heat effects and mitigation strategies. PACA Newsletter.

Hira G S (2009). Water management in northern states and the food security of India. *J Crop Impro*, 23: 136–157.

Hundal S S and Prabhjyot-Kaur (1996). Climatic change and its impacts on crop productivity in Punjab, India. pp 377-393. In: "*Climatic Variability and Agriculture*". Abrol, Y.P.et.al (Eds), Narosa Publishing House. New Delhi, India

IPCC (2007). Synthesis Report. Contribution of Working Groups I, II and III to the Fourth Assessment Report of the Intergovernmental Panel on Climate Change (eds Pachauri, R. K and Reisinger, A.), Geneva, Switzerland, pp 104.

Jat M L, Saharawat Y S and Gupta R (2011). Conservation agriculture in cereal systems of south Asia: Nutrient management perspectives. *Karnataka J Agric Sci*, 24: 100-105.

Kang M S and Banga S S (2013). Global agriculture and climate change: a perspective. In: Kang, M.S., Banga, S.S. (Eds.), Combating Climate Change: An Agricultural Perspective. CRC Press, Boca Raton, FL, pp. 11–25.

Kaur N and Prabhjyot Kaur (2016). Projected climate change under different scenarios in central region of Punjab, India. *J Agrometeorology*, 18: 88-92.

Kavikumar K S (2010). Climate sensitivity of Indian agriculture: role of technological development and information diffusion. In: *Lead Papers of National Symposium on Climate Change and Rainfed Agriculture*, February 18-20, 2010; pp. 192. Indian Society of Dryland Agriculture, CRIDA, Hyderabad, India.

Killman W (2008). Foreword. In: Climate Change and Food Security: A Framework Document. FAO of the United Nations, Rome, p. iii.

Kimbal B A, Kubayashi K and Bindi M (2002). Response of agricultural crops to free CO_2 enrichment. *Adv Agron*,77: 293-368.

Krishnan P, Swain D K, Bhaskar B C, Nayak S K and Dash R N (2007). Impact of elevated CO2 and temperature on rice yield and methods of adaptation as evaluated by crop simulation studies. *Agr Ecosyst Environ*, 122: 233–242.

Kumar S (2006). Climate change and crop breeding objectives in the twenty first century. *Curr Sci*, 90: 1053–1054.

Kumar S N, Aggarwal P K, Rani S, Jain S, Saxena R and Chauhan N (2011). Impact of climate change on crop productivity in Western Ghats, coastal and northeastern regions of India. *Curr sci*, 101: 332-341

Ladha J K, Pathak H, Padre A T, Dawe D and Gupta R K (2003) Productivity Trends in Intensive Rice-Wheat Cropping Systems in Asia. In: Ladha, J.K., Hill, J., Gupta, R.K., Duxbury, J. and Buresh, R.J., Eds., Improving the Productivity and Sustainability of Rice-Wheat Systems: Issues and Impacts, ASA Special Publication 65, American Society of Agronomy, Madison, 45-76.

Lal M, Singh K K, Srinivasan G, Rathore L S, Naidu D and Tripathi C N (1998). Growth and yield response of soybean in Madhya Pradesh, India to climatic variability and change. *Agr Forest Meteorol*, 89: 101–114.

Mahajan G, Bharaj T S and Timsina J (2009). Yield and water productivity of rice as affected by time of transplanting in Punjab, India. Agric. *Water Manage*, 96: 525–535.

Mahajan G, Timsina J and Singh K (2011). Performance and water use efficiency of rice relative to establishment methods in northwestern Indo-Gangetic Plains. *J Crop Imprv*, 25: 597–617

Majumdar P P (2008). Implication of climate change for sustainable water resources management in India. Physics and Chemistry of the Earth Parts A/B/C 33: 354-358.

Mittal S P and Singh P (1989). Intercropping field crops between rows of Leucaena leucocephala under rainfed conditions in northern India. *Agroforestry Sys*, 8: 165–172.

Mulchi C L, Slaughter L , Saleem M, Lee E H , Pausch R and Rowland R (1992) Growth and physiological characteristics of soybean in open-top chambers in response to ozone and increased atmospheric CO_2. *Agric Ecosyst Envir*, 38: 107-118.

Pandey S, Velasco L (2005). Trends in crop establishment methods in Asia and research issues. In "Rice Is Life: Scientific Perspectives for the 21st Century" Toriyama K, Heong KL, Hardy B (eds.)International Rice Research Institute, Los Ban˜ os, Philippines and Japan International Research Center for Agricultural Sciences, Tsukuba, Japan pp 178–181.

Pandey V, Patel H R and Patel V J (2007). Impact assessment of climate change on wheat yield in Gujrat using CERES-wheat model. *J Agrometerol,* 9: 149-157

Parry M L, Rosenzweig C, Lglesias A, Livermore M and Fischer G (2004) Effects of climate change on global food production under SRES emissions and socio-economic scenarios. *Global Environmental Change,* 14: 53-61.

Peng S, Huang J, Sheehy J E, Laza R C, Visperas R M, Zhong X, Centeno C S, Khush G S and Cassman K G (2004). Rice yields decline with higher night temperature from global warming. *Proceedings of the National Academy of Science* 101: 9971-75.

Prabhjyot-Kaur and Hundal S S (2006). Effect of possible futuristic climate change scenarios on productivity of some kharif and rabi crops in the central agroclimatic zone of Punjab. *J Agric Physics,* 6: 21-27

Prabhjyot-Kaur and Hundal S S (2010). Global climate change vis-à-vis crop productivity. *In* "Natural and anthropogenic disasters – Vulnerability, preparedness and mitigation" (Editor Jha, M.K.), Capital Publishing Company, New Delhi and Springer, The Netherlands pp. 413-431.

Ram H, Yadvinder Singh, Saini K S, Kler D S, Timsina J and Humphreys E J (2012). Agronomic and economic evaluation of permanent raised beds, no tillage and straw mulching for an irrigated maize-wheat system in northwest India. *Experimental Agriculture* 48: 21–38.

Rickman J F (2002). Manual for laser land leveling, Rice-wheat Consortium Technical Bulletin Series 5. New Delhi-12, India: Rice-wheat Consortium for the Indo-Gangetic Plains. pp. 24.

Rosenzweig C and Parry M L (1994). Potential impact of climate change on world food supply. *Nature,* 367: 133-138.

Seneweera S P and Conroy J P (1997). Growth, grain yield and quality of rice (Oryza sativa L.) in response to elevated CO2 and phosphorus nutrition. *Soil Sci Plant Nutr,* 43: 1131–1136.

Sharma P, Saini M, Sareen S, Verma A, Tyagi B S and Sharma I (2010). Assessing genetic variation for heat tolerance in synthetic wheat lines using phenotypic data and molecular markers. *Aust J Crop Sci,* 8: 515-22.

Singh H and Prabhjyot-Kaur (2018). Adaptive strategies to reduce the impact of terminal heat stress in wheat (*Triticum aestivum*) crop. *Journal of Agrometeorology* 20: 91-97

Sinha S K and Swaminathan M S (1991). "Deforestation, Climate Change and Sustainable Nutrition Security: A Case Study of India". *Climate Change,* 19: 201-209.

Sudhir Yadav G, Humphreys G E, Kukal S S, Walia U S (2010). Effect of water management on dry seeded and puddle transplanted rice. Part 1. Crop performance. *Field Crop Res,* 120:112–122.

Swaminathan M S and Kesavan P C (2012). Agricultural Research in an Era of Climate Change. *Agric Res,* 1: 3–11.

Tao F, Yokozawa M, Liu J, Zhang Z (2008). Climate crop yield relationship at province scale in china and the impact of recent climate trends. *Clim Res*, 38: 83-94.

Taub D R and Wang X Z (2008). Why are nitrogen concentrations in plant tissues lower under elevated CO2? A critical examination of the hypotheses. *J Integr Plant Biol*, 50: 1365–74

United Nations (UN) (2015). http://www.un.org/en/development/desa/news/population/2015-report.html

Yoshida S and Parao P T (1976). Climatic influence on yield and yield components of low land rice in tropics. *Proc Symposium on Climate and Rice.* Pp 471-94 IRRI, Manila.

Ziska L H, Namuco O, Moya T and Quilang J (1997). Growth and yield response of field-grown tropical rice to increasing carbon dioxide and air temperature. *Agron J,* 89: 45-53.

17

Climate Change Impacts in Agriculture over Jammu and Kashmir

Mahender Singh, Charu Sharma and Bharat Singh Ghanghas*

Agrometeorology Section, Division of Agronomy, SKUAST-J, Chatha Jammu-180004 (J&K), India
**CCSHAU, Hisar -125004 (Haryana)*

The Jammu and Kashmir State lies in the extreme North of the Himalayas and constitutes about 67.5 per cent of the North West Himalayan region. The State is predominantly a mountainous state with all the major Himalayan ranges and trans-Himalayas adequately represented. The State of Jammu and Kashmir located in the north Western corner of India, extends between 32°-17' and of 37°-50' North parallels of latitude and 73°-26' and 80°-30' East of meridians of longitudes and 81° East of Greenwich. The State on the basis of physiography may be divided into three main regions (i) outer Himalayas which comprise of Jammu province (ii) lesser Himalayas which comprises of Kashmir Valley and (iii) inner Himalayas which comprise of Ladakh province. Jammu, Kashmir and Ladakh regions have distinct agro climatic characteristics. Jammu region has two different climatic zones depending primarily on altitude. Lower hills & plains bear subtropical climate with hot dry summer lasting from April to July. The summer monsoon commences around first fortnight of July and fading away in early September. This is followed by dry spell from September to November. Winter is mild and temperature seldom touches freezing point. In the high reaches of Chenab valley, the climate is moist temperate, winter is severe and varied quantity of snow is received. The Kashmir valley with Pir Panjal Mountains on its South and Karakoram on its North receives precipitation in the form of snow due to western disturbances. The winter is severely cold and temperature often goes below 0°C. Summers are warm and dry and autumn is again cool and sometimes wet. Ladakh is situated in eastern mountain range of Kashmir. This is one of the highest ranges in the world. It is a cold desert receiving very little precipitation. The temperature remains below the freezing point during winter due to its high altitude.

The climate change is the biggest environmental emergency that the state faces today. This inevitable disaster has a multitude of serious implications for both environment and human society. Owing to its peculiar topography, rugged terrain, extreme weather conditions and underdeveloped economy, the State has suffered with frequent natural disasters in the changing climate. Mountain ecosystems, in particular are among the entities most vulnerable to climate change. The State has witnessed many natural disasters especially in the 19th and early 20th centuries. Hazards like earthquakes, floods, forest fires, droughts, avalanches and landslides often convert into disasters leading to loss of human lives as well as public and private property. Enhanced vulnerabilities of the built environment make the State highly prone to natural disasters. Scientific evidence suggests that the mountain ecosystems may provide an early warning of what may be the impacts in lowland environments.

The climate change has potential implications on different sectors including glaciers & permafrost, rivers, forests, soils, biogeochemistry, ecosystems, society and economy etc. The major climate change impacts includes, (a) warmer climates causing shifts season and amount of precipitation, (b) changes in annual freeze-thaw cycles, (c) erratic weather and prevalence of extreme climatic events, (d) changes in phenology of the crops, (e) degradation of quality of wildlife habitat and (f) loss of agricultural and horticultural production. Climatic change is affecting all regions of the globe which not only effects the production but it also disturbs the normal life in Kashmir valley. Due to the climatic change the occurrence of floods, draughts, heat and cold waves are common across the world and have adverse impact on the livelihood of billions of people who depend on land for most of their needs. As we know the weather of Kashmir valley is entirely different from other states of India. An unusual case of freezing rain and snowfall was reported from Jammu and Kashmir during the month of August 1996. It resulted in death of 241 pilgrims going to Amarnath while about 9000 pilgrims from different parts of India, were stranded in the area along the route. Due to the climatic change the occurrence of floods, droughts, heat and cold waves are common across the world and have adverse impact on the livelihood of billions of people who depend on land for most of their needs. Till 1999-2000 Kashmir was receiving well distributed precipitation in terms of rain and snow to the extent of 1000-1200 mm, but at present has decreased to 600-800 mm. The need of the hour is that such problems should be undertaken seriously in the states of Jammu and Kashmir as it is the backbone of economy, if it remains same the whole nation will suffer. The times ahead may be very challenging. The new generation will be in difficult situation. Even for us, it will be quite tough living. Living conditions will become harsher. The impacts of climate change are particularly worrying for the foothills region of western

Himalayan state of Jammu and Kashmir as agriculture is the mainstay of the people and it provides employment, directly or indirectly to about 70 per cent of the work force in the State.

17.1 Weather extremes

Extreme weather events might have taken six decades to hit Jammu & Kashmir but it may not now wait for that longer. Climate models, analyzed by experts, predict that India will be hit more and more by extreme rainfall events in future and J&K is certainly among the most vulnerable ones due to impact of global warming on Himalayan glaciers. 66 heat waves recorded in last 100 years across the State, resulting increase in temperature causing to increase evapotranspiration rates, reduce water availability and thereby altering the hydrological cycle. The maximum number of cold waves in Jammu & Kashmir is about 211 in last century. However, this was over a very small area over the mountainous region of Ladakh with the temperature going 19.7^{0}C below its normal. Snow storms occur during winter (December, January and February) over the northern regions. Sometimes early or late snowfall is also reported in November, March or April. In March and April, when spring starts, it coincides with snowing in mountains. Then it reaches in plains. It harms the Cherry, Almond and Peach blossoms in orchards. Crops are lost. For the last five years, 'untimely snowfall' has been devastating the crops in the area. Due to wide fluctuations in temperature hailstorms are experienced which cause damage to flowers and various stages of growth and development and damages plants directly. Data from 1973 to 2018 showed that there is a clear declining trend in precipitation amount from November to March. Less precipitation and snowfall closer to summer cause avalanches. Sudden cloudbursts or melting of glaciers add to the Siachen floods. The year 2010 was a landmark year for Ladakh because of the floods. It was a wake-up call for the locals and nature was not viewed in the same light. In August 2010, a cloudburst saw the Ladakh region receive more than 350 mm of rain, three and a half times its annual rain - in just two days. Rivers overflowed, villages and towns were inundated. Reports said 234 people died and 800 others went missing, perhaps washed away by raging rivers and waterways. Four years later, in the summer of 2014, another spell of heavy rainfall affected more than half a million people - most of them in Srinagar. Witnessing snowfall on April 6, 2017 shocked everyone, the reason is climate change. December, January and February were the apt months for snowfall. Snow would get frozen. But since last three to four years these three months, stay warm. Several media reports said this was perhaps the worst disaster recorded in over 50 years, leaving 280 dead. More than half a million people were trapped in their houses as the city was submerged under 18 feet of water for more than three weeks.

17.2 Climate change over J & K

The climate profile of Jammu and Kashmir indicates variation in climate characteristic over the years, there is increase in an average temperature in Jammu and Kashmir. The Kashmir valley has shown rise of 1.45 degree Celsius and Jammu region has shown a rise of 2.32 degree Celsius over last two decades. The maximum temperature has increased by 0.5 degree Celsius per year in Kashmir valley and 0.08 degree Celsius per year in Jammu region. Moreover, amount of snowfall has reduced over the years, the document said while laying stress on in-depth exercise to study the climate change projections in Jammu and Kashmir.

17.2.1 Climatic Change in Jammu Province

To analyze the impact of changing climate the region, long term climatic data of weather parameters *viz*., maximum & minimum temperature, morning & evening relative humidity and Rainfall of past thirty years of Jammu, Katra, Batote, Banihal and Bhaderwah covering sub-tropical, sub-temperate and temperate areas were used.

17.2.1.1 Temperature

Jammu: The trend shows that the non-significant (significance level of 0.05 & 0.01) variation in mean annual, *kharif* and *rabi* season's maximum temperature of Jammu decreases with the rate of 0.02, 0.034 and 0.039 °C per year, respectively. While the minimum temperature also showed non- significantly increasing trends (0.039°C) and decreasing trend with rate of 0.010and 0.043°C per year during annually and *rabi* season, respectively (Table 17.1).

Katra: The mean *annual* and *kharif* season maximum temperature showed non significantly decreasing trends in with the rate of 0.001 and 0.020°C per year, but non-significantly increasing during *rabi* season (0.004°C per year) in last 37 years data. Whereas, this station showed significantly (<0.05) increasing trends in mean *annual, kharif* and *rabi* season minimum temperature with the rate of 0.050, 0.055 and 0.055°C per year (Table 17.1).

Banihal: The mean *annual* & *rabi* season maximum and minimum temperature showed significantly (<0.05) increasing with the rate of 0.032 & 0.041°C and 0.021 & 0.024°C per year and non-significantly increasing at the rate of 0.010 and 0.012°C per year during *kharif* season, respectively (Table 17.1).

Batote: The mean annual and *rabi* season's maximum temperature non-significantly increasing (0.010 and 0.019°C per year, respectively) and decreasing during *kharif* season the rate 0.006°C per year. While showed significantly (<0.05) increasing trends in mean annual and *rabi* season minimum temperature

Table 17.1: Annual, crop seasonal temperatures and trend at different locations of Jammu region.

Station Name	*Kharif season*		*Rabi Season*		Annual	
	Temperature°C/ year	Trend °C/ year	Temperature°C/ year	Trend °C/ year	Temperature°C/ year	Trend °C/ year
Maximum temperature (°C)						
Jammu	34.8	-0.034	24.3	-0.039	29.7	-0.020
Katra	31.6	-0.024	21.7	0.004	26.6	-0.001
Banihal	27.6	0.010	16.7	0.041	21.3	0.032
Batote	26.5	-0.006	16.5	0.019	20.7	0.010
Bhaderwah	29.2	-0.020	14.1	0.015	22.5	0.000
Minimum temperature (°C)						
Jammu	23.7	0.039	10.2	-0.043	16.2	-0.010
Katra	21.3	0.055	10.0	0.055	15.2	0.054
Banihal	14.7	0.012	3.6	0.024	8.3	0.021
Batote	16.2	0.028	5.8	0.046	10.1	0.035
Bhaderwah	17.8	0.069	2.9	0.074	7.6	0.071
Mean temperature (°C)						
Jammu	29.3	-0.036	17.3	-0.001	23.0	-0.016
Katra	26.5	-0.009	15.9	0.355*	21.0	0.026
Banihal	21.1	0.026	10.2	0.019	14.8	0.426*
Batote	21.4	0.005	11.2	0.037	15.4	0.022
Bhaderwah	23.5	-0.003	8.5	0.372*	15.1	0.009

**Significance at 95 confidence level*

at the rate of 0.035 and 0.046°C per year. The mean *kharif* season minimum temperature fluctuates non- significantly with the rate of + 0.028°C per year (Table 17. 1).

Bhaderwah: The mean *annual* and *rabi* season maximum temperature showed non-significantly increasing with the rate of 0.002 and 0.015°C per year, respectively and decreasing 0.015°C per year during *kharif* season. The mean *annual, kharif* and *rabi* season maximum temperature showed significantly (<0.05) increasing trends at the rate of 0.71, 0.069 and 0.074°C per year (Table 17.1.

17.2.1.2. Rainfall

The annual rainfall in Jammu province is showing the non significant decreasing trends at all the locations except Banihal. The rates of decreasing in annual rainfall were found 1.690, 3.152, 6.121, and 5.583 mm per year at Jammu, Katra, Batote and Bhaderwah, respectively. While in Banihal the annual rainfall showed the increasing trend at the rate of 0.541 mm/year The crop season wise rainfall trends were analyzed in order to compare the inter seasonal rainfall variability and its change in pattern and is presented in Table 2. During *Kharif* season the rainfall showed the increasing trends at the rate of 1.954, 0.383, 2.804, 0.121 and 0.121 mm per year at Jammu, Samba, Bhaderwah, Batote and Banihal, respectively. At Katra station, the rainfall decreases at the rate of 0.213 mm per year during monsoon season (*Kharif* Season). But during *rabi* season the rainfall trends increasing at Jammu, Banihal and Batote with the rate of 1.252, 1.224 and 1.224 mm year, respectively, while decreasing in Katra, Samba & Bhaderwah at the rate of 3.306, 0.142 and 7.257 mm per year, respectively (Table 17.2).

17.2.1.3 Relative humidity

Jammu: The variation in mean annual, *kharif* and *rabi* season's maximum relative humidity at Jammu location showed significant (< 0.05) increase at the rate of 0.481, 0.252 and 0.558 percent per year with a regression coefficient of 0.59, 0.25 and 0.45, respectively. While the mean minimum relative humidity significantly (<0.05) decreasing trend in case of annually and *rabi* and with rate of 0.200 and 0.438 percent per year, respectively and during *kharif* season, non-significantly decreasing trends at the rate of 0.066 percent per year (Table 17.3).

Katra: The mean *annual* and *rabi* season maximum relative humidity showed non-significantly increasing trends at the rate of 0.063 and 0.081 percent per year, but non-significantly decreasing during *kharif* season (0.864°C per year). Whereas, this station showed non-significantly increasing trends in mean annual,

Table 17.2: Annual, crop seasonal rainfall and trend at different locations of Jammu region.

Station Name	*Kharif season*		*Rabi Season*		Annual	
	Mean Rainfall (mm)	Trend mm/ year	Mean Rainfall (mm)	Trend mm/ year	Mean Rainfall (mm)	Trendmm/ year
Jammu	868.9	+1.954	240.6	+1.252	1173.4	-1.125
Samba	882.7	+0.383*	211.5	-3.306	1143.6	-2.518*
Katra	1472.8	-0.213	503.3	-0.142	2077.8	-1.254
Bhaderwah	532.8	+2.804*	770.3	-7.257	1305.7	1.997*
Batote	627.1	+0.121*	998.4	+1.224	1624.2	-4.159*
Banihal	433.6	+0.121*	910.4	+1.224	1343.6	-4.714

*Significance at 95 confidence level

Table 17.3: Annual, crop seasonal relative humidity and trend at different locations of Jammu region.

Station Name	*Kharif season*		*Rabi Season*		Annual	
	Max. Relative Humiditymm/year	Trend mm/ year	Max. Relative Humiditymm/year	Trend mm/ year	Max. Relative Humiditymm/year	Trend mm/ year
Maximum Relative Humidity (%)						
Jammu	77	0.352*	83	0.558	76	0.482*
Katra	76	0.390*	67	0.081	73	0.064
Banihal	77	0.065	70	-0.026	71	0.002
Batote	68	0.067	67	-0.208	68	-0.110
Bhaderwah	75	0.140	73	0.078	73	0.095
Minimum Relative Humidity (%)						
Jammu	58	-0.066	55	-0.438	52	-0.200
Katra	70	0.048	63	0.023	58	0.010
Banihal	57	0.336*	51	-0.004	53	0.033
Batote	65	0.397*	59	-0.512	61	-0.065
Bhaderwah	59	0.069	58	-0.068	59	-0.021

**Significance at 95 confidence level*

kharif and *rabi* season minimum relative humidity with the rate of 0.009, 0.048 and 0.023 percent per year, respectively (Table 17.3)

Banihal: The mean *annual* maximum and minimum relative humidity showed non-significantly increasing with the rate of 0.002 & 0.033 percent per year, respectively. The *kharif* season mean maximum relative humidity showed non-significantly increasing with the rate of 0.065 percent per year, although mean minimum relative humidity during kharif season showed significantly increasing with the rate of 0.136 percent per year with regression coefficient of 0.28. The mean *rabi* season maximum and minimum relative humidity showed non-significantly decreasing with the rate of 0.026 & 0.004 percent per year, respectively (Table 17.3).

Batote: The mean annual maximum and minimum relative humidity showed non-significantly decreasing at the rate of 0.110 & 0.065 percent per year, respectively. The mean *rabi* season maximum relative humidity showed significant decreasing with the rate of 0.208 percent per year with regression coefficient of 0.21 and minimum relative humidity showed non-significantly decreasing at the rate of 0.152 percent per year. While, the mean annual maximum and minimum relative humidity showed non-significant increasing during *kharif* season at the rate 0.067 and 0.097 percent per year, respectively (Table 17.3).

Bhaderwah: The mean *annual, kharif* and *rabi* season maximum relative humidity showed non-significant increasing at the rate of 0.096, 0.140 & 0.078 percent per year, respectively. While, mean *annual* and *rabi* season minimum relative humidity non-significant decreasing at the rate of 0.021 and 0.068 percent per year, respectively and non-significant increasing (0.069 percent per year) during kharif season (Table 17.3) .

17.2.2 Climatic Change in Kashmir Province

The data for baseline (1980 – 2016) was procured from Indian Meteorological Department for six stations: Srinagar, Gulmarg, Kupwara, Pahalgam, Kokernag and Qazigund located in various districts. The maximum temperature of Srinagar, Qazigund, Pahalgam, Gulmarg, Kokernag and Kupwara station showed an increasing rate of about 1.18°C, 1.29°C, 1.38°C, 1.19°C, 1.80°C and 1.90°C during last year's, respectively. Whereas the minimum temperature showed an increase of about 0.18°C in Srinagar station, an increase at Qazigund station is about 0.06°C, Pahalgam station showed an increase of about 0.74°C, Gulmarg station showed an increase of about 0.24°C, Kokernag station showed an increase of about 0.32°C and Kupwara station showed an increase of about 0.24°C during these 30 years.

17.2.2.1 Rainfall

Rainfall showed a decreasing trend except two stations of Kupwara and Qazigund where the trend showed an increase from 1980 to 2010. The Srinagar station showed an decrease in rainfall from 918.7 mm to 779.2 mm, Qazigund station showed an increase from 368.6 mm to 1207.9 mm Pahalgam station showed an decrease in rainfall from 1274.2 mm to 1543.2 mm, Kupwara station showed an increase from 293.5 mm to 1216.3 mm, Kokernag station showed an decrease in rainfall from 1001.6 mm to 1226.5 mm and Gulmarg station showed an decrease in rainfall from 1598.9mm to 1380.4 mm.

17.2.2.2 Future Anomalies in Maximum Temperature

The predictions point towards the fact that annual maximum temperature will increase by 0.1 °C for the period 2021-2050 which is not quite significant. For the period 2051-2100, an overall decreasing trend in the annual maximum temperature will be observed with the annual maximum temperature plunging by 3°C and 3.5 °C during 2051-2075 and 2076-2100 respectively. The maximum temperature will decrease by about 0.9 °C during *Kharif*(2021-2050). However, for the same period, the maximum temperature for *Rabi* season will surge by 1.3 °C. For the period, 2051-2100, maximum temperature will increase during both *Kharif* and *Rabi* season. The deviations in monthly, seasonal, cropping season and annual maximum temperature with respect to the baseline period 1985-2015 for 2021-2050, 2051-2075 and 2076-2100 over the maximum temperature of the baseline period (1985-2015).

17.2.2.3 Future Anomalies in Minimum Temperature

The predictions point towards the fact that annual minimum temperature will not vary for the period 2021-2050. For the period 2051-2100, an overall decreasing trend in the annual minimum temperature will be observed with the annual minimum temperature plunging by 1°C and 1.2°C during 2051-2075 and 2076-2100 respectively. The minimum temperature will increase by about 0.1°C during both *Rabi* and *Kharif*(2021-2050). For the period, 2051-2100, minimum temperature will increase during both *Kharif* and *Rabi* season. The deviations in monthly, seasonal, cropping season and annual minimum temperature with respect to the baseline period 1985-2015 for 2021-2050, 2051-2075 and 2076-2100 over the minimum temperature of the baseline period (1985-2015).

17.2.2.4 Future Anomalies in Precipitation

The predictions point towards the fact that annual precipitation will increase by 35.2 mm during the period 2021-2050. For the period 2051-2100, increasing trend in the annual precipitation will be observed with the annual precipitation increasing by 38.7 mm and 32.5 mm during 2051-2075 and 2076-2100 respectively. The precipitation will increase by about 17.7mm during both *Kharif* and 45.3 mm during *Rabi* (2021-2050). For the period, 2051-2100, there precipitation varies significantly during both Kharif and Rabi season. The deviations in monthly, seasonal, cropping season and annual period with respect to the baseline period 1985-2015 for 2021-2050, 2051-2075 and 2076-2100 over the baseline period (1985-2015). The climate of Kashmir Valley has witnessed a different change in climate and climatologically variables as compared to the other parts of India. The maximum and minimum temperatures of the region did show no significant increasing or decreasing trends since last few decades. The amount of rainfall has however shown an increase in quantity during this period. The results obtained from the present study are therefore in agreement with the current conditions prevailing in the valley.

17.3 Climate change impact on irrigation

Irrigation plays an important role in the agriculture sector of J&K economy. The State does not receive rain throughout the year. In Jammu region, climatic conditions favour cultivation of crops throughout the year but due to insufficient water for irrigation is a major impediment. The model prepared by the University predicted that the crop water demand under subtropical condition of Jammu increases by 7 per cent if temperature alone rises by 3°C. The focus of attention now remains on extending irrigation to various parts of the state owing to the fact that self-sufficiency and food security are directly linked to it. Kashmir valley, receives rainfall mostly in winter when temperature is too low for plant growth. When the temperature starts rising in May and onwards, the rainfall decreases and except some showers of rainfall in July-August, most of the growing season remains dry. Since ages, the farm economy has been dependant on a single crop and the cultivator cannot take chances with it. The farmer always requires sufficient water supply for his field, therefore, he depends mostly upon canals for irrigation. Many snows fed streams, running down the slopes of the mountains, makes it very easy for the Government to construct small canals or pools, despite this only 60 per cent of the land in the valley is irrigated.

17.4 Climate change impact on Glaciers

The studies conducted during last three decades by the National Institute of Hydrology, Roorkee, reveal that about the changes observed in glaciers in Jammu and Kashmir, particularly in Ladakh, Zanskar and the Great Himalayan ranges of Jammu and Kashmir are generally receding and the glacier volume changes range between 3.6% and 97% with the majority of glaciers showing a degradation of 17%–25%". The 23-km long Drang-Drung glacier in the Zanskar valley is highly affected by western disturbances. Further, "the Nubra valley of Jammu and Kashmir has 114 small-sized glaciers varying between less than 5 km and 10 km in length. The glaciers of the valley, however, do not show much change in their length and area during the period 1989–2001. However, variable decline in the glacial area of the Siachin glacier has been observed with the area reducing from 994.99 km^2 in 1969 to 932.90 km^2 in 1989. Glacial melt in the Himalayas is projected to increase flooding, rock avalanches from destabilized slopes, and affect water resources and decreased river flows as the glaciers recede. However, small change in the area (932.90 km^2 to 930 km^2) has been noticed during the following decade (1989–2001) .The impact of rising temperature and reducing snowfall on glacier mass require a sound long-term database for precise climate change assessment.

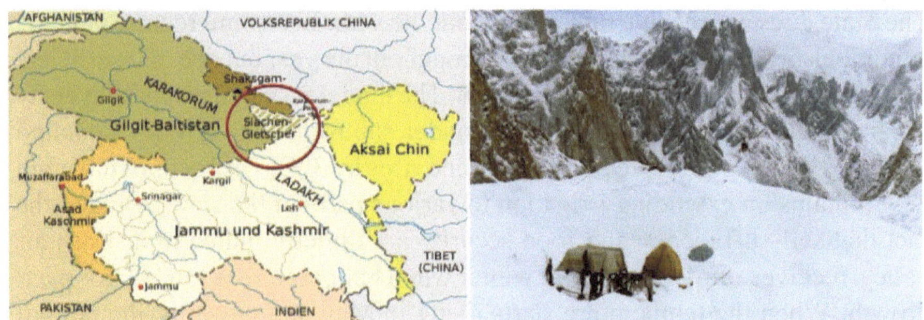

17.5 Climate change impact in horticultural crops

It is more important in Kashmir as the State's economy is entirely depending on agriculture and horticulture sector. The various climatic factors affect crops directly or indirectly due to which production. In addition, climate change also affects other factors of production agriculture, such as water availability, soil fertility, and pests. The state offers good scope for cultivation of horticultural crops covering a variety of temperate fruits like apple, pear, peach, plum, apricot, almond, cherry, and guava, citrus litchi and phalsa. Besides medicinal and aromatic plants floriculture, mushroom plantation crops and vegetables are cultivated in the state. State also has the honour of being amongst the world's few places where quality saffron (corcus sativus) is cultivated. The various reason behind the change in land use range from qualitative decrease in agricultural production and lack of quality agricultural seeds to change in land use by farmers besides extra income fetched by horticultural production. Thus temperature rise and rainfall decline, the walnut apricot etc are fast failing from several parts of the Kashmir. Because of general increase temperature and fewer availability of water deteriorating the production and quality of walnuts in Kashmir valley and in Jammu region in mid temperature. In Kashmir production and productivity grows very slow in rain-fed areas of Kashmir's Karewas because of abnormal windstorms and hailstorms in summer to different fruits like walnut, apricot, plum, apple and cherry are getting damage greatly. In latest years the style varies in Kashmir with snowfall pattern which affects all the stone and pome fruits. In so many years it has observed that flowering and snowfall are losing quantity and quality in great level. In Jammu and Kashmir, the area under walnut production has reduced with drastically due to high temperature and less rainfall in winter. Besides, medicinal and aromatic plants, floriculture, mushroom, plantation crops and vegetables are cultivated in the state. Apart from this, well known spices like saffron and black Zeera are also cultivated in some pockets of the state. Saffron which is one of the most important commercial commodities of Indian agriculture, J&K enjoys a virtual monopoly in the sub continent.

17.5.1 Walnut Production

The Jammu and Kashmir state cultivated the walnut about 86,263 tones within area 61,723 hectares in 2013.Walnut is the next major fruit occupying 26.86 percent among all other fruit. Walnut production is general in Anantanag, Kupwara, Kalgan, Baramulla, Budgam, Srinagar, Poonch, Badarwah, Ganderbal, Bandipora, Rajouri and other hilly areas. The walnut production favor temperate climate structure otherwise its trees are damaging in the cold winters and bloom flowering with immature nuts injures due to the spring frosts. In walnuts there is wind –pollination and different verities are planted in one orchard with close proximity. In Indian different states, walnut is cultivated but earlier it was found in Jammu and Kashmir. Where more than 92 percent of walnut production is exports to the Indian different markets and outside? Its production is cyclical and shows the fluctuation trend year to year due to the climate changes. Mostly walnut and apple are impacting due to the change of temperature where trees grow 2-3 weeks before but usually walnut trees grow in April. In mid march majority of trees began to open their buds. In March end, it becomes unusual extremely cold again and vulnerably to frost damage to the open buds. Due to these climatic changes, walnuts colour, taste and growing levels highly changes. Corm sprouting flower initiation and time of flowering are the critical stages that are influenced by environmental fluctuation in terms of temperate and availability of water. Day temperature of 23°C -25°C in the month of September is critical for corm sprouting and flowering starts when the day temperature reaches to 17°C with a night temperature around 10° C.

17.6 Climate change impact on pollination

The occasion of spring frost and low temperature at the time of flowering and fruit setting adversely affects production of fruits like saffron, apple and walnut. The area's most vulnerable to influence of low temperature are located at high altitudes where good spring season with adequate sunshine promote apple, walnut and saffron flowering. However due to fluctuating temperature during this critical period particularly rains accompanied low temperature inhibits the transfer of pollen because of restricted bees activity. Climate change has brought alternation in the pattern of blooming bearing and ultimately the fruit yield. The lack of early cold in December and January has adversely affected the production of various fruits. Apart from this, heavy rains wash away the pollen in plants during these months. Chilling affects the flowering and subsequent fruit setting quality as well as quantity in month of April and May.

17.6.1 Lack of sufficient chilling hours

The majority deciduous fruit trees require adequate accumulated chilling, or vernalisation to shatter winter dormancy insufficient chilling due to improved greenhouse warming may result in extended dormancy, leading to decreased fruit quality and productivity. The low down warming situation is less than 1°C is not likely, to affect the vernalisation of high-chill fruits like walnut, apple, apricot, cherry, almond varieties and if warming situation go over's 1.5°C and would considerably raise the risk of extended dormancy for both pome-fruit and stone-fruit. Temperatures of 7°C and below are affective in meeting the chilling requirements of different temperate fruits including apple, saffron and walnut. Sufficient chilling hours are not being met as there is lack of snowfall and rainfall during winter month especially in hilly areas of Kashmir. Hence results an erratic and poor flowering which is responsible for poor fruit set with low yields. The extinct plantation of apple in south Kashmir districts such a Shopian, Pulwama and Kulgam are the live examples of impact of climate change. The low snowfall and the less precipitation have affected the saffron production since last many years. It causes big problem of chilling necessity which is significant to quality and higher production of walnut.

17.7 Climate change impact on insect pest and diseases

Erratic variation in temperature and rainfall, guides to more incidences of pests, insect and diseases. In the previous some years, the attack of walnut scab, powdery mildew in walnut, flee beetle in approximately all the fruit crops has been enlarged. The less precipitation and high temperature records from last many years has damaged the crops, mostly apple, saffron and walnut and at the same time has caused diseases like scab, red might etc. that damages the fruits. Due to high weather and humidity the insect and germs like bacteria, fungus, algae grows more and affects the fruits. Due to low rainfall the growth and colour remain slow as in apple and saffron. Under the changing climate scenario existing fungi, pathogen and viruses may cause more damage to the crop. An increased number of sprays are now required for the routine control of pests. The number of sprays has increased from about four per year in 1980s to about 11 per year today. The disease like red mite and scab are common in each part of the valley due to the existing climate.

17.8 Climate Change impact in agriculture

Climate change impacts on agriculture are being witnessed all over the world, but countries like India are more vulnerable in view of the huge population dependent on agriculture, excessive pressure on natural resources and poor coping mechanisms. The projected impacts are likely to further aggravate yield fluctuations of many crops thus impacting food security. There are evidences already of negative impacts on yield of wheat and paddy in parts of India due to increased temperatures, increased water stress and reduction in number of rainy days. Significant negative impacts have been projected with medium-term (2010-2039) climate change, eg. yield reduction by 4.5 to 9 percent , depending on the magnitude and distribution of warming. Since agriculture makes up roughly 15 percent of India's GDP, a 4.5 to 9.0 per cent negative impact on production implies a cost of climate change to be roughly at 1.5 per cent of GDP per year. Enhancing agricultural productivity, therefore, is critical for ensuring food and nutritional security for all, particularly the resource poor small and marginal farmers who would be affected most. In the absence of mitigation and adaption strategies, the consequences of long term climate change could be even more severe on the livelihood security of the poor. Studies have shown that production and quality of fruit and vegetable crops are affected by high temperature and exposure to elevated levels of carbon dioxide and even ozone concentration.There is no doubt that climate change will increase the fluctuation of food production and have an important effect on food security. Climate change will affect all economic sectors to some degree, but the agricultural sector is perhaps the most sensitive and vulnerable. The impact of climate variability on agricultural production is important at local, regional, national, as well as global scales. Climate change leads to failure of crops and decreases production besides it also affects need and demand of particular crops in markets. The interaction between agricultural performance and weather has been a main area of research interest due to the direct dependence of the farmer on the latter. Climate generally affects cropping pattern directly as well as indirectly as of its direct association with growth and production of different crops, which mandates change in cropping pattern. Climate change affects cultivation of crops because of direct association of crops with climate of the region. Climate change affects agriculture in two ways—direct and indirect. Changes in climatic factors (for example, temperature, and rainfall) affect agricultural productivity through physiological changes in crops. However, research in this area has traditionally focused on the influence of yearly weather on agricultural production. With the threat of possible climate change gaining momentum, the focus has now shifted to the exploration of the relationship between agricultural performance and the long-term climate variables such as temperature, rainfall and atmospheric CO_2 concentration level. Understanding the relationship between climate and crop yield helps in enhancing resilience of agricultural production systems to climate change is getting pace. The cropping pattern of the farmers is dynamic and unique as of diverse topography of the region. The chapter throws light on various aspects of land use patterns and associated trends which are being followed by people. With regard to climate change the respondents uphold that non-seasonal snowfall compels them to change land use pattern. Climate plays an important role for every type of sector, especially in agriculture. Climate threat increases from various interactions between environments, economic, social, religious and political systems. It is more important in Jammu and Kashmir as the state's economy is entirely depending on agriculture and horticulture sector. Cropping is highly sensitive to climate, crops having a limited environment in which they are productive and profitable. The state offers good scope for cultivation of horticultural crops covering a variety of temperate fruits like apple, pear, peach, plum, apricot, almond, cherry, and guava, citrus litchi and phalsa. Besides medicinal and aromatic plants floriculture, mushroom plantation crops and vegetables are cultivated in the state. State also has the honour of being amongst the world's few places where quality saffron (corcus sativus) is cultivated. Usually critical months of September and October are dry and thus flowering is delayed due to delayed sprouting, In the latest induction of climate change an unpredicted hailstorm compiled with untimely rain has damaged the crop and causing loss of 2 billion rupees. The floods of 2014 make the destruction of major crops like, apple, pear, walnut, saffron, cherry etc. and causes loss of over 14.25 billion rupees. For the last many years the production of saffron is decreasing at high level, in 2015 the saffron production was below average. The less precipitation and high temperature records from last many years has damaged the crops, mostly apple, saffron and walnut and at the same time has caused diseases like scab, red might etc. that damages the fruits. Due to high weather and humidity the insect and germs like bacteria, fungus, algae grows more and affects the fruits. Due to low rainfall the growth and colour remain slow as in apple and saffron which results in the low price and also reduces its taste. The productions of various fruits of Jammu and Kashmir are badly affected by the challenging climatic conditions. Recently floods destroyed the apple quality and quantity in various parts of the state. The studies revealed that livelihood of people are in worse situation due to the present climate. The fruits such as apple, saffron, and walnut are the major sources of income for Jammu and Kashmir population are facing challenging movements day by day.

In the latest induction of climate change an unpredicted hailstorm compiled with untimely rain has damaged the crop and causing loss of 2 billion rupees. The floods of 2014 make the destruction of major crops like, apple, pear, walnut, saffron, cherry etc. and causes loss of over 14.25 billion rupees. For the last many years the production of saffron is decreasing at high level, in 2015 the saffron production was below average. The less precipitation and high temperature records from last many years has damaged the crops, mostly apple, saffron and walnut and at the same time has caused diseases like scab, red might etc. that damages the fruits. Due to high weather and humidity the insect and germs like bacteria, fungus, algae grows more and affects the fruits. Due to low rainfall the growth and colour remain slow as in apple and saffron which results in the low price and also reduces its taste. The productions of various fruits of Jammu and Kashmir are badly affected by the challenging climatic conditions. Recently floods destroyed the apple quality and quantity in various parts of the state. The studies revealed that livelihood of people are in worse situation due to the present climate. The fruits such as apple, saffron, and walnut are the major source of income for Kashmiri population and are facing challenging movements day-by-day due to climate change/variability.

17.8.1 Wheat

Rise in temperature during *rabi* season affect the wheat crop adversely because it shortens the duration of all development stages. It was estimated that the wheat yield reduces to the extent of 0.5 q/ha per rise of temperature by 1°C. The increase in maximum and minimum temperature by 2°C and 4°C without change in CO_2 concentration during the crop growing period the wheat maturity observed earlier by 17 and 30 days under temperate condition, while 10 and 18 days under subtropical condition, respectively. However increase in carbon dioxide levels to 350,450 and 750 ppm under normal temperature did not affect maturity of wheat crop under temperate condition of Jammu and Kashmir, but it enhances the wheat maturity by 3-5 day under sub tropical condition. Wheat crop matured in advance by 6 to 18 days under sub tropical condition, while 17-30 days in temperate condition. But when the temperature decreased to 2

°C and 4°C the ripeness of wheat crop was delayed by 20 to 38 days and 10 to 17 days under temperate and Sub tropical condition. The increase in maximum and minimum temperature by 2 °C and 4 °C throughout the crop growing periods then crop yield reduced by 29.09 and 37.29 per cent, respectively under temperate condition and 17.52 and 32.18 percent, respectively under subtropical condition. Increase in maximum and minimum temperature by 2°C under elevated CO_2 concentration 450 ppm abruptly decreased wheat yield by 16.82 per cent, while increased in temperature by 2°C with CO_2 concentration of 700 ppm enhanced grain yield by 13.14 per cent in temperate condition. Decrease in maximum and minimum temperature by 2 and 4°C with normal concentration of CO_2 decreased wheat yield by 7.07 and 26.96 per cent in temperate condition. In temperate condition increase in temperature (4°C) shortened the wheat crop duration about one month, which decline the wheat yield up to 37.29 percent, but in subtropical condition reduces seed yield by 32.18 per cent. The seed yield fluctuation found more in temperate condition compared to sub tropical condition of Jammu and Kashmir.

17.8.2 Maize

Rise in temperature showed a general tendency towards diminishing future maize yield in Jammu district of Jammu and Kashmir state. The testing and validation of CERES-maize model under subtropical condition predicted the significant reduction in maize yield by 34 percent under increase in temperature by 3°C. The overall grain yield of maize is going to decline by 3.66, 8.06, 16.29 and 26.14 where as biomass by 1.66, 8.64, 18.00 and 28.08 per cent from the production level of the year 2005 in 2020, 2050, 2080 and 2100, respectively. The elevated temperature up to 0.64°C did not show the marginal effect on grain yield where as significant yield reduction was observed at 2.6 and 3.2°C rise in temperature. The double CO_2 compensated the grain yield reduction due to the temperature rise up to 1.6°C. Beyond 1.6°C, the rising CO_2 could not completely mitigate the adverse effect of rising temperature in maize. Decrease of yield following rise in temperature up to 2.56 and 3.2°C coupled with doubling of CO_2 was 7.98 and 18.04 per cent, respectively. As far as bio mass is concerned, the adverse effect of rise in temperature is mitigated by the rising temperature up to 0.64°C.

17.8.3 Rice

The emissions, from the country's 42.21million hectares of land under rice cultivation, comprise about 2.07 Tg of methane [Tg is the unit of measurement of greenhouse gas emissions and is equivalent to 1012 grams], 0.19 Tg nitrous oxide and 72 Tg of carbon dioxide, annually. However, the emission levels vary from region to region depending on cultivation practices and the inherent carbon content of the soils. The methane outflow from the paddy fields of some districts in state has been observed to be relatively high. This is due to the presence of higher organic carbon content in the soils and the traditional practice of keeping rice fields constantly submerged under water. Similarly, emission of nitrous oxide is higher in paddy fields in northern States because of application of high doses of nitrogenous fertilizers. On the whole, the eastern and southern parts of the country have a relatively higher global warming potential because of higher discharge of methane and carbon dioxide and the predominance of rice cultivation in these regions.

17.8.4 Biodiversity

According to the "2005 Millennium Ecosystem Assessment", the climate change will cause loss of biodiversity by the end of this century. The significance and utility value of biodiversity for food and agricultural purpose will increase as and when climate changes. Genetic resources are the living materials that local communities, researchers and breeders use to develop high yielding crop varieties/strains that can adapt to changing needs. Maintaining and using this reservoir of genetic diversity will be the foundation for copping with climate change.

17.9 Climate risk management

Mitigation and Adaptation

- Agriculture is one of the major sources of greenhouse gas emissions. Climate change has been a cause of serious concern if the agricultural sector has to grow in the context of country's overall economic growth, to respond to rural households' livelihood, state's food security and poverty alleviation. It may take some years to fully experience the devastating effects of climate change on agriculture but the time is ripe for the Government, private sector and public to have adequate concern, commitment and accountability to mitigate the effects of climate change.
- Significantly investing in expanding, modernizing and equipping agricultural meteorology facilities in all agro-ecological regions to make it world class, thereby continuously improving weather and climate forecasting system.
- Evolving policy and programs to manage and mitigate risks due to climate change
    - Improving early warning systems followed by effective monitoring and evaluating its impact in the region

  - Developing climate impact modules that give a better understanding of how climate change may affect crop, livestock and fish farming and forestry at a local level in order to be well prepared
- Diversifying pattern of livelihoods and adapting agricultural, fishing and forestry practices to efficient water management and soil conservation practices and growing resilient crops and trees.
  - Developing a database on climate, soil and water use and crop yields to assess, map and monitor land-use performance under given technology conditions. Assessment of how vulnerable our food system is and how we can adapt agriculture, livestock, fisheries and forestry to future climate-related disasters.
  - Building sufficient resilience of the food systems to avoid enormous future economic losses in agriculture, livestock, fisheries and forestry.
  - Evolving comprehensive climate resilience strategies comprising risk assessment, development of varieties that can perform well in stressful conditions, better land, water and livestock management and bringing about specific changes in agricultural practices that can respond to climate change.
- Government and private sector will have to invest substantially in agricultural research on one hand and motivate/train farmers to take better advantage of the dry *rabi* season in the flood prone regions and help them supplement their income through non-farm activities on the other.
- Besides, following measures are necessary to better adapt to climate change impacts
  - To develop land use plans, food security programs, fisheries and forestry policies that can help farming community suitably adapt to climate changes
  - T o undertake cost/benefit analyses of climate change risks for irrigation or coastal protection and for investment decision
  - Promotion of "best crop-livestock-fish farming practices" through farmers' capacity building and networking
  - Conceptualization and Implementation of "National Adaptation Program of Action on Climate Change"
  - Developing contingency plans to cover new and evolving risk scenarios

- Due to decrease in rainfall during *rabi* season the changing planting dates could have a significant impact that planting wheat earlier than usual can help reduce climate change-induced damage with zero tillage and laser-based leveling can also help conserve water and land resources. Crop planning can be conducted as per the climatic zones of different regions, while utilizing better genotypes for rain-fed conditions with the decreasing in rainfall in different location of the region.

17.9.1 Participatory varietal selection and community based seed production

Initiatives were also taken for making farmers the seed producers and suppliers of various varieties of seeds. In this regard, Participatory Plant Breeding, a novel approach for up-scaling availability of location specific varieties/cultivars was followed, involving plant breeders and the farmers in partnership mode. The activity is being strengthened further both for identifying location specific varieties and strengthening the local seed system. A group of farmers is now engaged in producing quality seed under the scientific supervision. The seed production efforts were up-scaled and different approaches adopted to increase the out-reach of quality seed and planting material. Seed processing and packaging plants are now available at different constituent units and nurseries improved through laser levelling, irrigation, drainage etc. to expand production of quality planting material of fruit crops. Specific crop varieties recommended for the region were put in the production chain that included varieties recommended for release (rice, maize, wheat, moong, pea and masoor) by the State.

17.9.2 Conservation and utilization of local biodiversity

In order to harness the comparative advantage and mountain *Kashmiriyat* in agriculture, research focus on indigenous crops and varieties were achieved with local aromatic rice cultivars, *Mushkbudji* and *Kamad*, involving Sagam farmers of Anantnag district. Efforts were also started in 2011 with maize seed farmers of Lolab valley to augment the requirement of quality seed in maize. The blue print has been developed for upscaling these programmes in pa

17.9.3 Emphasis on micro propagation

Availability of quality planting material through tissue culture techniques was strengthened further. In the case of saffron, micro-propagation protocol has been standardized to achieve high corm multiplication rate. The technique has good potentialities for adoption under public-private partnership.

17.9.4 Production of quality planting material

Fruit nurseries expansion efforts resulted in production of one lakh saplings of different fruit crops particularly apple and walnut. University is now geared-up to increase the planting material production to 5 lakh plants by 2015. Models of High Density planting in apple for better productivity and quality have also been created. The adoption of high density orchards under valley conditions is expected to improve the apple production of farmers. During the same period, 15,000 saplings of three high yielding but difficult to root varieties of mulberry viz., Goshoerami, KNG and Ichinose were raised through poly house technology and supplied to various agencies.

17.9.5 Development of mother orchards

The mother orchards of quality bud wood have been established in apple (2 ha), cherry (1 ha), walnut (1.5 ha), almond (1 ha) and pear (0.75 ha) at different locations of the University to increase availability of adequate bud wood for production of quality planting material and distribution to the farming community.

17.9.6 Protocol for virus indexing of fruits

Protocol for virus indexing of fruit plants was standardized and a central facility created for marking virus free planting material and bud wood plants of apple and other fruits. Based on our investigations, four apple viruses viz. Apple Mosaic Virus (ApMV), Apple Chlorotic Leaf Spot Virus (ACLSV), Apple Stem Grooving Virus (ASGV) and Apple Stem Pitting Virus (ASPV) were detected in different samples. Molecular Pathology Lab has been set up as a centralized facility in the University. The facility also helped in identifying diversity of the fungus causing apple scab. It will help develop management strategies, through studies on host plant resistance. Four pathogenic races were identified for the first time through comprehensive molecular analysis, first of this kind in India.

17.9.7 Development of bio-control agents for wilt of solanaceous crops

For development of newly invented bio-pesticides, five bio-control agents viz., BA-18, BCA-20, BCA-16, BCA-31 and K-2 were identified for management of wilt/root rot of solanaceous crops. These bio-pesticides are being evaluated in the farmer's field for wider adoption by farmers.

Conservation and utilization of apricot and ambri

For conservation and utilization of apricot in Ladakh region, Kargil Apricot Research Station centre has been established at Kargil. Similarly for revival of Ambri, Apple Research Station has been established at Balpora, Shopian.

17.9.8 Commercial production of liquid biofertilisers and organic inputs

Efforts were initiated for strengthening the organic input production programme for eventual use in transition towards organic farming. Facilities for vermin-compost production and effective microbial consortia (EMs) production have been established to meet the expected requirement of the state in future. Efforts are underway for skill development and trainings in management of production systems organically.

17.9.9 Rural appraisal for livelihood systems

Scientific groups made extensive surveys in disadvantaged areas in the region particularly Changthang, Zanskar and Gurez. Based on the livelihood analysis of the population in the region, relevant technology interventions should now be possible with establishment of Highland Pastoral Systems Research and Extension Station, Nyoma, Changthang, Ladakh and Highland Agriculture Research and Extension station at Padum in Zanskar valley in Kargil district.

17.9.10 Integrated nutrient management

This will be based on soil health card and would move gradually from major fertilizer application to micro-nutrient. The adoption gap of fertilizers in the state from a study is given below to show there exists scope for improvement. The balancing will be done on climate induced soil characteristic deterioration like leaching, or deterioration of soil organic carbon. Sub-activities like organic cultivation, nutrient recycling and farm manure integration would be under this category.

17.9.11 Diversification of livelihoods

Livelihood action plans will be based on the livelihood shock during the climatic stress and actions proposed thereof and will be different from normal top down livelihood programmes and would be designed through a micro-planning process.e.g. introduction of small ruminants, introduction of green fodder will be under this category

17.9.12 Crop water balance and water use efficiency

The focus in this case will be based on the water requirement crops and the overall efficiency. This is in line with *Pradhan Mantri Krishi Sinchayee Yojana* (PMKSY). The micro-irrigation equipment will come from that source. However, the crop rotation, intercrop that will have best water use will be the mode of integration. In any case cereals and staples do not use drip. These systems will be given in areas where the vegetable cultivation or floriculture is feasible.

17.9.13 Specific Adaptation Activities to be implemented

- Specific adaptation activities to be implemented to reduce the climate variability / change vulnerability which are beyond the scope of business-as-usual are as follows.However, the overall scope of the project will be in line with the sector development focused along with adapting to the changing climatic situations.
- Vulnerability analysis of the existing cropping system and document deviations from recommended practices
- Supportive and diversified livelihoods system with mixed farming and advanced farming technique to cope with climate variability and increased output per unit of land
- Water scarcity situation managed bringing water use efficiency in rain-fed farming conditions
- Improved capacity of the farmers on climate resilient agricultural practices and its adoption in farming.
- Identifying priority crops for interventions
- Introduction of climate adaptive cropping system as per the assessment findings.
- Assessment of feasibility of inter-cropping / mixed cropping and its promotion.
- Soil testing and soil-moisture management system based on soil health card.
- On-Farm water management in rain-fed & water stress areas through drip/sprinkler systems.
- Farm ponds / water harvesting structures in the water stress / rain-fed areas.
- Training / Orientation of target farmers on climate resilient agriculture / horticulture.
- Demonstration of different package of practices that is adaptive to climate variability.
- Credit linkage and convergence of other existing schemes at farmer and area level.
- Promotion of Integrated Farming System, taking in to account livestock and agriculture.

- Monitoring and documentation climate benefit of adaptive practices and overall adaptation benefits and sharing the learnings.

Summary

In Jammu and Kashmir temperatures are increasing, with major increase of temperature in maximum level is 0.05 Celsius ever year. The increase in temperatures in Jammu and Kashmir has to do with the topography and high altitude. In Kashmir region average mean temperature has risen 1.45^0 and 2.32^0 Celsius rises in past three decade's years in Jammu region. For the period 2051-2100, an overall decreasing trend in the annual maximum temperature will be observed with the annual maximum temperature plunging by 3°C and 3.5 °C during 2051-2075 and 2076-2100, respectively. The maximum temperature has increased by 0.5 degree Celsius per year in Kashmir valley and 0.08 degree Celsius per year in Jammu region. Moreover, amount of snowfall has reduced over the years. Climate change and warming is a global trend, its impact on Kashmir is significant. The annual rainfall in Jammu province is showing the non significant decreasing trends at all the locations except Banihal. Thus temperature rise and rainfall decline, the walnut apricot etc are fast failing from several parts of the Kashmir. Because of general increase temperature and fewer availability of water deteriorating the production and quality of walnuts in Kashmir valley and in Jammu region in mid temperature. Due to these climatic changes, walnuts colour, taste and growing levels highly changes. This fluctuating temperature has reduced agriculture yield in Kashmir by 30 percent.

Some of climatic observed in area with the long term data are:

- Climatic Variability and Extreme Events
- Mean Temperature Rise: 0.2 degree Centigrade per decade
- Mean Summer Rainfall to increase, subsequently decrease during *rabi* seaon
- Frequent and Longer dry spells during south west monsoon season
- Shorter but Heavier Precipitation Events
- Hot Extremes, Heat Waves to be more common in sub tropical areas of the state
- Contraction of Snow Cover in hilly areas

It is concluded from the foregoing facts that the future climate and its impact could well trigger bloody wars fought over access to basic necessities like drinking water and food grains. It is estimated that greater loss is expected in *Rabi* as compared to *Kharif* crops and especially in productivity and quality apple. The growers have observed uneven distribution of winter, high temperature, etc. The area under food crops and fruits has shifted to off season crop cultivation. The change of climate is causing melting of glaciers which are precious reservoirs of clean water. The projected impacts of Climate Change would be an additional pressure that can significantly affect the future availability of goods and services from forests and hence would have a more profound impact on forest-dependent communities and their livelihoods. Certain measures are required to be taken to overcome this problem which include creation of laws; desilting, growing of pulses, millets and adoption of agro forestry.

References /Suggested Readings

Agriculture Production Department, (2013-14) "Result Framework Document (RFD), Horticulture Sector J&K".

Irrigated rice, wheat and mustard productions may be reduced by 6%, 4% and 4%, respectively.

Irrigated rice, wheat and mustard productions may be reduced by 6%, 4% and 4%, respectively.

Irrigated rice, wheat and mustard productions may be reduced by 6%, 4% and 4%, respectively.

IPCC (2014). Climate change 2014–impacts, adaptation and vulnerability: regional aspects. Cambridge University Press, Cambridge

IPCC (2015). Mitigation of climate change, vol 3. Cambridge University Press, Cambridge

Kumar V, Jain SK (2010) Trends in seasonal and annual rainfall and rainy days in Kashmir Valley in the last century. *Quatern Int*. 212(1):64–69

Lone PA, Bhardwaj AK, Shah KW (2015) Prenanthes violaefolia Decne (Asteraceae)—a new report from Kashmir Himalaya, India. *Asian Pac J Trop Biomed* 5(7):552–554

Rajendra, K. and Kitoh, A. (2008). Indian summer monsoon in future climate projection by a super high resolution global model. *Current Science*, 95(11): 1560-1569.

Sharma, S.K. and Chauhan, R. (2011).Climate change research initiative: Indian network for climate change assessment. *Current Science* (Bangalore), 101(3): 308-311.

18

Climate Change and Agriculture Scenario in Temperate and Cold Arid Region

Latief Ahmad

Agromet Field Unit-Srinagar, Division of Agronomy
Sher-e-Kashmir University of Agricultural Sciences and Technology of Kashmir
Shalimar, Jammu and Kashmir, India

Agriculture is very intensely influenced by weather and climate. While farmers are often compliant in managing weather and annual variability, there is however a high degree of adaptation to the local climate. This may be in the form of conventional infrastructure, native agricultural practices or personal experience. Climate change an in this way be relied upon to affect the agriculture, possibly threatening established characteristics of farming systems and provide chances for improvements. The role of climate change in influencing human activities and the natural environment is all around acknowledged. This may go from an extreme event enduring just a couple of hours (i.e. hurricane, thunderstorm and showers) through to the multi-year dry spells (Maracchi *et al*., 2005). Increase in concentration of greenhouse gases in the atmosphere further contribute to climate change. This results in a strong impact on human activities and natural environments. Essential areas, for example agriculture and forestry, will be more susceptible than auxiliary and tertiary segments, for example manufacturing and retailing (Parry *et al*., 1998). Agriculture is the fundamental activity by which people live and survive on the earth. Evaluating the impacts of climate change on agriculture is a vital job. In developed as well as developing nations, the influence of climate on crops and domesticated animals regardless of water system, enhanced plant and creature cross breeds and the developing utilization of synthetic manures. The continuous dependency of agricultural production on light, heat, water and other climatic factors, the reliance of a great part of the total populace on farming exercises, and the noteworthy extent and quick rates of conceivable atmosphere changes all consolidate to make the requirement for an exhaustive thought of the potential effects of atmosphere on worldwide agriculture (Rosenzweig, and Liverman, 1992).

Climate change impacts on agriculture differ regionally. Climate change impacts on subtropical and tropical areas will be predominately negative, particularly where agriculture is right now minimal as for high- temperature and moisture-deficit conditions. The most susceptible agricultural systems are present in arid, semi-arid, and dry sub-humid regions in the developing world, home to half of the world's malnourished populace, where high precipitation variability and recurring droughts and floods regularly disrupt food production and where poverty is unescapable. Agriculture in cold-limited (high-latitude and high-altitude) areas could benefit from humble levels of warming that adequately increment the length of the growing season (Padgham, 2009).

18.1 Climate Change and Agriculture

In light of a portion of the past encounters, effect of climate change on agriculture will be one of the real central components impacting the future sustenance security of humanity on the earth. Agriculture isn't just sensitive to climate change yet additionally one of the significant drivers for climate change. Understanding the weather changes over some stretch of time and modifying the management practices towards accomplishing better harvest are snags to the development of agricultural sector in general. The sensitivity of agriculture to climate is questionable, as there is territorial variability in precipitation, temperature, crops and cropping systems, soils and management practices. The inter-annual variations in temperature and precipitation have been much higher than the predicted changes in temperature and precipitation. The crop losses may increase if the predicted climate change increases the climate variability. Different crops differ in their response to global warming will have a complex impact. The tropics are more reliant on farming as 75% of total populace lives in tropics and 66% of these individuals' fundamental occupation is farming. With low levels of innovation, extensive variety of pests, diseases and weeds, land degradation, unequal land distribution and rapid population growth, any effect on tropical agriculture will influence their business. Rice, wheat, maize, sorghum, soybean and barley are the six chief crops in the world grown in 40% cropped area, and comprise 55% of non-meat calories and over 70% of animal feed (FAO, 1995). Subsequently, any impact on these crops would adversely disturb the food security.

The State of Jammu and Kashmir lies between 32°17' and of 37°5' North and 73°26' and 80°30' East. The State is located amidst three climatic regimes of Asia n its south fringe lies the frail rainstorm zone of Punjab. On the north-east the State is bounded by the vast arid plateau of Tibet while the North-west border areas confront the eastern furthest reaches of Mediterranean climatic district. This topographical position, combined with the differed physiography, gives the State a wide climatic variety. The State has been alienated into four broad macro-climatic zones (i) sub-tropical (ii) valley temperate (iii) intermediate, and (iv) cold-arid. The State has generally a mountainous area and occupies a central position in the Asian continent. The state comprises nearly two-thirds of the mountainous area of India. 2.3 million ha out of 3.5 million ha of mountainous area of India belong to Jammu and Kashmir. The State is limited on the north by Chinese and Russian regions, on the east by Tibet, on the south by Punjab (India) and on the West by Pakistan (Fig. 18.1). The topography of the state is highly uneven comprising of many snow-covered peaks having altitude range of 554 to 7077m amsl in North and North-west, which are followed towards the South by lower range of hills. Total geographical area of the State is 2,22,236 km^2 including 78,114 km^2 (35.15%) under Pakistan, and 42,735 km^2 (19.23%) under China. Ladakh is the largest hilly arid zone which involves 58321 km^2 (42.00%).

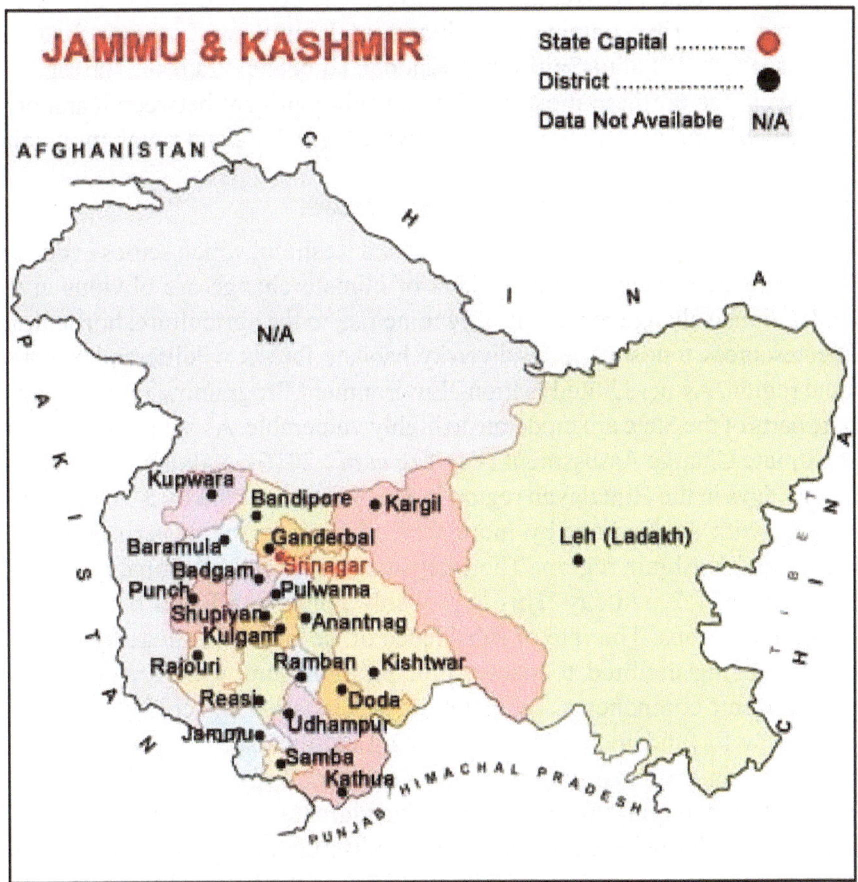

Fig. 18.1: Map showing different districts of J&K State and its neighboring areas

Kashmir area includes the temperate zone of the state of Jammu and Kashmir. his zone encounters wet and regularly extreme winters with frost, snow and rain and comparatively dry and warm summer. Snowfall, a critical type of precipitation, keeps up sufficient moisture supply amid summer when precipitation is sparse. The valley temperate zone involves the zones of varied relief. The plain valleys have an elevation of 1560 m amsl, which ascends to 1950 m in low altitude Karewas in mid belts, 2400 – 3000 m in the upper belts and to 4200 m in snow bound ranges. The soils of Kashmir valley are alluvial in nature having 62% of area under irrigation. The prominent meteorological features of temperate zone demonstrate that the zone receives annual precipitation of around 680 mm, of which nearly 70 per cent is received during winter and spring seasons (from December to May). The overall average temperature in different months varies from 1.2°C to 24.5°C with cold thermal index and humid hydric index. In India, arid zone encompasses 3,87,390 km 2 area of which 1,07,545 km 2 lie in the cold arid region of Western Himalayas. The cold arid region of Western Himalaya primarily contains Ladakh area of Jammu and Kashmir State and certain sections of Lahul-Spiti sub-division in Himachal Pradesh. The region in J&K lies in the northern most tip of Asian sub-continent between Karakoram and greater Himalayan ranges and is entwined with bare and rough mountains.

18.2 Climate Change in Kashmir and Ladakh

With regards to India, particularly Jammu and Kashmir, which settles in delicate Himalayan Ecosystem; there markers of climate change are obvious at this point. Climate change represents a genuine risk to the agriculture, horticulture, water resources, tourism, species diversity, habitats, forests, wildlife, and livelihood in the region. As per United Nations Environment Programme (UNEP) report some parts of the State are moderate to highly vulnerable. As per Indian Network on Climate Change Assessment (Parvaze *et al.*, 2016) evaluation the number of rainy days in the Himalayan region in 2030s may increase by 5-10 days on an average, with an increase by more than 15 days in the eastern part of the Jammu and Kashmir region. The intensity of rain fall is probably going to increment by 1-2 mm/day. This is probably going to affect a portion of the horticultural crops. The rate of subsidence of ice sheets is allegedly differing which is being credited to winter precipitation, atmosphere warming and anthropogenic components. Temperature, precipitation and cold wave are well on the way to fundamentally affect the agricultures sector. Scarcity in food production is growing in Jammu & Kashmir. With the diminishment in precipitation, the rain-fed agriculture will endure the most. Horticultural crops like apple are likewise demonstrating decrease underway especially because of decrease in snowfall. Around 34% and 39% of the forested frameworks are probably going to experience shifts in vegetation type with a trend towards increased occurrence of the wetter forest types.

The climate of Kashmir and Ladakh Valley has witnessed a different change in climate and climatological variables as compared to the other parts of India. During last few decades, the maximum and minimum temperatures of the region have shown no significant increasing or decreasing trends. The amount of rainfall has however shown an increase in quantity during this period (Parvaze *et al.*, 2016). Compared to other parts of the country which are witnessing externalities of climate change, long term annual rainfall in the state of Jammu and Kashmir showed positive growth momentum with annual growth rate of 1.39 percent during last 28 years. Overall, annual maximum and minimum temperature for the State almost exhibited a uniform trend during last 40 years (Jammu and Kashmir Agricultural Production Department). According to a report by Jammu and Kashmir Environmental Information System (J&KENVIS), "the climate change projections predict an increase from 0.9±0.6 °C to 2.6±0.7 °C in the 2030s. The net increase in temperature ranges from 1.7 °C to 2.2°C with respect to the1970s. Seasonal air temperatures also show a rise in all seasons. The annual rainfall in the Himalayan region is likely to vary between 1268±225.2 and 1604±175.2 mm in 2030s. The projected precipitation is likely to increase by 5% to 13% in 2030s with respect to 1970s. The intensity of rain fall is likely to increase by 1-2mm/ day. The water yield in the Himalayan region, mainly covered by the river Indus, is likely to increase by 5%–20% in most of the areas, with some areas of Jammu and Kashmir showing an increase of up to 50% with respect to the1970s" (J&KENVIS,2015).

18.3 Vulnerability of agriculture and allied activities due to climate sensitivity

Temperature, precipitation and cold wave significantly impacts the agricultures sector and enhances its vulnerability. This happens due to the early onset of rain and number of dry days. In western Himalaya, the traditional farming operation cultivating operation is an unpredictable result of yield farming crop husbandry, animal husbandry and forest resources constituting interlinked diversified production systems. Nonetheless, with changing land use, the area under cultivation of many traditional crops has been reduced and some others are at the brink of extinction. Deficit in food production is growing in recent times in Jammu & Kashmir. With the reduction in rainfall, the rain-fed agriculture will suffer the most. Plant crops like apple are additionally indicating decrease underway and a genuine scope especially because of decrease in snowfall.

18.4 Impact of Climate change in temperate Kashmir Valley

Crop yields have been reliably observed to be higher in temperate regions than in the tropics (FAO, 1995). Numerous factors contribute to this result. Soils in the humid tropics have a tendency to be profoundly drained of nutrients and are consequently unproductive because of high temperatures, intense rainfall, and erosion. Temperate soils are generally viewed as more favorable to agriculture than tropical soils because of higher nutrient levels. However, there are exceptions such as poorly developed and infertile soils in temperate regions. Solar radiation, which is basic to plant development, and whose intensity is controlled by the angle of the sun, day length, and cloudiness, is lower in winter and higher in summer in temperate zones. In the tropics, sunlight based radiation is regularly restricted by cloudiness amid the rainy seasons.

18.5 Impact on forests

Vegetation designs (circulation, structure and nature of woods) crosswise over globe are controlled for the most part by the climate. Even with global warming of 1-2 °C, significantly less than the latest projections of warming amid this century, most biological systems and landscapes will be affected through changes in species composition, productivity and biodiversity. It has been exhibited that upward movement of plants will occur in the warming scene. Because of increment in temperatures, change in vegetation, fast deforestation and shortage of drinking water, natural habitat devastation and corridor fragmentation may prompt an awesome risk to annihilation of wild flora and fauna. It is normal that with the expansion in dry spell cycles and corresponding increment in forest fires some pine species will infringe upon alternate species and will assume an essential part in diminishing the yield of non-timber forest products (NTFP) from these forests. The ecosystem services created would likewise modify. Unavoidably, any adjustment in the forest (distribution, density and species composition) under CC would massively impact economies like forestry, agriculture, livestock husbandry, NTFPs and medicinal plants based livelihoods. Some impacts of climate change on forests are listed below (J&KENVIS,2015):

- Marked extension (11%) in Temperate deciduous, cool mixed and conifer forests at the cost of alpine pastures which are probably going to shrivel.
- The uncommon patterns of winter movement of birds in the wetlands of Jammu, Kashmir and Ladakh
- Decline financially vital species like Deodar, Fir and spruce and increase in Blue Pine in Kashmir Valley and Chir Pine in Jammu.
- The spread of invasive alien species like Parthenium, Lantana, Ageratum, Xanthium, Anthemis, etc.,
- Decreasing tree density and forest fragmentation
- Increased net primary productivity.
- Increased incidences of forest fires.

18.6 Impact on agriculture

Agriculture is highly dependent on weather and changes in weather cycle have a major effect on crop yield and food supply. Mountain agriculture is mostly rainfed and driven by biomass energy of surrounding forests and confined to terraces carved out of hill slopes. Climate change may result in the following major impacts on agriculture in Kashmir valley (J&KENVIS,2015):

- Irrigated rice, wheat and mustard productions might be decreased by 6%, 4% and 4%, individually.
- The shortfall in food production in Kashmir region has touched 40 %, while the deficiency is 30 % in vegetable production and 69 % in oilseed production, putting food security at a more serious hazard.
- Invasion of weeds in the croplands and those are consistently removed by the farmers (e.g., Lantana camara, Parthenium sp.
- Increased rate of recurrence of insect-pest attacks
- Decline in crop yield.
- More and more paddy land being changed over to rainfed orchard or dry land. The huge chunk of paddy land has been converted into rain-fed dry land in many regions of the valley as of late.
- Area under apple cultivation increased however yield per hectare has has fundamentally declined amid the previous decades..

These factors have led to loss in agridiversity and change in crops and cropping patterns. Conventional farming in the Himalayan mountains has been a rich storehouse of agrobiodiversity and strong to crop maladies.

18.7 Impact on Agriculture in Ladakh

Agriculture is one of most delicate areas to effects of climate change due its high reliance on atmosphere and climate. Over hundreds of years, agriculturists in Ladakh have developed self-supported cultivating frameworks not withstanding in the midst of a climatically difficult condition. In fact, it has been the pillar of economy and gives nourishment and occupation security to the general population of Ladakh. However, recent environmental changes are harmful for the fabric of sustainability for the general population of Ladakh. The effect of environmental change in the course of the most recent couple of years has unfavorably influenced farming in this district in various ways. Snow and water from glaciers is the chief source of irrigation for nine-tenths of the farming community. Thus, glacial melt determines the productivity in the region to a large extent. Besides glaciers are the primary source of water for the farmers in the region. Excessive glacial melt has now become a cause of floods in the region, putting the lives of more than 80% farmers to risk. Continuous retreating of glaciers in the region will have an adverse effect on t agriculture. The economy and food security of the region will be threatened due to changes in agricultural produce. It is supposed that increase in temperature could modify agricultural produce. The change may be either positive or negative. The impact on agriculture and its outcomes on the general public are probably going to differ locally contingent upon the kind of environmental change that has occurred here and the choices accessible to farmers. It could cut down the yield of wheat. However, higher temperature may enhance the productivity of wheat in the high altitude areas. Warmer temperatures ay influence many crops to develop all the more rapidly, however hotter temperatures could likewise diminish yields. Crops have a tendency to grow quickly in warmer conditions. Nonetheless, for a few crops, (for example, grains), speedier development decreases the measure of time that seeds need for growth and maturity. Throughout the years, agriculturists have brought new plants into their farmlands. Crops tend to grow faster in warmer conditions. However, for some crops (such as grains), faster growth reduces the amount of time that seeds have to grow and mature. Over the years, farmers have introduced plants into their farmlands. Looking at the brighter side of the impact of climate change, it merits saying that farmers in Ladakh have profited from the current global warming as they can develop number of new vegetable products.

Climate change is likewise connected with an increase in the incidence of extreme climatic events, for example drought, cloudburst and floods. The nomads known as the Changpas living in the Changthang region, intensely feel environmental change. These individuals are reliant on the raising of animals, the pashmina goats for their fleece. As of late, the migration paths of the Changpas have changed because of decline in pasture land. Further, indeterminate snowfall has led loss of numerous domesticated animals in the district. In future, environmental change in Ladakh is probably going to influence agriculture and increment the danger of water for irrigation. The villagers could confront serious water shortage which is credited to overwhelming runoff from rapid snow melt. Besides, an expansion in the occurrence of pest attacks and weeds could also make destruction in the district.

Summary

Agriculture and forestry will be particularly sensitive to changes in mean climate and climate variability. Climate change will definitely impact agriculture and agricultural productivity especially with the introduction of new crop species and varieties, higher crop production and expansion of suitable areas for crop cultivation. However this may be accompanied by a rise in need for plant protection, risk of nutrient leaching and accelerated breakdown of soil organic matter. The rise in carbon-dioxide level may result in increased water use efficiency, which will recompense for some of the adverse effects of increasing water constraint and extreme weather events. This on the other hand will also lead to low harvest yields, higher variation in yield and reduction in areas suitable for growing traditional crops. In Kashmir Valley, the increased precipitation is expected to be large enough to compensate for the increased evapotranspiration. On the other hand, increased precipitation, cloudiness and rain days and the reduced duration of snow cover and soil frost may negatively affect the agricultural production.

References/ Suggested Readings

Food and Agriculture Organization. (1990). FAO Yearbook. Production. Vol. 44. Food and Agriculture Organization of the United Nations. Rome, 1991.

INCCA, (2010) Climate Change and India: A 4 x 4 Assessment – A Sectoral and Regional Analysis for 2030s.

J&KEVIS (2015). ENVIS Newsletter October – December, 2015 Climate Change & Concerns Of J&K, J&KENVIS Centre Department of Ecology, Environment & Remote Sensing Jammu & Kashmir.

Maracchi, G., Sirotenko, O. and Bindi, M., (2005). Impacts of present and future climate variability on agriculture and forestry in the temperate regions: Europe. *Increasing Climate Variability and Change*, 117-135.

Padgham, J., (2009). Agricultural development under a changing climate: opportunities and challenges for adaptation. World Bank.

Parry, M. L., Carter, T.R. and Knoijin, N. T. (eds.): (1988), 'The Impact of Climatic Variations on Agriculture', Kluwer, Dordrecht, 876.

Parvaze, S., Ahmad, L., Parvaze, S. and Kanth, R.H., (2017) Climate Change Projection in Kashmir Valley (J and K). *Current World Environment*, *12*(1), 107-115.

Rosenzweig, C. and Liverman, D., (1992). Predicted effects of climate change on agriculture: A comparison of temperate and tropical regions. Dalam Global Climate Change: Implications, Challenges, and Mitigation Measures. Dalam SK Majumdar (Ed.) The Pennsylvania Academy of Sciences. Pennsylvania, 342-61.

19

Climate Resilient Agriculture in Himachal Pradesh

Rajendra Prasad, Anupam Sharma and Manoj Kumar Negi

CSK Himachal Pradesh Krishi Vishvavidyalaya, Palampur, Himachal Pradesh, India

The State of Himachal Pradesh encompasses a geographical area of 55.7 lakh ha and is situated between 31°22'40" and 33°12'40"N latitude and 75°45'55" and 77°10'20"E longitude. The state is bordered by Jammu and Kashmir on the north, Punjab on the west, and Haryana on the south-west (Fig. 19.1). About 12 per cent of the state area is under agriculture which is mostly (80 per cent) rain fed. Wheat, maize, rice, and barley are the principal crops. In state, there are 12 districts, viz., Bilaspur, Chamba, Hamirpur, Kangra, Kinnaur, Kullu, Lahaul & Spiti, Mandi, Shimla, Solan, Sirmaur and Una. About 20 per cent of the area is under forest cover. It is rich in biodiversity of flora and fauna. Nearly 27 per cent of the area of the state is covered under permanent pastures. The state is endowed with a wide range of physiography/landform, climate, vegetation and geology which have influence on genesis of soils. The soils are mostly shallow, medium deep to deep, well/excessively drained, sandy, sandy-skeletal, loamy-skeletal, coarse-loamy and fine-loamy, calcareous as well as non-calcareous with low available water capacity. The State is the biggest montane ecosystem in India situated in northern part of the country. It is seventeenth largest state in terms of area and twentieth in population of which 65 per cent people are dependent on agriculture and 60 per cent of GDP comes from agriculture sector. Increasing temperature and reducing rainfall affecting the crop growth and crop production is a major issue everywhere in the world nowadays and this state has also not remained untouched by these prime issues of climate change. For sustainable development of agriculture in state, there was/is a need to identify and quantify climatic change where agriculture has a significant influence on both the economy and livelihood.

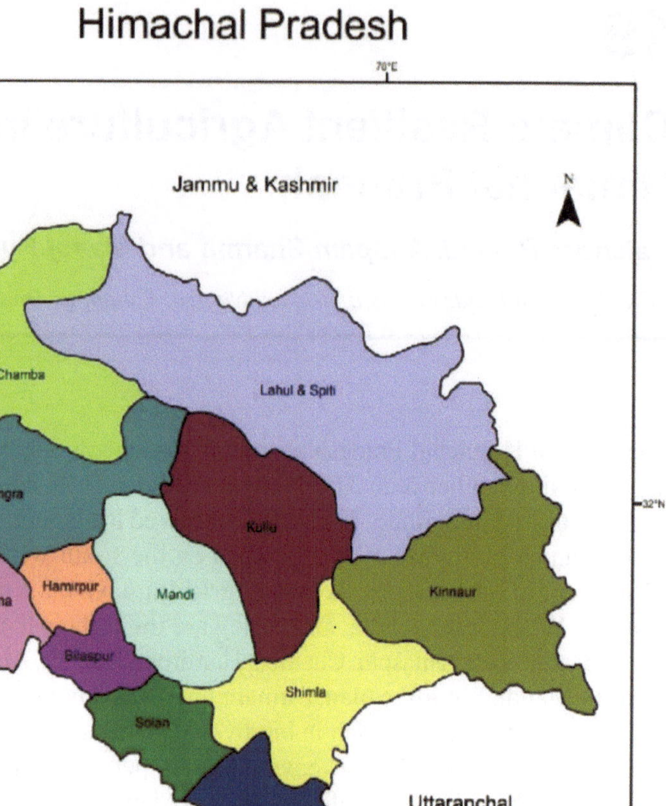

Fig 19.1: State map

19.1 Climate

This district falls in sub-humid sub-tropical zone. The annual mean temperature varies from 27-28°C. During summer maximum temperature varies from 29-32°C while during winter it fluctuates from 19-20°C. The minimum temperature in summers remains between 13-16°C while in winters from 5-7°C. The mean annual rainfall is 1313 mm (CV 23 per cent). The annual average number of rainy days is 71. Bimodal rainfall during winter, October to May (in association with the passage of western disturbances and thunder shower activity) and June to September is received in the state. A major portion (1075.2 mm) of precipitation was received during monsoon period from June to September. July and August are the wettest months. During south west monsoon the rainfall fluctuates from 501-1500 mm (CV 15-45 per cent) and number of rainy days varies from 31 to 60. In summer season (March to May) the rainfall deviates from 101-200 mm (CV 41-70 per cent) and 5-15 rainy days. The winter rains varies from 75-150 mm with (CV 30-70 per cent) and 8-9 rainy days. During post monsoon period, the rainfall fluctuates from 25-100 mm having (CV 76-100 per cent) and the number of rainy days ranging from 1 to 10. Himachal Pradesh has a very diverse biome and a very fragile ecosystem. According to Walia (2009), there will be an increase of temperature by 3°C in the world in the next 3 years. Rabindranath (2008) warns that Himachal Pradesh's forests, especially the higher altitude ones, are highly vulnerable to the effects of climate change. The forest type shifts, such as shifting of apple belt to higher altitudes are predicted to occur in greater than 80% of forested grids. Furthermore, a survey conducted by Pratap and Pratap (2007) in the Kullu valley of Himachal Pradesh concluded that climate change has forced the mountain farmers to abandon one set of crops. The mountain farmers then had to adopt new farming techniques to cope up with the unsuitable climatic conditions. The rising temperature has made it uneconomical and unprofitable for farmers to grow apples in the lower valley area; whereas, it has made it possible for farmers to grow them in the upper valley areas.

19.2 Climate Change

According to the survey conducted by Rana *et al.* (2010) in the Himachal districts of Kullu and Shimla, a large percentage of farmers perceived a definite reduction in percentage snowfall overtime during the winter season. Farmers reported that the onset of early snow in December and January has occurred more infrequently over time and the period of snowfall now extended through the months of February and March. As mentioned above, although these farmers might lack understanding of climate change concepts, but they were nonetheless conscious of the changes in the weather patterns due to their day-to-day experience. Since, the cropping patterns and productivity are influenced by spatial and temporal distribution of rainfall affecting the runoff, soil moisture and groundwater reserves; there was/is an urgent need of systematic documentation of characteristics and alterations in rainfall at regional/local level for developing prediction models and appropriate water management strategies in near future.

19.2.1 Trends in rainfall

On annual basis rainfall and rainy days were found to be significant in 43 and 40 per cent stations, respectively (Table 19.1). Though rainfall was found to be decreasing in 25 per cent stations and but incidentally the rainy days were found to be increasing at equal number of stations. The similar pattern was also observed in the south west monsoon. During post monsoon and winter, the rainfall was found to be significant in 28 and 13 per cent stations and rainy days 28 and 23 per cent, respectively. Compared to the stations showing decrease in rainfall or rainy days, more number of stations were observed with increasing rainfall or rainy days during these seasons. During summer, the trends in rainfall and rainy days were found to be significant in 23 and 40 per cent stations, respectively. Compared to the stations showing increasing rainfall, more number of stations were however observed with decreasing rainfall indicating hotter and drier summers in Himachal Pradesh. In nutshell, it shows that 20 per cent stations showed decreasing trend in SW Monsoon and summer rainfall. The analysis indicates that though there are regional increasing or decreasing trends but no trends (change) has also been found in >62 per cent stations for rainfall and >60 per cent stations for rainy days during different seasons in the state. This analysis indicates a mixed picture as far as changes in rainfall or rainy days are concerned in Himachal Pradesh.

19.2.2 Trends in heavy rainfall events

For heavy rainfall events of 75-100 mm and >100 mm category no trend was observed in the state during different seasons at most (>92 per cent) and (>82 per cent) of the stations, respectively (**Table 19.2**). Significant increase/decrease in extreme rainfall events (75-100 mm & >100 mm) were observed only at few stations during SW Monsoon with more stations showing decrease in rainfall then the stations showing increase in rainfall.

Table 19.1: Account of stations (in per cent) with increasing/decreasing trend of rainfall and rainy days in Himachal Pradesh

		Trend (in per cent stations)				
Names & total No. of districts	Names & total No. of stations	Annual (rainfall & rainy days)	Southwest monsoon (rainfall & rainy days)	Post monsoon (rainfall & rainy days)	Winter (rainfall & rainy days)	Summer (rainfall & rainy days)
Hamirpur, Kangra, Kinnaur, Kullu Lahaul & Spiti Mandi, Shimla, Sirmaur, Solan Una Salooni, Bhoranj,	Hamirpur, Nadaun, Dehra, Dharmashala, Nurpur, Kangra, Malan, Palampur, Nichar, Sangla, Banjar, Bajaura, Keylong, Udaipur, Chachiot, Jogindernagar, Karsog, Mandi, Sarkaghat, Sundarnagar, Shimla, Jubbal, Kothkhai, Kumarsain, Mashobra, Rohru, Theog, Dhaulakuan, Nahan, Pachhad, Paonta, Arki, Kandaghat, Kasauli, Akrot & Una	Rainfall: • No trend: 57 • Significant:43 • Increasing: 18 • Decreasing: 25 Rainy days: • No trend: 60 • Significant: 40 • Increasing: 25 • Decreasing: 15	Rainfall: • No trend:62 • Significant: 38 • Increasing: 18 • Decreasing: 20 Rainy days: • No trend:70 • Significant: 30 • Increasing: 25 • Decreasing: 5	Rainfall: • No trend: 72 • Significant: 28 • Increasing: 15 • Decreasing: 13 Rainy days: • No trend: 72 • Significant: 28 • Increasing: 23 • Decreasing: 5	Rainfall • No trend: 87 • Significant: 13 • Increasing: 8 • Decreasing: 5 Rainy days: • No trend: 77 • Significant 23 • Increasing: 13 • Decreasing: 10	Rainfall: • No trend: 77 • Significant: 23 • Increasing: 3 • Decreasing: 20 Rainy days: • No trend:60 • Significant:40 • Increasing: 17 • Decreasing: 23

Table 19.2: Trend in extreme rainfall at different locations in percentage

Season/Category/ Trend	75-100 mm			>100 mm		
	No Trend	Increasing	Decreasing	No Trend	Increasing	Decreasing
SW monsoon	92	3	5	82	8	10
Post monsoon	100	-	-	100	-	-
Winter season	97	-	3	100	-	-
Summer season	100	-	-	100	-	-

19.3 Climate change impacts

Agricultural scenario of 4 major cereals viz., wheat, maize, rice & barley and 2 important oilseed crops viz., rapeseed-mustard & sesame out of 24 field crops grown in Himachal Pradesh indicate that in addition to inter-annual fluctuations in the area, there was also reduction in area in five out of the six crops. The rapeseed mustard is the only crop showing increase in area, production and productivity. The impact of weather parameter and technology trend in productivity of wheat crop, which is one of the major cereal crops of the state, has been bifurcated for all the districts and it was found that there was decreasing trend in the productivity of three (Kangra, Una and Sirmaur) out of five major wheat growing districts of the state (Prasad *et al.*, 2014). An analysis to know the impact of temperature on major *rabi* crops, maximum temperature during the month of March for three years 2002, 2003 and 2004 at eight locations in the state was carried out by Prasad and Rana (2006). It was observed that the mean maximum temperature was 2-8°C higher during 2004 compared to 2003. The number of days with maximum temperature above a given value was also found to be more during 2004. Similar high temperatures were also observed during 2002. The temperature rise was more pronounced in the fruit and vegetable growing areas of Bajaura and Salooni. Rise in temperature and dry condition during the month of March reduced, green tea leaves (increase over February only by 6 times during 2004 against 29 times during 2002, 25 times during 2003 and 18 times during 2005), egg laying 20.9 per cent during April and reduced yield (wheat 21.3 per cent, rapeseed and mustard 59 per cent) compared to *rabi* season 2001-02.

19.4 Climate risk management

The application of the atmospheric knowledge for practical agricultural use is core objective of the agrometeorology. It promises great practical utility in protection against or avoidance of adverse climatic risks. It has tremendous application in improved management techniques based on sound interpretation of meteorological knowledge. These include irrigation; shelter from the wind or cold; shade from excessive sun; anti-frost measures including choice of site; anti-erosion measures; soil cover and mulching; animal housing and management; climate control in storage and transport; and efficient use of herbicides, insecticides and fertilizers and many more. Though in changing climate there are strong evidences of rising temperatures and soil moisture scarcity but the farm lands of hills and mountains continue to reel under sub optimal air and soil temperature conditions along with soil moisture scarcity having considerable impact on growth and productivity of different crops/ animals. The innovative farmers are resorting to various management practices which modify the microclimate and also accruing some economic benefit out of it. Such practices are required to be identified, documented, further improved upon and standardized through micrometeorological studies by measuring various components of radiation and water balance for dissemination among the farmers through Agromet advisories.

19.4.1 National Innovations on Climate Resilient Agriculture (NICRA)

National Innovations on Climate Resilient Agriculture (NICRA), a network project of the Indian Council of Agricultural Research (ICAR) launched in February, 2011 with the aims to enhance resilience of Indian agriculture to climate change and climate vulnerability through strategic research and technology demonstration. The research on adaptation and mitigation covers crops, livestock, fisheries and natural resource management. Through this project in NICRA District, Hamirpur, Himachal Pradesh we are trying to enable the farmers for weather-wise agriculture through: capitalizing the benevolent weather and minimizing the impact of malevolent weather. This will help increase the income margins of the farmers by reducing the input cost and increased efficiency of the production system. The Agro-climatic Characteristics of the district are given below:

19.4.2 Agro-climatic Characteristics of NICRA District, Hamirpur

Hamirpur district is located between 76°13" and 76°44" E longitudes and 31°52" and 31°58" N latitudes. This is the fifth smallest district as per population (4, 54,293) consisting of 6.63% of total population (2011 Census of India) and smallest district (1118 sq. km) as per the geographical area (Survey of India) of state of Himachal Pradesh. Its altitude varies from 400-1232 m. It is bound by Kangra district in the north, Mandi district in the north east, Bilaspur and Una districts in the south east and south respectively. From administrative view point, the district has been divided into six developmental blocks *viz*. Hamirpur, Bhoranj, Nadaun, Bijhri, Sujanpur Tira and Tauni Devi. There are 209 Panchayats covering all the villages which are 1617 in number. The KVK is 65 km away from university. The four villages have been selected in the NICRA district Hamirpur (Fig. 19. 2).

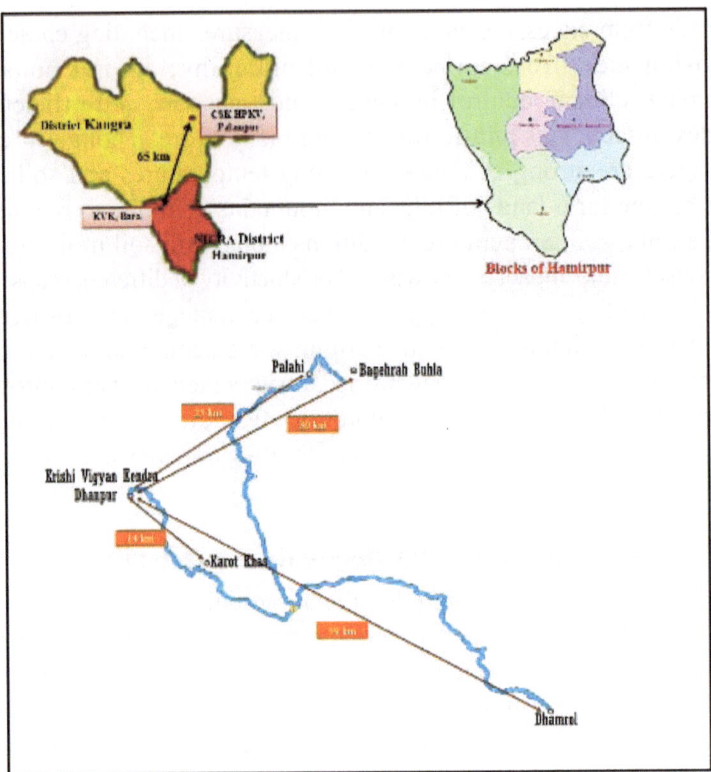

Fig. 19.2: Location map

19.4.2.1 Micro level agro climate

Block level rainfall of three blocks viz., Bhoranj, Hamirpur and Naduan shows that south west monsoon (June-September) season is the wettest season during which 900-1000 mm (CV 20-30 per cent) rainfall is received in 35-40 days. Winter (January – February) season is the second wettest season with 100-150 mm (CV 40-70 per cent) in 7-10 days. Summer (March-May) season is the third wettest season with 90-150 mm (CV 50-60 per cent) in 7-12 days. Post monsoon (October-December) season is the driest season with 40-70 mm (CV 70-130 per cent) in 3-5 days. Annually, 1000-1300 mm rainfall is received (CV 15-25 per cent) in 50-70 days. Since the district is rain fed (only 5 per cent area irrigation), the analysis of the rainfall probabilities for both the blocks viz., Bhoranj & Sujanpur was carried out and it was observed that there were variations in mean rainfall and probabilities of wet spells. The soils of Bhoranj are relatively deeper and levelled with better soil moisture storage capacity. The mean rainfall and the probability of getting wet spell, wet spell followed by wet spell by taking 10 mm as wet spell and consecutive wet spell of 2, 3 & 4 weeks duration by taking 20 mm as wet spell is shown in Fig. 19.3. The criteria for 10 & 20 mm rainfall have been taken for sowing rains.

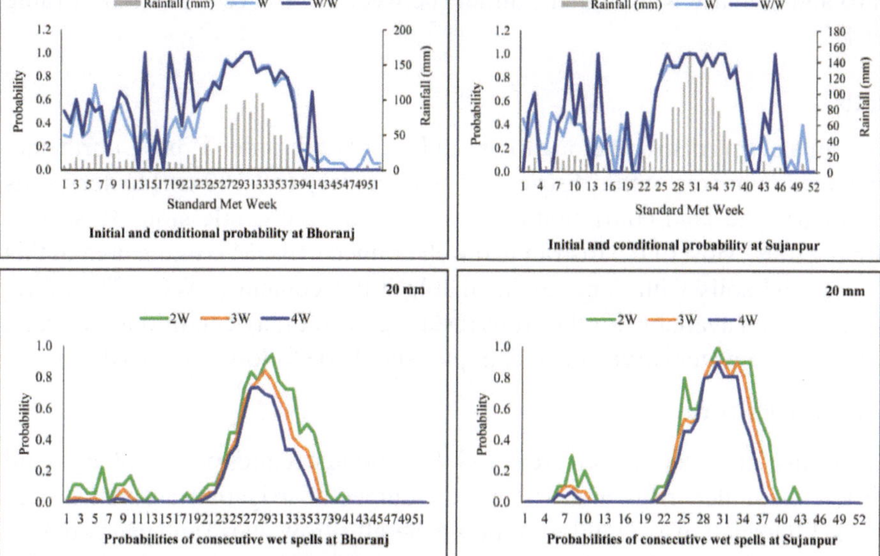

Fig.19.3: Mean rainfall and the probability of getting wet spells in Bhoranj and Sujanpur blocks

The extreme rainfall events showed that highest rainfall occurred at Bhoranj was 314.3 mm on 11th August, 2007 during last eighteen years (2000-2017). Whereas, it was 160.0 mm at Sujanpur during 1971-2017 (Table 19.3).

Table 19.3: Highest rainfall recorded in 24 hours period in different blocks of the district

Stations	Datum period	Date	Highest rainfall (mm) recorded in 24 hrs period
Bhoranj	2000-2017	11th August, 2007	314.3
Hamirpur	1971-2014	31st January, 1976	175.0
Nadaun	1991-2014	13th August, 2007	270.0
Sujanpur	1971-2017	30th July, 2012	160.0

The comparison of monthly and seasonal rainfall for the district during past two *kharif* seasons, showed that the rainfall distribution and amounts were better during July and August, 2017 (27 and 78 mm higher) compared to 2016. The early onset by 6 days (on 1st July against 25th June) and delayed withdrawal by 9 days (on 30th September against 21st September) interfered with sowing and maturity/harvesting of crops.

19.4.2.2 Extreme rainfall events

Compared to 2001-10, the extreme rainfall events are on rise during different seasons in recent decade (2011-15) as there is decrease in rainfall in more number of cases in events with lower rainfall (between 25-50 mm & 50-75 mm) and increase with higher rainfall (between 75-100 & > 100 mm) (Table 19.4).

Soils

The soils of this district belong to group Entisols sub-order Typic Udorthenta. The maximum area of this district had medium deep (50-100 cm) to deep soils (100 cm). The land configuration ranged from very gentle slope to steeply sloping hills. Most of the district had well drained soils. Maximum area of this district had soils with very low available water content (AWC). Therefore, water was not available for the crops resulting in draughty conditions. Dry land farming with water harvesting techniques should be followed in this district.

19.5 Agriculture

In this district, the cultivated area is 34.9 thousand hectares out of 1.8 thousand hectares is under irrigation. The major sources of irrigation are tanks, bore wells, lift irrigation and kuhls. Farmers generally grow maize and paddy during *kharif* season and wheat during *rabi* season. Some farmers having irrigated area have now switched over to the cash crops like fruits and vegetables. As compared to other districts, the productivity of these major crops is higher in this district thus possessing higher area and potential for improvement in productivity. The cropping intensity of the district is 198 per cent. The effect of El Niño on rainfall and crop productivity was studied in the district. It was found that there was positive impact of El Niño on rainfall. Compared to non - El Niño year, there was 11 per cent increase during SW monsoon and 8 per cent on annual basis with marginal decline during winter (2 per cent). There was reduction in maize and rice yield by 13 & 17 per cent, respectively. But there was positive effect on productivity of wheat and barley 18 and 2 per cent, respectively (Prasad *et al.*, 2018). The anomalies in seasonal rainfall for the years with weak, moderate and strong El Niño for other districts of the state indicated that during south west monsoon, rainfall deficiency mostly increased with strength of the El Niño while subsequent winter season received higher rainfall. Hence, El Niño had a compensatory effect on winter rains. The weak or strong El Niño attracted higher winter rains while the moderate El Niño caused reduction in south west monsoon rainfall (Prasad *et al.*, 2014). The maize yield of most of the districts was positively affected by the weak El Niño while moderate and strong El Niño years negatively affected it. In rice, wheat and barley the weak and strong El Niño positively affected the productivity in most of these crops growing districts while the moderate El Niño had negative impact. This indicates that moderate and strong El Niño years influence the productivity of the crop perhaps depending upon the location of the district.

Table 19.4: Increasing extreme rainfall events during 2011-15 as compared to 2001-10

Season	Rainfall (mm)											
	25<50			50< 75			75<100			>100		
	Hamirpur	Bhoranj	Nadaun	Hamirpur	Bhoranj	Nadaun	Hamirpur	Bhoranj	Nadaun	Hamirpur	Bhoranj	Nadaun
Winter (Jan-Feb)	D	I	D	D	D	D	NC	D	I	NC	NC	D
Summer (March-May)	D	D	D	I	D	D	NC	I	I	I	NC	I
Southwest monsoon (June –Sep)	D	D	D	D	D	D	I	D	D	I	D	D
Post monsoon (Oct-Dec)	D	D	D	D	D	D	NC	D	D	NC	NC	D

D- decreasing; I-increasing; NC- no change

Similar observation has also been made by Patel *et al.* (2014) for Gujarat state. The compensatory effect of El Niño on winter rains was tested during *rabi* season 2014-15 in NICRA district Hamirpur. 54.7 per cent higher rainfall than normal was received during the year following El Niño which caused water logging in wheat. This resulted in poor grain filling and excessively low 1000-grain weight. Perhaps the district has a geographical advantage due to which it receives generally 16 to 18 per cent higher rainfall during different seasons in El Nino years.

19.5.1 Comparison of district and block level forecast

The comparison of district and block level forecast showed considerable differences in various weather parameters, but frequent updates available on Doordarshan Kisan (DD Kisan) have been effectively utilized. The launching of DD Kisan during 2015 was a real landmark achievement in the history of agriculture in India. DD Kisan telecast comes at 8.30 AM & 7.30 PM. State level detailed weather forecast is regularly given in "Khas Khabar" means 'Special News'. The rainfall received during past 24 hours is given under "Mausam ki Chal". The other information is also covered under "Mausam ke Lihaj se Kheti" means 'Weather-Wise Agriculture', "Krishi Vishashagy ki Salah" means 'Advisory from Agricultural Experts' and "Krishi Vishashagy se Khaas Batcheet" means 'Special Interview with Agricultural Experts'. Generally package of practices of different crops are covered with very little information on weather based management practices. The available weather system can be easily tracked through these free to air channels. This Centre enthusiastically makes use of state/district/and block level forecast & takes available valuable weather based management input from this and advise NICRA farmers accordingly through mobile Short Messages Service (SMSs).

Orientation Workshop: An orientation workshop of Farmers representative from 10 villages at office of Subject Matter Specialist, Department of Agriculture (Sujanpur) was conducted on 28th March, 2018. In this workshop, all the points about DD Kisan along with the objectives of the project were discussed and explained to these farmer representatives.

Success rate (%) of block/district forecast: High success rate of forecast is indispensable for building confidence of the person who is issuing agromet advisories and for the farmers utilizing these advisories. In Bhoranj block the analysis indicated, improvement in 0 mm and 1-10 mm rainfall categories till 2016. But 2017 was a peculiar year for this state, with poor success rate i.e., less than 66 per cent. We deliberately did not issue any advisory during *rabi* 2017-18 as the success rate was exceptionally poor. There is urgent need for improvement in the Block level forecast.

19.6 Development and dissemination of AAS

The agromet advisories prepared by the project staff which are vetted and finalized by the Principal Investigator AICRPAM-NICRA. The contents of the SMS are approved by the Head of the Department to avoid any controversy. At present the SMSs are being sent for the farmers of 5 villages depending upon the significant weather event.

19.6.1 Weather based agromet advisories issued

Weather based SMS service has been started in ten districts viz., Hamirpur, Kangra, Bilaspur, Mandi, Una, Chamba, Kullu, Shimla, Sirmaur and Solan since the beginning of the project. A total of 3945 farmers were registered, out of which 1147 are from district Hamirpur and are being provided with weather based advisories through SMS. During 2018, 16 advisories were sent through SMS out of which 11 were on harvesting/thrashing of wheat, 3 on harvesting/ thrashing of mustard, 1 each on irrigation in vegetables and management of heat load in dairy animals. The formulation of the advisories was really a time taking and interesting experience. Suggesting harvesting windows for wheat & mustard during April, 2018 when there were continuous rains, was really a tough time and kept us on toes.

19.6.2 Farmer's awareness programme

Ten farmers' awareness programs on Climate Change in different districts of Himachal Pradesh *viz.*, Hamirpur, Bilaspur, Kangra and Una were successfully conducted since inception of the project. More than 3500 farmers participated who could be trained in these one day programmes. It was possible to communicate the message of climate change effectively to farmers and peoples' representatives through the experts drawn from different disciplines of Agriculture including Agrometeorology.

19.6.2.1 Economic impact/case study of micro level Agroment advisory services (AAS): B:C ratio in AAS and Non AAS farmer

In Bhoranj block

1. Economic impact assessment of AAS issued indicates that the timely sowing of maize done as per weather based advisory produced more yield 3960 kg/ha (4107 & 3750 kg during 2015 & 2016) yield as compared to 1480 kg/ha (800 & 2159 kg during 2015 & 2016) in late sown maize which was 62.6% lower than timely sown crop (Table 19.5). The gross income per ha from the timely sown maize crop was Rs 1, 17,300/- & 48,378/- whereas it was Rs.24, 480/- & 13,927/- in case of late sown

Table 19.5: Impact of forecast aided timely sowing in maize

Year /DOS	Timely sown				Late Sown			
	2015 (12-16th June) (five farmers)		2016 (12-16th June) (three farmers)		2015 (1-29th July) (seven farmers)		2016 (5-8th July) (three farmers)	
Particulars of selected farmers	Area	Production	Area	Production	Area	Production	Area	Production
	0.40+	1300+	0.20+	750+	0.40+	1800+	0.16+	320+
	0.20+	1000+	0.28+	1050+	0.40+	160+	0.12+	270+
	0.40+	1600+	0.40	1500	0.32+	300+	0.16	360=
	0.28+	1200+	=0.88	=3300	0.20+	300+	=0.44	950
	0.40	1800			0.20+	250+		
	=1.68	=6900			0.16+	150+		
					0.12	100		
					=1.80	=3300		
Total area (ha)	1.68		0.88		1.80		0.44	
Total production (kg)		6900		3300		1440		950
Yield (kg/ha)		4107		3750		800		2159
Gross income (Rs.)		1,17,300@ Rs 17.00/kg		48,378@ Rs14.66/kg		24,480@ Rs 17.00/kg		13,927@ Rs 14.66/kg

during 2015 & 2016. The further analysis indicated that the gross income was found to be dependent upon prevailing market rates, good weather captured by the forecast and forecast aided timely sowing undertaken by the farmers.

This is a case study reported by Mr. Sanjeev Mehar of Bhoranj block, an AAS farmers who grew wheat in 2400 m^2 area and obtained wheat yield of 900 kg with gross income of Rs. 20,000/-, net income of 8760/- with BC ratio of 0.78 (Table 19.6). The non AAS farmers obtained negative BC ratio of 0.35. The timings of the weather based operations and the report submitted by the farmer is presented in Table 19.7.

Table 19.6: B:C ratio in AAS and Non AAS farmers (Wheat crop) in Bhoranj villages of NICRA district Hamirpur (2017-18)

Particulars	AAS Farmer	Non AAS Farmer
Cost of cultivation (6 kanal=2400 m^2)	Rs. 11,240/-	Rs. 9,950
Wheat yield (kg)	900	350
Gross income	Grain:13,000+ Straw: 7,000= Rs. 20,000/-	Grain:5,056+ Straw: 1,444 = Rs. 6,500/-
Net income	Rs. 8,760	Rs.-3,450
Benefit Cost ratio	0.78	-0.35

Table 19.7: Timings of forecast based agricultural operations

Agriculture operations	(AAS Farmer)	Non AAS Farmer)
Field preparation	10th-17th Nov	01st-10th Nov
Sowing	18th Nov	12th Nov
Fertilizer application	With sowing	With sowing
Herbicide spray	19th Nov	13th Nov
Urea application & Interculture Operations	14th Jan	22nd Dec
Harvesting	01st & 10th May	26th April
Threshing	05th & 13th May	28th April
Irrigations applied(5 No. instead of 7 = saving of 2 irrigations i.e., Rs. 300/- saved@ Rs. 150/- per irrigation)	10th Nov, 7 Jan, 4 Feb, 10 March and 22 April	

Account of saving input cost/irrigation: As per the version of the farmer, he normally applies 7 irrigations at 25 days interval. But for this crop he applied only 5 irrigations and there was saving of 2 irrigations due to well captured forecast. According to him, there was nominal saving of Rs. 300/- @ Rs, 150/ - per irrigation, the account of which is given in the Table 19.8.

Table 19.8: Account of saving input cost/irrigations

Sr. No.	Forecast of rain & Advisory issued	Occurrence of rain	Irrigation: Applied/ Skipped
1	10.11.2017 (No rain) Sowing of wheat (18.11.2017)	-	Applied on 10.11.2017
2	12-13.12.2017 (Rain)	58.6 mm	Skipped (Saved Rs. 150)
3	07.01.2018 (No rain)	-	Applied on 07.01.2018
4	04.02.2108 (No rain)	-	Applied on 04.02.2018
5	12-13.02.2018 (Rain)	42 mm	-
6	10.03.2018 (No rain)	-	Applied on 10.03.2018
7	15.03.2018 (Rain)	15.2 mm	-
8	22.03.2018 (Rain)	12.4 mm	-
9	07.04.2018 (Rain)	07, 09, 11 April, 2018 (130.8 mm)	Skipped (Saved Rs. 150)
10	11.04.2018 advisory of proper management in view of benevolent weather cond.		
11	22.04.2018 (No rain)	-	Applied on 22.04.2018

19.6.2.2 Weather aberration and their impact on wheat crop

Rabi 2017-18 was a season with lot of weather aberration. The temperature remained above normal throughout the season with below normal rainfall. The district witnessed heavy unseasonal rains (about 80 mm rainfall) during month of April, 2018 which not only interfered with the harvesting of the crop but also caused grain/straw discoloration. The yield reduction in wheat varied from 9-75 per cent with lot of quality deterioration in NICRA villages (Table 19.9).

Table 19.9: Decrease in yield due to aberrant weather conditions during two *rabi* seasons

Sr.No.	Name of the farmer& area (Kanal)	Yield (q)		
		2016-17	2017-18	Per cent decrease
1.	Jai Ram (20)	5.5	4.5	9.1
2.	Rekha Devi (3)	2.0	0.5	75.0
3.	Desh Raj (10)	4.5	4.0	11.1

Discolouration in wheat (grain & straw)

The block witnessed 145 mm rainfall during April, 2018. The impact of weather aberration was considerable as evident by the press visibility of the impact of excessive rains. Reports in vernacular press mentioned: 'Hamirpur main gehun ki 35 per cent kam pedavaar', 'Kheton main kali padi gehun' which means that 35 per cent reduction in wheat yield in Hamirpur district and heavy discoloration of wheat. In our case study, we found that there was advisory of harvesting and safe storage but Mr. Chain Singh & Mr. Desh Raj though harvested their crop but did not ensure safe storage as per advice given on 21.04.2018 & 25.04.2018. Hence there was discoloration in grain and straw and the effect was more in low lying valley/irrigated areas having higher humidity.

19.7 Contingency Crop Plans

The district is prone to drought, cold wave, frost and pest and disease outbreak. The occasional hailstorm also affect the crops. District Agriculture Contingency Plans (DACPs) for eight districts including Hamirpur, were prepared and published (Prasad *et al.*, 2013). These plans are being reviewed for incorporating relevant changes in tune with the present time requirements. For avoiding crop losses due to adverse effect of irregular and erratic rains the following is the gist of contingency plans suggested:

1. In case of the normal onset of monsoon followed by 15-20 days dry spell after sowing the farmers are advised to carry out gap filling if plant population e”75 per cent in transplanted rice or >50 per cent in direct seeded rice. If plant population in direct seeded rice is < 50 per cent, re-sowing should be done with 25 per cent increased seed rate. The crop should be kept weed free, top dress with nitrogen, apply mulch and follow in situ moisture conservation by diverting water from adjoining areas.
2. For delayed onset of southwest monsoon by 4 weeks i.e., 3rd week of July, undertake intercropping/mixed cropping the fodder maize with soybean/black gram/finger millet/cowpea/sesame.
3. If the monsoon delays by 6 weeks than mixed cropping of Maize + Cowpea or Sorgham + Cowpca should bc adoptcd. Usc 40 kg seed for maize, 15 kg for cowpea and 45 kg for sorghum/ hectare. The vegetables like *kharif* onion, early cauliflower and early peas can be grown.
4. In case of mid-season drought or long dry spells, farmers are advised to follow the recommendations as suggested above. In addition, farmers are advised to keep the fields weed free, apply foliar spray of nitrogen and lifesaving irrigation where irrigation facilities are available, carry out inter culture operations to create soil mulch. If the crop stand is poor then use the crop as fodder. The farmers are also advised to sow radish/peas/toria as catch crop followed by early varieties of wheat.
5. In *rabi* season, if the rains are delayed by 2 weeks (onset on 1st week of January) sow late varieties of wheat and increase the seed rate by 25 per cent.
6. If the rains are delayed by 4 weeks (onset on 3rd week of January) farmers are advised to increase the seed rate and fertilizer by 25 per cent.

7. In case of heavy rainfall or unusual rainfall drain out excess water and carry out inter culture operations for soil aeration in vegetative stage, top dress with nitrogen at flowering stage and harvest the crop at physiological maturity if the crop is in maturity stage.
8. The outbreak of pests and diseases due to unseasonal rains leads to major losses to the farmers. Advised to drain the fields, routine preventive sprays for the control of insect/pests.

Summary

Sustainability of agricultural production systems is a core and major issue. Inadequate & ill distributed rains/soil moisture stress in 95 per cent rain fed area (prolonged period of drought) are frequent in the district. Large chunk of soils are shallow coarse textured with undulating topography. Low/high air/soil temperature coupled with soil moisture stress in lower Shivalik Hills during initial crop phases-gives stunted growth. People are giving up farming due to shrinking croplands & land holdings which are really inadequate for decent living. Heavy crop damage occurs due to stray/wild animals (Sharma, 2018). Due to being more people in government and private jobs, dependence for food and income on family farming declining. Availability of food grains at cheaper rates through Public Distribution System is also a reason for abandoning agriculture. Being non-remunerative, there are fewer takers of traditional agriculture. Having highest (94.4 per cent) literacy rate among all the districts, the aspiration of youth for white collar jobs seems to be very high. Consequently 50 per cent of arable land is lying vacant- increasingly infected by weeds. Organic agriculture for safe food is trending besides the visible shift to cash crops, small dairy farming and more demand for cultivated fodder other food crops. Villages are tending to turn into market towns. Weather products for these crops and peri-Urban Agriculture required for which we are ill equipped at present.

Planned adaptation is essential to increase the resilience of agricultural production to climate change. The majority of the farmers of the district are small and marginal. They are mostly affected due to their low adaptive capacity and risk bearing ability. To overcome the challenges the following strategies are proposed to enable the farmers for weather-wise agriculture through: capitalizing the benevolent weather and minimizing the impact of malevolent weather. This will help increase the income margins of the farmers by reducing the input cost and increased efficiency of the production system. Thus by incorporating various adaptation measures in the agriculture system we can increase resilience and adaptive capacity of the small land holders. Preparedness for the weather variations is critical. The targeted farmers are prepared to respond to climatic variations and minimize the risks. Hamirpur is the rain fed district and the difficulties faced by the famers are shorter sowing windows, poor germination, poor flowering, flower shedding, poor fruit set, and shorter harvesting windows in some cases. When we weigh the options available with the farmers, it is found that is the single and most important input which has inherent capacity to give resilience against all climatic/weather aberrations. All out efforts are required to be made to bring more area under irrigation minuscule (5 per cent) in this district. On the basis of district and blocks level MRWF issued by IMD, Pune agro-advisories are issued to the selected farmers of the district through toll free service Mobile Seva of GOI and economic impact of these advisories through case studies is being undertaken for past couple of years. As evident from the case studies, the selected farmers have really been benefitted who have adopted the weather based advice issued by this Centre.

References/Suggested Readings

Patel H. R., Lunagaria M.M., Pandey V., Sharma P. K., Bapuji Rao B. and Rao V.U.M. (2014). "El Niño episodes and agricultural productivity in Gujarat". Technical Bulletin 01/2014-15, Anand Agricultural University, Anand, pp. 22.

Prasad R and Rana R. (2006). A study on maximum temperature during March 2004 and its impact on *rabi* crops in Himachal Pradesh. *Journal of Agrometeorology*, 8(1): 91-99.

Prasad R., Rao V.U.M. and Bapuji Rao B. (2014). El Niño-Its Impact on Rainfall and Crop Productivity: A Case Study for Himachal Pradesh. CSKHPKV, Palampur, HP and CRIDA, Hyderabad, India. pp. 24.

Prasad R., Ravindra Chary G., Rao K.V., Prasad Y.G., Srinivasarao Ch., Ramana D.B.V., Sharma N.K., Rao V.U.M. and Venkateswarlu B. (2013). District Level Contingency Plans for Weather Aberrations in Himachal Pradesh. CSK HPKV, Palampur, HP and Central Research Institute for Dryland Agriculture. Hyderabad-500 059, India. pp. 222.

Prasad R., Sharma A. and Sood R. (2018). El Niño and seasonal rainfall dynamics vis-à-vis food grain productivity in Himachal Pradesh. *Journal of Agrometeorology* 20 (2): 157-160.

Pratap B and Pratap T. (2007). Climate Change Impact on Hill Agriculture and Farmers Adaptive Strategies. Department of Environment, Science & Technology, Government of Himachal Pradesh. State Strategy and Action Plan on Climate Change, pp. 249.

Rabindranath NH. (2008). Climate Change & Himachal Pradesh: Impacts, Adaptation & Mitigation Shimla, Himachal Pradesh, India.

Rana R.S., Bhagat R.M., Kalia V. and Lal H (2010). Impact of Climate Change on Shift of Apple Belt In Himachal Pradesh. ISPRS WG VIII/6 Archives XXXVIII-8/W3 Workshop proceedings; pp 131-137.

Sharma H.C. (2018). Strategies for Doubling Farmers Income in Himachal Pradesh. YSP University of Horticulture and Forestry, Nauni, Solan 173230, Himachal Pradesh. Website: http://www.yspuniversity.ac.in.

Walia J. S. (2009). Impact of climate change and Himachal Pradesh's action plan strategy on forestry. *Indian Forester* 135(9): 1155-1164.

20

Climate Change and Bihar Agriculture Assessment, Projection and Risk Management

A. Sattar[1] and Mithilesh Kumar[2]

[1]Agrometeorologist, [2] Director Research
Dr. Rajendra Prasad Central Agricultural University, Bihar, Pusa, Bihar, India

Bihar's economy is highly dominated by agriculture and allied sectors. Around 85 to 90 percent of the population still lives in rural areas, where agriculture along with animal husbandry, is the main source of their livelihood. In Bihar, the gross and net sown area is estimated at 8.26 and 5.64 million hectares, respectively. Nearly 60 per cent area under cultivation in the state is rainfed (Anon, 2008). The region comes under sub-humid subtropical monsoon climate. Most of the annual rainfall (almost 85 per cent) received from south-west monsoon, occurs during June to September (Sattar, 2010). There are three crop growing seasons in the state commonly known as *kharif, rabi* and *zaid.* The important *kharif* season crops, which are planted in May-June and harvested in September–October, are rice, maize, sesame, sorghum and bajra. Pigeon pea is an important long duration pulse crop sown during *kharif* season as sole crop or intercropped with maize. The important *rabi* crops which are sown in October–November and harvested in February–March are wheat, *rabi* maize, potato, chickpea, lentil, rapeseed, mustard, barley. *Zaid* crops, which are cultivated in short growing season extending from March to May in between *rabi* and *kharif* seasons, are also called summer crops, which include green gram, cowpea and sesame etc. Rice, wheat, maize, pulses and oilseeds are grown in all the districts. Sugarcane is also an important annual cash crop. However, the choice of crops, their varieties and crop rotation varies across different agro climatic zones. Rainfed crops during both *kharif* and *rabi* seasons in the state mostly experience moisture stress at different phases of growth and produce lesser yield. Like most parts of the country, crop production in Bihar is largely dependent on the performance of south-west monsoon.

The State is highly vulnerable to hydro-meteorological natural disasters, with North Bihar in general being highly flood-prone, and South Bihar being highly drought prone. Rainfall during the monsoon season (June-September) is the key element of Bihar's climate and continues to be the primary source of water for the large rain fed agricultural regions in the State. Erratic and high variability in monsoon rainfall in recent years is not only affecting rice and other *kharif* crop production, but also impacting the production potential of *rabi* crops. The occurrence of extreme temperatures during *rabi* season are also affecting the performance of wheat, maize, pulses, potato and other crops. On the basis of rainfall, temperature, terrain and soil characteristics, the state has been broadly divided into three agro climatic zones **(Fig.20.1)** viz. Zone I (North West Alluvial plains), Zone II (North East Alluvial Plains) and Zone III (South Bihar Alluvial Plains). Zone III is further subdivided into Zone IIIA and Zone IIIB. Each zone has its own unique characteristics in terms of agroclimatic features and the crop production. Zone I is situated in the northwestern part of the state whereas zone II is located in the northeastern part. Zones I and II are flood prone whereas zone III is drought prone. Zone IIIB is regarded as chronically drought prone region of Bihar.

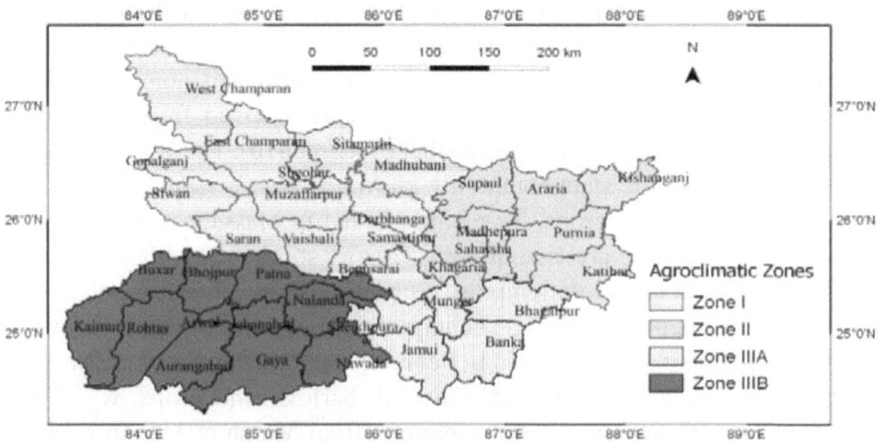

Fig. 20.1: Agroclimatic Zones of Bihar

In the backdrop of large inter and intra annual variability in monsoon rainfall, occurrence of extreme temperature particularly during *rabi* season, vast rainfed areas, climate sensitive geographical setting of the state and enormous rural population, there is an urgent need to address the challenges posed by climate change on agriculture and its allied sectors in the state. The overall strategy would be to transform agricultural production systems into a climate resilient and climate smart vibrant production system through various interventions in order to achieve sustainable food security amidst land degradation and over exploitation of natural resources.

20.1 Climate Change Impacts

Bihar is one of the most climate sensitive states of India on account of its unique geo-climatic conditions, high population density and poverty. It is besieged by all sorts of natural hazards. Drought, flood, heat wave, cold wave, extreme rainfall events and hailstorm are a common occurrence in Bihar. Being an agriculturally dependent state, the impacts of climate change and climate variability will be much more on livelihood options of the people and consequently on the state economy. Thus, threats posed by climate change and environmental degradation to sustainability, human prosperity and the state's economy are enormous and real. Climate change is one of the major challenges of agriculture in Bihar. Hence, overall strategy would be to transform agriculture and its allied sectors into climate-resilient and vibrant production system in order to ensure sustainable food and nutritional security. Climate change is now a reality in Bihar and could be felt in terms of increasing temperature, erratic rainfall pattern leading to frequent dry spell/ drought and related extreme weather events impacting agricultural production system in the state.

20.1.1 Temperature Trends

Many physical processes and chemical reactions which are vital to crop growth are greatly influenced by temperature. High temperatures shorten crop cycles by inducing early flowering and shortening the grain-filling period, thereby reducing crop productivity (Chattopadhyay, 2011). High day- and night-time temperatures would become more frequent in the near future and represent a tremendous environmental challenge to global food production and food security (Lobel *et al.*, 2011; Cairns *et al.*, 2012; Hijioka *et al.*, 2014). Like many regions in the world and India as well, average annual temperature tends to show an increasing trend in most of the districts in Bihar. In Zone I, data on daily temperatures of two locations located in Muzaffarpur and Saran districts indicated significantly increasing trend of temperature (Table 20.1). In Zone II, similar trend in temperature as that of Zone I was observed. Purnia district exhibited significantly increasing trend in temperature, while decreasing trend in temperature was observed in Supaul district. Bhagalpur district located in Zone IIIA recorded significant increasing trend in annual temperature. Annual temperatures of Gaya and Patna districts under Zone IIIB showed increasing trend. Considering the temperature during *rabi* season, analysis of long term data of daily maximum and minimum temperatures at Pusa (Samastipur), Bihar indicated significantly increasing trend of minimum temperature during *rabi* season and a decreasing trend of maximum temperature during November to March (Table 20.2).

Table 20.1: Time trends in average annual temperature in different districts of Bihar (Database: 1969-2014)

	Zone/district	Trend (^{0}C/year)	Significance test
Zone I			
	Muzaffarpur	0.021**	Significant
	Saran	0.039**	Significant
Zone II			
	Purnia	0.017**	Significant
	Supaul	- 0.013*	Significant
Zone IIIA			
	Bhagalpur	0.015*	Significant
Zone IIIB			
	Patna	0.014*	Significant
	Gaya	0.004	Non significant

*Significance at 5 % level, ** Significance at 1 % level

Table 20.2: Temperature trend during *rabi* season at Pusa, Bihar

Parameters	Nov	Dec	Jan	Feb	Mar	Annual
Max Temperature (^{0}C)	NS (D)	** (D)	**(D)	NS(D)	*(D)	**(D)
Min temperature (^{0}C)	*(I)	**(I)	**(I)	**(I)	*(I)	**(I)'

D: Decreasing Trend, I: Increasing Trend, NS: Non-significant (*): p=0.05, (**): p=0.01

20.1.2 Bright Sunshine

Almost all growth and development processes in plants are governed by solar radiation. The solar energy is converted into carbohydrates through the process of photosynthesis. The photosynthesis is largely affected by clear day or cloudy conditions. Any reduction in the intensity and duration of bright sunshine during the growing period of crops would affect the agricultural productivity. As for example, lesser sunshine or cloudy weather during flowering stage of rice affects its yield. Sreenivasan (1979) observed that more sun shine at the time of panicle emergence to fertilization leads to greater yield. Yamagata (1958) reported that the number of tillers and ears (panicles) increased with the intensity and quantity of light. Daily bright sunshine hour (BSH) of 7 to 8 hours at Samastipur (Bihar) was found to enhance the grain yield, while the yield decreased at BSH of less than 7 hours during flowering stage (Sattar *et al.*, 2017a). The duration of bright sun shine is showing a decreasing trend over north Bihar. As observed from the Fig.20.2, the bright sunshine duration has decreased significantly during the recent period 2001-15 compared to the period 1990-2000 (Anon, 2016). Reduced level of sun shine hours in the recent period may be due to atmospheric pollution, decreasing activity of winter rains in the region and wintry foggy weather.

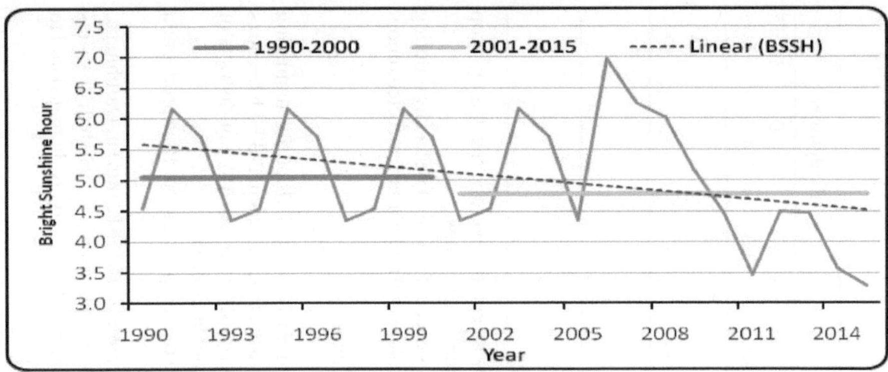

Fig. 20.2: Variability in bright sunshine duration during rabi season at Pusa, Samastipur

20.1.3 Rainfall Trends

In recent years, extreme weather events such as abnormal rise and fall of temperature, frequent dry spell/drought and floods have increased, posing serious challenges for sustainable and profitable agricultural production systems. In Bihar, climatic perturbations with respect to occurrence of extreme weather events have been detected as under:

- As evidenced from the results of analyses of some station rainfall data in Bihar, it indicated a slightly decreasing trend. With increasing number of heavy rainfall events, number of rainy days is decreasing. Consequently, the phenomena of extended dry spell/drought and flood are increasing. A case study related to increasing number of dry spells during *kharif* season at Pusa (Samastipur) region of Bihar presented in Table 3 shows that number of dry spell has increased in the recent years in this region (Sattar *et al.*, 2016a).
- A perusal of the data (Table 20.3) reveals that the month of June has become wetter during the recent period (1996-2015) compared to the period 1976-1995. On the other hand, dryness has increased alarmingly during the month of August. The months of July and September also showed increasing trends in the frequency of dry spells, which is worrisome for sustainable and profitable rice production.
- In recent years, the phenomenon of frequent dry spell/drought, high intensity rainfall in short period and flood have increased manifold, seriously affecting the production potential of *kharif* rice. Bihar experiences a lot of rainfall related anomalies. Some examples of rainfall anomalies during *kharif* season in the recent past are presented in the Table 20.4.

Table 20.3: Tracking climate change through dry spell analysis at Pusa, Bihar

Year	June			July			August			September		
	5-7 Days	8-10 Days	11-15 Days	5-7 Days	8-10 Days	11-15 Days	5-7 Days	8-10 Days	11-15 Days	5-7 Days	8-10 Days	11-15 Days
1976-1995	22	7	9	11	1	0	6	3	1	17	3	2
1996-2015	12	5	11	14	2	1	17	7	0	17	6	6

Table 20.4: Drought and flood in Bihar during recent past

Year	Total annual rainfall (mm)	Drought/flood	Agricultural Impacts
2005	637	Prolonged dry spell and drought	Worst drought, which affected severely the *kharif* crop cultivation and subsequent *rabi* crops.
2006	1066	Below normal, 70 % deficit in August	Rice and vegetable crops encountered high water stress during August. Higher temperature during February affected onion, pea and wheat
2007	2431	Devastating flood in Bihar, Course of Kosi river was changed	Severe damage to agriculture
2008	1534	Early season(June-July) flood	Heavy rainfall created early season flooding damaging vegetable crops and destroyed rice crops in 2-3 lakh hecatres
2009	910	Early season (June-July) and mid season disastrous drought	Rice cultivation was seriously affected. Most of districts of Bihar were declared drought affected
2010	985	Good rainfall in May and during 18-24 June and devastating drought thereafter	Severe drought affected rice crop productionRice production fell by 25-35 percent
2011	1428	Almost normal year. However, distribution remained highly erratic	Extremely very heavy rainfall (357- 412 mm) in Darbhnaga, Muzaffarpur, Shivhar, Samastipur and Vaishali districts during 26-27 September, which extensively damaged mid to early duration rice crops.

Contd.

Year	Total annual rainfall (mm)	Drought/flood	Agricultural Impacts
2012	880	Scanty rainfall in June and August.	Rice growth was seriously affected.
2013	865	Deficit rainfall. Only 45.5 mm rainfall was received in July against the normal of 295.5 mm	Rice transplanting was severely affected. Area under rice decreased drastically (40-50 %) in this year due to poor transplanting
2014	1117	Deficit rainfall in June	Rice seedlings were damaged in the nursery due to dry spell
2015	946	Highly deficit rainfall during *kharif* season. However, heavy rainfall event was also observed. 316 mm rainfall was recorded in consecutive two days in August.	Rice crop yields were seriously affected in the state due to both poor rainfall, dry spell and subsequent drought
2016	1016	Prolonged dry spell in June, August and September. August remained extremely dry. Only 5 rainy days. Two long dry spells (10 days and 14 days long)	Dry spell in June, August and October affected the growth of rice seedlings and rainfed rice crops
2017	1134	Very dry condition in June, poor distribution in July, Extreme dry condition in September and October	Rice seedlings got damaged, transplanting of rice was affected. Farmers had to employ diesel pump sets for frequent irrigation escalating the cost of production

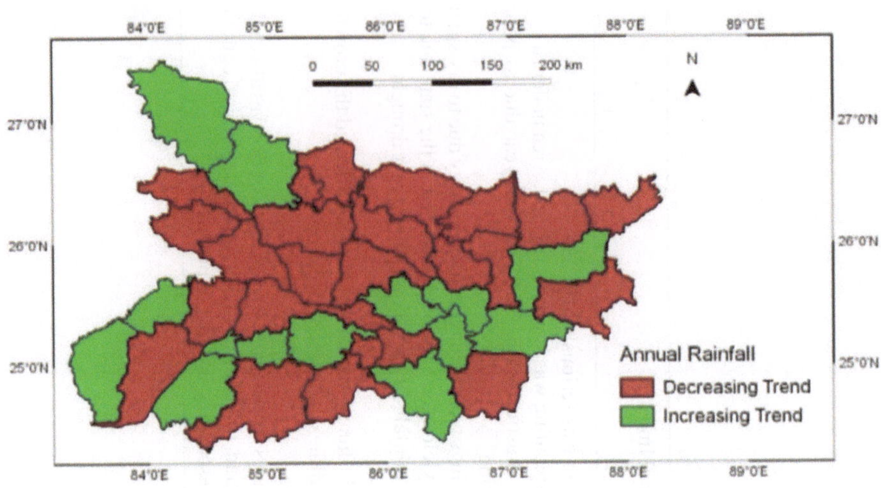

Fig. 20.3: Trend in annual rainfall over the districts of Bihar

Analysis of long term rainfall data suggests that most of the districts of Bihar are showing decreasing trends of rainfall. The decreasing trends are mostly observed in north Bihar districts (Fig.20.3), where amount of rainfall remains always higher than south Bihar districts. Seasonal as well as intra-annual variability in rainfall would continue to affect rainfed crop production potential in the state as about 60-70 % of the crop production takes place under rainfed condition. On the other hand, increasing trend of temperature will have considerable negative consequences on major crops by affecting crop phenology, physiology and plant-water relations during *rabi* season. Elevated temperature regime shortens crop duration by inducing early flowering and shortening the grain-filling period, thereby reducing crop productivity (Chattopadhayay, 2011).

20.1.4 Standardized Precipitation Index (SPI)

Standardized precipitation index (SPI) is an important analytical tool to track the change in rainfall climatology in a region. SPI values of monsoon rainfall during two periods *viz.* 2001-2010 and 2011-15 presented graphically in Fig. 20.4 (a-c), indicated that drier conditions as represented by higher negative values of SPI have been observed in the recent five years period across all the districts of different agroclimatic zones of Bihar. As presented in the figures, broken lines representing the variation in district-wise SPI values for the last five year period indicated alarming drying condition in the many districts of Bihar. Studies on drought incidence in Bihar based on SPI highlighted the vulnerability of several districts to drought incidence (Sattar *et al.*, 2013).

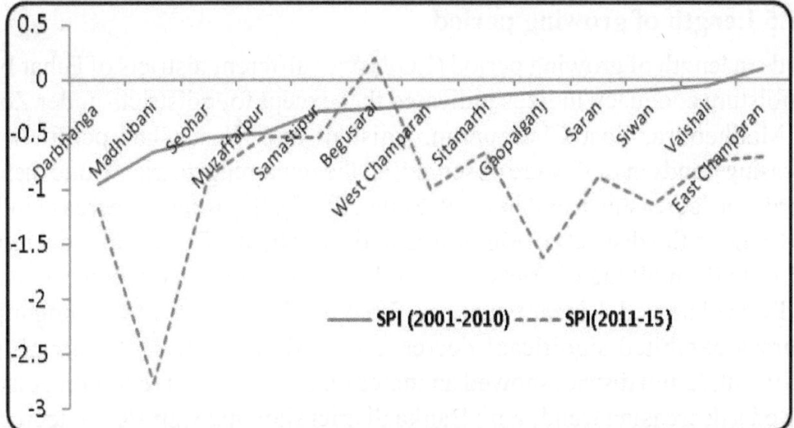

Fig. 20.4 (a): Variability in SPI values in Zone I

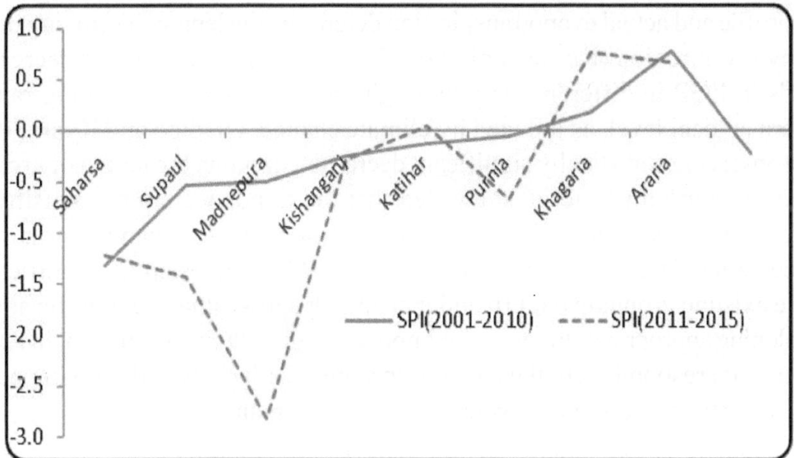

Fig.20.4 (b): Variability in SPI values in Zone II

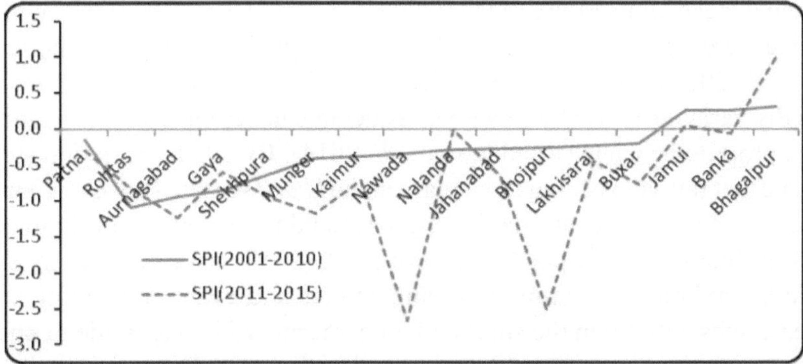

Fig. 20.4 (c): Variability in SPI values in Zone III

20.1.5 Length of growing period

Trends in length of growing period (LGP) over different districts of Bihar based on moisture adequacy indices indicated that except four districts under Zone I, *viz.* Madhepura, East Champaran, Vaishali and West Champaran, where increasing trends in LGP were observed, all the remaining districts (nine districts) recorded a decreasing trend in LGP (Sattar, 2015). Significant decrease in LGP was found in the districts of Begusarai and Gopalganj. There was a decreasing tend in LGP in all the districts of Zone II, except Khagaria, where increasing trend was observed. Linear equations fitted for the districts of Kishanganj and Saharsha exhibited significant decrease in LGP. In Zone IIIA, though non-significant, Jamui district showed an increasing trend and all remaining districts showed a decreasing trend, with Banka district showing significant decreasing trend at 5 % level. In Zone IIIB, six out of eleven districts, registered a decreasing trend in LGP. The amount and distribution of rainfall, moisture storage in the soil profile and actual evapotranspiration determine the length of growing period in any given region (Kokilavani *et al.,* 2012). Either increasing or decreasing trends in LGP in different districts might be in esponse to changing rainfall pattern at local level, as induced by climate change. Oladipo and Kyari (1993) after observing statistically significant decreasing trend in the length of growing season in northern Nigeria, suggested that the trends were more sensitive to large inter-annual fluctuations in the start of rainy season than to variation in the cessation dates. Subramaniam and Raju (1988) suggested for modification of the existing cropping pattern under changing agro-climatic condition and for developing appropriate strategies for improving agricultural production. In reality, farmers are responding to the changes in water availability and growing period and have started adopting new cultivars for risk management.

20.2 Climate change and crop projections

Studies on climate projections over the Indian sub-continent indicate an increase in annual mean temperature to the tune of 3.5 ^{0}C and 5.5 ^{0}C by 2080s with the relative increase to be less in *kharif* (monsoon) than in the *rabi* (winter) season (Lal *et al*., 2001). A review of historical rainfall trends suggests that trends are spatially variable across India with considerable uncertainty in the magnitude of the projected change (Randhawa *et al*., 2014). By 2050, projection studies indicate that rainfall is going to increase in the *kharif* season. However, it might decrease in the rabi season (Lal *et al*, 2001; Randhawa *et al*., 2014; Prabhakar, 2008) in Indo-Gangetic plains. Tesfaye *et al.* (2017) observed that in Bihar, rainfall, maximum and minimum temperatures and evapo-transpiration will increase substantially in the short and longer terms with magnitude of change varying with time, season and location within the state. The major challenges that came out of their study are heat stress during *rabi* season, high temperature regime during spring season and extreme rainfall and prolonged dry spell during *kharif* season. Thus, increasing level of temperature and erratic rainfall is likely to impact agricultural production endangering food security in the longer run. Bihar being an agriculturally dependent state, the impacts of climate change would be much more serious.

20.2.1 Projected Climate

Simulation studies based on GISS-E2-H-rcp4.5, GISS-E2-H-CC-RCP8.5 climate model indicate that till 2040 western Bihar and eastern Bihar would experience an increase of maximum temperature by 0.6 to 1°C and 0.2 to 0.4°C, respectively. Considering the minimum temperature, it is likely to increase by 1.2 to 2.0 °C in South Bihar, whereas, north Bihar may warm up to the extent of 1.0 to 1.6°C (http//:www.moef.gov.in). Simulations with regional climate models also project increasing temperature trends for future. During 2081-2100, the mean temperature may increase to the tune of 3.6°C for RCP 8.5 (IPCC, 2013). For Eastern India, the mean temperature increase may be to the tune of 1°C for 2020 and 3°C in 2050 for A2 scenario (Boomiraj *et al.*, 2010). Future projected change in rainfall during the months of June, July and August based on simulation of CCSM-RCP4.5, CCSM-recp8.5, CESM1-CAM5-rcp4.5 and CESM1-CAM5-rcp8.5 climate model shows that eastern and central Bihar would experience excess rainfall to the extent of 5-10% by 2040 from the base data of 1960-1990, whereas south and western Bihar might face rainfall deficiency by 5 percent. Singh and Sontakke (2002) pointed out a decreasing trend of summer monsoon rainfall of about 50 mm per 100 years during 1900–1984 followed by an increasing trend of 180 mm per 100 years during 1984–1999 over eastern Indo-Gangetic Plains. Das and Lohar (2005) detected a maximum 4% change in rainfall for all seasons during 2010–2039 over eastern part of India using six GCMs from Third Assessment Report of IPCC.

20.2.2 Crop Projections

Bihar being an ecologically vulnerable state, the impacts of climate change on agriculture would be much more pronounced. The direct impact of climate changes is that the *kharif* agriculture will become vulnerable due to increased frequency of extreme weather events such as erratic rainfall and frequent dry spell during monsoon (Table 4). The principal kharif crop of the region is rice, which is grown with receipt of monsoon rainfall. So, any change in rainfall pattern would be detrimental to sustainable rice production (Sattar *et al.*, 2016a). Rabi crop production may become comparatively more vulnerable due to increase in temperature, asymmetry of day and night temperature. Changes in temperature play an important role in determining crop productivity (Fiscus *et al.* 1997). Haris *et al.* (2013) observed decline in productivity of rice-wheat cropping system in Bihar owing to increase in temperature. Future scenarios indicated substitution of winter wheat by winter maize in regions unfavourable for wheat cultivation. Pathak *et al*. (2013) reported that decline in potential yield of wheat and rice was associated with negative trend in solar radiation and an increase in minimum temperature in the Indo-Gangetic Plains of India. In this background, it is necessary to have climate driven food production system in the eastern region of India to enhance resilience of agricultural production to the changing climate. This would bring stability in the farmers' income and contribute to rural prosperity. Studies conducted under All India Coordinated Research Project on Agrometeorology (ICAR) at Dr. Rajendra Prasad Central Agricultural University indicated that increasing trend of temperature during winter season is contributed majorly by minimum temperature (Sattar, 2010). Minimum temperature has a profound impact on crop productivity and increase in minimum temperature leads to increased respiration rate and its decrease increases the crop duration and yield. All available information suggests increase of average temperature by 1.5 to 2 C by 2050 with changes in rainfall intensity and distribution pattern. Three important food crops such as rice, wheat and maize would be seriously affected in Bihar under scenarios of climate change and global warming.

20.2.2.1 Wheat

It is likely to face harmful impact due to increasing trend of temperature. If current rate continues, simulation studies suggest that average productivity of wheat is likely to decrease by 3 to 5 percent by 2020 and almost 10 to 15 percent by 2050. Due to warmer nights during winter season, crop duration decreases leading to reduction in total biomass production (Anon, 2009). The studies on climate change impact on wheat and winter maize shows that simulated yield of wheat is possible to decrease in 2050s and 2080s by 3.6% and 14.1% respectively, for Pusa located under Zone I (North west alluvial plains). At Madhepura (Zone II, North east alluvial plains), a decrease in yield would be to the tune of 5%, 13% and 21% for the time periods 2020s, 2050s and 2080s, respectively, while Patna and Sabour in Zone III reported a decrease of 40 % in simulated yield by 2080 (Haris et. al., 2013). Considering crop duration, maximum decline of 26 days was observed at Madhepura. Number of grains increased marginally for Pusa and Patna during 2020; otherwise a decrease is noticed for other time periods from the baseline data. The study highlighted the suitability of North Bihar region (zones I and II) for wheat cultivation. However, due to frequent abnormal rise of day and night temperature during tillering and reproductive period of wheat crop in a particular season may lead to high reduction of yield. Wheat productivity was decreased by almost 20 to 30 percent during *rabi* season of 2008-09 owing to elevated ambient temperature condition during the month of December and January. It caused very high reduction in the number of tillering (20-35 %) and poor grain formation (Anon, 2009).

20.2.2.2 *Rabi* maize

Maize cultivation requires a temperature of 10 to 28°C for its proper growth. In Bihar, rabi maize encounters very low temperature during December and January leading to problem of pollen sterility. Growth and yield of the crop gets limited by cold sensitivity. Thus, if normal range of temperature during the month of January increases marginally, it will enhance the average productivity of the *rabi* maize in the longer run. Simulated yield of winter maize showed an increase in the range of 8.4–18.2%, 14.1–25.4% and 23.6–76.7% during 2020s, 2050s and 2080s respectively.

20.2.2.3 *kharif* rice

Monsoon rainfall, its amount and distribution plays crucial role in the successful raising of rainfed kharif crops particularly rice and maize crop. Long duration rice cultivars would be affected much more than medium and short duration cultivars. Delay in transplanting is another important contributing factor leading to poor performance of the crop. Transplanting of long and medium duration rice varieties beyond 25th July is not recommended. After this, rice productivity would be reduced significantly. A temperature below 23.9 °C & above 27.5 °C during 50 percent ear head to maturity of *kharif* rice was found detrimental for getting optimum yield under north Bihar condition (Sattar *et al.*, 2017a). Again, under future changing climatic condition, there is every possibility that incidence of droughts and extended dry spell during *kharif* rice growing season would tend to increase. This might severely affect the prospect of rainfed *kharif* rice cultivation in the state.

20.2.2.4 Potato and other horticultural crops

In future climate scenarios warming may ease the chilling conditions in these regions to favour potato productivity, while in other regions with cooler winter season the warming from current levels may prove detrimental. Based on INFOCROP-POTATO model (Singh et.al., 2005), it is found that potato production in 2020s and 2050s, will decrease by 3% and 11% over Bihar. It is worth noting that the agro-climatic conditions are very congenial for whole range of vegetables; a variety of roots and tubers crops; perennial fruit crops like mango, litchi, guava, and limes; fruit crops like banana, pineapple and papaya and spices like ginger, turmeric and chilly. Of late, floriculture is also showing excellent prospects. Under changing climatic scenario, horticultural crops have greater prospect and may offer an adaptation and mitigation option for the farmers.

20.2.2.5 Livestock and animal husbandry

Bihar is rich in livestock resources. The total livestock in the state has decreased from 53.742 million to 42.990 million between 1997 and 2003 showing a decrease of 20%. According to recent census (2012-13), the number has further got down to 32.938 millions, indicating a sharp decrease in the total livestock population in the recent period. Cattle, buffalows and goats are homotherms and are very sensitive to thermal stress, which reduces the animals ability to counter other stresses such as disease and parasites. Studies show that increase in maximum temperature during summer (> 4^0C) than the normal and fall in minimum temperature (> 3^0C) from the normal during winter was observed to reduce the milk production. The decline in yield varies from 10-30 percent in first lactation and 5-20 % in second/or third lactation (Rao *et al*., 2013). In climatologically vulnerable state, such as Bihar, the extreme weather events such as heat and cold waves are very frequents. Under projected climate change situation, such events are bound to aggravate, which may affect the livestock and animal performance. The poor farmers of the state, who account for more than 90 percent of the total farming community, would face the real threat of livelihood options dependent on animal rearing. Based on temperature humidity index (THI), stressful THI with 20 or more daily THI-hrs (THI >84) for several weeks affects animal performances. The economic losses may be very high in Bihar which produces crossbred cows, and the buffaloes would be affected more by climate change. Under climate change scenario, increased number of heat stress days with an elevated temperature regime and probable decline in water availability may further impact animal productivity and health.

20.3 Interventions and Risk Management

Climate change is now a reality and its impacts are clearly visible at multiple levels leading to crop damage and unsustainable crop production. In this context, several interventions which are cost effective, eco-friendly and practical solution oriented are required to be identified in the regime of climate change and brought into practice to make agricultural enterprises climate resilient and vibrant production system. Developing climate-resilient and climate smart agriculture requires identifying climate risks and opportunities at micro level followed by development and deployment of intervention mechanisms that help minimize the risks in order to harness maximum benefits.

Keeping the reality of climate change in mind, mechanisms for adapting to climate change could be developed by considering the agroclimatic zones with precise information of water availability period at local level considering the soil type, rainfall, potential evapotranspiration and growing period. This will lead to have agricultural sustainability and food security in the vulnerable agro-ecosystem in the state of Bihar, where occurrence of drought and flood is a recurring phenomenon (Anon, 2008). Among the four agroclimatic zones, agro-ecosystem of Zone IIIB is quite vulnerable due to having semi arid climate (**Fig. 20.5**). Drought resistant and low water requiring crops should be selected for planting under rainfed condition. Crops like maize (*Zea mays*), pigeon pea (*Cajanus cajan*) and millets during *kharif* seasons and chick pea (*Cicer arietinum*) during *rabi* season could do well in this zone. Rice (*Oryza sativa*) which is also an important crop in Bihar could perform well in moist sub-humid and humid climatic conditions. However, under rainfed condition, only short duration varieties of rice in moist sub-humid and long duration varieties in humid condition could be planted. Under dry sub-humid climate, which prevails over twenty one districts (Fig. 20.5), maize (*Zea mays*), pigeon pea (*Cajanus cajan*), sesame (*Sesamum indicum*) and chick pea (*Cicer arietinum*) could perform well. Usually rice (*Oryza sativa*) or maize (*Zea mays*) or pigeon pea (*Cajanus cajan*) or sunflower (*Helianthus annus*) during *kharif* season followed by rapeseed (*Brassica campestris*), or chick pea (*Cicer arietinum*), or lentil (*Lens culinaris*) during *rabi* season are grown under rainfed condition and they mostly experience moisture stress at different phases of growth and produce lesser yield. Instead of adopting long duration varieties, emphasis should be given for selection of appropriate short duration varieties for making the sequential cropping during *kharif* and *rabi* seasons more successful. As compared to long duration rice (*Oryza sativa*), upland rice (*Oryza sativa*) and maize (*Zea mays*) are of shorter duration and so the choice of improved short duration varieties of the latter crops is required to ensure the sowing of subsequent crops early in the

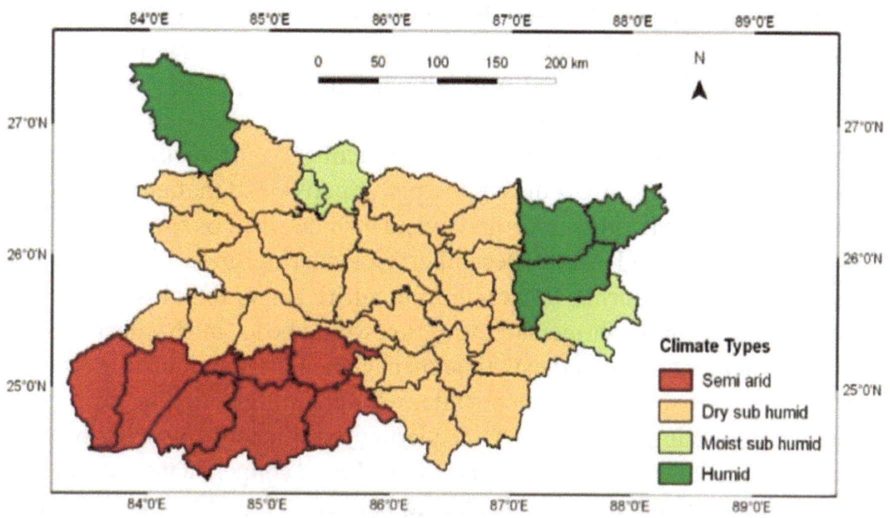

Fig. 20.5: Climate types of Bihar

season. Following the harvest of *kharif* crops, short duration varieties of *rabi* oil seeds (rapeseeds) and pulses (chickpea, lentil) could be grown with greater success. Sugarcane (*Saccharum officinarum*), which is a long duration (10-12 months) and high water requiring crop, could be grown in north-western districts (West Champaran, East Champaran) under Zone I and some districts (Kishanganj, Araria, Purnia) of Zone II owing to the availability of higher moisture availibility and longer growing period (Sattar, 2015).

Sustainable agricultural development should be the motto of agricultural planning and it is essential that fluctuations and uncertainty in productions due to floods, droughts, water logging and extreme temperature events are reduced and kept at minimum. In this context, promotion of environment friendly techniques like promoting organic farming, improvement of soil health, reuse of wastes and by products is called for under changing climatic scenario. Climate smart agricultural intervention should form the central point of development. Following are some important crop-climate based interventions which if integrated in the agricultural production system would enhance resilience and be helpful in boosting crop production and maintaining sustainability.

20.3.1 Adapting crops by shifting planting dates

Changing planting dates/sowing window of crops would be an important adaptation strategy in the wake of changing climatic scenario. Sowing of crops such as maize, wheat and rice during the period of traditional sowing windows is no longer climatologically sustainable. Hence, optimum sowing windows are essential to be identified to make the crops adaptive and high yielding to the new realms of climate change. The studies on crop-weather relationship conducted under All India Coordinated Research Project on Agrometeorology (ICAR) at Dr. Rajendra Central Agricultural University, Pusa (Samastipur), Bihar brought out some important information (Sattar *et al.*, 2017b), which revealed that that early sowing of wheat by November would enable the farmers to escape terminal heat stress during the reproductive and maturity phases of wheat and thereby realizing higher grain yield than those crops sown during last week of November and whole December. Considering grain yield vis-à-vis temperature regimes during flowering to maturity stage, the most important recommendation for the farmers of the region would be to finish wheat sowing before 25 November. The anthesis-time management by manipulating sowing dates could be a realistic adaptation strategy for attaining optimum grain yield under changing climate scenario. Considering the impact of weather on rabi maize, the study indicated that *rabi* maize would avail congenial thermal regime for proper growth and yield if it is planted between 10 to 30 November. The sowing window of the crop can be much better appropriated if planting is completed within 15-25 November. Similarly, the transplanting of rice needs to be finished by 25 July in order to properly harness the benefits of optimum crop-climatic normals for realizing desired crop yield.

20.3.2 Development of early warning system and climate smart farming

Crop plants remain exposed to the whims of weather right from germination to harvesting. Thus, yields of crops are directly dependent on weather. Even after using healthy seed and undertaking best management practices, we cannot expect desired crop yield if the prevailing weather remains inclement at any stage of crop growth. Thus, weather should be taken as an input in agricultural production. It is an important resource. In the context of climate change, the role of weather forecast based farm advisory among the farmers acquires new dimension. Actionable weather forecast based advisory can play a pivotal role in the mitigation of climatic risks in agriculture and help in the efforts towards sustainable agricultural production. The need of the hour is to adopt weather-wise farming practices. Farmers of the state are facing the brunt of climate variablities in terms of loss of production, weather related damage in crop production and above all un-sustainability due to frequent dry spell, flood, abnormal temperature, frost, heat and cold wave, and hailstorm. To cushion against such things, we need to have weather based farming practices. Since more than 50 percent of the cultivated land in Bihar is rainfed and agricultural production is heavily dependent on the monsoon rainfall, agriculture productivity is thus highly sensitive to climatic variability Thus inclusion of meteorological information in agricultural policy and decision making is necessary to cushion against weather induced variability in crop production. Feedbacks collected from farmers about the utility of weather based advisories point to the fact that they are using advisories issued from this centre on a massive scale and are regarding weather information as an input in crop production like seeds, fertilizers and irrigation. The total benefits in terms of yield stand at 7-12 percent apart from saving critical inputs.

20.3.3 Crop diversification strategy

Diversification of crops and allied activities are the need of hour in the wake of climate change and global warming scenario. Farmers tend to diversify their farming systems in order to minimize risks in crop production. In Bihar, rice is the major cereal crop, followed by wheat and maize. Mixed cropping is a common practice in upland areas during rainy season involving cereals and pulses. In winter, wheat, maize, chick pea, lentil, linseed and mustard are grown depending soil moisture and provision of irrigation. Rice-wheat is most dominant cropping system due to higher productivity. However, in recent years due to changing climatic condition, erratic monsoon is a cause of concern for the farmers as 70 percent of cultivated rice is grown under rainfed condition. On the other hand, increasing trend of temperature and extreme temperature events are affecting the performance of wheat and other *rabi* crops. In view of this, some crops and cropping system based technological interventions could prove to be effective in enhancing productivity and maintaining sustainable agriculture.

- Diversification of rice-wheat cropping system: Farmers are resorting to crop diversification by adapting rice-garlic-maize and rice-potato-oninon+maize relay cropping as an alternative replacing wheat from traditional rice-wheat-cropping system. Rice-potato-onion+maize relay cropping gave additional rice equivalent yield of 20.43 tonnes ha^{-1}, whereas rice-garlic-maize earned additional net returns of about Rs. 62,000 per ha per year over the existing rice-wheat system (Shekhar *et al.*, 2011).

- Under late sown irrigated condition, rice-winter maize is poised to be the most profitable cropping option than most common rice-wheat cropping system.

- Farmers are adopting intercropping of wheat and gram (5:1), which seems to be more remunerative than sole cropping of wheat and chick pea under limited irrigation (one irrigation) conditions.

- Under poor monsoon and erratic rainfall scenario, groundnut is gaining importance among the farmers of some districts (eg. Begusarai and Samastipur) of north Bihar particularly under medium and upland situations. This crop has bright prospect in central and south Bihar, where rainfall anomalies and occurrence of drought during kharif season are frequent. Groundnut has ample scope in replacing maize and rice during *kharif* season particularly in upland areas. Inter-cropping of groundnut either in kharif maize or arhar needs to be popularized among the farmers for enhanced income generation.

- As wheat is an important *rabi* crop in Bihar, farmers need to be advised to grow this crop by following several tested and proven technologies and interventions. The productivity of the crop can be increased by introducing timely sowing, use of zero tilled seed drills, discouraging late sowing, provision of assured irrigation and providing the farmers with wheat cultivars having high yield potential. Sowing the crop before 25 November would help it to avoid terminal heat stress, which is an important climatic constraint for wheat production in Bihar. Wheat-weather relationship indicated that the crop sown early by 15 November would ensure completion of their most sensitive flowering phase during 15- 20 February, beyond which increase in day temperature above 25 ^{0}C accompanied by dry westerly winds tend to interfere with pollination of wheat crop, thereby adversely affecting the grain setting (Sattar *et al,* 2017b). Hence, if a crop is planted earlier around 15 November, there is higher probability that the sensitive phases of reproductive and maturity periods would escape terminal heat stress.

- Adopting suitable crop sequence can help mitigate climatic risk in agriculture. Crop substitution can also play vital role in ensuring higher productivity under changing climatic scenario. In many areas of Zone I (North-west alluvial plain zone), there is ample scope of replacing kharif maize with yambean (Shekhar *et al.*, 2011).

- Besides traditional vegetables and fruit crops, farmers are turning over to non-traditional crops like aromatic and medicinal plants, flowers, spices, new species of fruits and vegetables. Vast areas under Zone I and II have marginal and sub-marginal lands. Diversion of low producing area into agri-horti system could be a win-win situation under changing climatic situation.

- Under changing climatic scenario, water is becoming a scarce commodity. To sustain long term production of rice, more efficient alternative methods are needed (Saharawat *et al.*, 2010). Under limited rainfall condition, farmers can adopt direct seeded rice. Direct seeded rice (DSR) method is water, labour and energy efficient along with eco-friendly characteristics and can be a potential alternative to traditionally puddled rice (Kumar and Lodha, 2011). DSR requires less water and is more tolerant to water stress compared to conventional rice. This has got impressive adaptive capacity under changing climatic condition. More adoption of DSR would tend to limit green house gas emission from the rice field. (Singh *et al.*, 2017). Farmers of Bihar are adopting this DSR in place of TPR (traditional puddled rice) in many districts due to uncertain monsoon rainfall. About 5-7 % of total area of rice (32 lakh hectares) is currently under DSR. This technology is gaining popularity among the farmers due to increased profits as a result of lesser cost of cultivation, since in this method there is no need of nursery preparation, thereby saving both labour and water. The districts where farmers are currently practicing DSR in medium and upland areas are West Champaran, East Champaran, Gopalganj, Muzaffarpur, Vaishali, Vaihslai, Bhagalpur, Banka, Nalanda, Nawada, Khagaria, Begusarai, Madhubani, Samastipur and Darbhanga (Singh, 2018).

- In assured production areas, location specific high yielding varieties need to be popularized. Rice should be grown in cropping system not as a mono crop. In case of kharif maize, emphasis should be given to sow it during 25 May to 10 June. Introduction of pre-kharif maize in diara areas and Naugachhia subdivision of Bhagalpur needs to be introduced for greater economic returns. As far as pulses are concerned, it should be grown as intercrop and short duration catch crop in intensive cropping system with cereals. Introducing rajmah during *rabi* season would be remunerative and profitable.

20.3.4 Stress tolerance crop development strategy

Development of heat and moisture stress tolerant crop varieties along with crop /crop varieties with higher albedo should be prioritized. There are some varieties of wheat which can tolerate heat stress during its growing period. Considering moisture stress as in the case of poor rainfall during monsoon season, aerobic rice has lot of potential. The university has developed a variety of aerobic rice *Rajendra Neelam* for its cultivation among the farmers. The variety matures in 110-115 days with yield potential of 40 quintals per hectare with only 2-3 irrigations. Such interventions could do wonders under limiting moisture and heat stress conditions.

20.3.5 Proper water management strategy

Water is a very important natural resource in agriculture. Under future changed condition, water would hold key to success of profitable and sustainable crop production. Proper rain water management along with concomitant use of ground and surface water resources should be properly chalked out. Drip and sprinkler irrigation systems should be popularized among the farmers. Emphasis on minimizing water loss and enhancing water productivity should be given in order to have sustainable water resources management.

20.3.6 Other need based agro-interventions

- Bio-energy plantations (eg Jatropha plantation) should be encouraged in large scale in Bihar in degraded and waste lands. This will obviously help increase the immensely low forest cover (6.8 %) in Bihar and at the same time will produce clean energy to meet the increasing energy demand. They may reduce nutrient leaching and soil erosion and generate additional environmental services such as soil carbon accumulation and improved soil fertility. It will also provide habitats for biodiversity in agricultural landscape.

- Tillage management (conservation and zero tillage) practices could encourage favorable impacts on environmental gases. In Bihar, this technology can prove a boon for sowing of *rabi* crops under delayed condition due to water stagnation in low land areas.

- Agroforestry and Agri-horticulture based cropping system will encourage carbon sequestration and at the same time will cushion against crop yield fluctuation under climate change scenario. The horticulture based system has great potential in Bihar. In all agroclimatic zones of Bihar, the potential and prospect of horticulture is enormous. This sector needs to be fully exploited to impart resilience in agricultural production system against the climate change impacts.
- In Bihar about 30-40% total geographical area gets periodically affected by flood. Alternative land use based on existing soil type, associated cropping pattern should be chalked out so that production potential of rabi crops does not suffer. If drought starts from first week of September, the preferred crops would be urd and toria in a given soil followed by normal *rabi* crop. Crop succession with time may be prepared as ready reference according to soil types, moisture deficit and pedo-ecosystem. As such crop calendar may be prepared on soil site basis (Choudhary *et al.*, 2009)
- Alternative interventions such as honey bee rearing, mushroom cultivation, fisheries, poultry, goatery and dairying hold a lot of promise for enhancing resilience in agricultural production systems and thus helping additional income generation among the farmers. Mushroom cultivation and honey bee rearing have caught the attention of the poor and landless farmers in Bihar. Many poor people are now involved in these ventures to earn their livelihoods.

20.3.7 Solar powered irrigation system

Solar energy use in agriculture is becoming increasingly popular. In Bihar, the use of solar energy based irrigation system can make a lot difference in enhancing farmers' income. Farmers of *diara* areas spread over about 15 lakh hectares in the state face the fury of drought and flood every year. In the absence of any permanent irrigation system, farmers find it difficult to irrigate their crops in these areas in spite of having a river flowing alongside their fields. Solar powered pumps can irrigate 5-6 acres of fields at a time. Dr. Rajendra Prasad central Agricultural University, Pusa, Bihar has installed several solar powered irrigation systems in the university farms and *dhab* areas and is in the process of popularizing the system among the farmers, which is very cheap compared to the cost incurred for electricity and diesel operated pumps. The system is eco-friendly and has enormous potential in enhancing crop resilience and food security.

20.3.8 Crop planning based on climatic risks and agroclimatic resources

Crop fitting based on AET/PET (index of moisture adequacy) values compared with water requirements during different growth phases of crops can help identify suitable crops and cropping systems in different districts of Bihar using long term rainfall data, available water holding capacity and PET values (Sattar, 2015; Sattar and Khan., 2016b). It helped (i) substituting the long duration rice crop variety with short duration rice and maize followed by toria/rapeseed, (ii) contribute substantially in delineating zone-specific efficient crops and cropping systems by matching with water availability period and (iii) adopt sustainable cropping systems based on local agroclimatic resources. The following information would be useful in enhancing agricultural resilience and mitigation of climatic risks.

- In Agroclimatic zone II, the study revealed that Araria and Katihar districts have better prospect for rainfed crop production than Khagaria district. In low rainfall district under Zone II, such as Khagaria, *kharif* maize could be successfully grown without any moisture stress, unlike *kharif* rice in Araria and Katihar districts. In *rabi* season, yellow *sarson* could be aptly fitted in the crop sequence in medium and fine textured soils.
- In Agroclimatic zone IIIA, it is apparent that rainfed crops in coarse textured soil experience higher moisture stress than medium and fine textured soils. In coarse textured soil, reproductive stage of toria faces severe moisture stress. In this zone, maize followed by toria or short duration rice followed by toria could be adopted under rainfed condition. When lentil is considered, the fitting analysis indicated that lentil has better prospect in Bhagalpur district.
- In Agroclimatic zone IIIB, the greater water stress is experienced by the rainfed crops grown in coarse textured soil. However, during *kharif* season, maize could be successfully cultivated without water stress at any phases of growth. In medium and fine textured soils, maize/short duration rice during *kharif* season followed by toria in *rabi* season could be adopted for potential productivity evaluation.

Summary

Bihar is one of the most climate sensitive states of India on account of its unique geo-climatic conditions, high population density and poverty. It is besieged by all sorts of natural hazards. Drought, flood, heat wave, cold wave, extreme rainfall events and hailstorm are a common occurrence in Bihar. It is a widely accepted fact that climate change is one of the major challenges of agriculture in Bihar. The analyses of past weather data revealed that the phenomenon of increasing temperature trend, extreme temperature events, frequent dry spell/drought, high intensity rainfall in short period and flood have increased manifold, seriously affecting the production potential of major food crops in the state. Available model predictions also suggest that these events would intensify and dominate the scenario of agricultural production systems. Hence, overall strategy would be to transform agriculture and its allied sectors into climate-resilient and vibrant production system in order to ensure sustainable food and nutritional security. As far as impact on crops are concerned, simulation studies suggest that average wheat productivity is likely to decrease by 3 to 5 percent by 2020 and almost 10 to 15 percent by 2050. The study highlighted the suitability of North Bihar (zones I and II) for wheat cultivation. However, due to frequent abnormal rise of day and night temperature during tillering and reproductive period of wheat crop in a particular season may lead to high reduction of yield. Simulated yield of winter maize showed an increase in the range of 8.4–18.2%, 14.1–25.4% and 23.6–76.7% during 2020s, 2050s and 2080s respectively. Under future changing climatic condition, there is every possibility that incidence of droughts and extended dry spell during *kharif* rice growing season would tend to increase. This might severely affect the prospect of rainfed *kharif* rice cultivation in the state. Amid all of this, horticulture has enormous potential in the state under changing climatic scenario. In this context, several interventions which are eco-friendly and practical solution oriented are required to be identified in the regime of climate change and brought into practice to make agricultural enterprises climate resilient and vibrant production system. Hence, developing climate-resilient and climate smart agriculture requires identifying climate risks and opportunities at micro level followed by development and deployment of intervention mechanisms that help minimize the risks in order to harness maximum benefits. Changing planting schedules of crops, adopting climate smart agriculture and weather based farm advisory services, crop diversification, changing cropping pattern, adapting to new crops, developing stress tolerant crop varieties, alternative sources of farm energy for irrigating crops and managing climatic risks based on local agroclimatic resources are some of the alternative and interventions that could be implemented in order to impart resilience in the production system and for enhancing farm productivity on sustainable basis under changing climatic scenario.

References/Suggested Readings

Anon. (2008) Bihar's Agriculture Development: Opportunities & Challenges. A Report of the special task force on Bihar. Govt. of India, New Delhi, pp.146.

Anon (2009) Annual Progress Report, AICRP on Agrometeorology, RAU, Pusa, Bihar, pp. 52.

Anon (2016) Annual Progress Report, AICRP on Agrometeorology, RPCAU, Pusa, Bihar, pp. 58.

Boomiraj, K., Chakrabarti, B., Aggarwal. P.K., Choudhary, R., Chander. S. (2010) Assessing the vulnerability of Indian mustard to climate change. *Agric Ecosys and Environ.* 138:265-273.

Cairns, J.E., Sonder, K., Zaidi, P.H., Verhulst, N., Mahuku, G., Babu, R., Nair, S.K., Das, B., Govaerts, B., Vinayan, M.T. (2012) Maize production in a changing climate: Impacts, adaptation, and mitigation strategies. *Adv. Agron.* 114, 1–57.

Chattopadhyay, N.(2011) Climate Change and Food security in India. In *Climate Change and Food Security in South Asia;* Lal, R., Sivakumar, M.V.K., Faiz, M.A., Rehman, A.H.M.M., Islam, K.R., Eds.; Springer: Berlin, 2011; pp.129-251.

Choudhary, B.C., Chandra, S.K., Prasad, S.N., Mohan, K. and Mishra, B.B. (2009) 25 Years' Perspective Plan of Land Use Planning of Bihar (2000-2025). Soil survey and land use planning scheme, Bihar Agricultural College, Sabour-813 210 (Bhagalpur, RAU, Pusa, Bihar, pp.221.

Das, L. and Lohar, D. (2005) Construction of climate change scenarios for a tropical monsoon region. *Climate Res.*, 30:39–52.

Fiscus, E. L., Reid, C. D., Miller, J.E. and Heagle, A.S. (1997) Elevated CO2 reduces O3 flux and O3-induced yield losses in soybeans: possible implications for elevated CO2 studies. *J. Exp. Bot.,* 48: 307–313.

Haris, A.V.A., Biswas, A., Chhabra, Vandana., Elanchezhian, E. and Bhatt, B.P. (2013) Impact of climate change on wheat and winter maize over a sub-humid climatic environment, *Current Sci.,* 104(2): 206-214.

Hijioka, Y., Lin, E., Pereira, J.J., Corlett, R.T., Cui, X., Insarov, G., Lasco, R., Lindgren, E., Akhilesh Surjan Asia. In Climate Change 2014: Impacts, Adaptation, and Vulnerability. Part B: Regional Aspects; Contribution of Working Group II to the Fifth Assessment Report of the Intergovernmental Panel on Climate Change; Barros, V.R., Field, C.B., Dokken, D.J., Mastrandrea, M.D., Mach, K.J., Bilir, T.E., Chatterjee, M., Ebi, K.L., Estrada, Y.O., Genova, R.C., *et al.*, Eds.; Cambridge University Press: Cambridge, UK; New York, NY, USA, pp. 1327–1370.

http://www.moef.gov.in/Bihar-State, GOB State Action Plan on Climate Change: Building Resilience through Development. Available online: http://www.moef.gov.in/Biharstate/default/files/ Bihar- StateActionPlanClimateChange.pdf (Accessed 5 February, 2018).

IPCC (2013) Summary for Policymakers. In: Climate Change 2013: The Physical Science Basis. Contribution of Working Group I to the Fifth Assessment Report of the Intergovernmental Panel on Climate Change [Stocker, T.F., D. Qin, G.-K. Plattner, M. Tignor, S.K. Allen, J. Boschung, A. Nauels, Y. Xia, V. Bex and P.M. Midgley (eds.)]. Cambridge University Press, Cambridge, United Kingdom and New York, NY, USA.

Kokilavani, S., Geethalaksmi, V., Arthirani,B. and Natarajan, S.K. (2012) Variability of length of growing period in Coimbatore district. *J. Agrometeorol.* **14:** 368-371.

Kumar, V. and Ladha, J.K. (2011) Direct Seeding of Rice: Recent developments and future research needs. *Advances in Agron.,* 111: 297-413.

Lal, M., Nozawa, T., Emori, S., Harasawa, H., Takahashi, K., Kimoto, M., Nakajima, T., Takemura, T. and Numaguti, A. (2001) Future climate change: Implications for Indian summer monsoon and its variability. *Current Sci.*, 81, 1196–1207.

Lobell, D.B., Schlenker, W. and Costa-Roberts, J. (2011) Climate trends and global crop production since 1980. *Science.* 333, 616–620.

Oladipo, E.O. and Kyari, J.D. (1993) Fluctuations in the onset, termination and length of the growing season in northern Nigeria. *Theor. Appl. Climatol.* 47: 241–250.

Pathak, H. *et al.* (2003) Climatic potential and on farm yield trends of rice and wheat in the Indo-Gangetic Plains. *Field Crops Res.,* 80: 223–234.

Prabhakar, S.V.R.K. and Shaw, R. (2008) Climate change adaptation implications for drought risk mitigation: A perspective for India. *Climate Change.*, 88, 113–130.

Randhawa, R.K., Singh, H. and Kang, M.S. (2014) Global warming and its possible impact on agriculture in India. *Adv. Agron.*, 123, 65–12.

Rao, V.U.M., Rao, A.V.M.S., Vijaya Kumar, P., Bapuji Rao, B. and Sastry, P.S.N. (2013) Agrometeorlogical Aspects of Extereme Weather Events, Central Research Institure for Dryland Agriculture, Santosnagar, Hyderabad, 303 pp.

Saharawat, Y.S., Singh, B., Malik, R.K., Ladha, J.K., Gathala, M., Jat, M.L. and Kumar, V. (2010) Evaluation of alternative tillage and crop establishment methods in rice-wheat rotation in North western IGP. *Field Crop Res.*, 116:260-67.

Sattar, A. (2010) Agroclimatic profiles of Zone I of Bihar.RAU, Pusa (Samastipur), pp.49.

Sattar, A. (2015) Agroclimatic characterization for assessment of crop potential under rainfed condition in Bihar. Ph.D. Thesis. Department of Agril. Meteorology & Physics, BCKV, Mohanpur, West Bengal.

Sattar, A., Kumar, Manish., Kumar, Mithilesh and Vijaya Kumar, P. (2016a) Agrometeorology of Rice in Bihar State of India, RPCAU, Pusa (Saamastipur), pp. 47.

Sattar, A and Khan, S.A. (2016b) AE/PE versus crop planning in North-West Alluvial Plain Zone of Bihar. *J. Agrometeorol.*, 18(2): 270-276.

Sattar, A., Kumar, Manish., Bobade, P.R. and Chandrawanshi, S. (2013) Assessing the length of growing period and drought incidence in Bihar, *J.Agrometeorol*, 15:192-194.

Sattar, A., Kumar, Manish., P Vijaya Kumar., and Khan, S.A. (2017a) Crop weather relationship in *kharif* rice under North-west Alluvial Plain Zone of Bihar. *J. Agrometeorol,* 19(1):71-74.

Sattar, A., Kumar, Manish., Kumar, Sumant and Kumar, Mithilesh (2017b) "Assessing crop weather relation in wheat for enhanced crop productivity in north Bihar" paper presented in the International Symposium "Stress Resilient Wheat in the Regime of Climate Change" held at RPCAU, Pusa during 22-24 April, 2017.

Shekhar, D., Kumar, Neeraj., Kumar, Rajesh., Singh, S.K. and Kumar, Uday (2011) Agricultural Technology Module, RAU, Pusa (Bihar), pp. 240.

Singh, N.K., Kumar, Rajesh., Kumar, A., Kumar, Santosh and Nilanjaya (2017) Direct seeded rice (DSR): Potent technology to mitigate green house gases from paddy fields in changing climatic scenario. Proceeding of National Conference on Climate Change and Agricultural Production: Adapting crops to Climate variability and Uncertainty held at BAU, Sabour (Bhagalpur) during April, 6-8, 2017.

Singh, N.K. (2018) Head, Department of Plant Breeding, RPCAU, Pusa, Bihar. (Personal Communication).

Singh, N and Sontakke, N.A. (2002) On climate fluctuations and environmental changes of the IndoGangetic plains in India. *Climate Change*, 52:287–313.

Singh, J.P., Govindakrishnan, P.M., Lal, S.S. and Agarwal, P.K. (2005) Increasing the efficiency of agronomy experiments using INFOCROP-POTATO model. *Potato Research.* 48:131-132.

Sreenivasan, P. S. (1979). Review of work done on crop, weather relation on rice in India. In: "Agrometeorology of rice crop". International Rice Research Institute, Manila, The Philippines, pp. 19-31.

Subramaniam, A.R. and Raju, P.A.N. (1988) Rainfall variability and crop production in north coastal Andhra. *Fertilizer News*. 33: 39-49.

Tesfaye, Kindie., Aggarwal , P.K., Mequanint , Fasil ., Shirsath, P.B., Stirling, Clare M Khatri-Chhetri Arun and Rahut, Dil Bahadur. (2017) Climate Variability and Change in Bihar, India: Challenges and Opportunities for Sustainable Crop Production. *Sustainability*, 9:198,1-22.

Yamagata, M. (1958). Studies on the limit of possibility of increase in number of leaves and grains of the main stem of rice. IV. Influence of sunlight intensity. *Bull. Fac. Agric.* Yamaguti, No. 9:1000-1010.

21

Climate Resilient Agriculture in Jharkhand

Pragyan Kumari and A.Wadood

Department of Agrometeorology & Environmental Science, BAU, Kanke Ranchi, Jharkhand, India

Climate change and its variability imposed a great threat to agriculture production and ultimately on food security. Understanding of anomalies in climatic variables is essential to make agriculture sector resilient. Jharkhand state, geographically located at 22^0 28' N – 25^0 30' N latitude and 83^0 22' – 87^0 40' E longitude with an altitude up to 1142 m above msl and having humid to sub-humid tropical monsoon type of climate, has a number of agro-climatic/physiographic constraints. The state has 3 agroclimatic sub-zones viz; Central and North Eastern Plateau sub zone (zone IV), Western Plateau sub-zone (zone V) and South Eastern Plateau sub-zone (zone VI) falling under agroecological regions XI, XII and XIII. Zone IV is having 15 districts; Zone V has 6 and Zone VI have 3 districts. Agriculture in Jharkhand is very tedious due to topographical, physiographic and climatic constraints. Vagaries of climate put greatest limitation before the agricultural production system in the State. The State enjoying nature's gift receives fairly a high amount of annual rainfall, on an average 1400 mm annually, of which more than 80% is instantly lost through high speed surface and sub-surface run-off causing loss of fertile soil and gully formation. The State's agriculture is virtually mono-cropped, major production being confined in kharif season (June-September) during which period 80-82% of annual rainfall is received. Raising a second crop (rabi season crop Oct-Nov. to March-April) in most part of Jharkhand is almost impossible because of a very meager irrigation potential (8-10% only). In absence of proper and adequate water management practices crop failure during rainy season in case of prolonged dry spells, virtually no crop during rabi season and drinking/domestic water crisis in the months of March to mid June have become a common feature in Jharkhand. Anomalies/irregularities in monsoon and its distribution (prolonged dry spells, agricultural drought etc.) have been affecting the Jharkhand agriculture since long. Now, in recent years, the effect of global climate change due to global warming has started appearing very prominently in Jharkhand. Increase in frequency and severity of extreme weather events like heat wave (unexpected rise in temperature in summer months, right from mid February onwards), hail storm, drastic decrease in pre-monsoon convectional rainfall and cold wave/frost in winter etc. are putting additional climatic constraints before the Jharkhand agriculture.

21.1 Climate Change

Study of temperature (maximum and minimum) showed significant increasing trend for more severe heat waves for almost whole state. Severe cold wave events were not observed at all, however, significant increase in cold wave (T Min < 4°C) was observed for Gumla, Latehar, Lohardaga and Ranchi districts (Table 21.1). Among 15 districts of zone IV, nine districts showed a decreasing trend in annual rainfall, which was significant for Giridih, Godda, Sahebganj and Pakur districts (Fig.21.1). Significant increasing trend was observed for Koderma, Dumka and Jamtara districts. Saraikela, East and West Singhbhum,

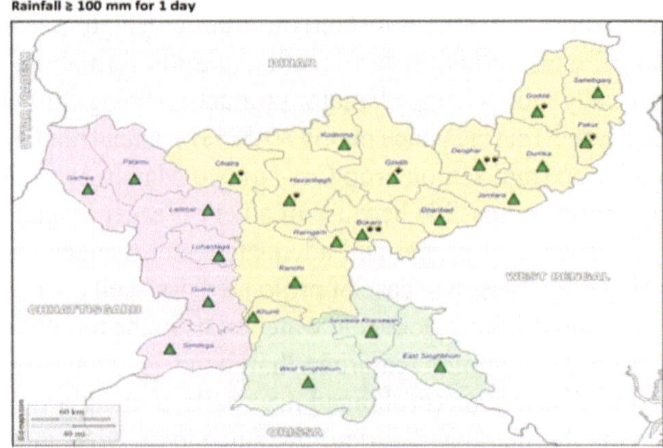

Note: : Increasing trend ▼decreasing trend + indicates 90 %* 95%; ** 99% and***99.9% confidence level

Fig. 21.1: Trend of annual rainfall and wet spells (rainfall > 100 mm) in Jharkhand

Table. 21.1: Increase (I)) in cold wave and heat wave with significance level at different districts of Jharkhand

SN	District	Cold Wave	Heat Wave (>40°C)	Severe Heat Wave (>45°C)	SN	District	Cold Wave	Heat Wave (>40°C)	Severe Heat Wave(>45°C)
1	Bokaro	I	I	I **	13	Khunti	I	I	I **
2	Chatra	I	I	I ***	14	Koderma	I	I	I **
3	Deoghar	I	I	I **	15	Latehar	I*	I	I
4	Dhanbad	I	I	I **	16	Loherdaga	I*	I	I
5	Dumka	I	I	I ***	17	Pakur	I	I	I ***
6	E. Singhbhum	I	I	I **	18	Palamu	I	I	I ***
7	Garhwa	I	I	I ***	19	Ramgarh	I	I	I **
8	Giridih	I	I	I **	20	Ranchi	I***	I	I
9	Godda	I	I	I ***	21	Sahebgunj	I	I	I **
10	Gumla	I^{+}	I	I	22	Saraikela	I	I	I **
11	Hazaribagh	I	I	I **	23	Simdega	I	I+	I **
12	Jamtatara	I	I	I **	24	W. Singhbhum	I	I	I **

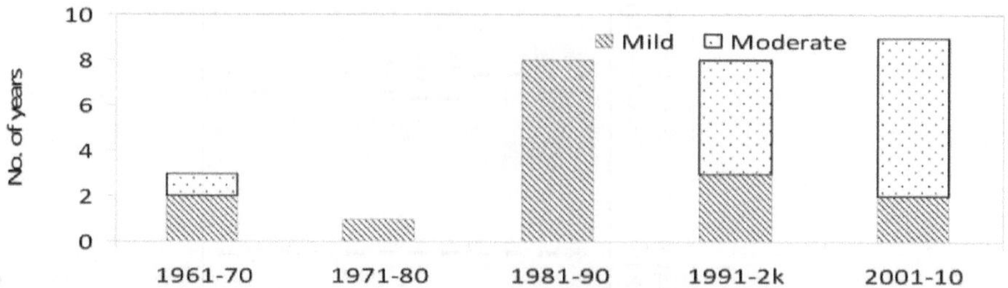

Fig. 21.2: Decade wise frequency and intensity of drought over Palamau (1961-2010)

which comes under zone VI showed an upward rainfall trend however, significant for Saraikela and East Singhbhum. Zone V was observed to be a deficit region in terms of rainfall. Extreme rainfall events (>100mm for 1 day) showed an increasing trend in all districts of Jharkhand (Fig. 21.2). Significant increase was observed in Chatra, Hazaribagh, Bokaro, Giridih, Deoghar, Godda and Pakur districts. As stated earlier, monsoon rainfall is decreasing in Giridih, Deoghar, Godda and Pakur districts but instance of extreme events increasing. Light texture soils of Jharkhand having shallow depth are very prone to erosion in case of high intensity rainfall under the undulating terrain condition. It is apprehended that such increase in number of erosive events may worsen the situation by leaving the top fertile soil barren and unproductive (Kumari 2016).

21.2 Climate Change Impacts

Climate change is already apparent in Jharkhand, there is reported rise in average rainfall in parts of state and this increase is not only undeniably steady but also significant and has potential of changing the agriculture pattern. Maximum and minimum temperature which is increasing in Jharkhand (during kharif season) has been found to have adverse effect on rice yield if it coincides with the flowering stage. Significant negative correlation between yield and minimum temperature at flowering stage was observed for upland paddy. Reduction with the unit increase in minimum temperature was reported at vegetative stage *i.e.* 608 kg ha-1 for Vandana and 293 kg ha-1 and 317 kg ha-1 for BVD-109 and BVD-111 at flowering stage, respectively. Increasing trend of maximum and minimum temperature are apprehended to reduce the rice yield in particular and yield of most of the other kharif crops in general. High rainfall at rice flowering stage has been found detrimental causing a yield reduction of up to 7 q ha-1 whereas at grain filling stage the high rainfall is beneficial causing yield increase up to 6.3 q/ ha (Kumari *et.al.*2017). High evening RH at emergence-flowering stage can cause rice yield reduction (up to 3 q ha-1) while variation in radiation seems to have no effect on rice yield. High Tmax with high evening RH during vegetative stage has been found to invite Brown spot disease and high rainfall at flowering stage causes chaffy grain of rice. Altogether, a yield reduction of 10-15 q ha-1 is reported. Above normal high maximum temperature coinciding with the flowering stage of wheat and other rabi crops has been found to cause pollen/flower sterility thereby causing appreciable yield reduction of rabi crops. However, as per a research result increase in both maximum and minimum temperature during the growing period of wheat has been found negatively correlated with yield (Kumari, *et.al.*2017). Appearance of new strains of disease/pests- Bristle Beetle in Arhar, Sheeth Blight and Rust in Kharif maize, Powdery mildew in Lentil, Alternaria Blight in Rapeseed-Mustard, Swarming caterpillar and Root Knot Nematode in Rice have been observed in Jharkhand due to changing climate scenario. Frequency and intensity of drought has been increased in recent years which affected production negatively and also changed the cropping pattern. Palamau, located in the Western Plateau sub-zone (V) of Jharkhand state, is a unique district with frequent occurrence of drought with uneven spatial and temporal distribution of rainfall. Study on rainfall and drought characteristics for crop planning in Palamau region of Jharkhand (Kumari *et.al.*2014) revealed that earlier this region was affected mostly by mild drought (1961-90) whereas in recent years frequency increased and intensity has been changed as moderate drought (Fig. 21.2). Meteorological as well as agricultural droughts in two consecutive years 2009 and 2010 restricted the rice crop coverage in this area thereby causing a heavy shortfall of nearly 50% in rice production.

21.3 Climate Risk Management

Jharkhand is divided into three agro-climatic zones, Central and North Eastern Plateau sub zone, western plateau Sub Zone and South Eastern Plateau Sub Zone. There is there need to have accurate weather information at least for each of the climatic zones. Further, there is need to have more micro level weather prediction system, sub district level weather information and weather advisory services are required to prepare farmers for climate change conditions. Majority of agriculture dependent communities are relatively poor residing in villages with limited access to resources hence there has to be a sound and supportive policy framework to facilitate adaptation process. The adaptation framework should address the issues of redesigning social sector with focus on vulnerable areas/populations, introduction of new credit instruments with deferred repayment liabilities during extreme weather events and weather insurance as a major vehicle to manage risk. The state government should encourage research and identify and prioritize adaptation options in areas with high impact on agriculture sector (micro level seasonal weather forecasting systems, water diversion, storage and distribution, agriculture planning and other infrastructure needs). The focus should be on integrating national development policies into a sustainable development framework that complements adaptation. The action plans based on this should accompany technological adaptation methods. In addition, the role of SHGs, rural banks and agricultural credit societies should be widened.

Table 21.2: Comparison of variability in Maize and Rice yield during recent decade (2000-2010).

Year	Maize yield (kg/ha)	Rice Yield(kg/ha)	Drought intensity
2000 01	1580	1235	Mild
2001 02	1445	1797	Mild
2002 03	1708	1498	Mild
2003 04	1600	1694	No Drought
2004 05	1444	1495	Mild
2005 06	1266	1150	Mild
2006 07	1504	1832	No Drought
2007 08	1500	2023	Mild
2008 09	1345	2035	Mild
2009-10	1280	1546	Moderate
S.D	142	302	
Mean	1467	1631	
CV%	9.7	18.5	

Globally, it has been established that many traditional knowledge is a valuable asset in observing and managing environmental change. Communities living in locations where climatic changes are fast and frequent have survived by traditional knowledge to adapt to biophysical changes in their environment for centuries (McNamara *et al.* 2010), and India is no exception. In this context, this is argued that traditional local knowledge such as local crop varieties (seeds and crops which can tolerate extreme weather conditions), decision on crop sowing time based on traditional understanding of weather pattern (rains, heat etc), use of intercropping to minimize crop failure risks, and use of plants/herds to control weed, insects etc are some very commonly practiced mechanism that are purely based on traditional knowledge system. Long term yield data of rice and maize, collected from Government of Jharkhand, showed that variability in productivity presented as CV per cent (Table 21.2) was found very less for maize (9.7 per cent) as compared to rice (18.5 percent) yield under the frequent occurrence of drought in recent decade. Maize crop having wider sowing window with lesser water requirement would be the better option for farmers as contingent food crop under the changing climatic condition of Jharkhand. A case study on recent droughts revealed that moderate drought hit the Palamau district of Jharkhand in year 2009 and 2010. As reported by Directorate of Agriculture, Jharkhand, rice cultivation worst affected by covering only 16 and 7.7 per cent of target area under cultivation whereas maize cultivation still achieved the target value by covering 83 and 87 per cent during 2009 and 2010, respectively. Adaptation of maize cultivation by farmers during drought years in Palamau region in particular and whole of Jharkhand in general, indicates the immediate adaptation measure against the climate variability.

Technique to reduce Urea use (FAO, 2012) Urea Deep Placement (UDP) technique, developed by the international Rice Research Institute (IRRI) and International Fertilizer Development Center (IFDC), is a good example of a climate-smart solution for rice systems. The usual technique for applying urea, the main nitrogen fertilizer for rice, is through a broadcast application. This is a very inefficient practice, with 60 to 70 per cent of the nitrogen applied being lost, and contributes to GHG emissions and water pollution. Further, the traditional practice to regulating and use of water bodies during the time of stress is something that has been found to work effectively. Under the scenario of changing climate monsoon has become very erratic causing agricultural droughts very often. In the event of early season drought or dry spell (mid June-mid July), which is less often, crop coverage suffers and farmers have little option to go for late sowing/transplantation of short duration and less water requiring crops. But in the event of mid season drought or dry spell, which is most often, well established standing kharif crops almost fail due to moisture stress and the farmers have no other alternative option except to lose the precious inputs like costly seed and labour leading to insecurity of livelihood of the farmers and their families. In such a situation had there been little water in small structures (Dovas), stored during the early monsoon season (mid June-mid July) when there was no drought/dry spell, it would have been utilized to save the standing kharif crop by providing onc or two lifc saving shallow irrigation. Experiences show that at least 5 shallow irrigations (5cm each) could be given to the standing crop in the catchment area out of the stored rain water from the same catchment area. Apart from saving of the suffering kharif crops the stored water could be utilized for raising a second crop (vegetables, mostly cucurbits with Thalla Technique) around the Dovas. It has been seen that the Dovas, if constructed at right site, remains full of water even after harvest of the kharif crop as such Dovas are continuously fed by seepage water.

Weather insurance plays an important role in mitigating climatic risks. But this strategy has worked successfully in regions having long term weather data, farmers have large holding and have a business approach for farming. In Jharkhand, the small holders are generally more prone to risks, the government sponsored crop insurance scheme has made some progress but it has a long way to go. Considering the climate trends being witnessed in recent years all over the country, weather based insurance appears to be a better alternative for mitigating risks in agriculture for Indian farmers. The state should work with agriculture research institutes and insurance companies to jointly develop crop wise data on weather sensitivity so that appropriate policies can be designed, which are friendly to farmers without affecting the viability of the insurance companies. Sporadic attempts for crops like lac, tasar have been made, besides other agriculture crops. Increasing resource base through development of wasteland: A pilot study is required to assess the reclamation potential of the waste land in the State. Hence, the use of indigenous knowledge and local coping strategies has to be taken as a base line and starting point of the adaptation planning.

In order to combat the climate change in the state a comprehensive action plan on climate change requires inclusion of some very important options for adaptations, well focused on climate resilient agriculture, and mitigations, having main focus on carbon sequestration. These would be; massive and repetitive Climate Awareness programmes, strengthening weather network and weather based Agroemt Advisory Services, strengthening Agricultural Information system, drought Assessment, promotion to dryland farming, identification of climate friendly and climate deteriorative cropping system/ pattern, disease/ pest-weather relationship and forewarning, soil & water conservation (mulching/ minimum or zero tillage etc), precision farming (sprinkler/drip irrigation, poly house, net-shed cultivation), crop improvement for drought/flashflood submergence and heat tolerance, crop Diversification and Agroforestry etc. under adaptations. Under mitigations the options would be; carbon sequestration, cultivable waste land to be brought under cultivation, promotion to Horticulture, NRM for Agriculture (Rain water & Soil management), minimal use of chemicals and inorganic fertilizers, organic farming, solar energy utilization and adherence to Legislations/Policies related to climate and environmental protection etc.

References/Suggested Readings

Kumari, P. (2016). Annual report of AICRPAM, BAU, Ranchi, Jharkhand.

Kumari, P. Kumar P. Vijay, Kumar, R. Wadood, A. and Tirkey, D. A. (2017) Effect of weather on grain yield of direct seeded upland rice varieties in Jharkhand, India. *Indian J. Agric. Res.*, 51(6): 562-567

Kumari, P., Wadood, A., Singh, C. S. and. Singh, S. K.(2017) Quantification of temperature and its effect on wheat yield in Jharkhand region, *The Bioscan.*12(1): 355-358,

Kumari, P., OJHA, R.K.A., Wadood, A., and KUMAR , R.(2014) Rainfall and drought characteristics for crop planning in Palamau region of Jharkhand. *Mausam*, 65, 1 : 67-72

McNamara, K.E., Westoby,R and Parnell, K. (2010) Elders' and Aunties'experiances of climate on Erub Island in the Torres Strait. Final Project report to the Marine and Tropical sciences Research Facility. Published online by the Reef and Rain Forest Research Centre Limited, Cairns (pp. 31)

22

Climate Change Mitigation Strategies in North Eastern Region

B. Goswami and P. Dutta

Department of Agro-Meteorology, Assam Agricultural University, Jorhat, Assam, India

The Northeast Region (NER) of India consists of Assam, Arunachal Pradesh, Manipur, Meghalaya, Mizoram, Nagaland, Sikkim and Tripura, covering an area of 255168 km^2. The area is characterized by rich biodiversity, heavy precipitation and high seismicity. Although the contribution from agricultural sector towards gross domestic product has declined, yet it remains the predominant sector for Indian Economy. Over 75% of the people of NE India directly depend on agriculture for their livelihood. Therefore, better performance of agriculture has a direct and multiplier effect across the economy of the region. Because of climate change the rain fed agriculture is facing a greater risk as it is resource poor and having limited flexibility for adaptation. Under changing scenarios there is increase in demand for fresh water and nutrient requirement, increase in soil erosion and nutrient losses affecting soil health, intensification of pest, disease and weed problems, reduction in nutritional quality and low input use efficiency. Continuous mono-cropping and the associated excessive tillage, inadequate erosion control measures and large scale slash and burn cultivation (Jhum) resulted in significant environmental damage and a non-sustainable crop production system in the region. The Jhum or slash- and- burn cultivation is very much prevalent in the region. This practice involves reckless slashing and burning of forest area for cultivation leading to release of carbon dioxide and significant amount of other Green House Gases such as nitrous oxide. The intensive ploughing of the cleared mountain makes the land very much susceptible for erosion leading to soil degradation and loss of soil organic carbon. The climate change is likely to add more burden to already struggling agricultural sector of the state. The climate change is a concern to researchers and farmers because it would lead to change in insect and disease dynamics i.e. the distribution, abundance and management of insect and pathogens. Moreover, there would be increase and variation of livestock disease and adaptation, and decline in soil health. The magnitude of impact varies with geographic location and is tailored to local circumstances and ecosystems. The erratic rainfall trend in the recent past lead to the failure in planning proper water management techniques and also makes the prediction of disease outbreak difficult in this region. Agro- climatic conditions of NEH are conductive for the development and multiplication of insect pests and their natural enemies. Due to climate change rice insect pests continue to be one of the major constraints to rice production, which causes about 20% yield losses. Maize is the second most important crop in the region. It is primarily grown in the Jhum land and terraced area in NEH region. In moisture stress area, occurrence of cutworm (*Agrotis flammatra* and *Agrotis segetum*) was found very serious. The important pulses grown in the region are pigeon pea, cowpea, rice beans, rajamash, Bengal gram, mung bean, lentil, fababean and lathyrus etc. Due to favourable climatic conditions, pulse crop are infested with number of pests throughout the cropping season.

22.1 Climate Change

NE region has a minimum/ maximum summer season temperature of 23/31°C and 9/24° C during winter. It receives an average 1000-2000 mm of rainfall during summer and about 200 mm during winter. Under changing climate the projected increase in minimum and maximum temperature in NE Region during summer is about 2°C by 2030. The summer rainfall is likely to reduce by about 10%. On the other hand, during winter projected increase in temperature is about 2.5°C and remains rainless mostly. Rainfall from 1871 to 2008 did not show any clear trend for the region as a whole, although there are seasonal trends across some hydro-meteorological subdivisions while rise was the trend in maximum, minimum, mean temperature and temperature range according to Jain *et al*., 2013. Notably for the post monsoon season, the Sen's estimator of slope (°C/year) was 0.019, 0.011 and 0.015 for the maximum, minimum and mean temperature, respectively. The impacts of climate variability on agriculture in Assam are already in terms of erratic monsoon, frequent flood, drought like situation and warmer winter. In three consecutive years since 2002, there were large floods in the State but the seasons of 2006, 2007 and 2009 witnessed deficient rainfall (IMD, 2006, 2007; IPCC 2007a). A study done at ICAR, Umiam has revealed that, the average monsoon rainfall/ rainy days in NE region have decreased from 900-3000 mm/ 65-91 days (1951-90) to 850-2350mm/ 57-85 days (1991-07). Significant decrease in rainfall is noticed in different parts of North- East India. The last decade (2001-2010) was the driest decade on the history of past 110 years in the Brahmaputra valley (Deka *et al*. 2012). Moreover, year to year variations in the onset, amount and distribution of monsoon rainfall are likely to be influenced under projected climate change scenarios in the region (Rupa Kumar *et al*. 2006).

22.2 Climate change impacts

The North Eastern Region is considered to be the richest reservoir of genetic variability of large number of horticultural crops. Horticulture is the backbone of agriculture sector of North East India and its potentiality is yet to be harnessed to its fullest extent. The simulation analysis indicates that the potato yields are likely to be marginally benefited up to 5% in upper parts of NE region due to climate change influence, but in the central part, the yields are projected to reduce by about 4% while in the southern parts of NE region, the negative impacts will be much higher. The heavy rainfall from April to October in NEH region causes heavy soil erosion and nutrient loss as well as results in manifestation of different types of insect- pest and diseases and as a result crop productivity is reduced to great extent. The long dry spell from November to March results in moisture stress in many of the horticultural crops and because of this the young orchard of khasi mandarin has started declining the entire NEH region. The incidence of fruit flies has become a serious concern in peach and guava since last three to four years. In underutilized crops, powdery mildew and blight are new record in *Prunus nepalensis*, which was never noticed until three years back. Cabbage butterfly is another serious problem in Cole crops during February- April, which was not recorded earlier. The late blight becomes a serious problem in tomato and potato during summer season (April- June) due to low temperature and heavy rains which provide congenial environment for disease development. As a result of global warming and climate change, glaciers in the Himalayas are retreating at an average rate of 15 meter per year, consistent with the rapid warming recorded at Himalayan climate stations since the 1970s. In the Himalayan region alone 67% of the glaciers have retreated in the past decade, e.g. the Gangotri glacier has been receding at a rate of 28 meter per year. Under a warming environment the Himalayan glaciers are expected to melt faster leading to increased summer flows and flooding initially for a few decades followed by progressive reduction in flow as the river feeding glaciers recede and disappear from the headstreams. Widespread water scarcity in the river basins like those of the Indus, the Ganga and the Brahmaputra is a serious consequence on the cards. Moreover, in case of glacier-fed rivers, glacial-melt runoff is seen to augment winter flow in the lean season. With glacial contribution deceasing over the years, in future lean season flow will decrease and water stress will increase in the Brahmaputra basin where large populations depend on agriculture for livelihoods. Glacial retreat in the Himalayas may lead to serious alterations in the hydrological regime of the Brahmaputra river system as the mainstream of the Brahmaputra (known as the Yarlung Jhangbo in Tibet, China) and some of its tributaries like the Subansiri and the Jia-Bharali are partly fed by snow-melt run-off. Projected increase in rainfall and accelerated summer flows may give rise to more intense flooding and flood hazards, but consequent retreat of glaciers may reduce flows in the long run (Das, 2009). Extreme precipitation events (heavy rain storm, cloud burst) may have their own impacts on the fragile geomorphology of the Himalayan part of the Brahmaputra basin causing more widespread landslides and soil erosion processes, and sedimentation in the Himalayan river basins could alter significantly due to climate change. Two extremely intense cloud bursts of unprecedented intensity- one in the western Meghalaya hills and Western Arunachal Pradesh in 2004 produced two devastating flash flood in the Goalpara and Sonitpur districts of Assam bordering Meghalaya and Arunachal respectively causing hundreds of deaths and enormous loss to the animals and agriculture. The most recent examples of such flash floods originating from extreme rainfall are two events that occurred in the north bank of the Brahmaputra River and caused significant damage to human life and property. The first of the two events occurred during the monsoon season on June 14th, 2008 due to heavy rainfall on the hills of Arunachal Pradesh north of Lakhimpur Districts causing flash floods in the rivers of Ranganadi, Singara, Dikrong and kakoi that killed at least 20 people and inundated more than 50 villages leading to displacement of more than 10,000 people. The other that occurred in the post monsoon season on October 26 affected a long strip of area of northern Assam valley adjoining foothills of Bhutan and Arunachal Pradesh causing flash flooding in the four major rivers (all are tributaries of the river Brahmaputra) and a number of smaller rivers. This episode of flash floods caused by heavy downpour originated from the Tropical Depression 'Rashmi' , (a depression over the west Central Bay of Bengal adjoining Andhra coast) and affected mainly the catchments of the rivers Puthimari (Assam- Bhutan border), Jia-Bharali (Assam- Arunachal border), Ranganadi (Assam Arunachal border), and the Subansiri (Assam – Arunachal border). The southern of Nagaon district in central Assam valley and the adjoining parts of Karbi Anglong from the rain shadow zone where annual rainfall is as low as 800-1200 mm. Water scarcities are a potential constraint for the people living in this rain shadow Zone and absence of effective irrigation systems or water harvesting practices adds to the vulnerability of the people. But what is of immediate concern is that rainfall in this zone is decreasing slowly as found in Lumding where rainfall is on the decline at a rate of 2.15 mm per year (Das, 2004). As a result water crisis might aggravate in this region in the coming years. Flood have caused mayhem in the region, especially in Assam, every year causing tremendous loss to crops, infrastructure, economy, livelihoods and lives of the people. During the period of 45 years between 1953 and 2004, the seven states of the region suffered together a loss of about 2 billion rupees due to flood damage to crops, houses and public utility while 1.25 million hectares of land were affected due to floods. In some year's floods have affected more than 3.8 million hectares of Assam's total area of 7.8 million hectares (WB,

2007). Floods inundate at least 2,000 villages every year in addition to destroying other infrastructure. The problem is further exacerbated by riverbank erosion, which destroys about 8,000 hectares of riparian land along the Brahmaputra annually. Vast areas in the region have been affected by erosion e.g. 1 million hectares in Assam; 815,000 hectares in Meghalaya; 508,000 hectares in Nagaland; 108,000 hectares in Tripura; and 14,000 hectares in Manipur (Venkatachary *et al.*, 2001).

22.3 Climate Change and crop projections

The present rapid climate change and its adverse effects are mostly governed by the rapidly increasing concentration of green house gases (GHG's), aerosols and other pollutants. These emissions are regulated by various natural and anthropogenic factors. Anthropogenic GHG emissions are mainly driven by population size, economic activity, lifestyle, energy use, land use patterns, technology and climate policy. The Representative Concentration Pathways (RCPs), which are used for making future climatic projections based on these factors. The The RCPs describe four different 21st century pathways of GHG emissions and atmospheric concentrations, air pollutant emissions and land use (IPCC, 2014). The RCPs include a stringent mitigation scenario (RCP 2.6), two intermediate scenarios (RCP 4.5 and RCP6.0) and one scenario with very high GHG emissions (RCP 8.5). Among different RCPs, RCP2.6 is representative of a scenario that aims to keep global warming likely below2°C above pre-industrial temperatures. Though these emission scenarios are highly uncertain but they serve as an important tool to analyze how various driving forces may influence future emission outcomes and possible impacts on the biological world. They also assist in the study of the location specific assessment of impacts of climate change amd potential adaptation and mitigation measures (Anon, 2015).

22.4 Rice yield under different climate change scenarios

Rainfed Agriculture in this region is predominant, covering about 80 per cent of the cropped area. Present and future climate change and variability potentially affect agricultural production by virtue of acute soil moisture deficit and lack of irrigation/water harvesting infrastructure. Hence, comprehensive studies on selection of suitable crops with multi stress tolerance ability, crop diversification and intensification are required in the context of present and future scenarios. Under these circumstances, a study was taken up to evaluate the impact of climate change on present and future kharif rice production scenarios in Jorhat, Assam, which is the major growing area in the state. Three transplanting windows were selected for transplanting of rice and they were maintained in the 3rd week of May, 2nd week of June and 1st week of July with near about 15 days interval between each transplanting. The simulated yields of different scenarios were compared with observed mean yield of 2009-2013 for their respective transplanting windows. The observed mean grain yield for first (DI), second (D2) an third (D3) dates of transplanting were 2600, 4594 and 3443 kg ha^{-1}, respectively, during 2009-2013 and rice yield under various RCP projections were compared with the present yield to estimate yield variability in future (**Table 22.1**). The results reveal that there is large negative deviation of projected rice in all scenarios compared to the mean observed grain yield. Under various scenarios, percent deviation in projected yield ranged from-12.7 (RCP 2.6) to - 28.8% (RCP 8.5); 34.2 (RCP 2.6) to-13.4% (RCP 8.5) and-23.9 (RCP 8.5) to- 30.9% (RCP 2.5) in first, second and third dates of transplanting, respectively, over observed mean yield of 2009- 2013. Considering different transplanting dates, the maximum mean yield reduction is likely in second transplanting date (-38.5%), followed by third (-28.8%) and first one (-23.3%) under various RCP projections at Jorhat. There is likely increase in maximum and minimum temperature at Jorhat during the rice growing season (Monsoon and post Monsoon). The projected magnitude of increase in maximum temperature ranged from 1.2 (2020) to 2.4 (2080) °C, during monsoon and 1.3 (2020) to 2.5, (2080) °C during post monsoon seasons. The minimum temperature has also shown similar trend of increase. Though there is likely increase in seasonal rainfall in 2020 and 2050, but by 2080 the rainfall is likely to decrease compared to present mean rainfall (Goswami *et al.*, 2016). These factors, associated with temperature and rainfall, may definitely be attributed for reduction in projected rice grain yield under various scenarios.

Table 22.1: Projected rice (Cv. Mahsuri) yield variability at Jorhat under different RCP scenarios and dates of sowing over current condition (2009-2013)

Periods		% Change in grain yield 2009-2013				
	observed grain yield (kg ha^{-1}) (2009-2013)	RCP2.6	RCP4.5	RCP6.0	RCP8.5	Mean
2020D1*		-14.4	-14.8	-27.5	-27.5	-21.0
2050D1		-14.5	-34.3	-17.1	-35.3	-25.0
2080D1		-9.3	-33.8	-27.0	-23.7	-23.0
Mean	2600	-12.7	-27.6	-23.9	-28.8	-23.3
2020D2		-27.6	-29.5	-36.7	-36.7	-33.0
2050D2		-43.9	-43.5	-44.1	-44.8	-44.0
2080D2		-31.1	-38.9	-35.9	-48.8	-39.0
Mean	4594	-34.2	-37.3	-38.9	-43.4	-38.5
2020D3		-21.1	-26.8	-14.0	-14.0	-19.0
2050D3		-56.6	-58.9	-64.0	-61.9	-60.0
2080D3		-14.9	-5.5	-12.2	-4.3	-7.0
Mean	3443	-30.9	-30.4	-30.1	-23.9	-28.8

*Dl (3rd week of May), D2 (2nd week of June) and D3 (1st week of July) are the dates of transplanting

22.5 Potato under climate change scenario

Potato is the most important non-grain field crop with a global production of 501 million tonnes in 2014 (FAO, 2014). However, potato production is likely to decline across many parts of the world by 2100. But, Stockle *et al.* (2010) indicated that, taking into account the positive effect of CO_2 and adaptation strategies on crop production might sustain the current production levels under future climate change conditions. Projected variations in fresh tuber yield of different cultivars of potato at Jorhat under different projection periods and dates of planting, considering all RCP together, is presented in Table 22.2. This is expressed as per cent difference over current observed mean yield (2014-15 and 2015-16). In the 1st planting period (third week of November), the mean observed fresh yield recorded for Kufri Jyoti (V1), Kufri Pokhraj (V2) and Kufri Himalini (V3) were 23.7, 19.8 and 12.8 t ha^{-1}, respectively. In this option of planting, cultivar V1 is likely to perform best in all projection periods, followed by V2. There is likely hike of fresh tuber yield by 33.8 (2020) to 18.1 per cent (2080) over the current mean yield for V1. Similarly cultivar V2 is likely to gain between 18.8 (2020) and 10.8 per cent (2080). In contrast, cultivar V3 is set to face reduction in its fresh tuber yield by 2.1 (2020) to 14.7 per cent (2080). In the 2nd option of planting (1st week of December), the mean tuber yields recorded were 27.2, 18.3 and 12.1 t ha^{-1} for V1, V2 and V3, respectively. Under projected scenarios, fresh tuber yield of cultivar V1 is likely to reduce by 7.1 (2020) to 30.1 per cent (2080). The yield of cultivar V3 will further go down by 37.6 (2020) to 52.7 per cent (2080). Whereas, cultivar V2 will maintain its gain in tuber yield by 9.8 (2020) to 10.7 per cent (2050), but is likely to record a loss of 16.1 per cent in 2080 over the current mean fresh tuber yield. If planting is further delayed to 3rd week of December, all these cultivars are likely to record drastic reduction in tuber yield. The V3 may record highest loss of 64 (2020) to 75 per cent (2080), followed by V1 (57.6% in 2020 to 71.5% in 2080) and V2 (45.2% in 2020 to 56.2% in 2080). All of the present cultivars may not remain sustainable under this option of planting period and tuber yield is likely to fall by over 50 per cent from the current mean values. The reduction in future tuber yield may be attributed to possible increase in temperature. High temperatures negatively affected the allocation of glucose, stimulated the growth of above ground biomass and simultaneously reduced tuber accumulation though not necessarily inhibit tuber initiation. Reduction in tuber yield of potato in Gujarat by 6-17 per cent due to increase in maximum and minimum temperatures during different months of potato growing.

Though the potato cultivars are not different much in current yield realization within themselves, even then if planting is delayed, Kufri Himalini may not remain suitable for planting beyond 2020 under all options of planting and Kufri

Jyoti beyond 2050 under December planting in this region. Kufri Pokhraj is likely to remain as a viable cultivar under subtropical of Jorhat till 2050 and can be planted up to 1st week of December.

Table 22.2: Projected variations in fresh tuber yield of potato cultivars under different projection periods and periods of planting over current mean observed yield (2014-15 and 2015-16) considering all RCPs together

Planting times (D)	Cultivars (V)	Mean observed yield (t ha^{-1})	Change in tuber yield over current mean yield (%)		
			2020	2050	2080
3rd Week of November	Kufri Jyoti	23.7	33.8	30.5	18.1
	Kufri Pokhraj	19.1	18.8	18.5	10.8
	Kufri Himalini	12.8	-2.1	-1.6	-14.7
1st Week of December	Kufri Jyoti	27.2	-7.1	-5.7	-30.1
	Kufri Pokhraj	18.3	9.8	10.7	-16.1
	Kufri Himalini	12.1	-37.6	-38.2	-52.7
3rd Week of December	Kufri Jyoti	23.0	-57.6	-60.7	-71.5
	Kufri Pokhraj	17.6	-45.2	-46.3	-56.2
	Kufri Himalini	10.2	-64.0	-65.7	-75.0

22.6 Climate risk management

As climate change is a complex subject and has many manifestations, it is very difficult to generalize possible remedial measures. Depending on the situation one has to observe and study critically to come out with possible sets of location specific adaptation and mitigation measures. Some of the general recommendations made by different workers are listed below:

1. A change of crop variety with different growth durations can mitigate the impact of extremes.
2. Changes in planting date may provide some adaptation
3. Seasonal weather forecasts as supportive measures to optimize planting and irrigation patterns.
4. Development of water/ temperature/ salinity stresses and flood tolerant rice varieties through extensive field screening.
5. Weed- pest-disease competition with crop has to be critically observed and corrective measures should be adapted as required by the situation.
6. System of rice intensification (SRI) method found as very effective to tackle water stress conditions. Even in high rainfall areas SRI method is giving very encouraging results in term of productivity.
7. Enhancement of irrigation facilities.

8. In situ rainwater harvesting through construction of farm ponds will provide lifesaving irrigation during the time of need. Rice cum fish culture may also be adopted to have additional source of livelihood.
9. Spraying of 2% urea can help the rice crop to overcome water stress very effectively.
10. Crop insurance programme, using weather derivatives, should be popularized to safe guard farmer's interest to the maximum possible extent. Explore feasibility of establishing feed, fodder, and seed banks.
11. Enhance climate literacy.
12. Encourage application of technologies, such as neem coated urea, to reduce GHG emission from rice fields.

22.7 Research needs

After observing 2009 as year of unprecedented drought, the monsoon season in 2010 is as far satisfactory by and large. In spite of that, extreme climatic events are causing disasters in different parts of South Asia, including India. Cloud burst of Leh., Flash Flood at Pakistan and China are some of the recent calamities causing extensive damage in all possible ways. The food security is always at threat under such circumstances, to mitigate the contingencies research and technological preparedness should always be prioritized.

IPCC Special Report on the Regional Impacts of Climate Change-An Assessment of Vulnerability explores the potential consequences of changes in climate for ten continental climate and sub-continental regions. The report underlines the potential for climate change to alter the ability of the earth's physical and biological systems (land, atmosphere and oceans) to provide goods and services essential .for sustainable economic development. The gap and deficiencies revealed in this special report suggest some priority areas for researchers to further work to help policymakers in mitigating the impact of possible climate change. These needs include:

- Collection/generation of better baseline data, both climate and socioeconomic Demographic change including Land-use change, Land degradation, Air and water pollution, Economic and social change, development and poverty.
- Better Scenarios, especially of precipitation, extreme events, sulfate aerosol effects and regional-scale changes.
- Better understanding of the ecological and physiological effects of increasing CO_2 concentrations, taking account of species competition and

migrations, soil and nutrients acclimation and partitioning between crop yields, roots, stems and leaves.

- Dynamic models of climate, biospheric processes and other socioeconomic factors to take account of the developing, time-varying nature of global change.
- Impact assessments across a range of scenarios and assumptions to enable the assessment of risk, particularly in regions comprised primarily of developing countries and small island states, where resources for research and assessment have been inadequate to date.
- Analysis of adaptation options, including the need for development of new technologies and opportunities for adapting existing technologies in new settings
- Integrated assessments across sectors, from climate change to economic or other costs, across countries and regions, including adaptations, and including other socio-economic changes.

22.8 Management options

In the wake of such a shift in climate in the region, there is an urgent need for reassessment of the agricultural practices. The foremost important thing i.e., aforestation programme should be taken by the community, Institutes, NGOs, schools and Governments. The agricultural practices that conserve natural resources and have the advantage to delay effect of stress should be given importance and necessary policy or subsidy may be encouraged. Some of the resource conserving technology suitable for the region are-

- Minimum tillage or zero tillage with residue cover in surface for improving soil quality and conserving soil moisture.
- Cover cropping, in-situ residue management and restoration of degraded lands for soil moisture conservation and improved C-sequestration.
- Agroforestry with multipurpose trees, crops and animal components for improving hydrology.
- Integrated farming systems and watershed development with animal, fishery and hedge row cropping for soil and moisture conservation and nutrient recycling.
- Screening short duration varieties for their drought resistance.
- Popularization of technologies like system of rice cultivation (SRI) and aerobic rice cultivation for water saving and mitigation of greenhouse gas (GHG) emission.

- Rain water harvesting: in-situ (land configuration, mulching etc.) and ex-situ (Ponds, micro water harvesting structure -jalkund etc).
- In-situ biomass management in shifting cultivation (cover about 1.6 million hectare in NER) instead of biomass burning to reduce CO2 emission and improve hydrology.
- Promotion of technologies that enhance biological N-fixation and improve nutrient and water use efficiency to reduce N_2O emission and dependence on non-renewable energy.
- Change in planning dates and crop varieties are another adaptive measure to reduce impacts of climate change to some extent. For example, the Indian Agricultural Research Institute study indicates that losses in wheat production in future can be reduced from 4-5 million tons 1-2 million tons if a large percentage of farmers could change to timely planting and changed to better adapted varieties.
- In North East India with the shift in rainfall pattern and rise in temperature it is important to relook at the present date of sowing and varieties. An example of this is that at mid-altitude of Meghalaya (950 m ASL), where it was not possible to grow a second crop of rice after Kharif (rainy season) rice due to early onset of winter, presently double cropping is possible at least at the experimental field with the adjustment of sowing dates and varieties.

Many of the strategy/ conservation agriculture practices mentioned here are proven and well tested for enhancing productivity and adaptive strategies to combat change. What is needed is the more data generation through intensive research and their adoption at farm level with proper extension and support policies.

Conclusion

Climate change is one of the major threats to food security. Since the supply of land is fixed, the continuous expansion of agricultural areas by bringing more areas under cultivation is not possible in the long run, further growth will have to rely on productivity increase. Meeting the food requirement of immensely growing population without any damage to the ecological aspects now is the topmost priority of agriculturists, environmentalists or development planners to focus on sustainable agricultural strategies for livelihood security especially in the context of climate variability and change. There is a need to develop and apply a standard methodology across the board for various studies related to climate change and agriculture.

References/Suggested Readings

Anonymous. (2015). Emission scenarios. Retrieved from WMO web site as on 15-09-2015. https://www.wmo.int/pages/themes/climate/emission_scenarios.php.

Bujarbarua, P., Baruah, S (2009). Vulnerability of fragile forest ecosystem of North East India in context with the global climate change: an ecological projection. IOP publishing Ltd.

Das, P.J., 2009. Water and climate induced vulnerability in northeast India: Concerns for environmental security. WATCH Research Report 1. AARANYAK. Guwahati, Assam, India. www.aaranayak.org (accessed October 25, 2009).

Das P.J., 2004. Rainfall Regime of Northeast India: A Hydro-meteorological Study with Special Emphasis on the Brahmaputra Basin. Unpublished Ph.D. Thesis Gauhati University (accessed October 25, 2009).

Deka R. L, Mohanta, C. Pathak, H., Nath, K. K. and Das, S (2012). Trends and fluctuations of rainfall regime in the Brahmaputra and Barak basins of Assam, India. TheorApplClimatol DOI 10.1007/ & 00704-012-0820-X. DES (2010) Directorate of Economics and Statistics, Department of Agriculture and Cooperation, Government of India.

FAO (2014). FAO Stat, http: //www.fao.org/faostat/en/data/QS as seen on 07-09-2017.

Goswami, B., Hussain, R., Rao, V. U. M and Saikia, U. S (2016). Impact of climate change on rice yield at Jorhat. *Journal of Agrometeorology* (2016). 18(2): 252-257.

IMD (2006, 2007) Southwest Monsoon end of season report for 2006 and 2007. IMD (India Meteorological Department), New Delhi (http:// www.imd.ernet.in).

IPCC (2007 a) climate change 2007: The physical Science Basis contribution of working Group I to the Fourth Assessment Report of the IPCC. In: Solomon S, Qm D, Manning M, eden Z, Marquies M, Averyt KB, Tignor M, Miller HL(eds). Combridge, UK: Cambridge University Press. 996.

IPCC. (2014). Climate Change 2014: Synthesis Report. Contribution of Working Groups I, II and III to the Fifth Assessment Report of the Intergovernmental Panel on Climate Change [Core Writing Team, R. K. Pachauri and L.A. Meyer (eds.)]. IPCC, Geneve, Switzerland, 151 pp.

Jain, S.K., Kumar, Vijay and Saharia, M. (2013). Analysis of rainfall and temperature trends in northeast India. *Int. J. Climatol.* 33:968-978.

Rupa Kumar K, Sahai AK, Kumar K K, Patwardhan SK, Mishra P K, Revadekar, JV, Kamala K, Pant GB (2006). High resolution climate change scenarios for India for the 21st century, *Curr.Sci* 90:334-345

Stockle, C.O., Nelson, R.L., Higgins, S., Brunner, J., Grove, G., Boydston, R., Whiting, M. and Kruger, C. (2010). Assessment of climate change impact on eastern Washington agriculture, Clim. CHANGE, 102;77-102.

Venkatacharya , K.V., Bandyopadhyay, K., Bhanumurthy, V., Rao, G.S., Sudhakar, S., Pal, D.K., Das, R.K., Sarma, U., Manikiam, B., Meena Rani, H.C. and Srivastava, S.K. 2001. Defining a space-based disaster management system for floods: A case study of Damage assessment due to 1998 Brahmaputra floods. *Current Science*, 80(3): 269-377.

WB. 2007. Development and growth in Northeast India: The Natural Resources, Water and Environment Nexus. World Bank report No. 36397-IN. the International Bank for Reconstruction and Development/The World Bank, DC, USA.

23

Rice Genotypes through Modeling under Projected Climate Change in Eastern India

***S. Pasupalak*[1] *and B. S. Rath*[2]**

[1]*Vice Chancellor, OUAT, Bhubaneswar*
[2]*Prof. & Head, Dept. of Agricultural Meteorology, CA, OUAT, Bhubaneswar, India*

Eastern India is frequently affected by weather hazards, such as drought, flood and heat wave. The region consists of Orissa, West Bengal, and Bihar. The coastal states of Orissa and West Bengal frequently face cyclones, while Bihar and eastern U.P. face cold wave. Due to unfavorable seasonal weather, crop productivity is poor. However, the recent change in climate induced by global warming has further threatened the agricultural productivity in the region. Rice is the principal crop of Eastern India and rice based cropping systems has long been practiced in the region. While rice-rice is the most dominant cropping system in Orissa and West Bengal, rice-wheat is in Bihar and eastern U.P. Besides, rice-groundnut occupies a considerable area in the region. It is now doubtful whether the resilience of these cropping systems would continue sustaining the productivity under the climate change. The major aim of the proposal was to identify suitable adaptation measures of crop production under projected climate change. Climate change is expected to alter precipitation pattern, increase temperature, change in humidity level and decrease sunshine. As a result it is like to influence right from land preparation to the harvest and post harvest process. Thus the project aim was to revisiting the existing package of practices of selected cropping systems by simulation modelling, deduce suitable adaptation measures and validate these by OTC experiments.

23.1 Trend analysis of climate change

- Increasing tendency in annual rainfall was observed in Eastern parts of Bihar, north-eastern Orissa and almost all regions of West Bengal except southeast n Orissa. Declining trend is noticed in western part of Odisha.

- Annual rainy days is showing increasing trend in central part of Orissa. Central and south eastern part of West Bengal and north-eastern trend in Bihar. It is decreasing in eastern Odisha.
- Frequency of occurrence of Tmax above 40°C is declining in west central part (Birbhum and Murshidabad district) of West Bengal during the month of April. In the remaining parts of three states, either no significant trend or no such event was noticed. During the month of May, only north eastern part (Supaul district) of Bihar showed decreasing trend.
- In the case of Tmin frequency (days with less than 10°C), declining trend is noticed in north eastern part of Bihar (Supaul, Araria and Kishanganj districts) and northern part (Darjeeling district) of West Bengal. No significant trend or no such event is seen in the remaining areas of eastern India.

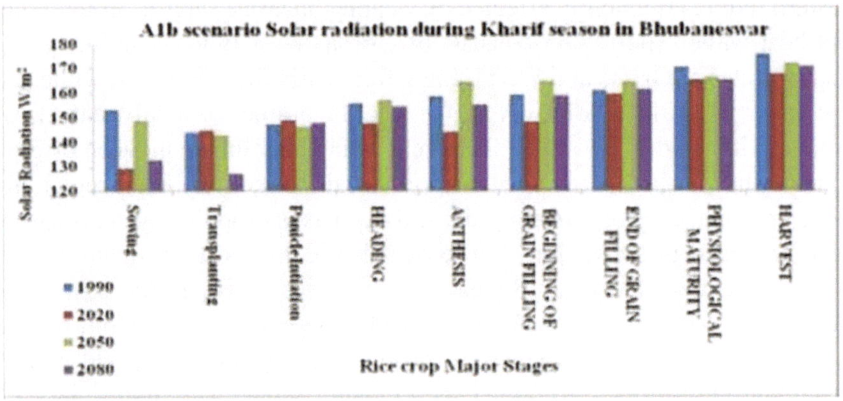

Fig. 23.1: A1b scenario: solar radiation variation during Kharif season

23.2 Climate change projection

From base line to middle and future scenarios, solar radiation is likely to increase in the coastal Orissa, West Bengal and Bihar during the SW monsoon season. In Base line most of the area received 130- 150 w/m2 , it may likely to be increased 150-160 w/m2 during the Middle and future scenarios.

23.2.1 Growth

Favourable growth under eCO2 was evidenced from taller plants, more tiller per hill (as high as 60%) and more root growth (as high as 80%). Increased biomass was due to increased net photosynthesis, despite increased photorespiration.

23.2.2 Development

The development in rice was accelerated by reducing the days to flowering by 3 to 5 in variety Swarna but the post flowering duration was prolonged by 3 to 5 days along with high LAD. The effect of eCO2 on rice yield was prominent (>15%) in summer season at all the places and less (<15%) in Kharif season.

23.2.3 Grain yield

- Grain yield of the cultivars Swarna and Swarna sub1 was decreased by 8 to 25% with increasing CO_2 level by 25% above the ambient CO_2 (390 ppm) during wet season in OTC due to decrease in filled grain.
- Increased grain sterility, which was in turn related to increased day temperature above 35°C coupled with high humidity (more than 75%). Chance of occurrence of such unfavourable

Adaptation

- eCO2 at 476 ppm required 25% more fertilizer for high yield in a rice-rice cropping system across varieties and places.
- The adverse effect of elevated CO2 on grain yield could be minimized through integrated management of chemical and organic sources of nutrients, and twice urea spray after PI.

References/Suggested Readings

Subash, N., S.S. Singh and Neha Priya. 2011. Variability of rainfall and effective onset and length of the monsoon season over a sub-humid climatic environment. *Atmospheric Research* 99:479-487.

N. Subash, S.S. Singh and Neha Priya. 2011. Extreme rainfall indices and its impact on rice productivity – a case study over sub-humid climatic environment. *Agricultural Water Management* 98, 1373– 1387.

24

Climate Change Impacts on Soil Microflora and Fauna

Suseelendra Desai[1], Arun K. Shanker[1], Chitra Shanker[2],Minakshi Grover[1], M. Srinivasa Rao[1], M. Prabhakar[1], Y.G. Prasad[1], J. Bhagyaraj[3]

[1]Division of Crop Sciences, ICAR-Central Research Institute for Dryland Agriculture, Santoshnagar, Hyderabad, India
[2]Division of Agricultural Entomology, ICAR-Directorate of Rice Research Rajendranagar, Hyderabad, India
[3]CNBRCD, 41, RBI Colony, Anand Nagar, Bengaluru, Karnataka, India

Climate change is a multifaceted alteration of climate which is perceptibly subtle and yet continuous, nevertheless it is extremely important due its consequences on whole biome that thrives under constant or relatively unchanged climates. The effects of climate change have reached such an extent that irreversible changes in the functioning of the planet are feared. Some of the main effects of climate change with specific reference to agriculture and food production especially during the last decade are: increased occurrence of storms and floods; increased incidence and severity of droughts and forest fires; steady spreading out of frost-free intervals and potential growing season; increased frequency of diseases and insect pest attacks; and vanishing habitats of plants and animals. Currently, the world is facing a challenge of producing sufficient food from dwindling natural resource base for the constantly increasing population. Strengthening of agricultural activities through improved productivity and effective resource management is one of the key options existing as competition for land and water are increasing from non-farm sectors. In other words, more production is required from fewer natural resources under future climate scenarios. Agriculture, human health and animal health are three important pillars of global livelihood security systems and they contribute to the developing nations' GDP significantly. While major crops and livestock world over contribute to the food security, the human resource has been a major driver of the growth in these sectors directly and indirectly. Across continents, climate influences these three pillars substantially. Climate –variability and -change are exerting additional pressure on developing countries of tropics and subtropics which are already at the threshold to cope with the increasing demand for food for increasing population pressure. Climatic disturbances in terms of increase/decrease in temperature, high intensity rainfalls, droughts, floods, hail storms and other extreme weather events in these areas will cause significant yield declines in major food grain crops. Several forecasts for the coming decades project changes in precipitation resulting in more frequent droughts and floods, increase in atmospheric CO_2 and temperature, widespread runoff leading to leaching of soil nutrients and reduction in freshwater availability. With the impending global climate change and the need to secure food supplies for the future, it becomes imperative to understand the consequences of climate change on the productivity of important agricultural crop species (Shanker *et al* 2014).

24.1 Climate–variability and Change

Climate -variability and –change have been the resultant of increasing global emissions. The increased temperature, precipitation and greenhouse gases (GHG) such as carbon dioxide, methane etc. are the major drivers of the climate change. During the past 100 years global average surface temperatures have increased by approximately 0.6°C (Houghton *et al.* 2001). Third IPCC report predicts that the global average surface temperature will increase further by 1.4 –5.8 °C by 2100 with the increased atmospheric carbon dioxide (CO_2) from 540-970 ppm over the same period. Concentrations of atmospheric CO_2 and temperature have both been rising very significantly since three decades and experienced across all sectors of agriculture. These changes affect the growth and development of the crop plants. Evaluation of Had CM3 model projections with measured mean monthly and annual rainfall data were studied. The projection data of rainfall for the year 2020 was taken from the output of Had CM3 model projections at various grid points and assumed that the projections follows the linear trend from 1990 to 2020 and calculated the values of year 2005 using the rain gauge stations located at or nearby grid point.The Spatial distribution of probability of occurrence of various categories of droughts viz., mild, moderate and severe at national level indicate that major part of the country can experience 3 to 4 drought years out of every ten years. Temperature trends shows that increasing trend in maximum temperature was noticed in 9 out of 12 stations in south zone followed by central (67%), east (60%) and west zone (57%). Minimum temperature showed increasing trend in 88 per cent of stations in central and east zone followed by north zone (80%), south zone (75%) and least in west zone (57%). Decreasing trend in minimum temperature was seen in 29 per cent of stations in west zone followed by 20 per cent and 8 per cent stations in north and south zone respectively. Mean annual temperature trends of 47 locations spread across the country indicated increasing trend in the central and southern parts and NE regions.

24.2 Crops, soil microflora and fauna, interactions

Modern commercial agricultural systems are constantly under threat of outbreak of pest and disease epidemics. In conventional agricultural systems, crops and their pests established a harmony and hence, equilibrium was established through a natural evolution process over decades. This equilibrium also included natural enemies which were parasitizing the pests and pathogens. This equilibrium helped in minimal crop losses due to diseases. However, increasing pressure on food security demanded increased agricultural output and there by disturbing the equilibrium. Hence, the global agriculture started experiencing the outbreak of epidemics of crop diseases leading to severe losses. While this development took place on one side, the environmental pollution has become additional variable to be reckoned with which had impacts on all biological systems. The elevated CO_2 levels coupled with increasing temperatures do affect host-pathogen interactions. Across global agricultural systems, the interactions of crops, soil microflora and fauna are very complex and involve multi-trophic interactions. Hence, it becomes often difficult to establish one-to-one relationship between two interacting organisms. The *in vitro* studies showing a clear relationship between a given crop and a pest often fail to show similar trends under field conditions where the interacting elements increase. These interactions are also significantly influenced by physical weather parameters which add a new dimension to the complexity.Soil biological activities are important in achieving crop productivity. Root exudates from growing plants provide an ideal source of readily available C which can enhance populations and activities of a diverse group of microbiota in the rhizosphere and interact with the plant with a particular emphasis on PGPRand also are involved in an extensive defense network by directly repelling, inhibiting, or killing pathogenic microorganisms (Baetz and Martinoia, 2014; Prashar *et al*., 2014) Ctinobacterial endophytes can colonise plants without disrupting the 'normal' endophytic populations, can produce antifungal antibiotics and plant growth hormones, and can also induce systemic disease resistance in plants (Gaiero *et al*., 2013;Conn *et al*. 2008).

Insectpests are known to inflict huge crop losses. While the lepidopteran pests are known to cause damage to plant parts by feeding on them,the sucking pests cause not only physical damage to plants but may also transmit virus diseases as vectors. Muralidharan and Pasalu (2006) reported that projected damage due to yellow stem borer *Scirpophaga incertulas* over rice ecosystems due to 1% dead heart as 6.4% yield loss. Pod borer in Pigeonpea caused yield loss of 14-100%, recently, the forecasted values of percent pod damage by this pest and productivity of Pigeonpea as been assessed 1137.40 kg ha^{-1} (Nath *et al*. 1977). Among oil seed crops, it is estimated that capitulum borer *Helicoverpa armigera* in Sunflower caused 30-60% damage (Dhaliwal and Arora, 1994) and in castor, *Achoea janata* caused the damage upto 30-50% and that of *Spodoptera litura* was 25-40% (Srinivasa Rao *et al.*2009). In vegetables (tomato and brinjal), fruit borers (*Helicoverpa armigera and Spodoptera litura)* put up a loss ranging from 15-46% (Singh, 1991; Kulkarni and Shekharappa, 2001). Babu and Azam (1989) reported that Grape mealy bugs caused the damage resulting in 90% yield loss. Plant diseases are caused by a variety of pathogens namely fungi, bacteria, viruses, viroids, nematodes, phytoplasma, and others. In nature, plants and their pathogens have co-evolved and established equilibrium,disease in populations results from an interaction among host, pathogen, and environment over a period of time. Depending on the type of plant pathogen, the disease development in populations varies. But pathogens still claim a considerable amount of global harvest. Like pests and pathogens, natural enemies of pests and pathogens have also co-evolved. Disturbing the fine-balance between plants, pests, pathogens and natural enemies would lead to outbreak of epidemics inflicting heavy crop losses. Microbial pesticides have been found to be effective in controlling pests and diseases; they essentially consist of bacteria, entomopathogenic fungi or viruses and sometimes also include the metabolic products of bacteria or fungi. Entomopathogenic nematodes are also often classed as microbial pesticides, even though they are multi-cellular. Some of the commercialized microbial pesticides include *Trichoderma, Pseudomonas, Bacillus,Bacillus thuringiensis, Nomuraea releyi,Verticillium leccani, Beauveria bassiana, Metarrhizium anisopliae* and *Paecilomyces fumosoroseus*. Trichoderma spp. are versatile opportunistic plant symbionts which can colonize the apoplast of plant roots and the ability ofTrichoderma spp. To colonise plant roots effectively is affected by various factors. In four soil types under different plant cover from Chambal ravines of Bhind in Madhya Pradesh, India, Trichoderma was frequently isolated and a significant positive correlation was noted between fungal population and carbon and nitrogen contents of soils, whereas clay contents of the soil and exchangeable calcium exhibited a significant negative relationship (Joshi and Chauhan, 1982). The ability of species of Trichoderma to colonize pine or maple litter differed with temperatures (Widden and Hsu, 1987). Thirteen species of *Trichoderma* were isolated from paddy fallows and populations in 90% of the samples ranged from 0.5 to 8.5 x 10^3 CFU/g of soil (Nagamani and Mew, 1987). There was a vide variation in distribution of microfungi in Turkey, at different altitudes of which *Trichoderma* formed 9.6% of the population (Karaoglu and Ulker, 2000). Similar variation in distribution pattern of Trichoderma was reported across North (Hebei) and South-West (Yunnan) areas of China. In addition, seasonal and habitat dependent variations in these endophytes has been observed in tea. *Pseudomonas* and *Bacillus* can tolerate a variety of physical conditions and this could be isolated from soil, marshes, coastal marine habitats, and plant and animal tissue. Boehm *et al.* (1993) found that the organic matter decomposition level of a mix significantly influenced the composition of bacterial species in the rhizosphere. *Pseudomonas* spp. capable of suppressing pythium damping-off predominated in the suppressive mix while in conducive- mix *Bacillus* spp. predominated. Although effective bacterial biocontrol agents were isolated from both the suppressive mixes and the conducive-mix, the majority were isolated from the less decomposed suppressive mixes. Arbuscular mycorrhizae (AM) fungi can not only protect plants from biotic stresses but also establish in heavy metal contaminated soils and act as a biological barrier in transfer of toxic metals to shoots (Gaur and Adholeya, 2004; Shanker *et al* 2005;Azcon *et al* 2013). Similarly, several studies have reported enhanced salinity tolerance various crops inoculated with AM fungi (Rabie and Almadini 2005; Estrada 2013;. Calvo-Polanco *et al* 2013; Aroca *et al* 2013) Eighty-nine species of AM fungi from 6 genera were detected in the iron stressed soils of Bailadila iron ore project in Baster region of Chattisgarh and the AM dependence of the host species was related to the establishment of the vegetation under metallo-toxic conditions.

24.3 Impacts of climate -variability and -change

Climate -variability and –change can influence the dynamics of crop production systems involving host-pest-pathogen-weather interactions. The predicted changes in climate have profound influence on the outcome of these complex interactions. Systems approach to understand these impacts needs understanding of individual responses and then see them in a systems perspective. The impacts at individual level sometimes may be significant but at systems level, it may be neutralized and vice-versa could also hold good. The impacts of climate change on soil microflora and fauna are described below.

24.3.1 Climate change and termites

Greenhouse gases (GHGs) greatly affect the temperature of the earth and causes the increase in atmospheric air temperature and it is predicted that temperature rises by 2-5 °C by 2050. methane, CO_2, nitrous oxide and other gases come under GHG and the source of emission of these gases is different between agriculture and industrial sector. Carbon dioxide emissions from agriculture are small; but other important GHGs are emitted from agriculture. Termites are a complex assemblage of species showing considerable variation in terms of social behavior and nutritional ecology. The digestive tract of termites normally paves a suitable niche for a variety of microorganisms to inhabit and multiply (Gomathi *et al* 2011) Agriculture accounts for about 60 percent of all nitrous oxide, mainly from fertilizer use and about 50 percent of methane mainly from natural and cultivated wetlands and enteric fermentation. Methane (CH_4)

has higher global warming potential than that of CO_2.Atmospheric CH_4 mainly comes from 70-80 % of biological sources. The contribution of termite in GHG emissions is also note worthy apart from rice cultivation, cattle production, mining and using of fossil fuels. The estimates of global annual CH_4 emission by termites vary between 5-40 % based on ecosystem. These termites generate 0.2-2 % global terrestrial CO_2 emissions. It is believed that termites are natural source of GHG and their emissions depend on temperature and feeding substrate. The soil feeding and fungus growing termites emit higher CH_4 than wood feeding termites whereas, wood feeding termite produce more CO_2. It is well known that termite gut would play significant role in activity and diversity of population of microbes for formation of GHGs. It is expected that the role of termites would be vital under future climate change scenarios as their population is driven by both temperature and soil moisture.Subterranean termites are ground dwelling social insects living in colonies. These termites have the ability to adjust the depth of their colony in soil depending on temperature and moisture requirements. The ground serves as a protection against extreme temperatures and provides a moisture reservoir. Understanding the effects of climate change on the subterranean ecosystem is indeed important because the terrestrial ecosystem is directly connected to the subterranean ecosystem. Termite identification is crucial to understanding termite distribution and their relationship to climate change. Termites are amongst the most difficult insects to study because of their cryptic behaviour and natural nesting habitat. It is well known that the population of termites *Microtermus*sp and *Odentotermus* sp increases during dry periods with low or no moisture conditions. The moist soil conditions discourage the multiplication of termites and thus adoption of suitable mulches with either sorghum straw or glyricidia retains the moisture and decreases the infestation of termites(Nyamadzawo *et al.*, 2012; Jamali *et al.*, 2013).

Some termite species are well known pests of agricultural crops, forest trees, wood products and timber-in-service causing considerable damage. In general, damage by termites is greater during dry periods or droughts than periods of regular rainfall, in lowland rather than highland areas, and in plants under stress (Wightman, 1988, 1991). With "global climate change expected" not only termite distribution will be affected and pest problems response change, but the effectiveness of termite management strategies will also change. As the majority of termites are soil dwellers; soil degradation and drought are the major factors affecting distribution and foraging behavior of termite in tropical and subtropical ecology. The potential for increases in economically devastating termite outbreaks in response to climate change in Africa is high and warrants intensive investigation into the effects of climate on crop pests (Ahmed *et al.,* 2011;Gomati *et al.*, 2011).

24.3.2 Climate change and white grubs

White grubs, *Holotrichia serrata* and *Lachnosterna consanguinea* are serious pests in Semi-arid tropics and the larvae feed on roots of the seedlings, resulting in withering of plants within a week. Climatic and soil conditions could be main factors responsible for high densities of white grubs and their feeding damages to the grass roots. Drought can intensify the effects of grub feeding to the crop plants by accelerating its withering (Hann *et al* 2008). Hot and arid climatic conditions as well as permeable soils are decisive for the emergence of grub damages. On-farm investigations performed in 2007 strengthened the hypothesis that drought and elevated soil temperatures might be the decisive factors for a strong development of grub populations and subsequent feeding damages. Additionally, drought can increase the effects of grub damage by delaying the regeneration of the damaged sward Hann *et al* 2008).The higher soil temperature enhances the development of high grub densities by reducing the egg and larval periods. Under climate change condition it is predicted that the occurrence of extreme weather events would be frequent and their influence on white grub is to be further investigated.

24.3.3 Climate change and Nematodes

Nematodes are the most numerous multicellular and unsegmented worms with elongated rounded body pointed at both ends. These are mostly free-living and some are parasitic and a handful of soil will contain thousands of the microscopic worms, many of them parasites of insects, plants or animals. As the competitive ability of different species changes with climatic conditions, raising temperature will alter the predominance of species in the nematode community at a given time of the year or soil depth, and may allow certain species to extend their occurrence northwards (Boag *et al.*, 1991). Warming is also affecting the soil fauna feeding on nematodes, like microarthropods and tardigrades (Briones *et al.*, 1997) and may desynchronise predator-prey interactions in the soil. Available information on effect of global warming on plant pathogenic nematodes though limited, indicate that nematodes show a neutral or positive response to CO_2 enrichment effects with some species showing the potential to build up rapidly and interfere with plant's response to global warming. Studies have also demonstrated that the geographical distribution range of plant pathogenic nematodes may expand with global warming spreading nematode problems to newer areas. Besides plant parasites, other trophic groups (microbial feeders, predators and insect parasites) of soil nematodes also shown to infl uence the plant productivity indirectly by regulating the key ecosystem processes including decomposition, nutrient mineralization, biological pest suppression and energy transfer in food webs (Somasekhar and Prasad 2012).Hence, climate induced stresses are influencing competition and trophic relationships in many ways and therefore are expected to alter soil biodiversity. Of the major groups of soil invertebrates, nematodes are particularly convenient for investigating impacts of CO_2 on soil food webs. Nematodes may be expected to respond both directly to changes in plant carbon allocation to the below-ground environment, and indirectly to changes in microbial communities that respond directly to plant carbon inputs. Elevated CO_2 clearly had a variety of effects on soil nematode communities. Impacts were also more pervasive and tended to be greater in soils of forest ecosystem (Neher *et al* 2004).

24.3.4 Climate change and micro-arthropods

Among soil fauna, the Collembola, oribatid mites (Acari: Oribatida) and enchytraeids (Oligochaeta: Enchytraeidae) are important in the ecosystem and come under category of micro-arthropods. These have significant role in decomposition, nutrient mobilisation, soil mixing and aggregate formation (Laakso and Setala 1999). In the study of Niklas 2003 examined possible effects of global change on soil fauna (Oribatida, Mesostigmata, Collembola, predatory macroarthropods and Enchytraeidae) and revealed that climate change resulting in frequent summer drought and irrigation caused the shifts in the community composition of Oribatida and Collembola, but drought affected soil fauna more than irrigation. Increased soil temperature during 5 years resulted in positive responses in several Oribatida species and negative in some Collembola species. Hovewer, temperature may directly affect survival and reproductive ability of collembola (van Straalen, 1994), indirect effects through changes in soil moisture may also be of higher importance (Sinclair and Stevens, 2006).The information on impact of climate change on terrestrial insects is well documented whereas, on subterranean insects is found to be scanty. It is understood that increase in atmospheric air temperature is also in turn influencing soil temperature and moisture and thus influencing the soil inhabiting insects substantially.

24.3.5 Elevated CO_2 and host-pathogen interactions

Under elevated CO_2 levels, the morpho-physiology of the crop plants is significantly influenced. The available data clearly suggests that atmospheric CO_2 enrichment asserts its greatest positive influence on infected as compared to healthy plants. This influence in turn will modulate the balance of co-evolution between the host and the pathogen as well as pathogen and biocontrol agent. Elevated carbon dioxide [ECO_2] and associated climate change have the potential to accelerate plant pathogen evolution, which may, in turn, affect virulence. Plant–pathogen interactions under increasing CO_2 concentrations have the potential to disrupt both agricultural and natural systems severely, yet the lack of experimental data and the subsequent ability to predict future outcomes constitutes a fundamental knowledge gap. Furthermore, nothing is known about the mechanistic bases of increasing pathogen aggressiveness. Under ECO_2 conditions, mobilization of resources into host resistance through various mechanisms such as reduced stomatal density and conductance, (Hibberd *et al.*, 1996a, 1996b); greater accumulation of carbohydrates in leaves; more waxes, extra layers of epidermal cells and increased fibre content (Owensby, 1994); production of papillae and accumulation of silicon at penetration sites (Hibberd *et al.*, 1996a); greater number of mesophyll cells (Bowes, 1993); and increased biosynthesis of phenolics (Hartley *et al.*, 2000), increased tannin content (Parsons *et al.* 2003) have been reported. Malmstrom and Field (1997) reported that CO_2 enrichment in oats may reduce losses of infected plants to drought and may enable yellow dwarf diseased plants to compete better with healthy neighbors. On the contrary, in tomato, the yields were at par (Jwa and Walling, 2001). Similarly, Tiedemann and Firsching (2000) reported yield enhancement in spring wheat infected with rust incubated under ECO_2 and ozone conditions. Chakraborty and Datta (2003) reported loss of aggressiveness of *Colletotrichum gloeosporioides* on *Stylosanthes scabra* over 25 infection cycles under ECO_2 conditions. On the contrary, pathogen fecundityincreased due to altered canopy environment. McElrone *et al.* (2005) found that exponential growth rates of *Phyllosticta minima* were 17% greater under ECO_2. Simultaneously, in the host *Acer rubrum*, the infection process was hampered due to reduction in stomatal conductance by 21-36% and thereby leading to smaller openings for infecting germ tubes and altered leaf chemistry. Reduced incidence of Potato virus Y on tobacco (Matros *et al.*, 2006), enhanced glyocellin accumulation (phytoalexins) after elicitation with ß-glucan in soybeans against stem canker (Braga *et al.*, 2006) and reduced leafspot in stiff goldenrod due to reduced leaf nitrogen content that imparted resistance (Strengbom and Reich, 2006) have been reported. Lake and Wade (2009) have shown that *Erysiphe cichoracearum* aggressiveness increased under elevated CO_2, together with changes in the leaf epidermal characteristics of the model plant *Arabidopsis thaliana*. Stomatal density, guard cell length, and trichome numbers on leaves developing post-infection increased under ECO_2 in direct contrast to non-infected responses. As many plant pathogens utilize epidermal features for successful infection, these responses provide a positive feedback mechanism facilitating an enhanced susceptibility of newly developed leaves to further pathogen attack. Furthermore, screening of resistant and susceptible ecotypes suggests inherent differences in epidermal responses to elevated CO_2. Gamper *et al.* (2004) noted that colonization levels of arbuscular mycorrhizae tended to be high on *Lolium perenne* and *Trifolium repens* grown under ECO_2 which may help in increased protection against stresses.

With the climate -variability and –change, two types of scenarios are possible that could have bearing on crop diseases. Under low moisture and high temperatures, organisms such as *Macrophomina phaseolina* that cause dry root rots may cause severe yield losses. Under low temperature and high humidity conditions, *Pythium* and *Phytophthora* could of serious concern for many field crops. On pigeonpea, it was observed that during high rainfall years, intensity of Phytophthora blight was also very severe. In sugarbeet, humid conditions favoured *Cercospora* and *Ramularia* significantly. Low temperature and high humidity may also promote wilt pathogens. Excessive irrigation, which can be compared to that of frequent and high intensity rainfall effects, led to the increased intensity of sugar beet leaf and crown wart. It was also presumed that Rhizomania virus, a major concern for sugar beet crop in the Europe, has spread due to irrigation, which can be equated with excess moisture conditions, because *Polymyxa betae* (a soil borne fungus) which is the vector for the beet necrotic yellow vein mosaic virus, prefers wet conditions. On the other hand, *Pleospora betae* causing blackleg was highest under drought conditions. Lane and Jarvis (2007) analyzed that with rising temperatures and change in the rainfall regime the global suitability for crops may not decrease *per se*, but could change geographically. Many instances have been recorded of species range shifts towards the poles or upward in altitude, and progressively earlier seasonal migrations and breeding (Parmesan, 2006). In relation to temperature, for each 1°C increase there will be a potential for range expansions of ~ 200 km pole ward and similarly, spread of the species ~170 m higher in elevation can be expected. Kiritani (2007) reported on the polar extension of several plant pests in Japan over the period 1965 to 2000. Lobell *et al.* (2008) further found a number of hotspots where key crops that contributed to the food and nutritional security of the local populations were predicted to be significantly negatively impacted. These include southern Africa where maize, a major staple, could be eliminated by 2050, and South Asia where productivity in groundnut, millet and rapeseed-mustard could be affected. Differential response of wheat leaf rust resistance genes was noticed at different temperatures. Minor or unknown pathogens may also become more aggressive such as alternaria blight of pigeonpea. Meaningful scenarios could be arrived at through a holistic analysis of response of host-pathogen-biocontrol system to the changing climate scenarios. Increasing temperature coupled with increasing moisture will favour the proliferation of *Ralstonia*, the causal agent of bacterial wilt across different crop species. The pathogen which is already playing havoc in S.E. Asia could be a threat to groundnut in India, where it is hitherto unknown.

Alternaria spp. and powdery mildew could be major concerns for various crops. Fungal species causing powdery mildews prefer dry and cool weather. In recent years, extended cloudy weather often prevails during crop season and these conditions are congenial for powdery mildew. It has been observed that *Alternaria*, hither too known as a minor pathogen on pigeonpea, could be an emerging disease in the years to come. Similarly, *Alternaria alternata* that causes leaf spot in many crops like groundnut, potato, cotton, castor becomes severe during long and hot dry spells. The fungus has been reported to sporulate three times more on plants cultivated under elevated CO_2 conditions. This increased spore density is not only harmful to the crops but also directly to human beings as an allergen. In recent years, increased occurrence of alternaria leafspot has been observed in groundnut. Rice plants grown in an elevated CO_2 concentration were more susceptible to leaf blast than those in ambient CO_2 as indicated by the increased number of leaf blast lesions. Higher number of tillers observed under elevated CO_2 concentrations might increase the chance for sclerotia of *Rhizoctonia solani* to adhere to the leaf sheath at the water surface. Consequently, the potential risks for infection of leaf blast and epidemics of sheath blight would increase in rice grown under elevated CO_2 concentration.

Hannukkala *et al* (2007) reported increased and early occurrence of epidemics of late blight of potato in Finland due to climate change and lack of crop rotation. One of the direct consequences of climate change in the pathogen-host relationship is the genetic resistance of plants to diseases. The ECO_2 conditions are known to alter host physiology and biochemistry which will have an impact on the host-pathogen interactions. It is possible that the genes imparting resistance in host may respond differentially. Though ECO_2 effects on host may give beneficial effects, the impacts on pathogen cycles in terms of increased spore production capacity and faster multiplication rates may nullify these positive effects. Indications have been that though a delay in initial infection and reduction in host penetration, post-establishment fast growth of pathogen in the host may result in increased disease intensity. Under drought stress, the disease symptoms may be reduced but at the same time the resistance of the host can also be modified thus leading to higher disease incidence. Drought impacted disease resistant plant types showed loss of resistance. Some pathogens could also enhance their ability to exhibit variability with which their fitness to the changed environment is enabled. Such kind of variability has also been suggested as an early indicator of environmental change because of their short generation times.

Limited studies have been conducted on effects of climate change on natural enemies. Warwick (2001) demonstrated that the rainfall regime and time of application of *Acremonium vittelinum* and *A. persicinum* for the control of the tar spot of coconut caused by *Catacauma torrendiella* and *C. palmicola*. Studies in this area will be important for understanding the interactions between host-pathogen-biocontrol interactions. The outcomes of such studies will be helpful to draw meaningful conclusions in a holistic systems approach. Such studies also will help in appropriate deployment of biocontrol agents either by encouraging natural populations to develop or through introduction of better strains. Effect of climate change on plant disease biological control is a complex issue and is currently based on indirect observations. Nevertheless, the vulnerability of biocontrol agents will surely be higher with climate change, since this is one of the problems when applying antagonists (Garrett *et al.*, 2006). Ghini & Bettiol (2008) opined that in general climate change may benefit the natural as well as introduced biological control agents as more emphasis will be laid on development of eco-friendly disease management strategies as compared to development of new fungicide molecules to reduce environmental pollution. The energy requirements for production of a unit of biocontrol agents is far less than that is required for production of one unit formulated chemical molecule. Rezácová *et al.* (2005) observed that under FACE conditions, *Chlonostachys rosea*, a biological control agent of *Botrytis*, and *Metarrhizium anisopliae*, an important entomopathogens, were strongly associated with the cover crop in a high CO_2concentration environment. The abundance of these fungi has been attributed to increased soil suppressiveness to phytopathogenic fungi.

24.3.6 Adaptation strategies

The adaptation strategies have to be multi-pronged as the extreme weather events can exert pressure on established plant-pest relationships of a given agro-ecological region. The focus must be laid on understanding host-pest dynamics under different climatic scenarios, deviations in outbreaks of pests, shifts in population dynamics of microflora & fauna, and invasive alien species. The adaptation strategies could be from different angles such as managing nutrient dynamics as well as microbe-mediated management of biotic- and abiotic-stresses. Plant growth is a net result of interaction of roots and shoots with the environment. Both roots and shoots have been shown extensively interacting at rhizosphere, rhizoplane and phyllosphere with microbial communities. Promoting rhizomicroorganisms (PGPR) such as *Rhizobium, Frankia, Pseudomonas, Bacillus, Trichoderma, Streptomyces* and mycorrhizal fungi have been shown to aggressively colonize roots and help in plant growth and development through nutrient uptake; induction of resistance mechanisms against invading pests; biological control of pests that attack the plants and plant growth promotion.

24.3.7 Microbe-mediated nutrient management

The importance of bacteria as key drivers of the nitrogen cycle is exemplified in both bulk soil and the rhizosphere (Delwiche, 1970; Rosswall, 1983). Many N_2 fixing bacteria have been found in rhizospheric and endophytic association but the transfer of biologically fixed nitrogen has been demonstrated only in few systems. Sugarcane has been a model in which biological nitrogen fixation (BNF) has been observed (Boddey, 1995; Yoneyama *et al.* 1997). Free nitrogen-fixing bacteria belong to a wide array of taxa; among the most relevant bacterial genera are *Azospirillum*, *Azotobacter*, *Burkholderia*, *Herbaspirillum* and *Bacillus* (Vessey, 2003). Sajjad Mirza *et al.* (2006) demonstrated nitrogen fixing ability of strains of *Pseudomonas.* Most cultivated tropical soils in India are reported to have relatively large populations (>100 cfu. g^{-1} dry soil) of rhizobia capable of nodulating the legumes and fixing nitrogen (Nambiar *et al.* 1988; Venkateswarlu, 1992; Khurana and Dudeja, 1997).Phosphorus is one of the major nutrients that limit plant growth. Most of the soils around the world are P deficient (Batjes, 1997). The use of rock phosphate as a phosphate fertilizer and its solubilization by microbes (Kang *et al*., 2002), through the production of organic acids (Maliha *et al*., 2004), has become a valid alternative to chemical fertilizers. Several studies have shown that phosphate soulubilizing microorganisms (PSMs) solubilize the fixed P in the soil resulting in higher crop yields (Gull *et al*., 2004). The combined application of *Rhizobium* and PSM (Perveen *et al*., 2002) or PSM and arbuscular mycorrhizal (AM) fungi (Zaidi *et al*., 2003) showed enhanced plant growth as compared to their individual inoculation in 'P' deficient soils. Phosphate solubilizing bacteria are ubiquitous (Gyaneshwar *et al*., 2002) and *Bacillus*, *Enterobacter, Erwinia* and *Pseudomonas* spp. are among the most potent strains.PSB are common in rhizospheres of crop plants and few examples of beneficial association of phosphate solubilizing PGPR and plants include *Azotobacter chroococcum* and wheat (Kumar and Narula, 1999), *Bacillus circulans* and wheat (Singh and Kapoor, 1998), *Enterobacter agglomerans* and tomato (Kim *et al.* 1998), *Pseudomonas chlororaphis* or *P. putida* and soybean (Cattelan *et al.* 1999), *Rhizobium* sp. and *Bradyrhizobium japonicum* and radish (Antoun *et al.* 1998). Zinc solubilizing potential of few bacterial genera has also been studied. Di Simine *et al.* reported (1998) a zinc solubilizing strain of *Pseudomonasfluorescens* from forest soil. Hutchins *et al.* (1986) reported that *Thiobacillus thioxidans*, *T. ferroxidans* and facultative thermophilic iron oxidizers solubilized zinc from sulphide ore (sphalerite). Saravanan *et al.* (2007) reported a strain of *Gluconoacteobacterdiazotrophicus* with zinc solubilization and its anti nematode activity against *Meloidogyneincognita.* Desai *et al* (2012) reported that strains of *Azospirillum* could support plant systems with P, Zn and N. Identification of stress tolerant strains of nutrient mobilizers/solubilizers will be of a great value under climate change scenarios as they can withstand low soil moisture conditions or high intense rainfall or floods or frost or high temperatures.Specific examples of tolerant strains surviving at higher stress levels include *Azospirillum brasilense* for salt (Tripathi *et al.*1998) *Azospirillum* *lipoferum* for temperature (Anand *et al.* 1999), *Azospirillum* and *Bradyrhizobium* for salt (Kulkarni and Nautiyal, 1999; Rao, 2001), *Bradyrhizobium* for moisture and temperature (Kulkarni and Nautiyal, 1999; Venkateswarlu and Katyal, 2001) and PSB for temperature tolerance (Rao, 2001). *Rhizobium, Azotobacter* and *Pseudomonas* produce copious exopolysaccharides (EPS) under natural environment but reports on the usefulness of EPS in desiccation tolerance are conflicting (Deaker *et al.*2004).Co-inoculation of *Rhizobium* and *Pseudomonas* can combat salt stress in maize by increasing the total uptake of K, P and Ca without affecting Na accumulation (Asghari andMussarat, 2009).

24.3.8 Microbe mediated biotic stress management

Biological control of soil borne- and foliar-pathogens through microbes has been commercially exploited. The biocontrol agents (BCAs) are known to produce chitinases, glucanases, cellulases, and proteases that are useful in biological control. Biological control in India is now gaining significance as there is no other economically viable option for controlling soil-borne phytopathogens. Biological control of rice diseases by *Pseudomonas* spp. under saline conditions has been demonstrated by Sunita Rangarajan *et al* (2003).

24.3.9 Invasive alien species management

The movement of plant pests, animal diseases and invasive alien aquatic organisms across physical and political boundaries threatens food security and creates a global public concern across all countries and all regions. Countries allocate large resources to limit the spread and control of transboundary pests and diseases. Examples of plant pests whose distribution is shifting in the United Kingdom and other parts of Europe, is most likely due to climatic factors. Farmers who closely monitor the occurrence of pests in their fields and keep records of the severity, frequency, and cost of managing pests over time will be in a better position to make decisions about whether it remains economical to continue to grow a particular crop or use a certain pest management technique. A regular survey and surveillance will help in understanding the changing pest scenario and thereby suggest suitable strategies of risk management. Awareness programs must be conducted regularly to all stakeholders about the emerging regional microflora and fauna scenarios especially those inciting damage to crops. Developing diagnostic tools for proper identification will help in reducing crop losses by recommending appropriate management strategies.

24.3.10 Remote sensing for detection of pest incidence

Early detection of incidence of pests helps in providing lead time to the farmers to pool their resources for management of risk. Remote sensing (RS) of biotic stress is based on the assumption that stresses interfere with photosynthesis and morphology organ level affecting the absorption of light energy and thereby altering reflectance spectrum (Riley, 1989). Vegetative spectral reflectance can help to gain a better understanding of the physiological, chemical and physical processes in plants due to pest and disease infestation. Depending on the spectral resolution and contiguous nature of spectral recording, the sensors can be either multispectral or hyperspectral.In multispectral RS, average spectral information over broadband widths is used to understand signatures resulting in loss of critical information available in specific narrow bands. Hyperspectral reflectance provides unique crop signatures with precision information covering these narrow band widths. The large amount of information in hyperspectral data provides the potential for more accurate and detailed information extraction than possible with any other type of remotely sensed data. Crop health conditions cause statistically significant variations of plant spectral signatures that allow not only discriminating between stressed and healthy vegetation but also to quantitatively describe the stress impact in terms of crop physiology and agronomy. Remote sensing and geospatial technology has been widely used and made greater strides in progress over the years. Some of the recent examples are detecting oil palms infected by basal stem rot disease and for mapping the disease in Indonesia (Santoso *et al.*, 2011), and monitoring the progress of wheat streak mosaic disease in Texas, USA using Landsat 5 TM imagery (Mirik *et al.*, 2013). In India also a few studies on ground based remote sensing for detection of blackgram yellow mosaic disease (Prabhakar *et al.*, 2013) and tracking of *Sclerotinia* rot disease in mustard (Bhattacharya *et al* 2013) using satellite data have been demonstrated. The recent reviews on this topic (Sindhuja *et al.*, 2010 and Prabhakar *et al.*, 2012) documented the possibility of using spectral reflectance for detection of several plant diseases and potential use of remote sensing of biotic stress in crop plants and its applications for pest management. Assessment of crop stress is crucial for implementing control strategies and realising the potential yield in crops. Ongoing pest surveillance programs through field scouts is often expensive, time consuming, laborious and many at times not very accurate. As RS gives a synoptic view of the area, non-destructive and non- invasive, this technology could be effective and provide timely information on spatial variability of pest damage for a large area. Thus can guide scouting efforts and crop protection advisory in a precise and effective manner. With the advancements in the communication, aviation and space technology, this could be a reality in the near future.Under climate change scenario, intra- and inter-seasonal extreme weather events alter crop-pest relationships. Often, extreme weather events occur suddenly and so by the time the interventions are planned, the crop is damaged. By using RS in conjunction with geospatial technology, such situations could be simulated for anticipated losses and thereby prepare contingency documents with suitable interventions. These techniques also could be used for tracking the invasive alien species that can affect crop production systems of a given agro-ecological region.

24.3.11 Pest simulation models and decision support systems

Simulation models based on mathematical descriptions of biological data as influenced by the environment are more easily applied across locations and environments. Computer programs or software to run these models facilitate the practical application of these models in understanding population dynamics and dissemination of pest forecasts for timely pest management decisions (Coulson and Saunders, 1987). Simulation approaches offer flexibility for testing, refinement and sensitivity analysis as well as field validation of developed models over a wide range of environmental conditions. Thorough descriptions of cropping systems being managed or studied are needed to explain the interactions among pests, plants and the environment (Colbach, 2010). Systems models or other prediction schemes can be used with appropriate biological, environmental, economic or other inputs to analyse the most effective management actions, based on acceptable control, sustainability and assessment of economic or other risks (Strand, 2000). In an effort to improve *Helicoverpa* management in Australia, a comprehensive population dynamics model (HEAPS: *Helicoverpa Armigera* and *Punctigera* Simulation) has been developed, which incorporates the spatial structure of the habitat and pest population and explicitly simulates the adult movement within a regional cropping system (Fitt *et al.*, 1995). This model incorporates modules based on adult movement, oviposition, development, survival and host phenology, and estimates the population in each unit of a grid (Dillon and Fitt, 1990). The EntomoLOGIC decision tool was derived from the SIRATAC decision support system deployed by the Australian cotton industry from 1976 to 1993 to reduce the risk associated with pest management using chemical pesticides. This was developed by the Commonwealth Scientific and Industrial Research Organisation (CSIRO) in collaboration with the University of Western Sydney, Australia (Hearn and Bange, 2002). Advances in hand-held computing have resulted in expanding the development of CottonLOGIC for use with Palm OS handhelds for widespread adoption by cotton growers in Australia (Bange *et al.*, 2004). A suite of predictive computer models called MORPH has been developed at the Horticulture Research International, UK (Phelps *et al.*, 1999), for use in fruit and vegetable crops. Using a multi-generation phenology model, ECAMON, Trnka *et al.* (2007) could explain 70% of the variation in the timing of key developmental stages based on daily weather data. ECAMON simulations correctly predicted the presence/absence of the European corn borer over a study region in Czech Republic during the 1961–

1990 reference period. It also helped to explain the sudden increase in maize infestation during the unusually warm periods of 1991–2000 and it also estimates that thispotential niche will expand within the next 20–30 years. RICEPEST, a model simulating yield loss due to several rice pests under a range of specific production situations in tropical Asia was developed by the International Rice Research Institute (IRRI) in the Philippines. Validation of the model under field experiments yielded promising results (Willocquet *et al.*, 2002). Web-based models and decision support systems are becoming popular and in future may become an absolute requirement for local, regional/area-wide and international implementation of IPM systems (Waheed *et al.*, 2003). In the USA and the Netherlands, commercial firms are applying mesoscale modelling techniques to forecast insect development and produce gridded products for regional and on-farm planning and pest management (Strand, 2000). A decision support system has been developed for forecasting black bean aphid (*Aphis fabae*) outbreaks in fields of springsown beans. The system takes into account the regional forecast and additionally information provided by the user on individual characteristics of the field and crop such as field shape, size, plant density and sowing date, which are used to downscale the area forecast to the specific field. The system also contains a module for the aphicides that are cleared for use on spring beans and calculates the economics of application (Knight and Cammell, 1994) SOPRA is applied as a decision support system for eight major insect pests of fruit orchards on a local and regional scale in Switzerland and southern Germany and has a wide range of possible applications in the alpine valleys and north of the Alps (Samietz *et al.*, 2008). Applying time varying distributed delay approaches, phenology models were developed driven by solar radiation, air temperature and soil temperature on an hourly basis. On the basis of local weather data, the age structure of the pest populations is simulated and crucial events for management activities are predicted by the SOPRA system. Phenology is directly linked to a detailed decision support system and to extended information about the pest insects, as well as to the registered plant protection products. Through a web interface, the simulation results are made available to consultants and growers (www.sopra.info). SIMLEP is a regional forecasting model used in practice for Colorado potato beetle (*Leptinotarsa decemlineata*) in Germany and Austria on a large scale and in the western part of Poland. The SIMLEP decision support system contributed significantly to the improvement of farmers' control measures for *L. decemlineata* (Jorg *et al.*, 2007) and later its use expanded to Slovenia. (Kos *et al.*, 2009). In the mid-1990s, CIPRA (Computer Centre for Agricultural Pest Forecasting) software was conceptualized, developed and implemented to access, in real-time, weather data from a network of automated stations. It allows the user to visualize forecasts for 13 insects, two diseases, and two storage disorders in addition to the apple crop phenology. These bioclimatic models, which have been developed, implemented and improved over the last 13 years, vary from a simple degree-days approach based on air temperature to more detailed epidemiological models based on air temperature, relative humidity and duration of leaf wetness. Many field specialists are using these model forecasts along with field pest scouting to provide valuable additional information for decision making in pest management and in apple storage strategies (Bourgeois *et al.*, 2008). Climate change impacts crop-pest interactions and hence well validated forewarning systems and DSSs will help to assess changing crop-pest dynamics in response to climate change and also will be useful for estimating the risks due to alien invasive pests. These models will also help to change crop production systems suitable to altered climatic scenarios thereby preventing outbreaks of pests.

24.3.12 Conservation Biological Control (CBC)

Conservation biological control has a key role under the changing climate scenario as an effective adaptation strategy which is has the qualities of being ecological sound, environmentally safe and economically viable. This approach has got considerable interest and also had fair amount of success in the last few years. This is commendable considering that the area is still in its infancy. The reasons for this is that (i) no new and exotic organisms are introduced, thereby addressing the major environmental concerns associated with classical biocontrol; and (ii) it would lay the foundation for sustainable, eco-friendly and low cost pest management which does not require the tedious and labour intensive mass rearing of natural enemies as in augmentative biocontrol. (Shanker *et al* 2012). Conservation biological control is defined as "the practice of enhancing natural enemy efficacy through modification of the environment or of existing pesticide practices" (Eilenberg *et al.* 2001). Conservation biological control is defined as "the practice of enhancing natural enemy efficacy through modification of the environment or of existing pesticide practices" (Eilenberg *et al.* 2001). In the last few years there has been a tremendous increase in this much neglected field of biological control. It must be however noted that the effects of conservation biological control are not as profound and evident as classical biological control and thus this forms a hurdle to gain true believers, although the advantages are manifold and one the important one being that it is low cost and requires smaller changes at the field level which even an individual grower can adopt. The best example for this is that in multiple natural enemy assemblages the negative impacts are mainly due to competition between species for resource directly or indirectly (interference). Dilution effects of biodiversity are rarely examined for predator communities. Intraguild predation is another major factor for negative effects of biodiversity. Predators often feed heavily on one another (Shanker *et al* 2012). Under the changing climate scenario, integrated pest management which takes into consideration the ecological principles for harnessing the *in-situ* biodiversity for pest suppression in the lightof the various factors that affect the presence of pests and its survival will be the future of sustainable management practices.

24.3.13 Microbe-mediated abiotic stress management

Many studies have been published on beneficial effects of bacterial inoculation on plant physiology and growth under abiotic stress conditions. Indeed, under stress conditions, including drought and salinity, the plant hormone ethylene endogenously regulates plant homeostasis and results in reduced root and shoot growth. It has been shown that plants produce ethylene at two different phases in response to stressful stimuli. In the first phase, the small amount of ethylene produced promotes the activity of stress-related genes. In the second phase (1–3 days after stimulus application) the larger amount of ethylene produced lead to inhibition of growth and harmful effects on plants including senescence, chlorosis, and abscission (Glick *et al*. 2007). Mayak *et al*. (2004a) reported that *Achromobacter piechaudii* having ACC deaminase activity significantly increased the fresh and dry weights of tomato seedlings grown in the presence of NaCl salt (up to 172 mM). Mayak *et al*. (2004b) reported that ACC deaminase PGPR *Achromobacter piechaudii* ARV8 significantly increased the fresh and dry weights of both tomato and pepper seedlings exposed to transient water stress. Growth promotion of cotton under salt stress by a strain of *Pseudomonas putida* was reported by Lixia *et al.* (2010).Bensalim *et al.* (1998) revealed that *Burkholderia phytofirmans* strain PsJN helped potato plants in maintaining normal growth under heat stress. Recently, Barka *et al.* (2006) reported ACC deaminase activity *in vitro* and inoculation of grapevine (*Vitis vinifera* L.) explants with the same bacterium (*Burkholderia phytofirmans* strain PsJN) enhanced plant growth and physiological activity at both ambient (26°C) and low (4°C) temperatures. Ali *et al.* (2009) reported that *Pseudomonas* strain AKM-P6 enhanced tolerance of sorghum seedlings to high temperatures (50°C).Degradation of the ethylene precursor ACC into 2-oxobutanoate and ammonia by bacterial ACC-deaminase lowers the ethylene concentration in plant roots, relieves the ethylene repression of auxin response factors synthesis, and indirectly increases plant growth (Glick *et al.* 2007; Kang *et al.* 2010). Siddikee *et al*. (2010) have also confirmed that inoculation with 14 halotolerant bacterial strains ameliorate salt stress in canola plants through the reduction of ethylene production *via* ACC deaminase activity. Inoculation of maize plants with *Pseudomonas fluorescens* containing ACC deaminase boosted root elongation and fresh weight significantly under saline conditions (Kausar and Shahzad, 2006). Inoculation with *Pseudomonas* spp. containing ACC deaminase partially eliminates the effects of drought stress on growth, yield, and ripening of pea (*Pisum sativum* L.) (Arshad *et al.* 2008). Nadeem *et al.* (2010) reported that rhizobacteria capable of producing ACC deaminase mitigate salt stress in wheat. Several studies correlated accumulation of nitrogen-containing compounds (NCC) with drought and salt tolerance in plants (Parida and Das, 2005). The most frequently accumulating NCC includes amino acids, amides, amino acids, proteins, quaternary ammonium compounds and polyamines. Very high accumulation of cellular proline (up to 80% of the amino acids pool under stress and 5% under normal conditions) due to increased synthesis and decreased degradation under a variety of stress conditions such as salt and drought has been documented in many plant species (Szabados and Savourè, 2009).Under abiotic stress conditions, increased proline biosynthesis was observed for various plant species inoculated with different PGPR (Barka *et al.* 2006; Jha *et al.* 2010; Kohler *et al.* 2009; Sandhya *et al.* 2010; Vardharajula *et al.* 2011). Exopolysacchrides produced by *Pseudomonas putida* have been implicated in imparting drought tolerance in sunflower (Sandhya *et al.*, 2009).Antioxidants such as ascorbic acid and glutathione, which are found in high concentration in chloroplasts and other cellular compartments, are also crucial for plant defence against oxidative stress (Miller *et al.* 2010). Induction of antioxidant enzymes (catalase and total peroxidase) is involved in the alleviation of salinity stress in lettuce plants inoculated with PGPR strains (Kohler *et al.* 2010).

24.4 Modelling crop growth and yield

Crop production system involving complete input management with recommended package of practice including scheduling irrigation, nutrient application, and crop protection is cost intensive for experimental data generation especially when done at different sowing dates and multiple locations. Appropriately validated crop simulation models could be used to test many such combinations in a brief time with limited expense. Such simulations can adequately describe relative trends in yields caused by environmental variation (Satapathy *et al* 2014, Penning de Vries *et al.*, 1989). Mechanistic models predict crop productivities based on quantitative functional relationships essentially fundamental to the processes studied such as - photosynthesis, water relations, source sink relationship and their interaction with changing environmental variables. This makes these models suitable for projecting the impact of predicted climate scenarios on crop productivity; in addition, these models can be used at multiple levels which include experimental field, regional, national and global. In addition, local-gridded approaches to crop modeling using climate change scenarios can assist to identify suitable adaptation and mitigation measures. Recently, application of the crop yield simulation systems approach to climate change adaptation and mitigation initiatives has been gaining popularity due to the increasing information on the processes that are affected by changing climate which in turn is helping us device possible adaptation measures (Nelson Shively 2014; Warren 2011; Falloon and Betts 2010). In addition to growth and yield prediction models econometric model are used to explain yield as a function of climatic and economic variables. They infer that yields will rise with increasing temperatures and longer growing seasons despite greater variability in rainfall, which in principle negatively influences yields (Weersink *et al* 2010). Crop growth simulation models are used to analyze the effect of numerous climatic factors on crop production in terms of growth and yield taking into account the relations with edaphic, biotic and agronomic factors. Numerous studies show that there is a need to develop an integrated valuation of climate variability and climate change on regional and global scale using a dynamic modelling framework. This suggests that there is an urgent need to develop an integrated modelling framework for crop models not only to predict yield but also to suggest and assess the various adaptation and mitigation practices that have been tested in filed conditions. The technique of linking crop model and climate models needs to be systematically assessed in order to arrive at meaningful conclusions from climate change impacts on crops yields. Many studies have found that simulated productions of grain crop are prone to uncertainty from using different climate models, emission scenarios and the crops simulation models. Moreover, it is seen that many studies use different spatial resolutions, and different methods for general circulation model (GCM) downscaling which tend to increase the uncertainty for regional impacts assessment (Ju *et al* 2013). Varied methods, scenarios, and models are being used to illustrate the potential impacts of climate change on crop yield and other related characteristics of agricultural production. There is a degree of predictability as a result of the uncertainties over projected changes in CO2 and climate and different methodologies are being tested with varying degree of accuracy in results. However, there is limited work on integration of soil fauna especially micro fauna in this area. Incorporation of this data in the modelling framework will benefit research on potential impacts of climate change and will help in establishing a coordinated resource for crop management, climate, and soil data.

24.5 Research gaps

Host-pathogen-natural enemy interactions are not only complex but also highly weather driven. Hence, considerable changes in these interactions are expected due to climate variability and climate change. The net sum of the positive and negative impacts of the changing climate on individual organisms will decide about the crop-microflora and fauna interactions. A process-based mechanisitic approach is required to quantify the impacts to adopt timely management strategies. The projections for the future depict that appropriate adaptation and mitigation strategies should be developed to meet worst possible scenarios.

After understanding these impacts, suitable adaptation strategies could be developed to reduce the negative impacts. The research efforts therefore must be focused on

a. Understanding the tritrophic interactions and develop future scenarios
b. Understand host-pest-pathogen dynamics
c. Thrust on epidemiology
d. Relationships of interactive weather components on insect-pests pathogens, and natural enemies through simulation modeling and real-time groundthruthing
e. Increasing awareness about the climate change impacts among farmers and
f. Capturing the practices adopted by farmers to overcome the climate change impacts
g. Development of policy framework to promote R&D and develop adaptation strategies
h. Research on real time basis for different situations to understand tri-trophic interactions to use them in modeling and also development of decision support systems.

For assessing the impacts of climate change on pest occurrence, simulation models for spatial and temporal distribution using future climate scenarios as inputs must be tried. For vector-transmitted pathogens the risk analysis will also include behavior of the vector population as discussed by Harrington (2002) for the barley yellow dwarf disease.

Conclusions

Climate change as well as climate variability has been demonstrated beyond doubt and hence it becomes imperative to study its impacts on agriculture and allied sectors. In agriculture, host-pest and host-pathogen interactions are very vital and these may be influenced by climate change. To cope with the scenarios of climate change in terms of elevated CO_2 levels and associated effects like drought, cold, heat and other adverse environmental conditions, indepth study of the interaction of soil microflora and fauna with biotic and abiotic stresses it is of paramount importance to understand plant responses to these stresses that disturb the homeostatic equilibrium.Hence, an interdisciplinary approach among plant physiologists, entomologists, pathologists, molecular biologists and computer modelers in a network mode is needed to measure, predict, and assess the impacts climate change on the susceptibility of agriculture to insects, diseases and weeds and to understand the dynamics as influenced by climate change. To address these varied and diverse goals a variety of facilities needs to be utilized in this approach. This understanding is very crucial to develop adaptation strategies for future to meet the food security demands in the years to come.

References/Suggested Readings

Aroca R, Ruiz-Lozano JM, Zamarreño ÁM, Paz JA, García-Mina JM, Pozo MJ, López-Ráez JA(2013) Arbuscular mycorrhizal symbiosis influences strigolactone production under salinity and alleviates salt stress in lettuce plants. *Journal of Plant Physiology* 170:47-55

Azcon, R., Medina, A., Aroca, R. and Lozano, J.M. 2013. Abiotic stress remediation by the arbuscular mycorrhizal symbiosis and rhizosphere bacteria/yeast interactions. *Molecular Microbial Ecology of the Rhizosphere* 1:2991-1002.

Babu TR, Azam KM(1989) Biological control of grape mealy bug, *Maconellicoccus hirsutus* (Green).*J. Pl. Protec. Sci.*, 17:123-126.

Baetz U, Martinoia E (2014) Root exudates: the hidden part of plant defense. *Trends in Plant Science* 19:90-98.

Boehm MJ, Madden LV Hoitink HAJ (1993) Effect of organic matter decomposition level on bacterial species diversity and composition in relationship to pythium damping-off severity. *Appl. Environ. Microbiol.* 59:4171–4179.

Calvo-Polanco M, Sánchez-Romera B, Aroca R (2013) Arbuscular Mycorrhizal Fungi and the Tolerance of Plants to Drought and Salinity.In Symbiotic Endophytes (pp. 271-288). Springer Berlin Heidelberg.

Conn VM, Walker AR, Franco CMM (2008) Endophytic actinobacteria induce defense pathways in *Arabidopsis thaliana. Mol Plant Microbe Interact* 21:208-218, 23:137–146.

Dhaliwal GS Arora R (1994) Trends in agricultural insect pest management. Commonwealth Publishers, New Delhi.

Estrada B, Barea JM, Aroca R,Ruiz-Lozano JM (2013)A native Glomus intraradices strain from a Mediterranean saline area exhibits salt tolerance and enhanced symbiotic efficiency with maize plants under salt stress conditions. *Plant and Soil* 366:333-349.

Gaiero JR, McCallCA, Thompson KA, Day NJ, Best AS DunfieldKE(2013) Inside the root microbiome: Bacterial root endophytes and plant growth promotion. *American Journal of Botany* 100:1738-1750.

Gomati V, Ramasamy K, Kumar K, Sivaramaiah N, Mula RVR (2011) Green house gas emissions from termite ecosystem. *African Journal of Environmental Science and Technology*, 5: 56-64.

Houghton JT, Ding Y, Griggs DJ, Noquer M, Van der Linden, PJ, Xiaosu, D (2001) Climate change 2001:The scientific basis. Cambridge University Press, Cambridge 944 pp.

Joshi IJ. Chauhan RKS (1982) Distribution of soil microfungi in various soil types of Chambal Ravines. *Proc. Indian Natl. Sci. Acad.* B 48:525-533.

Karaoglu SA, Ulker S (2000) Isolation, identification and seasonal distribution of soil borne fungi in tea growing areas of Iyidere-Ikizdere vicinity (Rize-Turkey). *J. Basic Microbiol.* 46:208-218.

Kulkarni KA, Shekarappa(2001) Integrated management of chilli fruit borer *Helicoverpa armigera* Hub.*Proceedings of II National Symposium on Integrated Pest Management (IPM) in Horticulture Crops, New Molecules, Biopesticides and Environment,* Bangalore, 17-19th October, pp. 59-60.

Muralidharan K, Pasalu IC (2006) Assessments of crop losses in rice ecosystems due to stem borer damage (Lepidoptera: Pyralidae). *Crop Protec.* 25:409-417.

Nagamani A, Mew TW (1987)*Trichoderma* in Philippine rice field soils. International-Rice-Research-Newsletter (Philippines) 12: 25.

Nath DK, Sen B Pal SR (1977) Insect pests occurring in Sunflower, *Helianthus annuus* in west bengal. *Sci. Cult.* 43:180-181.

Prashar P, Kapoor N, Sachdeva S (2014) Rhizosphere: its structure, bacterial diversity and significance. *Rev Environ Sci Biotechnol* 13:63-77.

Rabie GH, Almadini AM (2005) Role of bioinoculants in development of salt-tolerance of *Vicia faba* plants under salinity stress. *African J of Biotechnology* 4:210-222.

Singh D(1991) Pests of solanaceous vegetables in "A decade of research on pests of horticultural crops, 1980 – 1990", Group discussion of entomologists working on coordinated prioject of horticultural crops held at CIHNP, Luknow.

Widden P, Hsu D (1987) Competition between *Trichoderma* species: effects of temperature and litter type. *Soil Biology and Biochemistry* 19: 89-93.

25

Importance of Monitoring Weather Parameters in Crop Insurance Schemes

G.G.S.N. Rao

Former Project Coordinator (Agri. Met), CRIDA & Senior Consultant, NCML Hyderabad, India

Agricultural production in India is still vulnerable to vagaries of monsoon rainfall in spite of rapid technological advancements like 1) Availability of high yielding varieties 2) Improved crop management practices and increased irrigation availability and effective control of pests and diseases and etc, over last few decades. Majority of Indian farming community which broadly comprising of small and medium farmers are often subject to climate extremes such as droughts, floods, heavy rainfall events, heat and cold waves, frost, hail storms during the crop growing season. The probabilities of occurrence of such climate extremes and associated crop damages and production losses are common in tropical countries like India. Compensation to the resource crunched farmers in the form of crop insurance seems to be the best option for speedy recovery from the losses incurred during the season. It provides security and stability to farm income by protecting the farmers' investment in crop production and then improves their risk bearing capacities. Crop insurance scheme is based on the fundamental principle of insurance business in which risk is distributed across space and time. For successful implementation of any crop insurance scheme operating presently in the country, proper monitoring of accurate weather parameters at the identified locations throughout the crop life cycle is important. Since these ongoing schemes requires recording weather data right at the locations the existing weather network stations operated by the national agencies which are mostly located at the district head quarters are not adequate to cover the entire cropped areas, where crop is grown and insured. In order to avoid such difficulty, the government allowed few private entrepreneurs for monitoring and supply of daily weather data of high quality by adopting strictly the norms laid out by world Meteorological Organization (WMO) and India Meteorological Department (IMD). The process of monitoring,collection and supply of weather data to the insurance companies on a regular basis involves the following:

1) Site selection for establishment of either manual or self recording Automatic Weather Stations (AWS)
2) Proper exposure of weather sensors to record true weather conditions
3) Collection of timely weather data, filling of missing weather data and quality checking for identification of faulty data and replace them with corrected values
4) Supply of processed weather data to the insurance companies that covers crop risks against extreme weather conditions.

Unless good quality weather data is not supplied, the farmers shall be subjected to great financial loss in spite of being covered under the schemes. For example, the farmer is entitled to receive 50 percent of premium value of sum assured whenever a rainfall of above 10 mm is received during flowering stage. If the area covered under this scheme receives actually rainfall above 10mm and if the rain gauge does not show any rainfall due to its poor maintenance or wrong calibration etc, the farmers shall be recording very low crop yields due to flower drop that fetches low income from agricultural produce and simultaneously are not compensated by the insurance companies. Therefore there is a great need for proper maintenance of weather stations and adaptation of strict quality control procedures as prescribed by National and International organization for effective implementation of crop insurance schemes.

25.1 Weather data Net work in India

A good network of weather observatories is important for monitoring droughts and other calamities. In the past, data collected at other than the surface observatories (Hydro meteorological observatories), often were not processed for long time. With the advancement in communication technologies, real time weather is monitored for applications in various fields *viz.*, Agriculture, Floods, Tourism, Disaster management such as, landslides and avalanches etc. With the availability of data from many locations even from remote areas on real time basis makes it possible to develop Contingency plans and Decision Support System (DSS) for sustainable agricultural production. With the increasing awareness on climate and its changes over time, the importance of collection of weather data has gained movement around globe. In recent times the developing countries in Africa and in Asia have improved the observational network considerably over the years. India has the largest network of weather observatories of various categories in Asia. There are 25 types of atmospheric monitoring networks that are operated and coordinated by India Meteorological Department (IMD). It has 559 surface meteorological observatories, 35 radiosonde and 64 pilot balloon stations, 219 Agrometeorological and 45 radiation observatories (Attri and Tyagi, 2010). IMD in its massive modernization of existing network of observatories to improve the numerical prediction capabilities, it had established 550 Automatic Weather Stations (AWS) at different locations in the country. Meteorological and Oceanographic Satellite Data Archival Centre (MOSDAC) of Indian Space Research Organization (ISRO) also has established about 1098 AWS across the country to facilitate and promote the exchange of data to the academic and research needs.

Some of the state governments in India has also started establishing their own network of AWS across the states, with the objectives to monitor the weather for better management of climate resources. The erstwhile State of Andhra Pradesh took the lead in this regard by establishing 1100 AWS at each Mandal Headquarters (Fig. 25.1). Similarly the State Government of Tamil Nadu also established AWS at Taluka level and the State of Karnataka established a network of Telemetric Rain gauge stations at Taluka level. The states of Gujarat and Maharashtra are also establishing shortly AWS at block level to monitor the weather for improving the decision making ability of the Government.

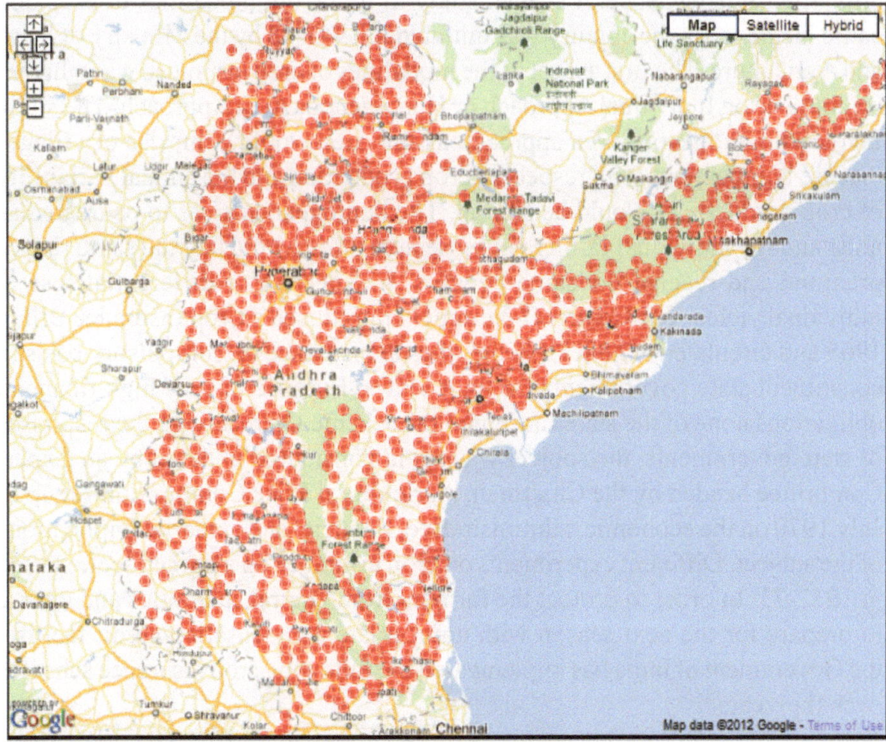

Fig. 25.1: Distribution of AWS network in Andhra Pradesh State
(*Source:* ttp://apsdms.cgg.gov.in)

Data from AWS, Automatic Rain gauge (ARG), Doppler weather Radars (DWR) and Satellite (INSAT-3D) are available on real time basis and is being used by IMD for running number of global and regional NWP models for weather forecast purpose. The National Agricultural Research System (NARS) in the country consisting of Research Institutes of Indian Council of Agricultural Research (ICAR) and state Agricultural Universities (SAU) and their regional center have also established Agro meteorological observatories spread across the country and generatedata for crop planning and agricultural operations on real time basis. In order to monitor the extreme weather events and to prepare contingency plans, the AICRPAM (All India Coordinated Research Project on Agro meteorology) at CRIDA, Hyderabad had established 100 AWS at different Krishi Vigyan Kendras (KVK) across the country to improve dissemination of Agromet Advisory Services (AAS) to sub district level (CRIDA, 2011).

25.2 Crop insurance program in the country

Though the concept of crop insurance program was designed way back in 1920 in India, its effective implementation by the Government of India has started only after independence. A special study to frame out modalities of crop and cattle insurance in the country was initiated in 1947-48 by the Ministry of Food and Agriculture. The first aspect of the modalities of crop insurance was whether it should be on Individual Approach or Homogenous Area Approach. The study favoured homogenous area approach as this area would comprise of villages that are homogenous with respect to crop production and the annual variability of crop productivity would be similar. The homogenous areas to be treated as units and the individual farmers in those area units would pay the same rate of premium and receive the same benefits, irrespective of differential loss in individual yields. The Central Government introduced a Crop Insurance Bill in 1965 and circulated a model scheme of crop insurance on compulsory basis to constituent state governments for their views. Because of very high financial obligations none of the states accepted the scheme.On receiving the responses of state governments, the subject was again considered in detail by an Expert Committee headed by the Chairman, Agricultural Price Commission set up in July 1970 on the economic, administrative, financial and actuarial implications of the subject. Different experiments on crop insurance on a limited scale started in 1972-73. In order to protect the farmers from natural calamities and ensure to prepare for the next season with appropriate compensation/credit facility; the Government of India has implemented the following crop insurance schemes over the years.

25.3 First Ever-Individual Approach Scheme

In 1972-73, the General Insurance Department of Life Insurance Corporation of India introduced a Crop Insurance Scheme on H-4 cotton. The Scheme was based on "Individual Approach". Subsequently the scheme included groundnut, wheat, potato and gram and was implemented in the states of Gujarat, Maharashtra, Tamilnadu, Andhra Pradesh, Karnataka and West Bengal. The schemes continued till 1978-79.

25.4 Pilot Crop Insurance Scheme (PCIS) – 1979

In the background and experience of the aforesaid experimental schemes for crop insurance, a study was commissioned by GIC in 1979 to evaluate the schemes. Based on the findings, the Pilot Crop Insurance Scheme (PCIS) was launched in 1979 and continued till 1984-85. The scheme was implemented in 13 states

25.5 Comprehensive crop insurance scheme (CCIS)

This scheme was operated on the basis of experience gained from implementation ofPCIS, a Comprehensive Crop Insurance Scheme (CCIS) was introduced with effect from 1st April 1985 by the Government of India with the active participation of State Governments on a voluntary basis covering 15states, 2UT. About 5 million farmers were the beneficiaries and around 8-9 M. ha were annually covered by this scheme. It was operated on the basis of crop yield losses in a region. If the actual yield of the crop in any area covered by the scheme fell short of the guaranteed yield. Then the farmers were entitled to an indemnity on the compensation to the extent of shortfall of yield. This scheme was abandoned in 1997 due to majority of claims went to one single crop like ground nut.

25.6 Experimental crop insurance scheme (ECIS)

This scheme was introduced for 1 year 1997-1998 covering non-loanee small and marginal farmers growing specified crop in few selected districts. This scheme was implemented in 14 districts of five states. Because the claims were much more than the premium collected from the farmers, this scheme discontinued within the same year.

25.7 Farm income insurance scheme (FIIS)

This scheme was introduced by GOI during 2003-2004. The scheme targeted two critical components of farmers' income viz. yield and price. The insured farmers were assured with minimum guaranteed income (**average yield X minimum support price**). If the actual income of the farmer was less compared with guaranteed income; the insurer shall be compensated to an extent of the shortfall by the insurance company. All the farmers availing loan facility shall be covered automatically under this scheme. This was withdrawn in 2004.

25.8 National agriculture insurance scheme (NAIS)

With the failure of comprehensive Insurance scheme in 1997 the GOI came up with a new scheme 'National agricultural insurance scheme', during 1999-2000.It was planned to cover all food crops (cereals and pulses), oil seeds, horticultural and commercial crops. This scheme was extended to all farmers, both loanees and non- loanees. Comprehensive risk insurance is provided to cover yield losses due to natural calamities which cannot be prevented. It operates on the basis of 1) area approach and on 2) individual basis. Area approach defines the area for each notified crop for wide spread calamities and where as individual basis covers for localized calamities such as hailstorms, landslides, cyclones and floods. The premium rates vary from 1.5% to 3.5% of sum assured for food crops. For horticulture and commercial crops, actual rates are charged. A subsidy of 50% the premium was given to small and marginal farmers. The subsidy on premium shall be equally shared by State and Central Governments and the subsidy has to be phased out over a five year period. The scheme was largely unsuccessful with low coverage and high claims to premium ratio. At present it is operating in a small cropped area covering about 15% of farmers.

25.9 Weather based crop insurance scheme (WBCIS)

Crop yields are mostly dependent on weather conditions during the season in addition to soils, seeds, fertilizers etc. Therefore an innovative weather insurance scheme was introduced in 2003 as the index based rainfall scheme. From *kharif* 2007, Weather Based Crop Insurance scheme (WBCIS) has been implemented across India. The scheme operates on the principle of "area approach" in selected notified reference unit area. Weather triggers for different crop growth stages of important food and commercial crops for different agro climatic region have been identified from research findings of Agricultural Universities and of ICAR Research Institutes. Weather data is monitored throughout the growing period by deploying Automatic Weather Stations (AWS) in the notified unit area. Farmers are eligible to receive payouts of the sum assured for that crop stage whenever the weather parameters exceeds the trigger value fixed for that stage. The major limitation of this system is that whenever a farmer lost his entire crop during a particular stage, the compensation according to the farmers may not be the maximum sum assured, but it will be the sum assured of the weather insurance covers operative during the stage. Under this scheme, overall claims ratio of few insurance companies was nearly 77%. The major advantage of the system in that payoff to the farmers is quick and they need not wait till the crop harvest is completed. As of today, many states like Maharashtra, Rajasthan, parts of M.P, Jharkhand, N.E regions, Karnataka, etc. have been implementing this scheme. Since, India meteorological department cannot deploy AWS in the notified areas; many private entrepreneurs have entered in to the weather market in way of providing quality data to the insurance companies. Even in the remote places and inhospitable terrains, AWS have been established and 24 hr data is being recorded and supplied to various stake holders.

25.10 Varsha Bima-2005

Varsha Bima covers anticipated shortfall in crop yield on account of deficit rainfall. It is voluntary for all classes of cultivators who stand to lose financially upon adverse incidence of rainfall. The insurance operates during June to September for short duration crops; June to October for medium duration crops; and June to November for longer duration crops. There is no necessity for submission of loss information or claims intimation by insured cultivator. Normally claims are paid on the basis of actual rainfall data within a month from end of indemnity period

25.11 Pradhan Mantri Fasal Bhima Yojna (PMFBY)

Each crop insurance scheme has its own limitations and the actual sufferers are not adequately and timely compensated. Studies have indicated that only 23% of the farmers have been covered under these schemes .In view of the short comings of the exciting crop insurance scheme, Pradhan Mantra Fasal Bhima Yojna (PMFBY) which has been launched recently is expected to cover 50% of farmers in the next 5 years period and shall replace all the other crop insurance schemes which integrate all the benefits in one single yojna. The premium rates in the existing crop insurance schemes have increased by about 25% of the sum assured in recent years. The PMFBY has drastically reduced the premium amount to 2% for *kharif* crops and 1.5% for rabi crops. It shall be 5% for commercial or horticultural crops. Farmers shall derive maximum benefits by paying minimal premium. The scheme entails immediate payment of 25% of the due compensation and the money will go directly to the bank accounts of the farmers. It also provides compensation for loss of seed plants and post harvest damages. It shall also assess the damage due to local calamities like hail storm, unseasonal rains, landslides and inundation of crop fields., Latest technologies like utilization of smart phones, remote sensing and drones shall be used to estimate the crop losses even at a remote place without delay. This scheme will be effective from 2016 *kharif* season onward.

25.12 Farmer needs for sustainable crop production

In view of changing climate, introduction of unconventional crops, exported oriented horticultural and floricultural crops, improved availability of water resources, chemical fertilizers, pesticides,liberalized financial assistances, intensive fish farming and animal husbandry, developments in communication technology and increased expenditure on social obligations , the requirements of farmers in the recent past have rapidly changed. Some of them are listed below

1) Assured water supply on demand
2) Technological inputs for modern irrigation methods to save water in high water demand crops
3) New crops and improved crop varieties with resistance to pests and diseases
4) Assured supply of quality seeds of all varieties to meet the variability in rainfall
4) Micro regional weather forecast with special emphasis on extreme weather events along with agro advisories to all levels of farmers
5) Climate related risk management strategies
6) Crop insurance facility against all natural calamities and extreme weather events to all the farmers including tenant farmers
7) Online market information to get better price for the produce
8) Liberal credit facilities and increased subsidies on inputs
9) Free accessibility of internet and mobile for receiving advisories
10) Frequent training trainings and interactions with subject matter specialists.

25.12.1 Strategies to enhance farm production

1) Crop diversification to meet the changing climate and its variability
2) Improved cropping techniques (Fertilization, micro irrigation practices, soil and water conservation methods, Farm implements for better crop establishment)
3) Rain water harvesting through farm ponds for life saving irrigation
4) Improve the efficiency of farm ponds through periodical maintenance

5) Enhance rural livelihoods by improved farming systems productivity Viz: better livestock management, growing vegetables,, fish farming where ever runoff water is plenty.

6) Weather forecast for next 4 to 5 days and weather alerts to cope up with weather extremes. Real time weather based advisories through internet and SMS

7) Micro regional crop plans, crop contingencies for all types of climatic situations like , 1)Droughts, 2)floods, 3)temperature extremes, 4) frost, 5) hail storm 6)Cyclones etc.

8) Improving the efficiency of fresh water resources in the sandy tracts of coastal areas.

9) Development of infrastructure (Roads, irrigation systems, major and medium irrigation project).

Conclusions

The improved capacity to monitor micro regional weather and the forecast shall benefit all the farming community by providing adequate safe guards towards weather uncertainties. Comprehensive planning for agricultural crops at regional scale and timely weather based agro advisories shall sustain agricultural production, rural economy and meet the challenges posed by changing weather.

References/Suggested Readings

Ramakrishna, Y.S., Rao, G.G.S.N., Rao, S.G. and Vijayakumar, P. (2006). Impact of climate change in Agriculture. In: Environment and Agriculture (eds. Chadha, K.L. and Swaminathan, M.S. Malhotra Publishing House, New Delhi, pp. 1-30.

Rao G.G.S.N., Rao V.U.M., Rao A.V.M.S. and Saikia U.S. (2006). All India Coordinated Research Project on Agrometeorology, Bi-Annual Report (2004-06). Central Research Institute for Dryland Agriculture, Hyderabad. Pp. 170.

Rao G.G.S.N., Rao V.U.M., Rao A.V.M.S. and Saikia U.S. (2007). All India Coordinated Research Project on Agrometeorology, Annual Report (2006-07). Central Research Institute for Dryland Agriculture, Hyderabad.Pp. 162.

Rao G.G.S.N., Vijaya Kumar, P., Rao V.U.M. and Rao A.V.M.S. (2009). All India Coordinated Research Project on Agrometeorology, Annual Report (2009). Central Research Institute for Dryland Agriculture, Hyderabad.Pp. 242.

Rao V.U.M., Bapuji Rao, B., Rajkumar Dhakar, Sarath Chandran, M.A., Vijaya Kumar, P. and Rao A.V.M.S. (2013). All India Coordinated Research Project on Agrometeorology, Annual Report (2012-13). Central Research Institute for Dryland Agriculture, Hyderabad.Pp. 127.

Rao V.U.M., Bapuji Rao, B., Rajkumar Dhakar, Vijaya Kumar, P. and Rao A.V.M.S. (2012). All India Coordinated Research Project on Agrometeorology, Annual Report (2011-12). Central Research Institute for Dryland Agriculture, Hyderabad.Pp. 188.

Rao V.U.M., Bapuji Rao, B., Sarath Chandran, M.A., Vijaya Kumar, P. and Rao A.V.M.S.. (2015). All India Coordinated Research Project on Agrometeorology, Annual Report (2014-15). Central Research Institute for Dryland Agriculture, Hyderabad.Pp. 132.

Rao V.U.M., Bapuji Rao, B., Sarath Chandran, M.A., Vijaya Kumar, P. and Rao A.V.M.S.. (2014). All India Coordinated Research Project on Agrometeorology, Annual Report (2013-14). Central Research Institute for Dryland Agriculture, Hyderabad.Pp. 116.

Rao V.U.M., Vijaya Kumar, P., Bapuji Rao, B. and Rao A.V.M.S. (2010). All India Coordinated Research Project on Agrometeorology, Annual Report (2010). Central Research Institute for Dryland Agriculture, Hyderabad.Pp. 180.

Rao, G.G.S.N. (2015). Droughts and Agricultural Production Monitoring and Management. Agriculture. New India Publishing Agency, New Delhi, 386p.

Rao, G.S.L.H.V.P, Rao, G.G.S.N. and Rao, V.U.M. (2010). Climate Change and Agriculture over India. PHI Learning Private Limited, New Delhi, India.328p.

Rao, G.S.L.H.V.P. (2011). Climate Change Adaptation Strategies in Agriculture and Allied Sectors. Scientific Publishers (India) Ltd. Jodhpur, India. 336p.

Rao, K.N. and Rao, G.S. (2017). Weather Based Livestock Insurance – Present and Future: Chapter 22. In. Climate Resilient Animal Agriculture. New India Publishing Agency, New Delhi, 460p.

Rathore, L.S. and Parvinder Maine. (2008). Economic Impact Assessment of Agrometeorological Advisory Service of NCMRWF. Report NO. NMRF/PR/1/2008. NCMRWF, MoES, Govt. of India, New Delhi.160p.

Shukla, P.R., Sharma, S.K. and Ramana, V.P. (2002). Climate Change and India - Issues, Concerns and Opportunities. Tata – McGraw-Hill Publishing Company Ltd, New Delhi, 314p.

26

Strategies for Improved Agro-advisory Services to Farmers

Y.S. Ramakrishna

Former Director, CRIDA, Hyderabad, India

Increasing evidences over the past few decades indicate that significant changes in climate are occurring across various regions of the globe in the form of increased occurrence of extreme weather events. The significant cause to this sustained increase in climate variability, known as climate change has been a resultant of increased level of greenhouse gases (GHGs) in the atmosphere. The reason for this increase is mainly ascribed to enhanced human activities such as industrialization, uncontrolled burning of fossil fuels, increased use of refrigerants and changed land use patterns and practices. Most significantly, the concentrations of GHGs like CO_2 has already crossed the 400 ppm mark, while other potential contributors like methane and nitrous oxide also show significant rise. As a result of the heat absorbing capacity of these gases, there has been a considerable rise in global mean temperatures, more so in the recent decade and more. The increase in mean air temperatures of the globe was 0.76^{0}C during the past 100 years. This rate of increase has been higher during the past two decades and the recent period recorded the warmest years in sequence, the latest being the 2017. The impacts of the climate change are visible across all regions of the globe, in the form of increased occurrence of extreme weather events like severe droughts, floods, storms, heat waves, cold waves, high wind regimes etc. Besides the physical risks to life and increased discomfort, these events directly influence the quantity and quality of agricultural production adversely. They increase the production risks and threaten the food security of nations. Agriculture is one of the main sectors that is likely to be impacted by climate change and in a country like India, the consequences can be harsh with projections of decline in yields of major food crops due to warmer temperatures and increased drought risk in the coming decades. Thus risk characterization and risk management are critical issues in agricultural productivity management. The risks are higher in rainfed areas which are mainly dependent on the quantum and distribution of rainfall during the monsoon season of June to September over major part of the country. As against the national average of 40% of the total cropped area being irrigated, 60% of the total cropped area is still rainfed and is strongly dependent on uncertainties of monsoon. This shows that the dependency of Indian agriculture on climate is very large and as such the Indian agriculture is called a gamble in monsoon.

With about 60 per cent of the Indian Agriculture being rainfed and is dominated by small and marginal farmers, the challenges to crop production have become all the more daunted by the increased vagaries of weather that have become not only a regular feature but also been increasing in intensity, over the years. The small and marginal rainfed farmers of the country, who are dependent on rainfed agriculture are less endowed in terms of financial, physical, human and social capital limiting their capacity to adapt to the changing climate. Research studies show that increased climate variability is already impacting Indian agriculture. Unseasonal rain and erratic weather, which have affected Indian farmers, agriculture, economy, and politics, is no more an aberration, but an indication of climate change according to many research studies. These studies also brought out that though the annual rainfall may remain about the same, an increase in high rainfall events and increase in dry spell durations during the cropping season, have been visible in many regions across India. Further, the last few years, especially 2015 witnessed El Nino impact with large part of the country experiencing drought conditions, more so in Maharashtra, peninsular states and in parts of AP and Telangana. As many as 15 states were affected and 33% of the cropped areas were damaged. The losses to farmers were in excess of '20,000 crore. Prolonged drought conditions are still continuing in states like Tamil Nadu and parts of Kerala. Under these increasing occurrences of adverse weather conditions, there is an urgent need for taking up systematic adaptation measures to increase the resilience and adaptive capacity of the farming communities.

From agricultural point of view weather plays a major role in influencing the decisions of the farmer, starting from land preparation, sowing to harvesting, post harvest management including storage and transport. For the rainfed farmer, all his farm operations begin with the onset of rains and his choice of crop and cropping system depends on the long range forecast of weather expected during the coming monsoon season. Once he decides on his crop, the next would be the land preparation, which depends on the quantum of the pre monsoon showers. In case these showers are poor to wet the top soil layers, he has to wait till the monsoon onset to carry out both the land preparation, followed by sowing after sufficient moisture accumulates in the top soil layers that can enable germination of the crop and support seedling growth till the next rains are expected. Thus from sowing onwards he depends on the medium and short range weather forecasts to manage his crop and to decide on the various farm operations he had to adopt for better plant growth and in managing any moisture stress that is caused by the inter spell duration of rain events. The timely availability of weather information thus becomes crucial to the farmer for efficient management of the rainwater for crop growth and better crop productivity. Good weather code thus becomes his motto, where he attempts to take best advantage of good weather situations and can manage to reduce the impacts of bad weather conditions on his crop growth and performance. Besides the rainfall regime his interest extend to other important weather variables like the temperature and humidity regimes which mainly control the evaporative demand of the crop and also influences the micro-climate and pest and disease incidence in his crop. Weather information across the country is being recorded mainly by IMD and also being supported by various other agencies and institutions. With the initiatives taken up various state government agencies, good net work of rain gauges and also many networks of Automatic weather stations (AWS) have come up across the country. Also remote sensing agencies have been collecting data from various satellites on vegetation status and monitoring drought and flood situations. Thus a large source of weather and related parametric data are being generated regularly. Thus data analytics of this various sources of information can help in analyzing long term weather patterns of a region and also in field level monitoring of state, region, district, block and even up to village level agricultural conditions on a regular basis. Looking to the importance of weather information needs of farmers, IMD and ICAR have initiated Agro advisory services to farmers in 1945 and have been improving upon the quality of the information disseminated to the farming community. From a regional level forecasts IMD now is providing short and medium range weather forecasts upto a block level and based on these weather forecasts both ICAR and IMD are providing district and block level forecasts on weather conditions and crop management strategies to farming communities. Thus weather forecasts and agro advisories form an integral part of improved crop management and farmers especially the small and marginal, need to sustained support through improved knowledge management for empowering them to meet the vagaries of the weather and take up adaptative measures to meet the challenges of increased climatic variability as a result of climate change. Access to improved knowledge can help them minimize the climate change risks and enable them in the efficient use of existing infrastructure and natural resources. It is here that the information and Communication Technology has come to the aid of the farming community in a great way. With the fast improvement in mobile technologies and accessibility of mobiles even in remote regions, it has now become possible to transfer knowledge to larger number of farming community on efficient crop management approaches and in adopting climate smart agriculture. In addition the crop insurance agencies are interested to generate information on crop status during different phases of growth and impacts of extreme weather events like high temperatures high intensity rains, floods, drought, high wind regime, pest damage etc. causing crop losses, so as to settle weather related insurance claims quickly and with confidence. Further many of the state governments are interested to generate accurate information on crop booking and area sown under each crop in space and time, with geo-references. However, the current approach to providing agro advisories is not fully satisfactory to the farming community. Analysis by extension agencies indicate that farmers are not keenly following up the agro advisories as they feel that they are general in nature and does not specifically deal to their situation or meet their individual need. For example, even though the same crop is grown in a region, some farmers might have sown early, some during the normal rowing season and some sown it late. Thus their crop are at different pheno-phases when a weather event occurs (rain or critical temperature) and so their crops response would be different. This has to be taken into account and also they would expect a personal approach to their problem and not a general approach to all with the same advice. Thus, agro advisories have reaped success mainly in those regions where the extension officials have adopted those farmers and are in regular contact with them / visit to their fields. Thus one of the main lacunae in the current approach is ' addressing the last mile' - that is including the farmer as a major player (provider and consumer) in the information exchange and in suggesting him the suitable alternate strategies. Any approach adopted to provide such ' smart' and improved agro advisories has to take many measures to synthesize and integrate all the relevant information to enable to provide more accurate advisory to the farmer. For this they have to bring on to one platform, the data from satellites, weather stations (regular and AWS) and integrate this data with simulation models and with ground information on crop status collected from village level officials and progressive (and other concerned) farmers on crop, phenophase, crop health through photo based information (in the form of pictures taken on crop condition and uploaded to the cloud) for analysis, integration and interpretation using machine language (to handle the large data base).

States like Andhra Pradesh have taken steps to monitor the individual farmers activities across the state, by mapping their survey numbers with latitude and longitude coordinates. Further they have appointed 5000 MPEOs (multi-purpose extension officers) who are to visit two villages a day and interact with 20 farmers each in their farms. This innovative approach provides many opportunities for monitoring individual farmers fields and provide them agro advisories individually. With a unique field ID tagged to each field and the farmers provided with a picture based App, the farmer takes a picture of his field when he sows his crop and fills a small proforma providing information on the crop, variety, date sown and inputs provided. The App saves this information and uploads it to the cloud when net is available, providing information on from which field it is taken and the mobile number of the farmer. Such information uploaded from individual farmers fields enables the tagging the information to the field pixel enabling easy identification of not only the crop but also whether it was sown early, normal or late. Any further photographs taken at later periods by the farmer and also the MPEO's during their field visits, can be tagged together to monitor the progress of growth and health of the crop, specific to the corresponding farmer. Through this process the agriculture department will get accurate crop booking information, enabling the mapping of the area sown under each crop in time and space in that region through geo references and the unique ID tagged to the each field visited by the MPEO's and extrapolated using satellite data to come up with aggregated statistics on area sown under different crops and at different periods. This will help in assessing the current individual crop condition in the geo tagged farmers fields and also to estimate the area covered by a specific crop, cropping pattern in the region. Further visits by these officials as well as the concerned farmer and also any individual (like the progressive farmer of the village) taking photographs at that field will keep on getting uploaded by the App (along with Lat and Long) and tagged to the concerned farmers database. Thus during the crop growing period, crop health assessments, other relevant crop information from individual fields, along with corresponding satellite information on a weekly basis can be compiled and provided to the farmer and officials regarding the crop condition, pest/disease/ moisture deficit (drought) or moisture surplus (flood) status. This will also lead to continuous and efficient performance of the MPEOs in data collection and information dissemination to government and the advisory agency in 'near real time'. Such farmer specific crop information will enable in providing improved/ smart crop specific agro advisories to individual farmers on a regular basis during the crop growing period. These advisories would be for three crop conditions viz; early, normal and late sown crop. Individual farmers will receive location specific crop relevant agro weather advisories on their mobiles.

The currently available District /Mandal level weather forecasts provided by IMD (using 9x9km grid WRF model) have to be down scaled (3x3 km) and using additional data from local sources, and data from AWS network, satellite information and weather radar data to improve the accuracy of advisories at the micro level. The weather advisory will be improved, to be applicable to the respective location, crop type, crop stage, soil properties, current soil moisture and will be used to provide location and crop specific agro advisories, to be of better use to the individual farmer at his farm. Such improved agriculture services, available at the door step of the farmer will not only boost his confidence to adopt innovative methods in agriculture management but also enhance the quality and productivity of his crop leading to improved livelihoods. Any specific query on the agro advisory in respect to any pest/ disease issue will, however, be provided to the farmers in a 'cluster approach' (one advisory to all those farmers having similar crop conditions/ phenophase) but will be received by the concerned farmer and also by other farmers on individual basis as per his crop status. In case of any adverse weather condition damaging the crop fully or partially, the farmer can take a photograph and upload it. This information can be cross checked by the insurance company through weather data and satellite data. Once confirmed of a likely damage, it can form a committee of MPEO, local Panchayati member and any other member to represent them and these members along with the concerned farmer, visit the field, assess the damage and come up with an agreement on the percent damage with the farmer. This information on percent damage in a proforma along with photo (provides the Lat/Long of the field from where the photo is taken) of the damage crop area are uploaded from the farmers field. Such a procedure can enable the insurance companies to judge the authenticity of the damage and clear the insurance payment rapidly, in 48 hours. Such seamless insurance payments by the concerned agency during extreme weather event damages will enhance the confidence of the farmers enabling to come forward in large numbers to avail crop insurance proposals. In addition, improving access of the farmers to vital market information, supporting them to adopt group dynamics to get good quality seeds and inputs will lead to better Finance-Production-Market linkages (through bankers, mobile trading, tendering, bartering platforms) can help reduce transaction costs and middle men interference thereby providing higher profits, boosting farmers income. The rural youth can play a big role in this Endeavour, because of their agility and marketing abilities. This will also help in providing self created job opportunities to the village youth and will also enable them to return back from the cities to the villages.

The major players in this smart/ improved agro advisory system would be the Ministry of Science and Technology, Earth sciences and Ministry of Agriculture and Farmers welfare, from the central government, while at the state level the concerned Ministry of Agriculture, Department of Agriculture, the concerned Agricultural universities, ICAR institutes and other institutes in the state providing technical backstop on strategies for crop production and contingency planning at district level. For technology transfer to the farmers, at the state level the main player would be the department of agriculture, while at the district level the concerned JDA/DDA/ATMA/KVK's / DATTC and other Mandated agencies. At Mandal level the ADA/AEO's/MPEO's will lead, while at the Village/ cluster of villages level, the VDO's/NGO's/Farmer societies/ Panchayati will play the major roles. Since this would need a lot of computer time, cloud computing, remote sensing data analysis, modelling, monitoring crop data, extreme weather events, assisting in rapid settlement of insurance claims and providing smart agro advisories to individual farmer's mobiles and attending to their queries. This would mean that the state department would need support of expertise from private agencies to take the lead and join forces with agriculture related central and state agencies and agriculture universities, in a P-P-P mode to take up such an integrated Farmer specific approach, to collate and analyze such large data and reach such a voluminous group of farmers with diverse cropping systems. This approach can be tried on an experimental basis in few districts and once proved successful to enthuse larger number of farmers, can be up scaled to other districts. Such an approach can not only make these improved agro advisory services more valuable to individual farmers providing them more confidence in managing their agriculture and achieve enhanced quality productivity believe in government's committed support for their welfare. It can also help farmers in future to come forward to pay for the advisory services in future, leading to a business model for a P-P-P. Also the experiences gained during such an experimental phase will enable the field level officials to carry forward the technological adoptions by farmers in future, in a seamless manner. In turn organizations like ATMA can take these strategies forward at ground level with the help of MPEOs, other field staff. KVKs can also help in the successful demonstration of the technologies and up-scaling them at village and block level. Ultimately the department of Agriculture can converge the efforts of public, private and NGO organizations to avoid duplications and to follow a clear strategy for a planned multi-stakeholder involvement. This will lead to improved benefits to farming community and increased productivity and profitability. Providing access to such adaptative technologies can help farmers in following resilient cropping strategies and achieving sustained agricultural productivity and overcoming the impacts of climate change, a win-win situation.

References/Suggested Readings

Prasad, R and Rana, R. 2006. A study on maximum temperature during March 2004 and its impact on rabi crops in Himachal Pradesh. *J. of Agrometeorology*, 8(1): 91-99

Ramakrishna, Y.S., Rao, G.G.S.N., Rao, S.G. and Vijayakumar, P.2006. Impact of climate change in Agriculture. In: Environment and Agriculture (eds. Chadha, K.L. and Swaminathan, M.S.). Malhotra Publishing House, New Delhi, pp. 1-30

Rao, G.S.L.H.V.P. 2011. Climate Change Adaptation Strategies in Agriculture and Allied Sectors. Scientific Publishers (India) Ltd. Jodhpur, India. 336p

Rao, G.S.L.H.V.P, Rao, G.G.S.N. and Rao, V.U.M. 2010. Climate Change and Agriculture over India. PHI Learning Private Limited, New Delhi, India.328p

Rathore, L.S. and Parvinder Maine. 2008. Economic Impact Assessment of Agrometeorological Advisory Service of NCMRWF. Report NO. NMRF/PR/1/2008. NCMRWF, MoES, Govt. of India, New Delhi.160p

Shukla, P.R., Sharma, S.K and Ramana, V.P. 2002. Climate Change and India - Issues, Concerns and Opportunities. Tata – McGraw-Hill Publishing Company Ltd, New Delhi, 314p

27

Agrometeorological Products in Agroadvisory Services

G.S.L.H.V. Prasada Rao

Founder Academy of Climate Change Education and Research
Kerala Agricultural University, Vellanikkara, Thrissur-680 656. Kerala, India

The District-wise weekly agroadvisory service based on medium range weather forecasting (3 to 10 days) has been made operational by the State Agricultural Universities in collaboration with National Centre for Medium Range Weather Forecasting (NCMRWF), India Meteorological Department. The Agroadvisory Bulletin (AAB) contains three parts. The first part of AAB consists of weather events occurred during the past week and weather forecast for three days ahead. These forecasts provide expected weather on cloud amount, rainfall, average wind speed, wind direction, maximum and minimum temperatures and cumulative rainfall. The second part contains the factual information on the varieties of crops, their state and stage, ongoing agricultural operations and occurrence of insect pest and diseases of crops. The final and third part of the AAB provides information on agroadvisory to be taken up as farm operations during the ensuing week. The AABs are prepared by the Agrometeorological Field Units (AMFUs) in consultation with the scientists belonging to various disciplines and Agricultural Officers of the locality. Such AABs are being prepared and issued in English as well in local languages. The AMFUs functioning across the Country (more than 130 AMFUs) attempt the weekly agroadvisory service very effective and operational from time to time with simple models developed in the concerned universities for sustenance of crop production against the ill effects of climate change/variability to a considerable extent. Agroadvisory service is based on weather forecasting and agromet products play a major role in AABs for the benefit of farmers. As such the AABs revolve mostly around agronomic practices rather than based on agromet products. Some of the typical tools prepared in Kerala Agricultural University are quoted for guidance and such models may be implemented wherever they are relevant for betterment of Agromet Advisory Services as a part of climate risk management in Agriculture.

27.1 Onset of monsoon over Kerala

The normal onset of monsoon over Kerala is 1st June, varing from 25th May to 8th June. The onset of monsoon is categorized into three viz., early, normal and late and this information can be utilized in Agro Advisory Bulletins. It is useful in the case of rice cultivation across the State of Kerala. Rice varieties suggested for cultivation accordingly during *Kharif* and the feedback obtained from farmers are given in **Table 27.1.** Based on the forecast of onset of monsoon, short or medium or long duration varieties can be planned for cultivation so as to maintain crop calendars intact. The varieties suggested may be there or not, if the idea mooted here is followed based on onset of monsoon the crop losses can be minimized to some extent in case of rice due to excess/shortage of water during crop growing season.

27.2 Crop simulation modelling

An attempt has been made to suggest optimum time of planting for Thrissur and Palakkad districts of Kerala during rabi season based on crop growth simulation models in the case of paddy (IBSNAT-DSSAT Rice model). The simulated grain yield in paddy across the Palakkad and Thrissur districts is relatively less if the transplanting is taken up after 20th September (Fig. 27.1). Accordingly, the time of transplanting is recommended. The farmers who followed the advisory got advantage and obtained better yield in the case of paddy during *Rabi.*

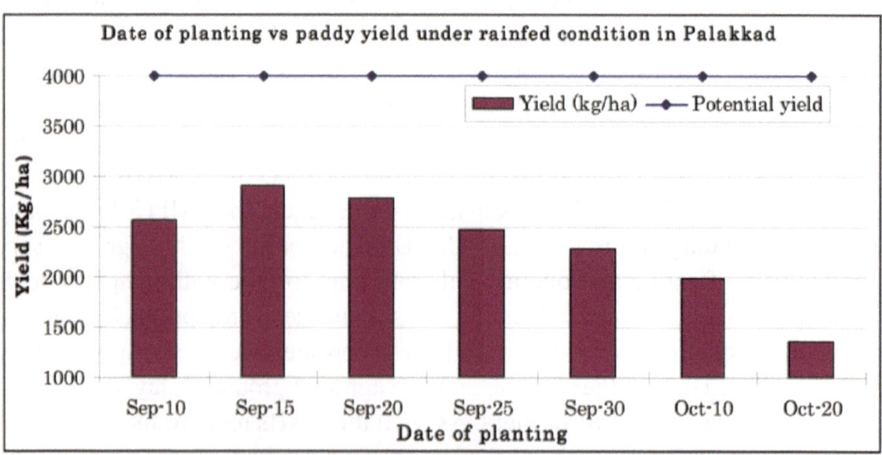

Fig 27.1: Date of transplanting and paddy yield across the Palakkad district

Table 27.1: Onset of monsoon versus suggested paddy varieties

Onset of monsoon	Package/advisory	Feedback	Remarks
Early (before 25th May)	Long duration varieties (Ponmani, Dhanya, Reshmi, Kumbham)	Very few farmers selected long duration varieties. Most of the farmers choose medium and short duration varieties	Normally, farmers select varieties based on their experience or yield potential. The choice of variety may be either medium or short duration irrespective of date of onset of monsoon. Some farmers choose local cultivars due to their comparative tolerance against pest and diseases.
Normal (25th May To 8th June)	Medium duration varieties (Jaya, Aathira, Aiswarya, Panchami, Karishma)	Most of the farmers selected medium duration and short duration varieties	
Late (after 8th June)	Short duration varieties (Jyothy, Mattatriveni, Kanchana, Remanika)	Most of the farmers selected medium and short duration varieties	

Advisory	Feedback	Remarks
Sow/transplant before September 20th, so as to overcome water stress during maturity period.	Most of the farmers transplanted before September 20th	The farmers who started cultivation before 20th September, got more yield than the non-followers

27.3 Predictive model for water discharge through canals

Peechi dam of Thrissur district of Kerala is 213.36 m long and 40.8 m height built structure across Manali river with a water spread of 1295 acres and a water storage of 109.88 M.cum capacity. Besides a supply of 0.055 M. cum of water daily for drinking purposes, irrigation is assured to an area of 18,400 ha through its two canal systems viz., left bank main canal (LBMC) and right bank main canal (RBMC). LBMC being positioned at a higher elevation of 66.6 m compared to RBMC (55.8 m), the discharge through LBMC stops whenever water level fall below 66.0 m and makes the command area of 8,000 ha spread in nine villages viz., Panachery, Mulayam, Kainoor, Puthur, Trichur, Kallor, Varadarapalli, Amballur and Manamangalam prone to drought situations. A regression model was developed to predict the water discharge through the left bank main canal (LBMC) of the Peechi dam of Kerala using the water storage of the dam during November. The predicted and actual discharge through LBMC of Peechi dam is given in Fig. 27.2. An advance forecast about the irrigation through the LBMC will enable the farmers to have necessary drought management strategies well in advance. The said model was developed using data from 1994-2005. The failure of rainfall coupled with lack of irrigation during January (0.314 M.cum) and February 2004 resulted in severe crop loss to the tune of 9.0 ha of Coconut, 14.0 ha of Arecanut, 6.0 ha of Banana, 80.0 ha of Black pepper and 6.0 ha of Nutmeg from the catchment area of Pananchery, Puthur, Varadarapalli and Puthukadu panchayats. The farmers were compelled to increase the irrigation interval from three days to one week during February and then to 10-12 days during March 2004. The regression equation is given below:

LBMC discharge = 5.393 + 0.241 x Storage in November ($R^2 = 0.71$)

Based on the result obtained through model, it has been forecasted that, there will be normal discharge of water during summer 2005 and 2006. Hence, farmers could irrigate their crops and good return was obtained from crops like banana, coconut, arecanut, nutmeg and vegetables.

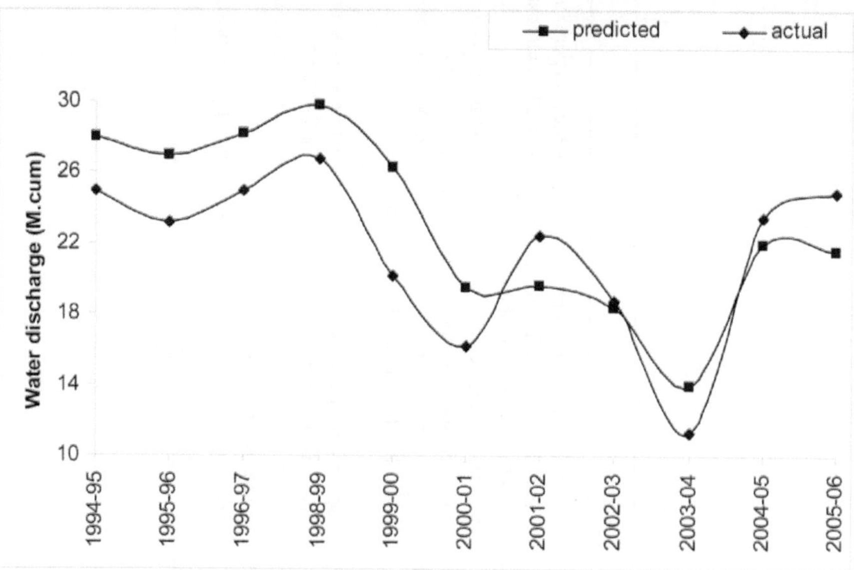

Fig. 27.2: Predicted and actual discharge through LBMC of Peechi dam

27.4 Rainfall analysis for crop planning and Agro-Advisory Services

Lack of dependable rainfall from December to March at Vellanikkara demands the need for soil and water conservation measures in this region. High coefficient of variation of rainfall during the fag end of the southwest monsoon in September (73.7%) also indicates the possibility of extreme rainfall events and uncertainty. Chances of getting 5 mm of weekly rainfall are below 50% from 12th November to 6th May. Based on the trends in the wet and dry spell probability, a plan of advisory on crop management was prepared and presented in Table 27.2.

27.5 Hopkins bioclimatic law

Hopkins (1938) attempted to express the importance of latitude, longitude and altitude in the distribution and rate of development of plants by means of a "Bio climatic law". It may be stated as "A biotic event in North America will, in general, show a lag of four days for each degree of latitude, five degree of longitude and 400 feet of altitude, northward, eastward and upward in spring and early summer". In case of mango, the India Meteorological Department also established that there is a delay in flowering from south to north of India, which generally followed Hopkins' Bio climatic law. In case of cashew also, all the biotic events of cashew in the three test varieties delayed with increase in latitude and altitude from south to north and northeast of Kerala. It is evident that the delay in biotic events such as bud break, flowering, harvest and the total crop duration of cashew depend on latitude and altitude if the distribution of rainfall is uniform and genotype is alike under better crop management

Table 27.2: A plan of advisory on crop management based on wet / dry spell probability of rainfall

Crop	Management practices	Period	Remarks
Coconut	Mulching/ water conservation	November 5 - December 16	Operations are to be completed between the said periods since the probability of getting a wet spell is almost zero from 10th December
	Crown cleaning to reduce bud rot	April 9 – May 20	Operations are to be completed between pre-monsoon showers and South west monsoon (April – May)
Coconut and other plantation crops	Fertilizer application	First dose:-June 25 – July 8 Second dose:- October 22 – November 4	High rainfall from 4 – 24 June and 1 -28 October, may cause leaching of applied nutrients
Black pepper	Shade regulation	April 9 – May 20	Operations are to be completed between pre-monsoon showers and South west monsoon (April –May)
	Prophylactic drenching against phytophthora	April 23 –May20	operations are to be completed just before the onset of South west monsoon
Ginger and Turmeric	Drainage against waterlogging	June 4 – August 12 & October 1 – October 8	Precaution must be taken to avoid water logging during this time

situation in rainfed conditions (Fig.27.3). This information is used in selecting suitable variety, manipulation of flowering and prophylactic crop protection measures in the case of cashew.

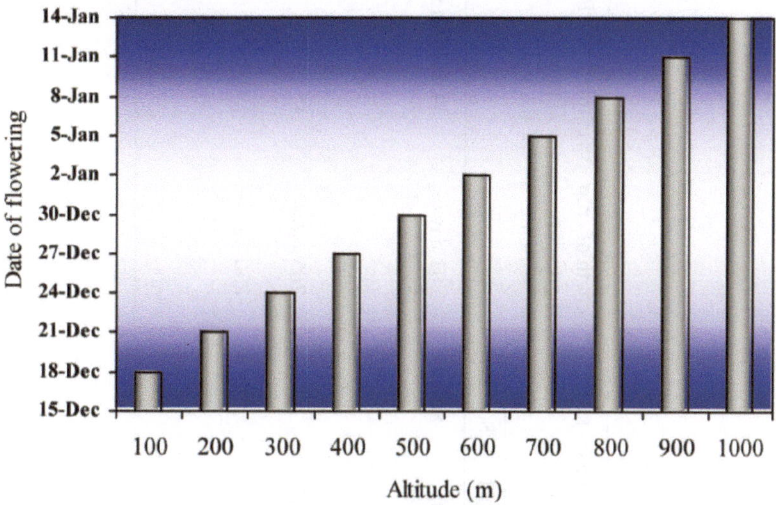

Fig. 27.3: Influence of altitude on cashew flowering

27.6 Agro-climatic zones of small cardamom and cashew

Studies on agroclimatic zonation of small cardamom revealed that the production of zone I is relatively better (>200 kg/ha) when compared to that of zone II and III across the western ghats, where the length of crop growing season is more than 300 days with annual Ima of more than 90%. In addition, the annual temperature range is very low and optimum across the zone I. the production potential of small cardamom is low (100-150kg/ha) over Zone III (Karnataka), where the length of growing season is less than 250 days with annual Ima of 70-75 %. The annual temperature range is also very high, which may be detrimental to cardamom production in the zone III. The zone II falls under intermediate category (150-200 kg/ha), where the length of crop growing season is more than 250 days with annual Ima varied between 80 and 85% (Table 27.3). Similar package is given in the case of Cashew also (Table 27.4).

27.7 Forecasting tea mosquito bug in cashew

Tea mosquito bug is a very severe pest in cashew and the incidence is seen during the flushing and flowering phases, which coincide from September to February depending upon genotype and geographical co-ordinates. The night minimum surface air temperature plays a major role in incidence of pest population (Table 27.5). The hot spot areas of tea mosquito bug across the

Table 27.3: Agro-climatic zones of small cardamom

Zone	Location	Soil type	Ima (%)		Length of crop growing season	Temperature range (°C)	Productivity (Kg/ha)
			Annual	Summer			
I	Vandanmettu, Ayyapan koil, Pampadumpara, Udumbanchola, Santhanpara villages in Idukki district of Kerala (an ideal Cardamom Hill Reserve) and Sethur hills, Meghamalai, Agasthiamalai, Palani hills of Tamil Nadu	Clay and fine loam	<90	80-90	301 days	14.1	More than 200
II	Wayanad district of Kerala and Gudallur district of Tamil Nadu (Nilgiri Biosphere)	Fine loam	80-85	>70	265 days	15.6	150-200
III	Virajpet, Madikeri< Somwarpet, sakleashpur and Mudigere taluks of Karnataka (Coorg, chickmagalur and Hassn districts of Karnataka)	Red loam	70-75	50	230 days	19.0	100-150

Table 27.4: Cashew and its environment across the west and east coast of India

Environment	West coast	East coast
PLANT		
1. Planting material	Pedigree & senile	Relatively new plantations
2. Crop management including crop protection measures	Poor	Relatively better
3. Dormant phase	No moisture stress	Soil moisture stress
4. Reproductive phase	Soil moisture stress	Severe soil moisture stress
5. Flowering	December/January	February/March
SOIL		
1. Soil	Laterite	Littoral sand and sandy loam
2. Topography	Hill slopes	Plains
CLIMATE		
1. Coastal climate	Same	Same (except Chintamani & Jagadalpur)
2. Annual rainfall	2000 to 3600 mm	600 to 2000 mm
3. Max. temperature	< 35°C (30-35°C) may be optimum	>35°C (Occasionally shoots up to 42°C) – may not be optimum - not conducive for nut development
4. Min. temperature		
5. Soil temperature		
6. Sunshine	> 15°C (15-20°C) may be optimum	<15°C (Occasionally goes below 10°C in north – may not be conducive for flowering)
7. Relative humidity	Moderate	High
8. Cyclones	2700 to 3000 hrs	2700 to 3000 hrs
	70 to 90%	50 to 70%
	Rare	Not uncommon
IRRIGATION	No summer irrigation	Irrigation given in some areas
PEST COMPLEX	Incidence is common	Incidence is not common
Tea mosquito bug	Moderate to severe	Low

cashew tract of whole country have been demarcated taking into account the optimum night temperature during the flushing and flowering stages. The criteria adopted for intensity of pest based on night temperature is follows:

Table 27.5: Intensity of tea mosquito bug versus night surface air temperature

Minimum temperature (°C)	TMB population (No/q/tree)	Intensity of TMB
>24	<4	Low
22 and 24	4 to 8	Low to Moderate
15 and 22	8 to 20	Moderate to severe
12 and 15	4 to 8	Low to Moderate
10 and 12	<4	Low
<10	0	Nil

27.8 Crop weather diagram and Crop weather calendars (Crop pest calendar and crop weather cultivation diagrams)

It gives season-wise information on the crop husbandry (tillage to harvest), actual weather and normal weather month/week-wise and information on pest and disease incidence. It may help in understanding favourable weather conditions that are responsible for better crop yields and the vice versa if they are prepared for more number of years. Using the crop weather diagrams, attempts can be made for obtaining better crop yields through agronomic manipulations and also one can predict crop yields qualitatively based on weather conditions. Such crop weather diagrams are prepared in the case of cashew and rice. They can be incorporated in the agro-advisory service.

27.9 Climate Risk Management

The State economy has adversely been affected to a considerable extent due to weather related disasters which are not uncommon in the recent past due to global warming and climate change. Summer drought and heavy rains during monsoon including waterlogging is a threat to plantation crops to a considerable extent. Waterlogging and floods due to excess rains could be minimised by providing surface and subsurface drainage facilities in plantation gardens which will improve the soil health and nutritional status, in turn yields can be increased in plantation crops. Floods are mostly location /region specific due to heavy monsoon rains or cloud bursts during pre-and-post monsoon seasons while effects of drought could be relatively seen in larger areas due to non receipt of rains. In the case of occurrence of summer droughts, drought management techniques could be adopted with irrigation and without irrigation under rainfed conditions. Irrigation may not be possible during summer due to scarcity of water, where plantation crops grow predominantly. However, the available water could be effectively used through various micro-irrigation systems and crop management and protection practices. At the same time, pro-active crop management practices are the only option to mitigate the ill effects of drought on plantation crops under rainfed conditions.

27.9.1 Drought management

The northern districts of Kerala and Palakkad experience prolonged dry spell from December to May, if pre-monsoon showers fail, leading to soil moisture stress. The occurrence of drought for the State as a whole was noticed during summer 1983, 2004 and 2013. It is the lack of emphasis on conservation measures which results in the recurring of drought situation. Rainwater during the monsoon should be harnessed with small investments. If careful arrangements for conserving of water are not done in time, it could result in the drying up of even bore wells and tube wells. In order to keep rivers and wells adequately replenished with water during the dry season, small sub surface dams and anicuts up to 40-odd rivers originating from the Western Ghats and flowing into the Arabian sea are necessary. Implementation of irrigation and hydel development projects in all the valleys would automatically take care of domestic water supply, sanitation and partial flood control. As even partial measures have not been undertaken in all the river systems, the water potential in the State remains poorly harnessed. Watershed based participatory natural resource development and management schemes such as tanks, vented cross bars, diversion weirs and lift irrigation can increase and sustain irrigation, ground water recharge, crop production and drinking water supply considerably. Even in the worst drought conditions, Kerala's rainfall statistics shows annual average as 1600 mm. This much of water if properly conserved and used would be plenty even to spare. The rivers of the State except one or two cease their flow with the cessation of the northeast monsoon and thereafter practically all the rivers dry up and remain dry until June for six months. It is during this period that the State tends to become drought prone recurrently. In addition, drought management practices have to be followed under field conditions. Cultivation of drought tolerant crop varieties, proper soil and water conservation practices like ploughing before the receipt of southwest monsoon and immediately before the cessation of the northeast monsoon, opening basins of crops like coconut before north east monsoon, organic mulching and husk burial can definitely conserve the available moisture in the soil and use water judiciously. Judicious use of water in day-to-day life and for agriculture purpose can reduce the impact of drought in Kerala. Awareness on drought management practices is the need of the hour to sustain plantation crop production in the State against the weather abnormalities. Drought may be a relative term used in case of deficit rainfall or soil moisture stress. Breaks or failure in monsoon may lead to droughts during *kharif*season across the Country. The *kharif* foodgrains production in India is dependent on the monsoon rainfall and its distribution under rainfed conditions as mentioned in *chapter 3*. At the same time, it is the northeast monsoon and pre-monsoon showers, which is important in the case of plantation crops. In the Tropical Monsoon Climates where monsoon is assured, it is the summer rainfall, which is also critical in the case of plantation crops as they are perennial in nature and the availability of soil moisture round the year is very important for survival of the crops.

27.9.2 Scarce water resources management

It is understood that the plantation crops that are grown in the low-and-mid-lands of Kasaragod and Kannur Districts experience severe soil moisture stress almost in half of the years. It was probably one of the reasons, why, the fluctuations in plantation crop production was noticed very often. In the recent past, the occurrence of droughts in summer adversely affected the plantation crop production to a great extent. Therefore, Irrigation schedule during the summer months under scarce water resources management as a part of climate risk management is one of the tools to mitigate the ill effects of summer drought through various ways under scarce water resources. If the irrigation requirement is worked out using reference evapotranspiration and crop coefficient, the irrigation water requirement is comparatively less and soil moisture status can be improved to certain extent and in turn plantation crop yields can be increased. To quote an example, the irrigation requirement of coconut was worked out at RARS, Pilicode using the reference evapotranspiration (modified Penman method has been generally accepted as a reliable method to determine evapotranspiration from climatic data) and crop coefficient. The irrigation requirement varied from 1106 litres/palm/month in December to 1488 litres/palm/month in April. The total irrigation requirement from December to May was worked out as 7807 litres (Table 27.6). Coconut growers can very well adopt the irrigation quantities during summer as recommended above wherever water is available which will improve coconut yield significantly as the yield per palm is well below the normal under rainfed conditions due to prolonged dry spell from November to May. In another experiment, a field experiment was taken up at RARS, Pilicode, Kasaragod where the palms were provided basin irrigation at the rate of 450 litres/palm/week till the water is available during summer and leaving the palms without watering further till the onset of monsoon. The yield was better in the above treatments when compared to that of pre-treatments' yield (Table 27.7). The quantity was arrived at 450 litres/palm/week, taking the evapotranspiration of coconut as 5 mm per day.

Table 27.6: Irrigation requirement of coconut

Month	ETo(mm)	Water requirement (mm) (ETo x 0.75)	Irrigation requirement (l) (pr^2h)
December	3.79	2.84	1,106.0
January	3.95	2.96	1,154.0
February	4.56	3.42	1,204.0
March	5.01	3.76	1,464.0
April	5.26	3.95	1,488.0
May	4.76	3.57	1,391.0
Total			7,807.0

ETo-Reference evapotranspiration; r- Radius of coconut basin- 2 metres; 0.75 – Crop coefficient of coconut

Table 27.7: Duration of soil moisture stress on coconut yield of WCT

Treatment	Pre-treatment yield(1976-89) (A)	Post-treatment yield(1991-97) (B)	Difference in yield (A – B)	% increase over pre-treatment yield	Whether significant over pre-treatment yield
T1	103	122	19	18.4	S
T2	103	112	9	8.7	S
T3	94	107	13	13.8	S
T4	87	120	33	37.9	S
T5	92	115	23	25.0	S
T7	91	95	04	4.4	S
T6	60	82	22	36.7	S

T1- Irrigating the palms @450l/palm/week during December and January
T2- Irrigating the palms @ 450l/palm/week from December to February
T3 – Irrigating the palms @450l/palm/week from December to March
T4 – Irrigating the palms @ 450l/palm/week from December to April
T5 – Irrigating the palms @ 450l/palm/week from December to May
T6 – Irrigating the palms as per climatic water balance procedure
(150 l/palm/week in December, 200 l/palm/week in January, 300 l/palm/week in February, 350 l/palm/week in March, 400 l/palm/week in April, 450 l/palm/week in May)

27.9.3 Rainwater harvesting /groundwater recharging

Efforts are needed for effective rainwater harvesting in farm lands. Wherever water is running we should make it walk. This is done by terracing, contour bunding, and increasing the soil vegetative cover. All walking water should be lied down. Rain water harvesting is unique and traditional system of recharging of the under- groundwater bodies for sustained water availability. There were several techniques in rainwater harvesting which can be practiced in house hold (roof top rainwater harvesting) as well as fields. Traditional sources of water like very shallow, small diameter wells known as 'keni' in Wayanad district, horizontal wells/tunnels called 'surangams' in Kasaragod district, natural wells, ponds, tanks, springs, lakes and wind breaks and shelter belts should form an integral part of the watershed development programme. Groundwater utilization has to be based on location specific investigations and over exploitation has to be avoided.

27.9.4 Soil health

The indiscriminate use of fertilizers and plant protection chemical and weedicides are polluting the soil, air and water beyond the tolerable limit. This is true in case of certain areas of Kerala where the practice is creating problem. There was drastic decline in the use of organic manure for the past years due to several reasons. In the process, organic content of the soils is very poor and water-holding capacity of the soils became very weak, resulting in frequent droughts. Therefore, it is essential to improve the organic content of the soils to retain the original status of the soils. Awareness has been created on the use of organic manures among the farmers with reasonable use of chemicals without affecting productivity. There are certain crops and areas where organic farming can be practiced in Kerala. Organic farming also plays an important role as an alternative food production process. It is a crop production system that avoids or largely excludes the use of synthetically produced fertilizers, pesticides, growth regulators and livestock feed additives. As far as possible it relies on crop rotations, animal manures, legumes, green manures, organic wastes, bio-fertilizers and aspects of biological pest control to maintain soil productivity and tillage to supply plant nutrients and to control insects, weeds and other pests. Input availability, productivity, premium price and marketing aspects have to be considered. In other areas, good agricultural practice with the minimum dependence will definitely reduce the pollution of soil, air and water and reduce its influence in greenhouse gas emissions.

27.9.5 Weather forecasting

The India Meteorological Department, popularly known as MausamBhavan, Government of India is one of the oldest departments in the Country, located in New Delhi. It was established in 1875. It has a network of approximately 6000 rain gauge stations and more than 550 surface meteorological observatories plus about 300 stations which include hydro-agro-meteorological, radiation, air pollution and pilot-balloon stations spread over the entire Country. The whole Country has been divided into 36 meteorological subdivisions based on homogeneity in its rainfall distribution. The aim of the department is to forecast daily weather on meteorological subdivision-wise and disseminate the same on time to the users. All the State Capitals in India also have State Meteorological Centres (SMCs) to serve the weather forecast needs of the States and these centres issue Farmers' Weather Bulletins (FWBs). The bulletins contain a forecast on weather elements valid for 36 hours and an outlook for the subsequent two days. There is a vast improvement in precise short range forecasting with the advent of supercomputers along with the radar and satellite information. Cyclone forecast with a network of weather radars along the East Coast of India is a classical example for success of short range weather forecasting. One of the prime objectives of providing weather forecast is to disseminate information to the farmers and users on real time basis so that they can utilize the beneficial aspects of weather information to increase crop and livestock production or to minimise the damage, which may be caused directly or indirectly by unfavourable weather. Incidence of crop pest and diseases and animal diseases could also be forewarned based on weather forecast if information on relationship between weather and specific pest and disease incidence is known in advance. The output of agriculture (includes animal agriculture) could be increased and recurring losses minimized if reliable forecast on incidence of pest and diseases is given timely based on weather variables. Hence, weather forecast plays a key role in sustaining agricultural/livestock production of the Country. This could be possible if information is provided to the farmers, in advance, on the type of weather situations likely to be encountered and method to be adopted for efficient management of inputs. Probably, this is one of the best tools to mitigate the ill effects of climate change/climate variability in plantation crops' production.

27.9.5.1 Long Range Weather Forecasting (LRF)

It is a forecast for more than 10 days, a month and for a season. The India Meteorological Department restarted issuing the long range weather forecasting since the year 1988 onwards on total monsoon rainfall of the country by the end of May (25th May) initially. Now, it is available by 15th April, 45 days ahead of onset of monsoon. This year's seasonal forecast of monsoon rainfall was released to the Press on 24th April and predicted as 97% of LPA with +/- 5% model error. The predicted and actual long period average (LPA) of monsoon (June-September) rainfall of the Country were broadly in agreement except significant deviations in 1994, 1997, 1999, 2002, 2004, 2007, 2009, 2012, 2013 and 2014 and 2018 (Fig. 27.4). Though the total monsoon rainfall tallies with predicted in majority of the years, the distribution of rainfall within the season is a very complex one to predict and it is the one which decides agriculture production and to some extent animal agriculture too. Therefore, the forecast need to be improved in many angles including zonal scale. The long range weather forecast can be used for predicting likely trends in foodgrains production of India before beginning of the kharif season as the foodgrains production depends mostly on the distribution and amount of monsoon rains across the

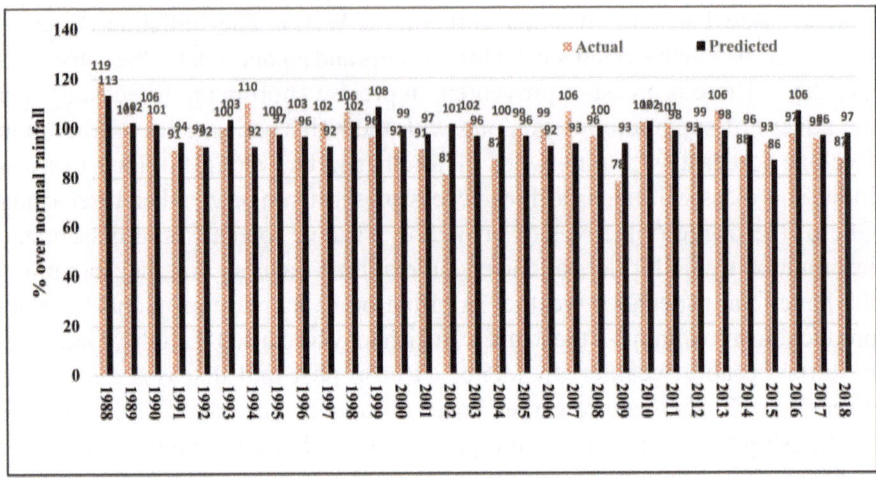

Fig. 27.4: Actual and predicted monsoon rainfall over India from 1988 to 2018 (Long Range Forecasting)

Country under rainfed conditions. These forecasts can hold the foodgrain prices in check through buffer stock operations. In good monsoon years when prices fall, the Government may step in and buy, and in bad years when prices tend to rise, it may unload a part of what it had purchased in the past. This will help in price stability, less financial losses, better distribution of the available foodgrains and its import and export policies. With this in view, the crop weather watch groups (CWWGs) in the Agriculture Ministry, Government of India closely follows the forecast and its impact on foodgrains production during the kharifseason.The operational Long Range Forecasting (LRF) system in India is based on the statistical methods. IMD issues operational long range forecasts for rainfall/ precipitation during winter (Jan- March), southwest monsoon (Jun to Sept) and northeast monsoon (October-December) seasons. Among these, forecast for monsoon season is most important as the rainfall received during this season accounts for 70-90% of the annual rainfall over most parts of the country and due strong positive association of monsoon season rainfall with both Kharif and rabi crop production in the country. India Meteorological Department (IMD) issues various monthly and seasonal operational forecasts for the south-west monsoon season using models based on latest statistical techniques with useful skill. Operational models are reviewed regularly and improved through in house research activities. Details of the present operational forecasting system used for long range forecasting for southwest monsoon rainfall can be from Pai *et al.* (2011). Operational forecast for southwest monsoon season rainfall over the country as a whole is issued in two stages; the first forecast in April and update for the April forecast in June. In June, along with the update forecast for the season rainfall, forecasts for the monthly rainfall (for the months of July and August) over the country as a whole and that for season rainfall over the four geographical regions of the country are also issued. In addition, forecast outlook for the rainfall during second half (August-September) of the monsoon season (issued in July) and that for the September rainfall (issued in August) over the country as a whole are also issued. Separate models are used for preparing each of these forecasts. While preparing, the operational forecasts, the experimental and operational forecasts from various research institutions from India and abroad were also taken into account. Along with the operational forecasts, experimental forecasts of monthly and seasonal rainfall over Indian region based on Seasonal Forecasting Model (SFM) are also prepared. In addition to the above long range forecasts, an operational forecast for the monsoon onset over Kerala is also issued in May.

27.9.5.2 Climatic/agroclimatic Forecasting/extended forecasting

It requires past meteorological data for a good number of years (say 30-50 years). The trends in rainfall and its variability, probability on distribution of rainfall over a season can be determined weekly using the past data on rainfall for a given location. This information is useful to crop planners and farmers as crop growth periods can be adjusted under rainfed conditions depending upon rainfall probabilities. The climatic trends also will help in understanding the impact of climatic variability on animal agricultural production over a period of time. The periods of water surplus and deficit could be worked out for a particular location based on past meteorological data (rainfall and evaporation) through the water balance procedure. This information can be used for selection of crops and adjusting crop growing periods, irrigation scheduling during dry spells and for providing drainage in excess of water surplus. The crop weather relationships could also be well understood based on past biological and meteorological data, which in turn will help in mitigating the ill effects of weather aberrations through better agronomic/livestock practices. In this direction, the research programmes should be strengthened in the field of agricultural/livestock meteorology at the ICAR, SAUs and SVASUs levels with technical personnel having the subject background.

27.9.5.3 Farmers' Weather Bulletins (FWBs)

A reliable system of short range and medium range weather forecasts is needed for effective on farm management practices such as cultivation of land, preparation of seed bed, planting, choice of crops in case of prolonged monsoon breaks, fertilizer management, harvesting of crops, post-harvest storage and transportation to markets. India Meteorological Department started issuing Farmers' Weather Bulletins (FWBs) on a regular basis since 1945 and later it was entrusted to Regional Meteorological Centres (New Delhi, Mumbai, Calcutta,

Chennai and Nagpur), which contain a forecast of expected weather for the next 36 hours and an outlook for another two days. In due course, meteorological centres at State headquarters started to issue FWBs. All India Radio used to broadcast FWBs every evening in regional languages to reach the most remote corner. However, they are found to fall short of the requirements of farmers in terms of duration of forecast validity and capability for small areas, likely impact on agronomic operations. The crop weather calendars prepared for main crops lost their significance with fast changing varieties. As they contain limited information, there was a need to frame them in relation to crops and to specific areas. Thus came the concept of Agro-meteorological Advisory Services (AAS) and it is in operation since 1977. However, the utility of the same is very much limited. Realizing the defects and make the AAS more meaningful, National Centre for Medium Range Weather Forecasting (NCMRWF) took up the task in 1988 based on medium range weather forecasting.

27.9.5.4 Dissemination of Agromet Advisory Services

The agro-advisory bulletins prepared weekly/biweekly by the AMFUs could be communicated on the same day/next day to the selected farmers through a special messenger. The AIR, DD and the other communication facilities can be effectively used to disseminate AAS. It is being done in almost all parts of the Country. The agro-advisory bulletins are prepared at the AMFUs by the panel of scientists based on the "Package of Practices Recommendations" developed by the State Agricultural Universities for the respective zones. In the absence of a separate package based on R&D in operational agricultural meteorology in the form of agromet products, it is difficult to improve the quality of AAS. The essence of agro-advisory is to make weather forecasting awareness among the farmers for day -to -day farm operations for sustenance of agricultural production. This is possible if the interaction between weather variables and crop growth and its development is well understood for which crop growth simulation models have to be revalidated/developed for major crops. In similar lines, an interdisciplinary team should work to develop models on pest and disease dynamics. It needs strengthening of research and development activities of the AMFUs in every agro-climatic zone. In addition to that, the NCMRWF in collaboration with the ICAR and SAUs should try to reorient the various research programmes ongoing in the field of agricultural meteorology to tune to AAS. Simultaneously, the skills of medium range weather forecasting must be improved further by the NCMRWF as of now the forecast skill is very poor. The utility of agro-advisory in the field of agriculture based on Medium Range Weather Forecasting will be fruitful only if the forecast skill is improved for at least one week ahead along with improvement in quality of AAS. The crop weather watch groups (CWWGs) should be established at the district level on similar lines what we have today at the Centre to integrate the various activities related to weather and agriculture. The CWWGs should be made operational to minimize the crop losses under projected climate change scenario.

27.9.5.5 Economic impact of AAS

A reliable and clearly disseminated weather forecast is a very important tool for forewarning crop managers of potential weather hazards. The India Meteorological Department is constantly working to improve forecast skill and help disseminate the forecast in a suitable form to aid farm level decisions. The economic impact of a weekly AAS based on medium range weather forecasting has been assessed for different crops and regions (Table 27.8). This showed that the percentage increase in yield varied from 6.4 – 19% depending upon the crop in the case of AAS farmers compared to the non-AAS farmers. Furthermore, it indicated that seasonal crops need intensive advisory, followed by less intensive for biennials and perennials. This is one of the effective tools to mitigate the ill effects of climate change/variability on crops. Reliability of weather forecast is very important as it declines with increase in time and decrease in space, for which the India Meteorological Department is working to improve the forecast skill. Even the forecast skill is better, the dissemination at farm level is very important. There should be a mechanism to disseminate information effectively through Information Communication Technology such that every farmer and needy people should get weather service at his/her doorstep. The Village Resource Centres under ISRO programme must be linked with online network to disseminate agro advisory based on weather forecasting. It is very pertinent during weather calamities to minimize crop losses. In an event of unusual summer rain, Kuttanad and Thrissur paddy farmers suffered to a large extent. If advance weather service is extended at community level, crop loss would have been minimized to some extent by covering the paddy grains as per the farmers' view. Another way of helping the farmers in such weather calamities is to implement weather insurance. As a long term measure, drainage facilities should be improved at micro-level so that flood damage is minimized in addition to development of flood tolerance/saline tolerant paddy varieties. As the paddy/wetlands act as water sink during the rainy season and water source during summer, such lands should be protected as the area under wetlands is declining vary fast in addition to deforestation reduction across the Western Ghats so as to reduce the run-off and degradation of soils.

Table 27.8: Impact of AAS on crop yields from 2003-04 to 2006-07

Season	Crop	Yield (t/ha)		% increase over in yield over non -AAS farmers	% increase in net return over non -AAS farmers
Rabi 2003-04	Paddy	3.7	3.0	19.0	30.0
	Banana	31.8	27.9	12.2	11.5
	Coconut*	13460	11025	18.1	13.2
Kharif 2004	Paddy	2.8	2.6	7.1	31.6
	Banana	22.8	20.5	10.1	7.9
Rabi 2004-05	Paddy	3.0	2.7	7.5	34.1
	Banana	31.3	29.3	6.4	11.9
Kharif 2005	Paddy	2.7	2.5	6.5	38.5
	Banana	25.3	22.4	11.2	7.1
Rabi 2005-06	Paddy	3.3	2.9	13.6	36.0
	Banana	27.8	24.5	11.8	12.3
Kharif 2006	Paddy	2.8	2.5	9.0	29.4
	Banana	24.5	21.6	11.9	10.0

27.9.5.6 Weather insurance

Weather extremes like flood, drought, landslides, lightening, cloud bursts and hail storms, heat and cold waves are not uncommon now-in-days under the projected global warming and climate change scenario. Both crop and farmer are to be safeguarded against these calamities. Often, the farmer who takes loan and invests in farming cannot repay it due to crop failures in the context of weather extremes. It leads to hardship and rural livelihood is in threat. Farmer suicides are frequent in India and such incidents were noticed in Maharastra, Telengana, Andhra Pradesh and Karnataka. It was also reported in Wayanad District of Kerala in 2004 due to unprecedented crop loss because of summer drought. Weather insurance is a different one when compared to other crop insurance scheme. If crop loss is noticed due to weather extremes like unusual summer rains in case of paddy, coffee and cashew, monsoon breaks in case of black pepper, occurrence of floods and droughts and heat and cold waves, the farmers need to be paid compensation. Farm level expected income should be assured in an event weather calamity so as to sustain rural livelihoods. Therefore, insuring farmers as well as crop becomes necessary against weather abnormalities. In this context, precise and accurate information on weather change and impacts are necessary. Short, medium and long range forecasting will definitely help in weather tune farming so as to minimize the adverse effect of weather abnormalities taking suitable precautionary measures to lessen the harmful effects. The meteorological net work of observatories has to be strengthened to disseminate agro-advisory service based on integrated weather forecasting. Pro-active measures against weather aberrations will go a long way in minimizing the crop losses against the weather and climate disasters.

27.9.5.7 Climate Risk Managers

Climate change awareness programmes should be organized at various levels, commencing from farmers to planners. These programmes should focus on climate change impacts in agriculture and site specific adaptation and mitigation strategies. The farmers who are trained in all the relevant aspects of climate change and its impacts in agriculture are known as "Climate Risk Managers". They are supposed to disseminate the information and knowledge gained by them through frequent trainings to the neighbouring farmers so as to sustain agriculture against the ill effects of climate change with their wisdom (Plate 27.1). Though ill-effects of abnormal weather events are well understood in plantation sector at the field level, it is difficult to quantify the same at the zonal/ state level based on secondary data. Hence, a realistic scientific method is to be evolved as such no technique is available to project future crop trends in plantation crops based on climate variability/change. Warming with decline in rainfall and increase in temperature range is projected at some of the locations across the high ranges along the West Coast, where cardamom, coffee, tea and black pepper are predominantly grown under the influence of forest-agro-ecosystems. Deforestation influences the crop boundaries to a considerable extent. Crop growth simulation models indicate that the cropped area and yields

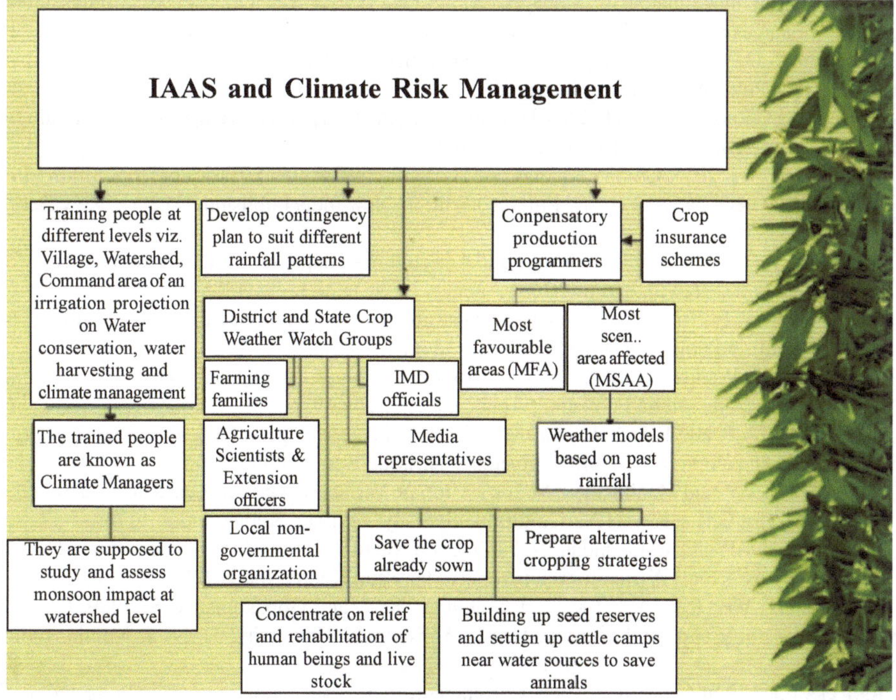

Plate 27.1:

are likely to be reduced due to increase in temperature in the case of coffee, tea and cocoa as well in other crops. Though the models indicate reduced crop yields due to long term climate changes (increase in temperature), it is in fact not seen such impacts as on today, but at the same time the adverse impacts of climate variability (occurrence of heat and cold waves and floods and droughts) on crop yields is significant. The GCM models indicate that the frequency of weather abnormalities is likely to increase in ensuing decades under the projected climate change scenario. Therefore, the adverse impact of climate change on crops is seen only through climate variability rather than the direct long term effects of climate change, which is doubtful. Therefore, short term crop management strategies need to be worked out as part of climate resilience agriculture to combat the ill effects of climate variability/climate change for sustenance of crop production in a changed climate scenario.

References/Suggested Readings

Gopakumar, C.S., Krishnakumar, K.N. and Rao, G.S.L.H.V. P. 2011. Climate and Agriculture (In Malayalam). Current Books.Thrissur, Kerala, India. 233p

Martinez, R., Hemming, D., Malone, L., Bermudez, N., Cockfield, G., Mushtaq, S.,Diongue,A. Hansen, J., Hildebrand, A., Ingram, K., Jakeman, G., Kadi, M., McGregor, G.R. Rao, P., Pulwarty, R., Ndiaye, O., Srinivasan, G., Eh.R., Seck, White, Nand Zougmore, R. 2012. Improving Climate Risk Management at Local Level – Techniques, Case Studies, Good Practices and Guidelines for World Meteorological Organization Members-Chapter 21 (477-532): In Risk Management-Current Issues and Challenges (Ed: Nerija Banaitiena). In Tech Publishers. Rijeka, Croatia,584p

Nybe, E.V. and Rao, G.S.L.H.V.P. 2005. Drought management in plantation crops. Kerala Agricultural University. Thrissur, Kerala, India. 176p.

Rao, G.S.L.H.V.P. 2015. Climate Change and Horticulture-Chapter-5. Basics of Horticulture (Ed. K.V.Peter). 2nd Revised and Enlarged Edition. New India Publishing Agency. New Delhi, India. 704p.

Rao, G.S.L.H.V.P. 2015. Climate Change and Agriculture - Chapter-1. Climate Resilient Crops for the Future. New India Publishing Agency. New Delhi, India. 450p.

Rao, G.S.L.H.V. P. 2011. Climate change adaptation strategies in agriculture and allied sectors. Scientific Publishers (Jodhpur), India, 336p.

Rao, G.S.L.H.V.P. and Gopakumar, C.S. 2016. Climate change and plantation crops in the Humid Tropics. New India Publishing Agency, New Delhi,492p

Rao, G.S.L.H.V.P. and Krishnakumar, K.N. 2011. Climate change and coconut productivity in Kerala. Kerala Agricultural University. Thrissur, Kerala, India. 219p.

Rao, G.S.L.H.V.P. and Manikandan, N. 2009. Climate Variability and small cardamom across the Western Ghats. Kerala Agricultural University. Thrissur, Kerala, India. 117p.

Rao, G.S.L.H.V.P. and Manikandan, N. 2009. Crop weather relationships of cocoa in the Humid Tropics. Kerala Agricultural University. Thrissur, Kerala, India. 117p.

Rao, G.S.L.H.V.P., Rao, A.V.R.K., Krishnakumar, K.N. and Gopakumar, C.S. 2009. Climate variability and food security. Kerala Agricultural University. Thrissur, Kerala, India. 117p.

Rao, G.S.L.H.V.P. and Nair, R.R. 1988. Agrometeorology of plantation crops. Kerala Agricultural University. Thrissur, Kerala, India. 187p.

Rao, G.S.L.H.V.P., Rao, G.G.S.N. and Rao. V.U.M. 2010. Climate change and agriculture over India. PHI Learning Private Limited, New Delhi, India. 320p.

Rao, G.S.L.H.V.P. 2015. Agricultural Meteorology (Third printing). PHI Learning Private Limited, New Delhi, India. 352p. 9.6 Climatic normals for Crop Production

Rao, G. S.L.H.V.P., Varma, G.G. and Beena, V. 2017. Livestock Meteorology, New India Publishing Agency. New Delhi, 532 p.

Rao, G. S.L.H.V.P., Varma, G.G. and Beena, V. 2018. Climate Resilient Animal Agriculture, New India Publishing Agency. New Delhi, 468 p.

28

Geospatial Technologies for Impact Assessment of Extreme Weather Events on Agriculture

K.V. Ramana, K.V.V. Ramesh, Sudheer Tiwari and P. Srikanth*

Andhra Pradesh Space Application Centre, Planning Department
Govt of AP, Vijayawada., India
**National Remote Sensing Centre, ISRO, Hyderabad, India*

Natural disasters are extreme events within the earth's system (lithosphere, hydrosphere, biosphere or atmosphere) which differs substantially from the mean, resulting in death or injury to humans, and damage or loss of valuable good, such as buildings, communication systems, agricultural land, forest, natural environment. They are a profound impact of the natural environment upon the socio-economic system (Alexander, 1993). This impact may be rapid, as in the case of earthquakes, or slow as in the case of drought. It is important to distinguish between the terms disaster and hazard. A potentially damaging phenomenon (hazard), such as an earthquake by itself is not considered a disaster when it occurs in uninhabited areas. It is called a disaster when it occurs in a populated area, and brings damage, loss or destruction. Natural disasters occur in many parts of the world, although each type of disaster is restricted to certain regions. Natural disasters are events which are caused by purely natural phenomena and bring damage to human societies (such as earthquakes, volcanic eruptions, hurricanes). Mitigation of natural disasters can be successful only when detailed knowledge is obtained about the expected frequency, character, and magnitude of hazardous events in an area. Many types of information that are needed in natural disaster management have an important spatial component. Spatial data are data with a geographic component, such as maps, aerial photography, satellite imagery, GPS data, rainfall data, borehole data etc. Many of these data will have a different projection and co-ordinate system, and need to be brought to a common map-basis, in order to superimpose them. We now have access to information gathering and organising technologies like remote sensing and geographic information systems (GIS), which have proven their usefulness in disaster management. Remote sensing and GIS provides a data base from which the evidence left behind by disasters that have occurred before can be interpreted, and combined with other information to arrive at hazard maps, indicating which areas are potentially dangerous. The zonation of hazard must be the basis for any disaster management project and should supply planners and decision-makers with adequate and understandable information. Remote sensing data, such as satellite images and aerial photos allow us to map the variabilities of terrain properties, such as vegetation, water, and geology, both in space and time. Satellite images give a synoptic overview and provide very useful environmental information, for a wide range of scales, from entire continents to details of a few metres. Secondly, many types of disasters, such as floods, drought, cyclones, volcanic eruptions, etc. will have certain precursors. The satellites can detect the early stages of these events as anomalies in a time series. Images are available at regular short time intervals, and can be used for the prediction of both rapid and slow disasters. When a disaster occurs, the speed of information collection from air and space borne platforms and the possibility of information dissemination with a matching swiftness make it possible to monitor the occurrence of the disaster. Many disasters may affect large areas and no other tool than remote sensing would provide a matching spatial coverage. Remote sensing also allows monitoring the event during the time of occurrence while the forces are in full swing. The vantage position of satellites makes it ideal for us to think of, plan for and operationally monitor the event. GIS is used as a tool for the planning of evacuation routes, for the design of centres for emergency operations, and for integration of satellite data with other relevant data in the design of disaster warning systems. GIS is extremely useful in combination with Global Positioning Systems (GPS) in search and rescue operations in areas that have been devastated and where it is difficult to orientate. The impact and departure of the disaster event leaves behind an area of immense devastation. Remote sensing can assist in damage assessment and aftermath monitoring, providing a quantitative base for relief operations. In the disaster rehabilitation phase GIS is used to organise the damage information and the post-disaster census information, and in the evaluation of sites for reconstruction. Remote sensing is used to map the new situation and update the databases used for the reconstruction of an area, and can help to prevent that such a disaster occurs again. The case studies of extreme weather events - floods due to cyclones, heavy rains and drought were presented hereunder.

28.1 Agriculture damage assessment due to flood in Andhra Pradesh

28.1.1 Study of rice inundation – September, 2009

The incessant rains, both in the upper and lower catchments of the Karnataka rivers, since the second half of September 2009 and especially from 29th Sep to

4th October, 2009 caused serious flooding in the 3 districts of Andhra Pradesh namely Kurnool, Krishna and Guntur districts all along the river Krishna. The flood event and the rice crop inundation were monitored using the pre and post event microwave and optical remote sensing data. The Radarsat1 data of Aug15, Sep 08 and Oct 02, 2009 over Krishna and Guntur districts; Aug 01, Aug 25 and Sep 18, 2009 over Kurnool districts were used for mapping the rice cropped area. The flood layer from DSC, NRSC was overlaid on the rice cropped layer and the areas of rice crop inundation were mapped (Figs. 28.1 and 28.2). Apart from the inundation, sand casting was observed over large areas damaging the crop severely. A total of 25085 ha of rice crop was submerged in the above mentioned three districts.

28.1.2 Mapping and estimation of inundated crop area – 2012 during NILAM cyclone in Andhra Pradesh

Heavy rains lashed parts of Andhra Pradesh and Tamil Nadu states due to the impact of strong cyclonic storm - 'NILAM' which made the landfall near Mahabalipuram on October 31. Satellite data has been analyzed to study the extent of total inundation and cropped area inundated in two states. The satellite data from Radarsat-2 and Resourcesat 2 was acquired and analyzed. The fortnightly time composited NDVI mosaic from AWiFS and MODIS were also used for estimating the cropped area.

Flood inundation layer generated using Radarsat-2 data was intersected with the *kharif* cropped area derived from AWiFS/MODIS data and generated the inundated cropped area. In the initial period (1 and 2nd Nov, 2012), the inundated

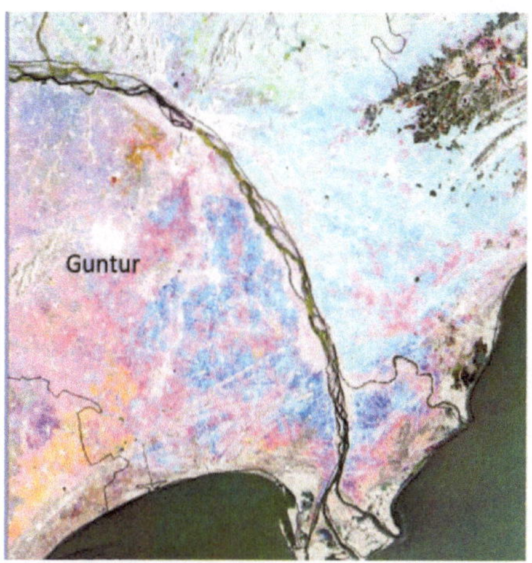

Radarsat1 FCC Aug 15, Sep 08, Oct 02 2009

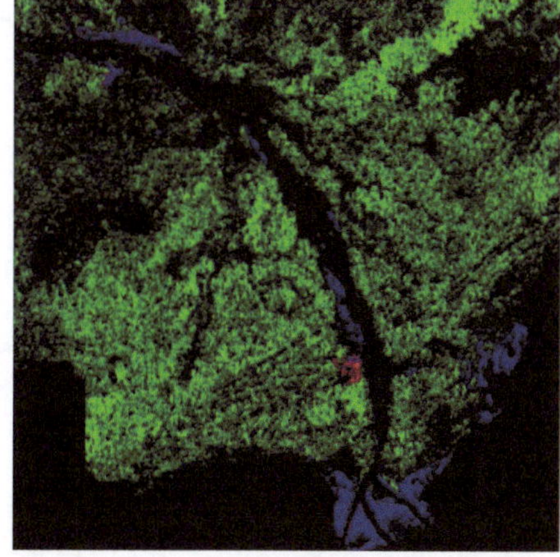

Rice cropped area overlaid with flood layer

Fig. 28.1: Satellite and the corresponding rice crop distribution overlaid with flooded areas

areas were concentrated in Southern AP and Tamilnadu (Table 28.1) and subsequently, the areas north of Guntur district were found to be inundated. The inundated areas were mostly found to be cropped areas only.

Table 28.1: District statistics of crop inundation in Andhra Pradesh.

S No	District	01 Nov, 2012 Cropped area Inundated (Ha)	02 Nov, 2012 Cropped area Inundated (Ha)	05 Nov, 2012 Cropped area Inundated (Ha)	06, Nov 2012 Cropped area Inundated (Ha)
1.	Prakasam	45160	16509		
2.	Nellore	30122	1166		
3.	Chittoor	4273			
4.	Guntur		482		
5.	West Godavari		939		
6.	Krishna		1129	1436	
7.	West Godavari			3139	2232
8.	East Godavari			5879	6039
9.	Visakhapatnam			5209	3701
10.	Vizianagaram			126	
11.	Srikakulum			441	1431
12.	Khammam			70	
	Total	79557	20225	16300	13403

Fig. 28. 2: Spatial depiction of crop inundation in Andhra Pradesh

28.3 Crop inundation due to Rainfall in parts of Coastal Andhra Pradesh - 2013

Heavy rainfall in the aftermath of Cyclone Phailin left parts of the Indian states of Odisha and Andhra Pradesh under flood water. High intensity rains ranging from 10 to 30 cm lasted over many parts of Andhra Pradesh for 4 to 5 days from 22nd to 28th October, 2013. The severely affected districts included were the Coastal area – Srikakulam, Visakhapatnam, East and West Godavari, Krishna, Guntur and Prakasam districts of Andhra Pradesh. An attempt was made to study the extent of crop inundation and crop damage using geospatial techniques in the above districts. The extent of crop inundation was found to be, 3152, 1138, 10687 and 13267 ha for Srikakulum, Visakhapatnam, Guntur and Prakasam districts respectively (**Fig. 28.3**).

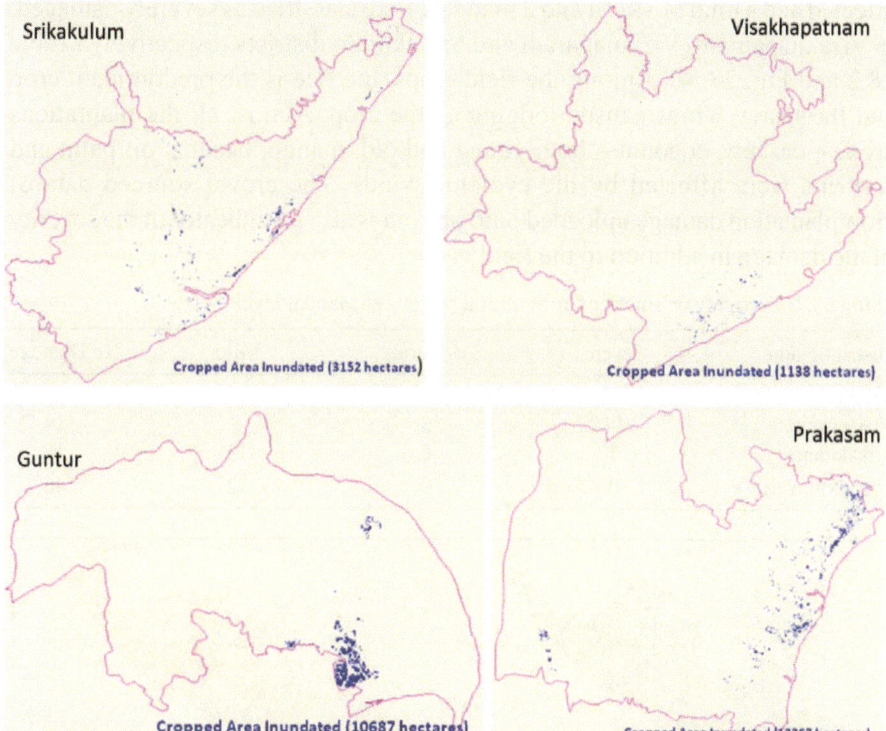

Fig. 28.3: Crop inundation in Visakhapatnam, Visakhapatnam, Guntur and Prakasam districts.

28.4 Agriculture damage assessment due to *Hudhud* cyclone (Wind and Heavy rain) in Andhra Pradesh

The northern coastal districts of Andhra Pradesh were hit by the Very Severe Cyclonic Storm (VSCS)-Hudhud on 12th October, 2014. The storm crossed north Andhra Pradesh coast over Visakhapatnam between 1200 and 1300 hrs

IST of 12th Oct. Due to the heavy gales and cyclonic rains, paddy and other rainfed crops apart from horticulture plantations have suffered extensive damage. The damage is caused due to the combined effect of heavy winds and rainfall and the subsequent inundation in the low lying areas. The maximum cumulative rainfall of around 30 cm received on the day of cyclone crossing the AP coast (12th Oct, 2014) and by the strong wind speeds of over 180 kmph caused extensive damage. A rapid assessment was carried out to study the effect on crops and horticulture plantations for prioritization of effected mandals. The wind speed and the rainfall data collected from AWS and the satellite data of LISS-IV, Landsat 8 OLI and MODIS data were used for assessing the damage. The wind speeds and the total rainfall were used for assessing the risk of crops for getting damaged at 1 km grids and the pre and post event satellite data was used for studying the impact. The Visakhapatnam district is the severe most affected and a total of 19, 14 and 2 mandals were classified as severely damaged, in Visakhapatnam, Vizianagaram and Srikakulum districts respectively (Table 28.2 and Fig. 28. 4). Among the field crops, the rice is the predominant crop and the heavy winds caused lodging of the crop. Almost all the plantations crops – cashew, coconut – both young and old, mango, banana, oil palm and casurina were affected by the cyclonic winds. The crowd sourced data of crop/plantation damage uploaded onto bhuvan is also an indicator of the severity of the damage in addition to the field visit.

Table 28.2: Impact assessment of agricultural regions at Mandal level

District Name	Severe	Moderate	Mild	No Damage
Visakhapatnam	19	7	14	3
Vizianagaram	14	3	3	14
Srikakulam	2	6	13	16
East Godavari	0	0	12	46

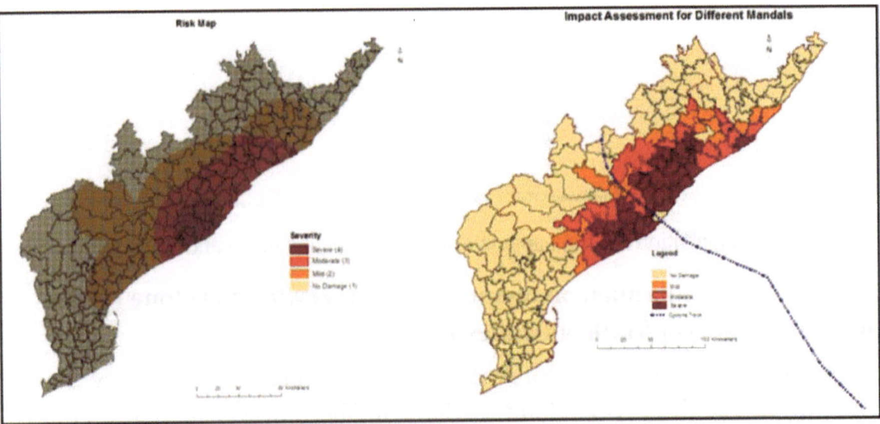

Fig. 28.4: Risk and impact assessment over agricultural regions

28.5 Agriculture damage assessment due to Drought in Andhra Pradesh

In Andhra Pradesh total 670 mandals were observed fortnightly using fifteen years (2002 to 2016) time series data for the month of June to October of Kharif season. Normalised Difference Vegetation Index (NDVI), Normalized Difference Wetness Index (NDWI) and Vegetation Condition Index were generated and cumulative fortnightly rainfall were computed using daily rainfall observed by 1169 Automatic Weather Stations (AWS). The analysis integrated all these parameters NDVI, NDWI, VCI, Dry spell and rainfall (Figs. 28.5 and 28.6). The assessment was divided in two part, first part "Drought Warning" was from June to August and grouped into Normal, Watch and Alert situation of each mandal and Second part "Drought Declaration" was grouped into Normal, Mild, Moderate and Severe for confirmation of agriculture drought situation. In *kharif* 2016 the mandals of Andhra Pradesh categorized into Normal (378), Mild (175), Moderate (98) and Severe (21) - (Figs. 28.7 and 28.8). Ground truth was also collected from field and compared and updated the analysis (Table 28.3).

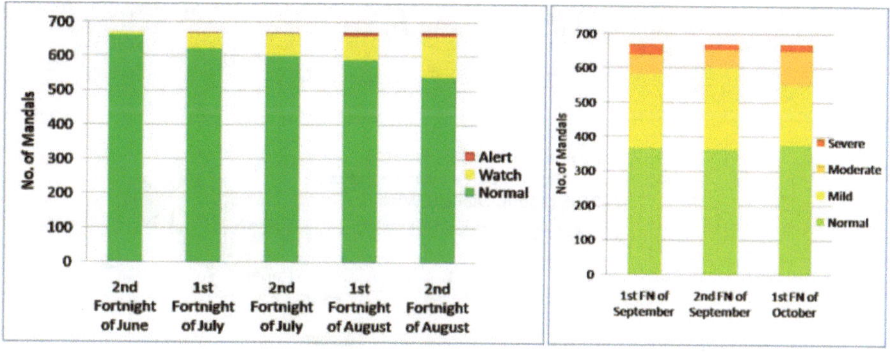

Fig. 28.5: Statistical graph representing mandals under draught situation

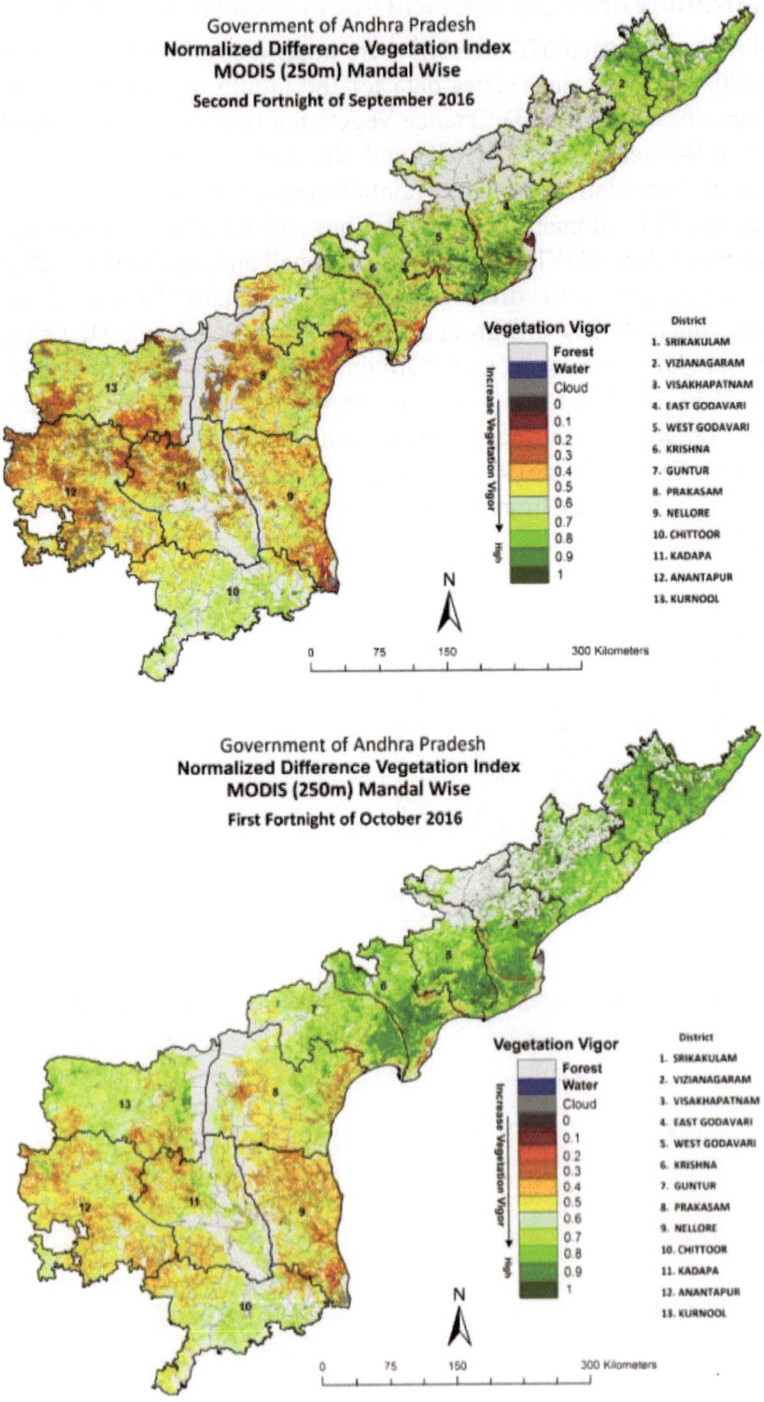

Government of Andhra Pradesh
Normalized Difference Vegetation Index
MODIS (250m) Mandal Wise
Second Fortnight of September 2016

Vegetation Vigor
Forest
Water
Cloud
0
0.1
0.2
0.3
0.4
0.5
0.6
0.7
0.8
0.9
1
Increase Vegetation Vigor
High

District
1. SRIKAKULAM
2. VIZIANAGARAM
3. VISAKHAPATNAM
4. EAST GODAVARI
5. WEST GODAVARI
6. KRISHNA
7. GUNTUR
8. PRAKASAM
9. NELLORE
10. CHITTOOR
11. KADAPA
12. ANANTAPUR
13. KURNOOL

N
0 75 150 300 Kilometers

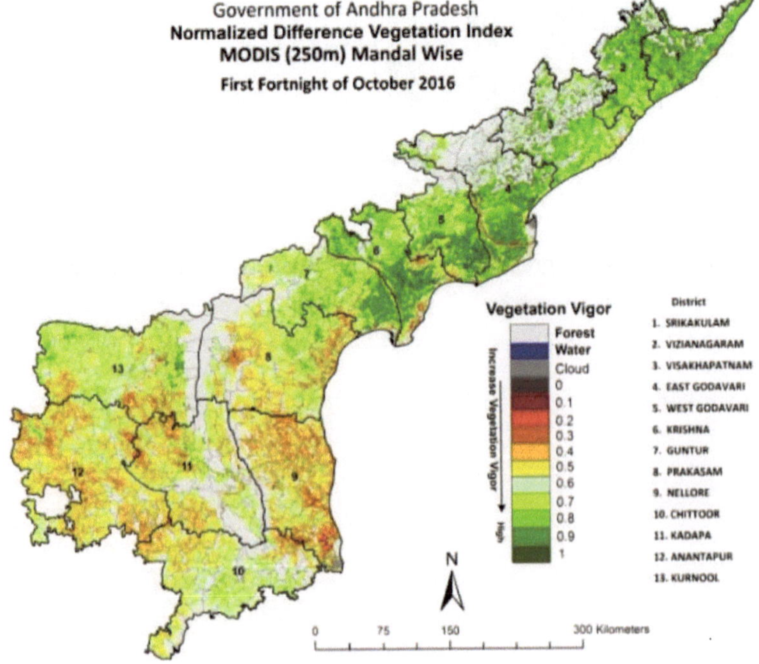

Fig. 28.6: Normalized Difference Vegetation Index (NDVI)

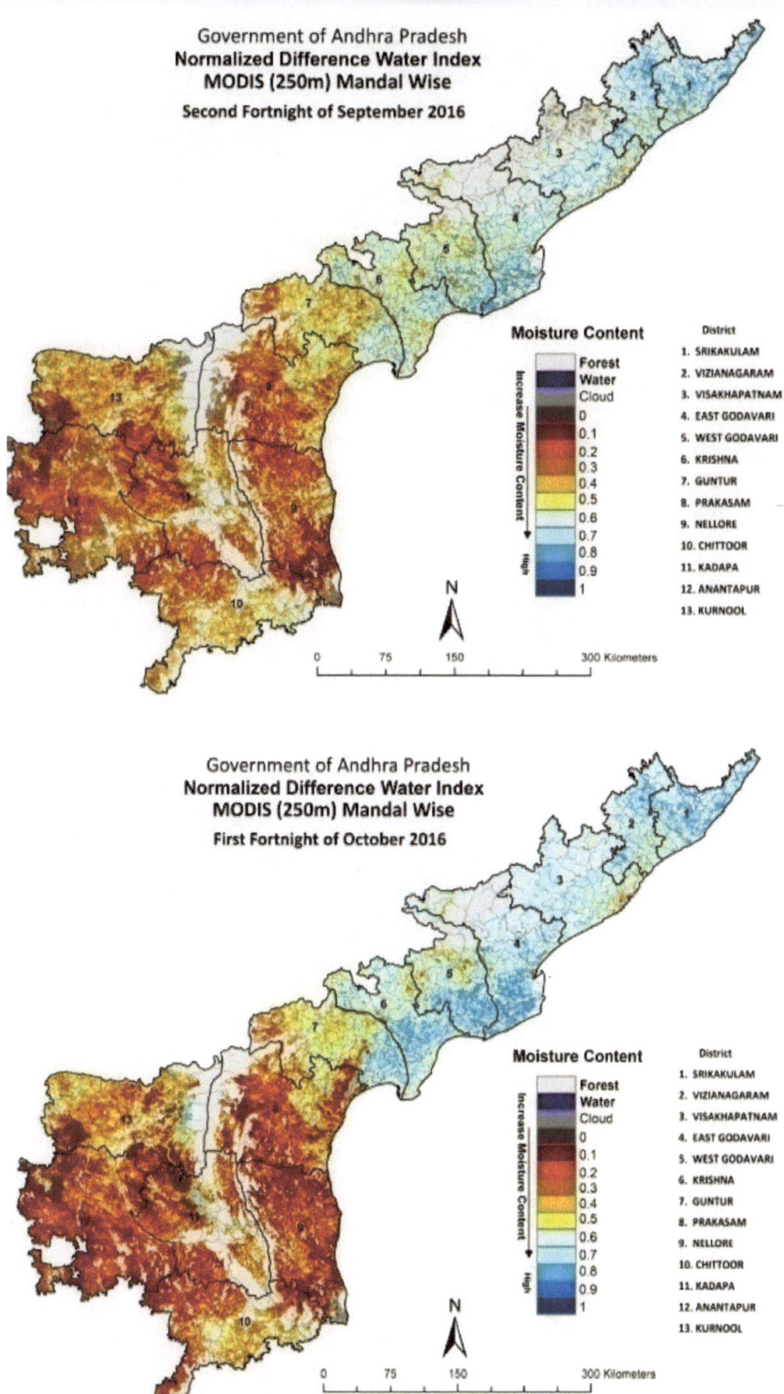

Fig. 28.7: Normalized Difference Wetness Index (NDWI)

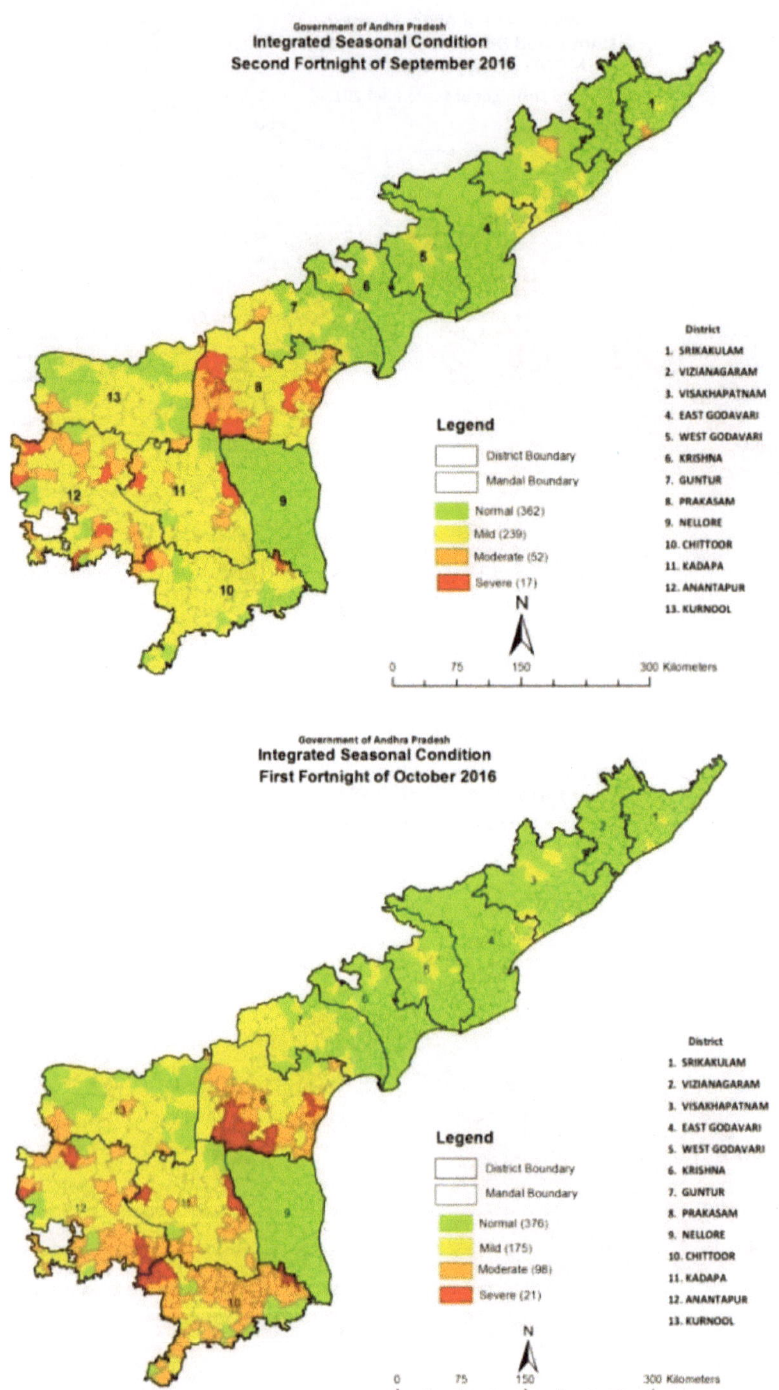

Fig. 28.8: Mandal level drought assessment

Table 28.3: Statistics of mandals under agriculture drought situation

Agriculture drought situation	2nd FN June, 16	1st FN July, 16	2nd FN July, 16	1st FN Aug, 16	2nd FN Aug, 16	Agriculture drought situation	1st FN Sept, 16	2nd FN Sept, 16	1st FN Oct, 16
Normal	395	622	599	588	539	Normal	366	362	376
Watch	275	45	68	73	121	Mild	215	239	175
Alert	00	03	03	09	10	Moderate	57	52	98
						Severe	32	17	21

28.5.1 Statistics of Agriculture drought affected mandals

Summary

Flood, Cyclone and Drought are the most devastating natural hazards in Andhra Pradesh, claiming more lives and causing extensive damage to agriculture, vegetation, human and wild life and local economies. The remote sensing and GIS technology significantly contributes in the damage assessment due to natural calamities. Remote sensing is used operationally in the warning and monitoring of cyclones, drought and floods. Here these studies prove and justifies the usefulness of remote sensing and GIS for damage assessment from natural disasters. The case study of the utilization of satellite data for studying the extent of flood damage to agricultural areas during the year 2009 due to flash floods, NILAM cyclone in 2012, Hudhud cyclone in 2014 and floods due to high intensity rains in 2016 was discussed. During the *Hudhud* cyclone, the combined effect of wind and rainfall is studied. Also the extent of area suffered due to deficit rainfall in Kharif, 2016 was assessed from the satellite born spectral indices – NDVI, NDWI and VCI. The extent of drought extent is manifested in 294 mandals other than in Nellore district.

References/Suggested Readings

A.P.M. Cunha, R.C. Alvalá, C.A. Nobre, M.A. Carvalho, 2015. Monitoring vegetative drought dynamics in the Brazilian semiarid region.*Agricultural and Forest Meteorology,* 214–215, 15 December 2015, Pages 494-505.

Alexander, D. (1993). Natural disasters. UCL Press Ltd., University College London. 632 pp.

Gao, B.C., 1996. NDWI –A Normalized Difference Water Index for remote sensing of vegetation liquid water from space. *Remote Sens. Environ.*58**:** 257-266.

Gibbs Wj and Meher JV, 1967, Rainfall deciles as drought indicators.Bureau of Meteorology, Melbourn, Austalia, Bulletin, 48, pp 33.

Kogan FN, 2001, Operational Space Technology for Global Vegetation Assessment, *Bulletin of the American Meteorological Society*, 82: 1949-1964.

Kogan, F.N., 1990. Remote Sensing of weather impacts on vegetation in non-homogeneous areas. *Int. J. Remote Sens.*,11**:**1405-1419.

Malingreau JP, 1986, Global vegetation dynamics: Satellite observation over Asia, International *Journal of Remote Sensing*, 7: 1121-1146.

Murthy CS, Chakraborty A, Sai MVRS, Roy PS (2011). Spatio-Temporal Analysis of the Droughts of Kharif 2009 and 2002. *Current Science*, 100: 1785-1788.

Liu, W.T.and Kogan, F.N.,1996. Monitoring regional drought using the vegetation condition index.*Int. J. Remote Sens.*, 17: 2761-2782.

29

Role of Satellites in Climate Change Monitoring

B. Manikiam

Indian Space Research Organization (Rtd.), Bangalore, Karnataka, India

The United Nations Framework Convention on Climate Change (UNFCCC) defines Climate Change as a change the weather parameters attributed directly or indirectly to human activity, altering the composition of the global atmosphere. Human activity includes the pollution that arises from industrial activity and other sources that produce greenhouse gases. These gases, such as carbon dioxide, have the ability to absorb the spectrum of infrared light and contribute to the warming of our atmosphere. Once produced, these gases can remain trapped in the atmosphere for tens or hundreds of years.

29.1 Causes of Climate Change

The several factors that contribute to climate change impacts are –

- Raising emissions to the atmosphere due to burning coal, oil and gas produces carbon dioxide and nitrous oxide.
- Cutting down of trees in forests leading to deforestation. Trees help to regulate the climate by absorbing CO_2 from the atmosphere. So when they are cut down, that beneficial effect is lost and the carbon stored in the trees is released into the atmosphere adding to the greenhouse effect.
- **Increasing livestock farming of** Cows and sheep that produce large amounts of methane when they digest their food.
- **Fertilisers containing nitrogen** produce nitrous oxide emissions to the atmosphere.
- **Fluorinated gases** produce a very strong warming effect, up to 23 000 times greater than CO_2. Thankfully these are released in smaller quantities and are being phased down by EU regulation.

The components of climate change are shown in Fig. 29.1.

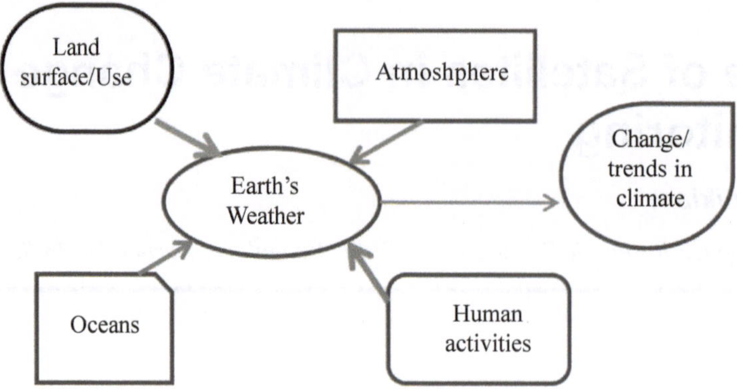

Fig. 29.1: Components of Climate Change

The role of green house gases in the atmosphere are summerised as follows:

29.1.1 Carbon dioxide

The burning of fossil fuels (oil, natural gas, and coal), solid waste, and trees produce large amount of carbon dioxide emission to the atmosphere. Deforestation and soil degradation also add carbon dioxide to the atmosphere. Some of the carbon remains in the atmosphere for thousands of years, due to the slow process by which carbon is transferred to ocean sediments. Human activities currently release over 30 billion tons of CO_2 into the atmosphere every year. The annual variation of atmospheric CO_2 is shown in Fig. 29.2.

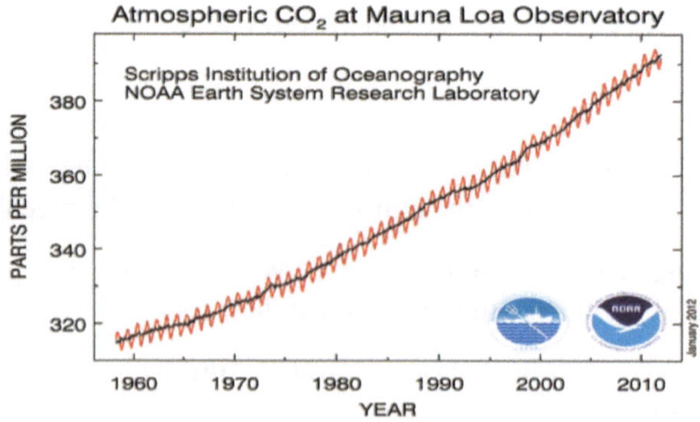

Fig. 29.2: Annual variation of CO_2

29.1.2 Nitrous oxide

It is emitted during agricultural and industrial activities, as well as during combustion of fossil fuels and solid waste. Its lifetime in the atmosphere stands at 121 years

29.1.3 Fluorinated gases

The hydrofluorocarbons, perfluorocarbons, and sulfur hexafluoride gases emitted from a variety of industrial processes, commercial and household uses have long life time.

29.1.4. Methane

Methane emissions are due to production of coal, natural gas and oil and also from livestock, agricultural practices, anaerobic decay of organic waste etc. with an average lifetime of over 12 years

The term greenhouse effect coined in 1827 by Joseph Fourier, a French mathematician and physicist, has envisioned the warming process of the Earth due to the greenhouse gases trapped in the atmosphere. The greenhouse gases have been increasing in the atmosphere due to various human activities. The relative contributions of green house gases to global warming have been as follows:

- CO_2 is responsible for 64% of man-made global warming.
- Methane is responsible for 17% of man-made global warming, nitrous oxide for 6%.

29.2 Role of Earth's reflectivity

Human changes in land use and land cover have changed Earth's reflectivity. Processes such as deforestation, reforestation, desertification, and urbanization often contribute to changes in climate in the places they occur. These effects may be significant regionally, but are smaller when averaged over the entire globe. In addition, human activities have generally increased the number of aerosol particles in the atmosphere. Overall, human-generated aerosols have a net cooling effect offsetting about one-third of the total warming effect associated with human greenhouse gas emissions. Reductions in overall aerosol emissions can therefore lead to more warming. However, targeted reductions in black carbon emissions can reduce warming.

29.2.1 Other climate forcers

Particles and aerosols in the atmosphere can also affect climate. Human activities such as burning fossil fuels and biomass contribute to emissions of these substances, although some aerosols also come from natural sources such as volcanoes and marine plankton. Black carbon (BC) is a solid particle or aerosol, not a gas, but it also contributes to warming of the atmosphere. Unlike GHGs, BC can directly absorb incoming and reflected sunlight in addition to absorbing infrared radiation. BC can also be deposited on snow and ice, darkening the surface and thereby increasing the snow's absorption of sunlight and accelerating melt. Sulfates, organic carbon, and other aerosols can cause cooling by reflecting sunlight. The current global average temperature is 0.85°C higher than it was in the late 19th century. Each of the past three decades has been warmer than any preceding decade since records began in 1850. An increase of 2°C compared to the temperature in pre-industrial times is seen by scientists as the threshold beyond which there is a much higher risk that dangerous and possibly catastrophic changes in the global environment will occur. For this reason, the international community has recognised the need to keep warming below 2°C.

The trend in the Earth's temperature is as shown in Fig. 29.3.

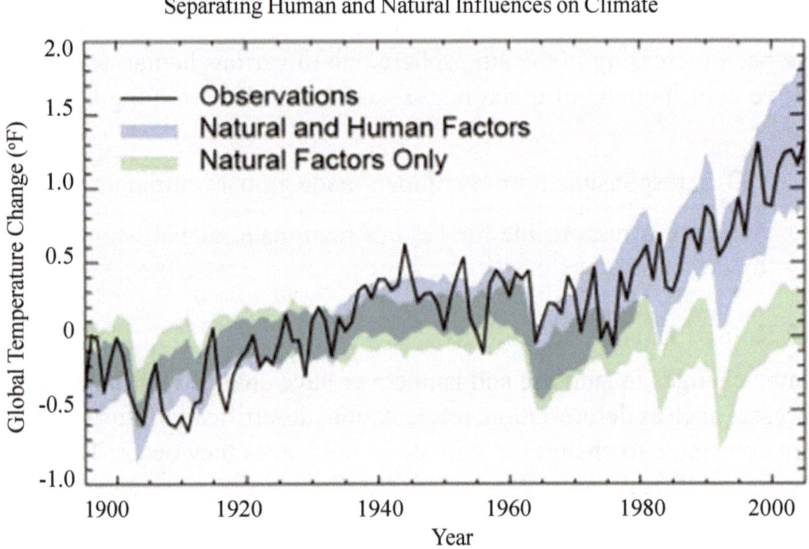

Fig. 29.3: Relative contributions to global temperature

Besides the green house gases, the other factors that can lead to warming or cooling of earth's atmosphere are variations in the sun's energy reaching Earth and changes in the reflectivity of Earth's atmosphere and surface. There are several studies dating back hundreds of thousands of years (and, in some cases, millions or hundreds of millions of years), by analyzing a number of indirect measures of climate such as ice cores, tree rings, glacier lengths, pollen remains, and ocean sediments. Models that account only for the effects of natural processes are not able to explain the warming observed over the past century. Some of the results of past studies are as shown in Fig. 29.4.

Besides direct impacts of green house gases on warming, there are feedback mechanisms that contribute to climate change. Water vapor appears to cause the most important positive feedback. As Earth warms, the rate of evaporation and the ability of air to hold water vapor both rise, increasing the amount of water vapor in the air. Because water vapor is a greenhouse gas, this leads to further warming. The melting of Arctic sea ice is another example of a positive climate feedback. As temperatures rise, sea ice retreats. The loss of ice exposes the underlying sea surface, which is darker and absorbs more sunlight than ice, increasing the total amount of warming. Some types of clouds cause a negative feedback. Warming temperatures can increase the amount or reflectivity of these clouds, reflecting more sunlight back into space, cooling the surface of the planet. Other types of clouds, however, contribute a positive feedback.

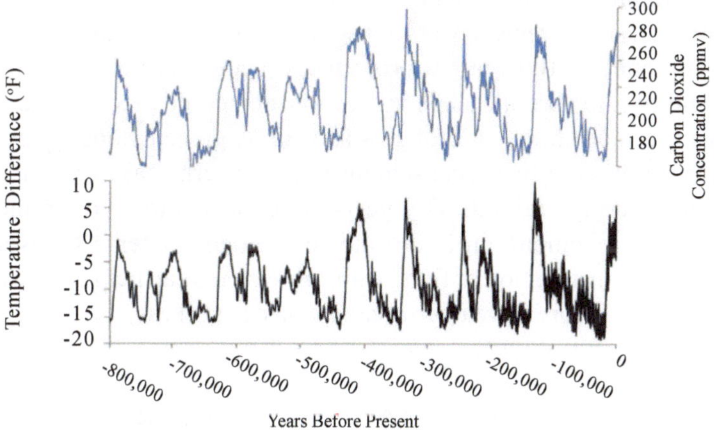

Fig. 29.4: Paleo studies of temperature

29.3 Impacts of Climate Change

The changing climate impacts society and ecosystems in a broad variety of ways. For example, climate change can alter rainfall, influence crop yields, affect human health, cause changes to forests and other ecosystems, and even impact our energy supply. Climate-related impacts are occurring across the country and over many sectors of our economy. There have been observed changes in precipitation, but not all areas have data over long periods. Rainfall has increased in the mid-latitudes of the northern hemisphere since the beginning of the 20th century. There are also changes between seasons in different regions. For example, the UK's summer rainfall is decreasing on average, while winter rainfall is increasing. There is also evidence that heavy rainfall events have become more intensive, especially over North America.

29.3.1 Impacts of Climate Change in India

29.3.1.1. Extreme Heat

India is already experiencing a warming climate. Unusual and unprecedented spells of hot weather are expected to occur far more frequently and cover much larger areas. Under 4°C warming, the west coast and southern India are projected to shift to new, high-temperature climatic regimes with significant impacts on agriculture. With built-up urban areas rapidly becoming "heat-islands", urban planners will need to adopt measures to counteract this effect.

29.3.1.2 Changing Rainfall Patterns

A decline in monsoon rainfall since the 1950s has already been observed. The frequency of heavy rainfall events has also increased. A 2°C rise in the world's average temperatures will make India's summer monsoon highly unpredictable. At 4°C warming, an extremely wet monsoon that currently has a chance of occurring only once in 100 years is projected to occur every 10 years by the end of the century. An abrupt change in the monsoon could precipitate a major crisis, triggering more frequent droughts as well as greater flooding in large parts of India. India's northwest coast to the south eastern coastal region could see higher than average rainfall. Dry years are expected to be drier and wet years wetter. Improvements in hydro-meteorological systems for weather forecasting and the installation of flood warning systems can help people move out of harm's way before a weather-related disaster strikes.

29.3.1.3 Droughts

Evidence indicates that parts of South Asia have become drier since the 1970s with an increase in the number of droughts. Droughts have major consequences. In 1987 and 2002-2003, droughts affected more than half of India's crop area and led to a huge fall in crop production. Droughts are expected to be more frequent in some areas, especially in north-western India, Jharkhand, Orissa and Chhattisgarh. Crop yields are expected to fall significantly because of extreme heat by the 2040s. Investments in R&D for the development of drought-resistant crops can help reduce some of the negative impacts.

29.3.1.4 Groundwater

More than 60% of India's agriculture is rain-fed, making the country highly dependent on groundwater. Even without climate change, 15% of India's groundwater resources are overexploited. Although it is difficult to predict future ground water levels, falling water tables can be expected to reduce further on account of increasing demand for water from a growing population, more affluent life styles, as well as from the services sector and industry. The efficient use of ground water resources will need to be incentivized.

29.3.1.5 Sea level rise

Mumbai has the world's largest population exposed to coastal flooding, with large parts of the city built on reclaimed land, below the high-tide mark. Rapid and unplanned urbanization further increases the risks of sea water intrusion. With India close to the equator, the sub-continent would see much higher rises in sea levels than higher latitudes. Sea-level rise and storm surges would lead to saltwater intrusion in the coastal areas, impacting agriculture, degrading groundwater quality, contaminating drinking water, and possibly causing a rise in diarrhea cases and cholera outbreaks, as the cholera bacterium survives longer in saline water. Kolkata and Mumbai, both densely populated cities, are particularly vulnerable to the impacts of sea-level rise, tropical cyclones, and riverine flooding. Building codes will need to be strictly enforced and urban planning will need to prepare for climate-related disasters. Coastal embankments will need to be built where necessary and Coastal Regulation Zone codes enforced strictly.

29.4 Impacts on Agriculture and food security

Even without climate change, world food prices are expected to increase due to growing populations and rising incomes, as well as a greater demand for biofuels.

Rice: While overall rice yields have increased, rising temperatures with lower rainfall at the end of the growing season have caused a significant loss in India's rice production. Without climate change, average rice yields could have been almost 6% higher (75 million tons in absolute terms).

Wheat: Recent studies shows that wheat yields peaked in India and Bangladesh around 2001 and have not increased since despite increasing fertilizer applications. Observations show that extremely high temperatures in northern India - above 34°C - have had a substantial negative effect on wheat yields, and rising temperatures can only aggravate the situation.

Seasonal water scarcity, rising temperatures, and intrusion of sea water would threaten crop yields, jeopardizing the country's food security.

Should current trends persist, substantial yield reductions in both rice and wheat can be expected in the near and medium term.

Under 2°C warming by the 2050s, the country may need to import more than twice the amount of food-grain than would be required without climate change. Crop diversification, more efficient water use, and improved soil management practices, together with the development of drought-resistant crops can help reduce some of the negative impacts.

Energy Security

Climate-related impacts on water resources can undermine the two dominant forms of power generation in India - hydropower and thermal power generation - both of which depend on adequate water supplies to function effectively. To function at full efficiency, thermal power plants need a constant supply of fresh cool water to maintain their cooling systems.

The increasing variability and long-term decreases in river flows can pose a major challenge to hydropower plants and increase the risk of physical damage from landslides, flash floods, glacial lake outbursts, and other climate-related natural disasters.

Decreases in the availability of water and increases in temperature will pose major risk factors to thermal power gener*ation.* Projects will need to be planed taking into account climatic risks.

Water Security

Many parts of India are already experiencing water stress. Even without climate change, satisfying future demand for water will be a major challenge.

Urbanization, population growth, economic development, and increasing demand for water from agriculture and industry are likely to aggravate the situation further.

An increase in variability of monsoon rainfall is expected to increase water shortages in some areas.

Studies have found that the threat to water security is very high over central India, along the mountain ranges of the Western Ghats, and in India's northeastern states. Improvements in irrigation systems, water harvesting techniques, and more-efficient agricultural water management can offset some of these risks.

Impacts on Health

Climate change is expected to have major health impacts in India- increasing malnutrition and related health disorders such as child stunting - with the poor likely to be affected most severely. Malaria and other vector-borne diseases, along with and diarrheal infections which are a major cause of child mortality, are likely to spread into areas where colder temperatures had previously limited transmission.

29.5 Adapting to Climate Change

"Adaptation" refers to the adjustments that societies or ecosystems make to limit the negative effects of climate change or to take advantage of opportunities provided by a changing climate. Adaptation can range from a farmer planting more drought-resistant crops to a coastal community evaluating how best to protect its infrastructure from rising sea level. While reducing greenhouse gas emissions is required to avoid the worst impacts of climate change, a certain amount of global warming is inevitable, due to the long-lasting nature of greenhouse gases already in the atmosphere, and to heat already stored in the oceans.

Adaptation actions include responding to conditions that have already changed and planning for climate change before impacts have occurred. Adaptation measures are already in place in many areas, as the examples below show. These actions can be expanded or modified to prepare for climate change. But additional measures, such as new technologies and policies, may also be needed. Such actions could require time and resources to carry out, so planning now is important.

The various proactive actions planned are:

- Develop crop varieties that are more tolerant of heat, drought, or flooding from heavy rains.
- Preserve wetlands and open spaces to protect coastal communities from flooding and erosion from storms and sea level rise.
- Improve evacuation planning for low-lying areas to prepare for increased storm surge and flooding.
- Reduce pollution, habitat loss, and other stressors that make ecosystems more vulnerable to climate change.
- Increase energy efficiency to help offset rises in energy consumption, such as from more air conditioning use as temperatures warm.
- Set up early warning systems and emergency response plans to prepare for more extreme weather events.
- Plant trees and expand green spaces in cities to reduce the "urban heat island" effect.
- Improve water use efficiency, and build additional water storage capacity by restoring stream and river banks.

29.6 Role of Satellite Data in Climate Change Studies

The potential climate changes and possible adverse impacts on the economy and society at large are causing concern. Observations indicate the possibility of more frequent severe weather events and high intense rainfall events and the submergence of low-lying islands, coastal areas, frequent flooding, coastal degradation and melting of glaciers in the Indian Himalayas. The various parameters associated with climate change need to be monitored and analysed. Indian agriculture being heavily dependent on the monsoon and the potential impact of climate change on monsoon dynamics through ocean warming needs attention. While studying the climate change scenarios, it is essential to build up necessary strategies at local level to reduce the adverse impacts especially on agriculture and water management. It will be a challenging task to counter the effect of climate change through scientific means.

29.6.1 Applications of Satellite Data

Over the past three decades, satellite remote sensing has grown in terms of technology and application potential. Multi-frequency, high resolution sensors have been developed and made operational. India is operating both INSAT and IRS series of satellites (Joshi *et al*, 2003) for land, ocean and atmospheric applications. The advanced sensors in operation include Atmospheric Sounder in INSAT 3D, OCM in Oceansat 2, Wind Scatterometer in Oceansat 2, WiFS in Resourcesat. These sensors are capable of capturing the dynamic aspects of land processes, ocean dynamics and atmospheric conditions.

Following sections highlight applications of satellite data for climate change studies.

Land applications: Due to rapid urbanization, the land use is getting changed at a rapid pace and this need to be monitored. The Indian Remote Sensing (IRS) satellites have high resolution sensors operating in visible and near IR region. These are useful in mapping land cover over large areas and quantify the changes. Fig. 29.5 depicts the land cover and vegetation changes for the months of June and September. The vegetation layer provides important information required by high resolution weather models.

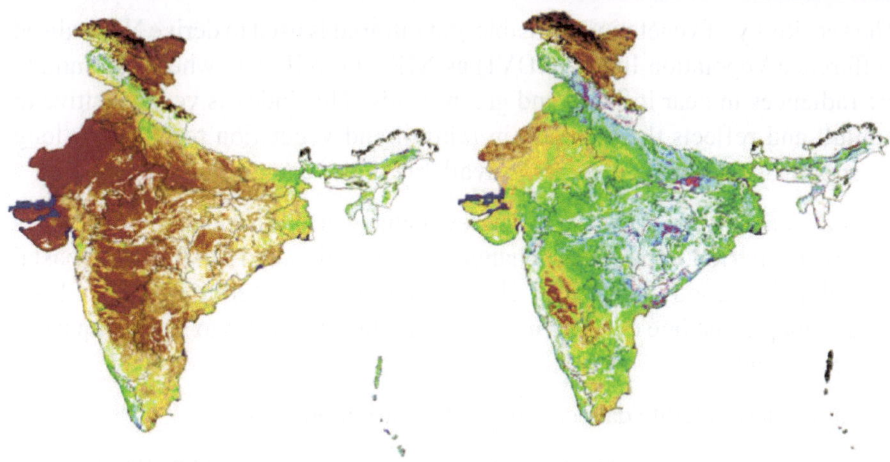

Fig. 29.5: Vegetation Map from IRS 1 B data-June 2016 (left) & September 2016 (right)

Forest cover across the globe is being monitored with great detail using multiple satellite sensors. The data provides details such as vegetation type, density, vegetation vigour and forest fire damages. The forest cover over parts of western Ghats in Karnataka state, India is shown in Fig. 29.6.

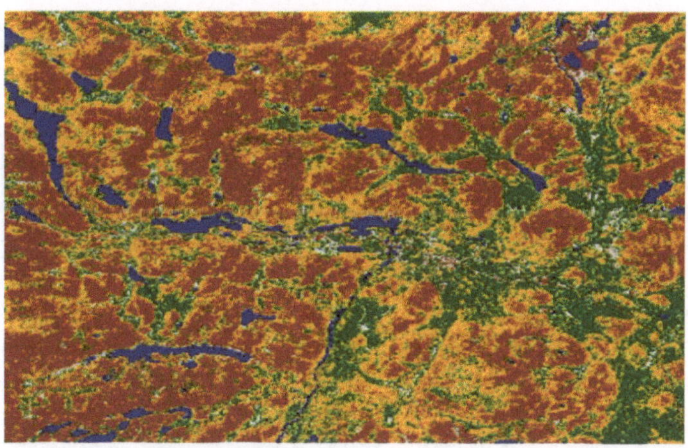

	Hardwood	Softwood	Mixed	Water	Open	User's Accuracy
Hardwood	59	1	8	0	11	0.75
Softwood	6	9	7	0	0	0.41
Mixed	17	4	31	0	0	0.60
Water	0	D	0	20	0	1.00
Open	0	2	1	1	3	0.67
Pro du cer's Accu racy	072	Q.5G	0.66	D.9E	0.42	

OveralF Accuracy: 0.G9

Fig. 29.6: Forest monitoring using IRS WiFS data

The sensitivity of vegetation in visible and infrared is used to derive Normalised Difference Vegetation Index (NDVI) as NIR - G /NIR + G where NIR and G are radiances in near infrared and green bands. This index is very sensitive to rainfall and reflects the anomaly in rainfall and vegetation response. A long term database of about 20-years is available in the country.

Coastal zone is very sensitive to sea level changes and frequency of cyclones and over a period experiences changes in coast line and landforms. Coastal zone thus is a good indicator of climate change IRS satellite data have been used to map coast line changes in yearly and decadal scales to study impact of sea level changes.

The multi date satellite data can give information on –

- Mapping and monitoring coral reefs, mangroves, coastal wetlands
- Primary productivity
- Monitoring of coast lines

Using satellite data, it is now possible to identify "hotspots" where large change in coastal line is taking place (Fig. 29.7).

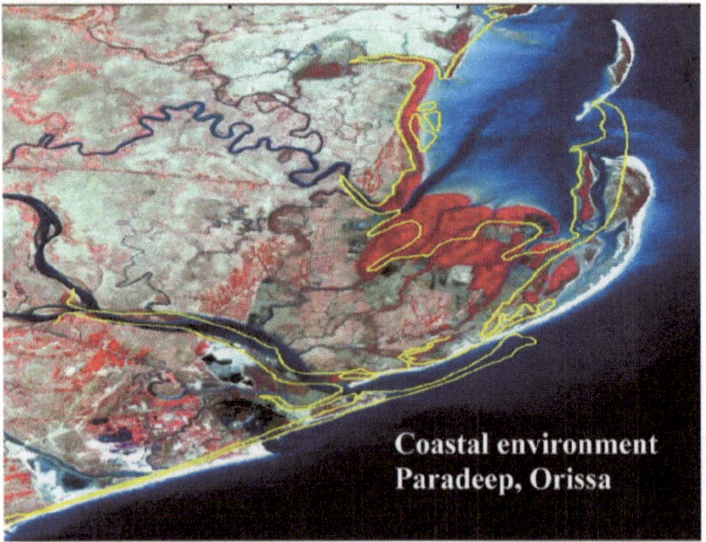

Fig. 29.7: Coastal Environment monitoring through satellites

Another potential application of satellite data is to inventory, monitoring and retreat of Himalayan glaciers using satellite data. Glacial inventory has been carried out for the glacial regions of Indus, Ganga and Brahmaputra Basins covered approximately in 1:50,000 scale. Specific measurements of mapped glacier features are used as inputs for generating the glacier inventory data.

Glacier mass balance studies have been taken up for 10 selected medium sized sub-basins spread over in Indian Himalayan region (Indus, Ganga and Brahmaputra Rivers). The IRS LISS III scenes are used to ascertain the boundary of glaciers using higher resolution data. Glacier boundaries are overlaid on all AWiFS scenes sequentially. Glacier Retreat mapping was done for the 14 selected basins spread over in Indian Himalayan Region. This involves, creation of database of spatial extent of all glaciers and overlaying of the boundaries; checking of the boundaries in the accumulation zone based on shadow of the ridges & on the drainages; identification of glacier retreat and estimation of glacier area evacuated and verification in the field the location of snout using GPS observations. Monitoring of snow cover of the Indian Himalayan region covering Indus, Ganga and Brahmaputra river basins have been carried out using multidate AWiFS data from Resourcesat-1 and snow cover atlases have been prepared. Careful monitoring is essential to assess any reduction in snow cover and retreat of snow lines due to climate change (Fig. 29.8).

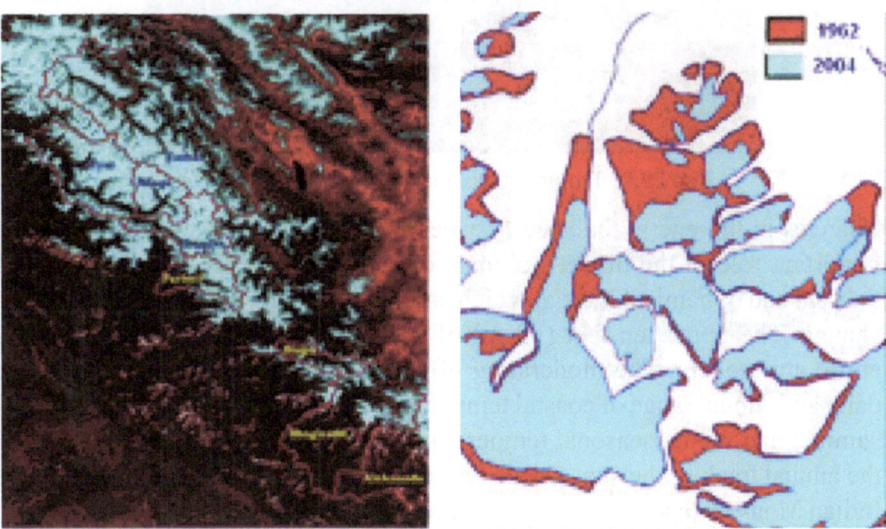

Fig. 29.8: Glacier Inventory of Ganga & Brahmaputra Basins 1962–2004 using IRS WiFS data (33 sub basins monitored and a loss of 16% has seen detected)

The oceans and atmosphere store and exchange energy in the form of heat, moisture, and momentum. The oceans are obviously the Earth's largest reservoir of moisture. They also absorb heat more effectively than land and ice surfaces, and store heat more efficiently than land. Oceanic heat is released more slowly than on land, keeping coastal areas more temperate.

Further, meteorological data along with satellites are valuable for monitoring and forecasting of cyclones. INSAT/VHRR images are being used to identify cloud systems over the oceans , where no observational data are available, as well as for cyclone tracking, intensity assessment and prediction of storm surges (Fig. 29.9). Current research around the globe is concentrating on use of meso-scale models with satellite data inputs to improve the cyclone intensity and track prediction.

Fig. 29.9: Tropical Cyclone study using INSAT data

Changes in the energy balance between the oceans and atmosphere play an important role in the planet's climate change. The oceans have tremendous thermal and dynamical inertia, which can slow and dampen the rate of climate change. The upper ten feet (~3 m) of the ocean holds as much heat as the entire atmosphere. The moderating effect of ocean temperature reduces the daily and annual range of coastal temperatures, and results in a lag of the global summer and winter seasonal temperatures extremes by several weeks behind the annual track of the sun. Ocean temperatures play a very crucial role in the Indian Monsoon and in the ocean productivity in terms of fishery potential and chlorophyll production. The satellite data from NOAA and INSAT geostationary satellites are used to prepare weekly Sea Surface Temperature (SST) maps. The accuracy achieved is in the region of ± 0.5 °C (Fig.29.10). These maps form the basic data for analysis of El Nino/La Nina conditions in Pacific Ocean and Cyclogenesis conditions in the Indian oceans. Long and short seasonal variations of SST are being monitored on a regular basis. The ocean stores and transports not only heat but also carbon dioxide (CO_2), a potential major source of global warming. About half of the totalCO_2 added to the atmosphere during the past century by human activity – mostly from the use of fossil fuels and deforestation– has been absorbed by the ocean.

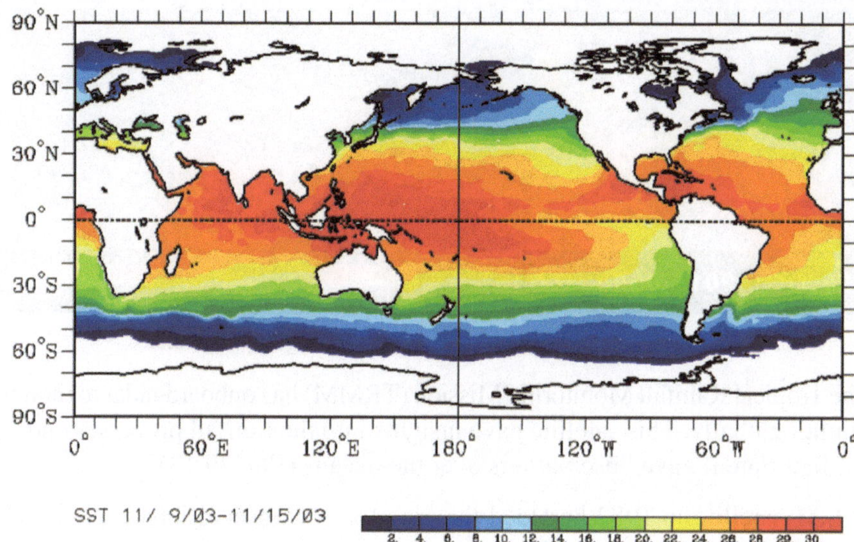

Fig. 29.10: Global Sea Surface Temperature using NOAA data

Most of the CO_2 emitted into the atmosphere will eventually be absorbed by the ocean. However, this process will take several decades to centuries. Phytoplankton also stores carbon dioxide from the upper layers of the ocean. Many species sequestered CO_2 in their carbonate shells, which eventually sink to the ocean floor, or long-term removal from the carbon cycle. Ocean Colour Monitor on-board Oceansat 2 has 8 narrow channels in the green and blue region of visible spectrum. This makes it very sensitive to ocean colour due to phytoplankton and sedimentation. Similarly MODIS satellite has multi channels sensitive to ocean productivity (Fig. 29.11).

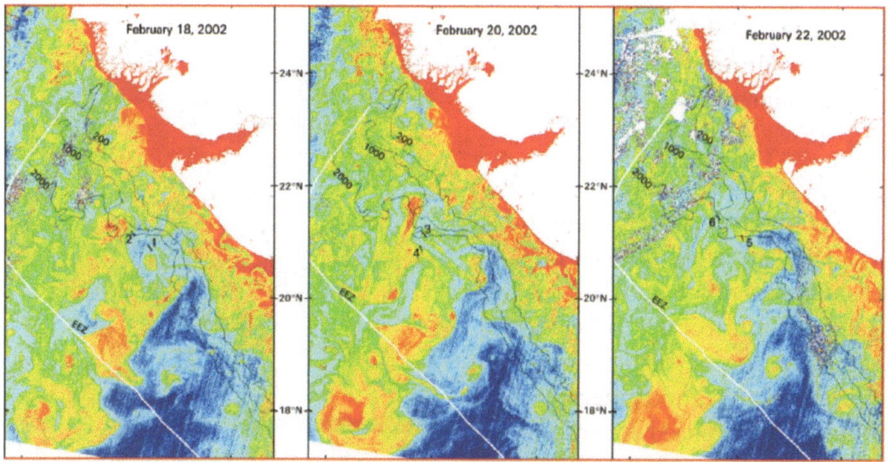

Fig. 29.11: Spatial and seasonal (monsoon) variability of the chlorophyll-a concentration in NW Atlantic and Indian Oceans.

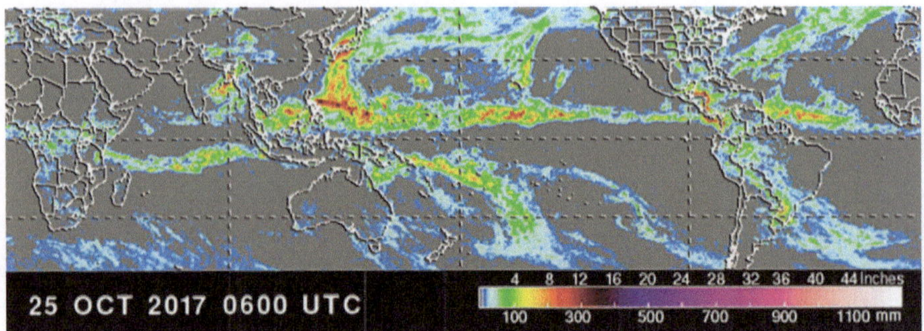

Fig. 29.12: TRMM based rainfall over global oceans.

The Tropical Rainfall Monitoring Mission (TRMM) has onboard radar to monitor rainfall globally. This satellite gave insights into the rainfall processes and for the first time it gave observations over the oceans (Fig. 29.12).

INSAT satellite in 1982 heralded the era of space observations and gave the first glimpses of the dynamic cloud systems over India. Since the launch of INSAT 1A in 1982, a series of satellites are in operation and IMD has developed a long term series of cloud cover, OLR, rainfall etc. The INSAT products will form an important data set in climate change analysis at local and regional level. Currently INSAT satellite is providing global and regional observations. It carries Very High resolution Radiometer (VHRR) payload which operated in two spectral bands – visible [0.55-0.75 ìm] and thermal infrared [10.5-12.5 ìm]. The INSAT satellite gives hourly weather imagery of the country showing the cloud systems, their movement and potential severe weather events. The quantitative products available from INSAT data give the numerical products such as Cloud Motion Vectors (CMVs), Quantitative Precipitation Estimates (QPEs), Outgoing Long-wave Radiation (OLR), Vertical Temperature Profiles (VTPRs) and SST (Fig. 29.13).

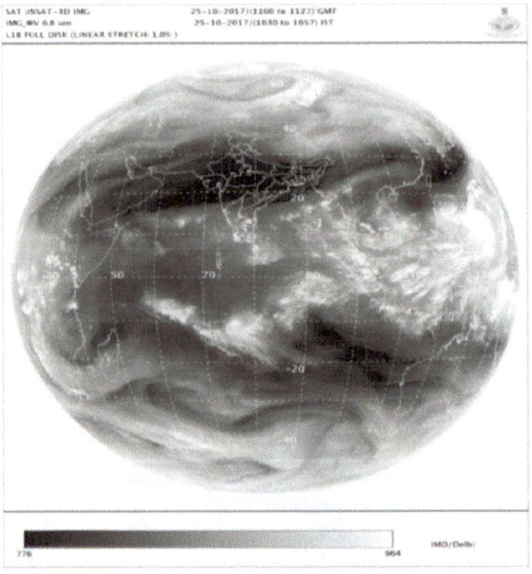

Fig. 29.13: INSAT Water Vapour Image.

The data from GPS satellites are used to derive atmospheric conditions, based on the principle that the time delay in arrival of signals from GPS satellite is directly linked to the moisture content and temperature profile of the atmosphere. A new stability index based on atmospheric refractivity at ~500hPa level and surface measurements of temperature, pressure and humidity is formulated. The new index named here as Refractivity based Lifted Index (RLI) is designed to give similar results as traditionally used lifted index derived from radiosonde profiles of temperature, pressure and humidity (Fig. 29.14).

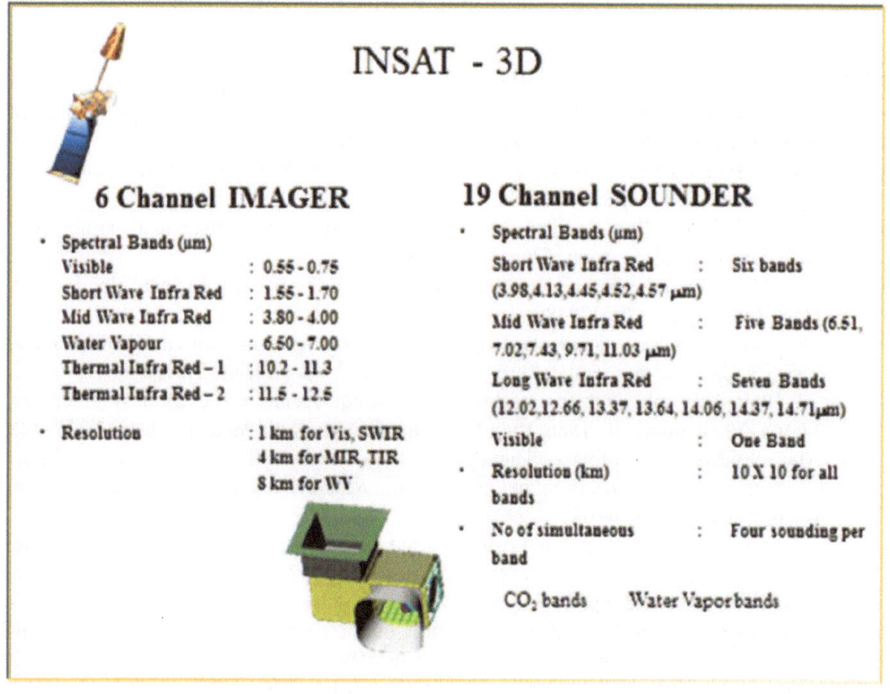

Fig. 29.14: INSAT 3D Advanced Sensors

The formulation of the stability index and its comparison with the traditional temperature profile based lifted index (LI) is discussed. The index is tested on COSMIC radio occultation derived refractivity profiles over Indian region. The forecast potential of the new index for rainfall on 2°×2° latitude– longitude spatial scale with lead time of 3–24 hours indicate that the refractivity based lifted index works better than the traditional temperature based lifted index for the Indian monsoon region. Decreasing values of RLI tend to give increasing rainfall probabilities. 4. Summary The satellites have a great role to play in the study of climate change. Climate change modelling requires accurate and long term observations of land surface, oceans and atmosphere. The satellites with advanced sensors are capable of measuring several critical parameters. Table-1 summarises the capabilities of satellites. The immediate task is to build a long term series of well calibrated data using satellite and ground/ ocean observations to look for signal of climate change.

Conclusions

The weather and climate play a key role in human endeavours. The recent trends in climate change is having short and long term effects and needs a clear scientific assessment and understanding. The new tools of satellites are providing valuable data on several key parameters which will help in continuous monitoring of the climate change and develop suitable mitigation measures.

References/Suggested Readings

Alexander GR, Keshavamurty N, De US, Chellappa R, Das SK, Pillai PV. 1978. Fluctuations of monsoon activity. *Indian J. Meteor. Geophys*. 29:76–87

Anderson DLT, Sarachik ES, Webster PJ, Rothstein LM, eds. 1998. A special issue on the TOGA decade. *J. Geophys. Res.* 103 (No. C7):14169–240

Annamalai H, Slingo JM. 2001. Active/break cycles: diagnosis of the intraseasonal variability of the Asian summer monsoon. *Clim. Dyn.* 18:85–102

Anthes RA. 1977. A cumulus parameterization scheme utilizing a one-dimensional cloud model. *Mon. Weather Rev.* 105:270–86

Bhat GS, Srinivasan J, Sulochana G. 1996. Tropical deep convection, convective available potential energy and sea surface temperature. *J. Meteorol. Soc. Jpn.* 74(No. 2): 155–66

Bjerknes J. 1969. Atmospheric teleconnections from the equatorial Pacific. Mon. Weather Rev. 97:163–72 Blanford HF. 1886. Rainfall of India. Mem. India Meteorol. Dep. 2:217–448

Carlin, Alan, "A Multi-disciplinary, Science-Based Approach to the Economics of Climate Change," *International Journal of Environmental Research and Public Health*, Vol. 8, No. 4, April 2011, pp. 985-1031.

Chen, Liqi, "The Role of the Arctic and Antarctic and their Impact on Global Climate Change: Further Findings Since the Release of IPCC AR4, 2007," *Advances in Polar Science*, (2013) Vol. 24, No. 2, pp. 79-85.

Chunhan, Jin, Liu, Jian, and Wang Zhiyuan., "Study on the characteristics and causes of inter-decadal temperature changes in China during the MWP," Quaternary Sciences, Volume 36, Issue 4, April 2016, pp. 970-982. http:// html. rhhz.net/ DSJYJ /20160418.htm

Christidis, N., P.A. Stott, and S. Brown, 2011: The role of human activity in the recent warming of extremely warm daytime temperatures,. *Journal of Climate* doi:10.1175/ 2011JCLI4150.1 http://journals.ametsoc.org/doi/abs/10.1175/2011JCLI4150.1

Deveerappa Jagadheesha, Kamsali Nagaraja and Balakrishnan Manikiam, 2014, 'Improved Technique for Retrieval of Temperature and Humidity from Neutral Atmospheric Refractivity Profiles', Mapana J. Sci., 13 (1), 75-83 ISSN 0975-3303|doi:10.12723/ mjs.28.5

Gadgil S. 1988. Recent advances in monsoon research with particular reference to the Indian monsoon. *Aust. Meteorol. Mag*. 36:193– 204

Gadgil S. 2000. Monsoon-ocean coupling. Curr. Sci. 78:309–23 Gadgil S, Abrol YP, Rao PRS. 1999. On growth and fluctuation of Indian foodgrain production. *Curr. Sci.* 76(No. 4):548–56

Gadgil S, Guruprasad A, Srinivasan J. 1992. Systematic bias in the NOAA outgoing longwave radiation dataset? *J. Clim.* 5:867–75

Gadgil S, Srinivasan J. 1990. Low frequency variation of tropical convergence zones. *Meteorol. Atmos. Phys*. 44:119–32

Gadgil S, Surendran S. 1998. Monsoon precipitation in the AMIP runs. Clim. Dyn. 14: 659–89

Gadgil S, Srinivasan J, Nanjundiah RS, Kumar KK, Munot AA. Rupakumar AA. 2002. On forecasting the Indian summer monsoon: the intriguing season of 2002. *Curr. Sci.* 83 (No. 4):394–403

IPCC AR4 WG1 *(2007). Solomon, S.; Qin, D.; Manning, M.; Chen, Z.; Marquis, M.; Averyt, K.B.; Tignor, M.; Miller, H.L., eds.* Climate Change 2007: The Physical Science Basis. *Contribution of Working Group I to the* Fourth Assessment Report *of the Intergovernmental Panel on Climate Change. Cambridge University Press.* (pb: 978-0-521-70596-7).

IPCC AR4 SYR *(2007). Core Writing Team; Pachauri, R.K; Reisinger, A., eds.* Climate Change 2007: Synthesis Report. *Contribution of Working Groups I, II and III to the* Fourth Assessment Report *of the Intergovernmental Panel on Climate Change. IPCC.* ISBN 92-9169-122-4.

Joshi. P.C., Narayanan. M.S. and Manikiam. B, 2003, 'Evolution of Indian satellite meteorological programme', *Mausam*, 54 (1), pp. 1-12.

Knutson, T., J. McBride, J. Chan, K. Emanuel, G. Holland, C. Landsea, I. Held, J. Kossin, A. Srivastava, and M. Sugi, 2010: Tropical cyclones and climate change. *Nature Geosci* 2010 http://dx.doi.org /10.1038/ ngeo779 http:// www. nature.com /ngeo/journal/v3/n3/suppinfo/ ngeo779_S1.html

Madden RA, Julian PR. 1972. Description of global-scale circulation cells in the tropics with a 40–50 day period. *J. Atmos. Sci.* 29:1109–23.

Manikiam. B and Sridharan. R, 1983, 'A statistical study of Bay of Bengal disturbances (1961-80) using satellite imagery', Mausam, 34, pp 219-222.

Manikiam. B, 2014, 'Applications of IRS and INSAT Data with Specific Case Studies', *Mapana J Sci,* 13, 1 ISSN 0975-3303 98.

Trenberth, K. E., 2011: Changes in precipitation with climate change. *Climate Research.* doi:10.3354/cr00953.

30

NABARD's Initiatives and Interventions in Climate Change

D.V. Subba Rao

NABARD Chair Professor, Department of Agricultural Economics
Agricultural College, Acharya N.G.. Ranga Agricultural University, Bapatla, India

Climate change has become real and tangible affecting people. It is a major challenge for agriculture, food security and rural livelihoods. Climate change impacts on vulnerable groups have social, economic, political, and environmental implications. The impacts of climate change are global. India is more vulnerable in view of the high population depending on agriculture, which in turn is dependent on the seasonal monsoons. India is confronted with the challenge of sustaining rapid economic growth amidst the increasing global threat of climate change. Government of India has accorded top priority for addressing climate change related concerns.

The Government formulated the National Action Plan on Climate Change (NAPCC) in 2008. Climate change is a complex policy issue with major implications in terms of finance. All actions and solutions to address the adverse impact of climate change ultimately involve costs. Funding is vital in developing countries to design and implement adaptation and mitigation projects. NABARD has already taken various initiatives in addressing the challenges posed by Climate Change particularly in the areas of agriculture and rural livelihood sectors. In this direction NABARD aims to channelize national, international and private finances for adaptation and mitigation activities in India. The Green Climate Fund has been designated as an operating entity of the financial mechanism of the UNFCCC. In the context of sustainable development, the Fund aims to promote a paradigm shift towards low emission and climate-resilient development pathways by providing support to developing countries to limit or reduce their greenhouse gas emissions and to adapt to the impacts of climate change, taking into account the needs of those developing countries particularly vulnerable to the adverse effects of climate change. The Fund will play a key role in channelizing new, additional, adequate and predictable financial resources to developing countries and will catalyze climate finance, both public and private

levels. The finance is expected to meet the agreed full and incremental costs for activities to enable and support enhanced action on adaptation, mitigation, technology development and transfer (including carbon capture and storage), capacity-building and the preparation of national reports by developing countries. NABARD is accredited as National Implementing Entity in 2015 on fast track. NABARD is eligible to submit large size projects having outlay of more than USD 250 million. The National Adaptation Fund for Climate Change (NAFCC) was established in 2015 to meet the cost of adaptation to climate change in India, particularly vulnerable to the adverse effects of climate change. In keeping with the commitment of NABARD to address impact of climate change the "Climate Change Fund" was created out of the profit of NABARD during 2016-17 for facilitating attempts to address impacts of climate change especially towards fostering sustainable development. NABARD contributes annually from its profit towards the corpus of the fund. Institution of the "Climate Change Fund" is a unique initiative of NABARD as a Development Financial Institution to foster sustainable development and contribute meaningfully towards national priorities. The objective is to promote and support activities aimed towards addressing climate change impacts, adaptation and mitigation measures, awareness generation, knowledge sharing and facilitate sustainable development.

30.1 Climate Change initiatives of NABARD in Agriculture

Climate change has become real and tangible, affecting people, particularly poor. It is a major challenge for agriculture, food security and rural livelihoods. Climate change impacts on vulnerable groups have social, economic, political, and environmental implications. The impacts of climate change are global. However, India is more vulnerable in view of the high population depending on agriculture, which in turn is dependent on the seasonal monsoons. India is confronted with the challenge of sustaining rapid economic growth amidst the increasing global threat of climate change. Government of India has accorded top priority for addressing climate change related concerns. Climate change is a complex policy issue with major implications in terms of finance. All actions and solutions to address the adverse impact of climate change ultimately involve costs. Funding is vital in developing countries to design and implement adaptation and mitigation projects. NABARD has taken various initiatives in addressing the challenges posed by Climate Change particularly in the areas of agriculture and rural livelihood sectors. In this direction NABARD aims to channelize national, international and private finances for adaptation and mitigation activities in India. The National Adaptation Fund for Climate Change (NAFCC) was established in 2015 to meet the cost of adaptation to climate change in India that are particularly vulnerable to the adverse effects of climate change. In keeping with the commitment of NABARD to address impact of climate change the "Climate Change Fund" was created out of the profit of NABARD during 2016-17 for facilitating attempts to address impacts of climate change especially towards fostering sustainable development. NABARD contributes annually from its profit towards the corpus of the fund. Institution of the "Climate Change Fund" is a unique initiative of NABARD as a Development Financial Institution to foster sustainable development and contribute meaningfully towards national priorities. The objective is to promote and support activities aimed towards addressing climate change impacts, adaptation and mitigation measures, awareness generation, knowledge sharing and facilitate sustainable development. Several pilot projects were initiated to deal with the problems of climate change through Climate Change Adaptation (CCA) fund. One such project in Ahmednagar district of Maharashtra is funded to the extent of US$ 3.5 million. It was taken up with an objective to develop the knowledge, strategies, and approaches that enable vulnerable communities to cope with Climate Change and adapt to the impending impacts. The impacts of climate change are very much evident in terms of delays in onset and withdrawal of monsoon, changes in temperature and unseasonal rainfall. Creating successful demonstration projects on community based adaptation practices is a key challenge to address these adverse impacts.

Sustainability of public programs, investments, or agricultural value-chains are at risk due to climate change. NABARD is also piloting projects on climate proofing of watersheds in all the states to address such risks. Besides, opportunities of reducing climate risks through interventions, increasing the capacities of communities and institutions to deal with climate change are explored through climate proofing. Further, NABARD is supporting projects many of which can be classified under climate finance. Over 28% of NABARD's cumulative disbursements have links with climate change adaptation and mitigation. The NABARD's areas of forestry, agriculture, animal husbandry, land development, minor irrigation, etc., have components with emission reduction potential. Further, the areas like farm mechanisation, SHGs, Storage and Market Yards, etc. have some kind of linkages with climate change adaptation.

Climate Change Adaptation (CCA) project in Ahmednagar district of Maharashtra was taken up with an objective to develop the knowledge, strategies, and approaches that enable vulnerable communities to cope with climate change and adapt to the impending impacts. The impacts of climate change are very much evident in terms of delays in onset and withdrawal of monsoon, changes in temperature and unseasonal rainfall. Creating successful demonstration projects on community based adaptation practices is a key challenge to address these adverse impacts. The CCA project in Maharashtra in collaboration with SDC precisely aims at creating a model which is replicable and scalable. Sustainability of public programs, investments, or agricultural value-chains are at risk due to climate change. NABARD is also piloting projects on climate proofing of watersheds in Tamil Nadu and Rajasthan in collaboration with GIZ to address such risks. Additionally, opportunities of reducing climate risks through investment interventions, increasing the capacities of communities and institutions to deal with climate change are explored through climate proofing. Scaling-up climate resilient agriculture practices towards climate smart villages (CSVS) in Haryana and scaling climate smart agriculture through mainstreaming climate smart villages in Bihar are planned foe wider coverage.

Rain water harvesting in hilly areas in Darjeeling as an adaptive measure to potential climate change impacts, ecosystem management of Kaziranga National Park, Climate Change Adaptation for Natural Resource Dependent Communities in Kachchh, Gujarat and Bundelkhand Region of Uttar Pradesh are supported by NABARD. The other projects include management of by creating climate resilient livelihood for vulnerable communities through organic farming and pond based pisciculture, addressing water sector at Gram Panchayat level in drought prone areas of Sikkim, gene pool conservation of indigenous rice varieties under traditional integrated rotational farming system (jhum optimization) for promoting livelihood and food security as climate change adaptation strategy in Nagaland, climate resilience building in rural areas through crop residue management are a few to mention supported by NABARD.

30.2 Linkages of NABARD projects to Climate Change Initiatives

NABARD is supporting projects many of which can be classified under climate finance. Over 28% of NABARD's cumulative disbursements have links with climate change adaptation and mitigation. Specifically, NABARD's thematic areas of forestry, agriculture, animal husbandry, land development, minor irrigation, etc., have components with emission reduction potential. Apart from these, areas like farm mechanisation, Self Help Groups, storage and market yards, etc. have some linkages with climate change adaptation. NABARD's five programmes are Rural Infrastructure Development Fund (RIDF), Tribal Development Fund (TDF), Integrated Watershed Development Programme, Umbrella Programme on Natural Resource Management (UPNRM), NABARD Infrastructure Development Assistance (NIDA), These programmes have supported projects with emission reduction potential including forestry, biogas digesters, rural energy management (pump replacement, HVDS, etc.), renewable energy and improving energy efficiency.

30.3 Climate Change initiatives of NABARD in Livestock and Aquaculture

Towards climate resilient livestock and aquaculture production systems, NABARD has been supporting projects in Punjab and Odisha, climate resilient interventions in dairy sector in coastal and arid areas in Andhra Pradesh and conservation and management of indigenous varieties of livestock (cattle and sheep) in Karnataka. Building adaptation capacity of small inland fisherman, coastal resource conservation, climate adaptation for wetland and integrated farming system programmes have been initiated in different states. Conservation and management of coastal resources as a potential adaptation strategy for sea level rise is financed in Andhra Pradesh. Building adaptive capacities of small inland fishermen for climate resilience and livelihood security undr adaptation fund has been initiated in Madhya Pradesh as adaptive measures to address rainfall variability, building resilience, knowledge generation and management with focus on gender issues. Fisheries and aquaculture sector in India is 3rd largest fish producer, 6% of world production(10% Inland and 3 % marine) contributing 1 % of GDP and 5.58% of agricultural GDP. It privided livelihood to 14 million fishermen, mainly unorganised while earning foreign exchange. The per capita fish consumption is around 9.13 kg/annum (20 Kg Global). The declining fish catch is observed in both marine and culture fisheries due to Drought, floods, cyclones, diseases leading to loss of diversity, changes in species composition. Share of Marine and Inland sectors in fish production in India in 1980 is Inland is 35%, and 65% in Marine sources changed to 64 and 36 % respectively in 2014. The change has increased the vulnerability of the ecosystem services. The impact of Climate Change in India is evident, ncrease in mean temperature by 0.60°C in the last 100 years. Fisheries sector is most vulnerable to climate change[1] which employs 15 million people reinforces the need for huge climate finance.

Building adaptation capacity of small inland fisherman, coastal resource conservation, climate adaptation for wetland and integrated farming system programmes have been initiated in different states. Climate Change Projects under Fisheries Sector are financed by Adaptation Fund and National Adaptation Fund for Climate Change (NAFCC). Rehabilitation of Coral Habitats

Building Adaptive Capacities of Small Inland Fishers for Climate Resilience and Livelihood Security in Madhya Pradesh undr adaptation fund initiated as adaptive measures to address rainfall variability, adaptive measures to address warmer climatic regime, building resilience for climate adaptation and knowledge generation and management with focus on gender issues.

Conservation and Management of Coastal Resources as a Potential Adaptation Strategy for Sea Level Rise is financed in Andhra Pradesh with an amount of USD 0.69 Million.

Conservation and Management of Coastal Resources as a Potential Adaptation Strategy for Sea Level Rise-Andhra Pradesh, has been initiated in2015 with USD 2.51 Million. Interventions linked to promoting climate adaptive and resilient livelihood systems through diversification, technology adoption and natural resource management.Preparation of land and water use master plan, reducing climate risks through timely and appropriate early warning, climate resilient technology transfer for enhancing the adaptive capacity of the community, learning and knowledge management, integration of aquaculture with agriculture and water management.

In order to address food, fodder, fuel insecurity and aquaculture micro planning for enhancing adaptive capacity and increasing resilience of small and marginal farmers in Purulia and Bankura districts, West Bengal, NABARD has initiated two projects. Climate adaptation in wetlands along the Mahanadi River Catchment Area in Chhattisgarh, management and rehabilitation of coastal habitats and bio diversity for climate change adaptation and sustainable livelihood in Gulf of Mannar, Tamil Nadu and integrated farming system of paddy and shrimp/fish (Kaipad and Pokkali) in coastal wetlands of Kerala for mitigating the adverse impacts of climate change on chilka lake. Several projects for restoration of mangrove cover in 4,628 Km^2 area and to improve livelihood have been initiated in several states.

Conclusions

NABARD took the lead to support various activities in the areas of climate change, particularly in agriculture, livestock, aquaculture and rural livelihood sectors and to facilitate for sustainable development. Several pilot projects were initiated to deal with the problems of climate change and also piloting projects on climate proofing of watersheds in all the states to address such risks. Besides, opportunities of reducing climate risks through interventions, increasing the capacities of communities and institutions to deal with climate change are explored through climate proofing. Towards climate resilient livestock production systems, NABARD supported projects in dairy sector and conservation and management of indigenous varieties of livestock. Building adaptation capacity of small inland fisherman, coastal resource conservation, climate adaptation for wetland and integrated farming system programmes have been initiated in different states. Conservation and management of coastal resources as a potential adaptation strategy for sea level rise. In this connection, there is a need to integrate climate change concerns with agriculture, livestock, fisheries and natural resource sectors in relation to livelihoods. Greater investments are required for climate proofing of all these sectors in view of high vulnerabilities. In the absence of the support the implementation of adaptation measures is extremely difficult. Considering the adaptation and mitigation measures, the peoples' participation of various stakeholders is to be ensured. Inter institutional collaboration and transfer of technology shall be promoted with all possible options by considering the needs of local people. Capacity building initiatives, are required to be conducted regularly to generate The knowledge base of different sectors and their inter-relations need to be strengthened.

NABARD has established the National Adaptation Fund for Climate Change (NAFCC) in 2015 to meet the cost of adaptation to the adverse effects of climate change. Further, NABARD has created "Climate Change Fund" out of the profit of NABARD during 2016-17 to address impacts of climate change especially towards fostering sustainable development. NABARD promote and support the activities aimed towards addressing climate change impacts, adaptation and mitigation measures, awareness generation, knowledge sharing and facilitate sustainable development. Several pilot projects were initiated to deal with the problems of climate change through Climate Change Adaptation (CCA) fund. One such project is in Ahmednagar district of Maharashtra. It was taken up with an objective to develop the knowledge, strategies, and approaches that enable vulnerable communities to cope with climate change and adapt to the impending impacts. The impacts of climate change are evident in terms of delays in onset and withdrawal of monsoon, changes in temperature and unseasonal rainfall. Creating successful demonstration projects on community based adaptation practices is a key challenge to address these adverse impacts. Sustainability of public programs, investments, or agricultural value-chains are at risk due to climate change. NABARD is also piloting projects on climate proofing of watersheds in all the states to address such risks. Besides, opportunities of reducing climate risks through interventions, increasing the capacities of communities and institutions to deal with climate change are explored through climate proofing. Towards climate resilient livestock production systems, NABARD supported projects in Punjab and Odisha, climate resilient interventions in dairy sector in coastal and arid areas in Andhra Pradesh and conservation and management of indigenous varieties of livestock (cattle and sheep) in Karnataka. Projects as building adaptation capacity of small inland fisherman, coastal resource conservation, climate adaptation for wetland and integrated farming system programmes have been initiated in different states. Conservation and management of coastal resources as a potential adaptation strategy for sea level rise is financed in Andhra Pradesh. Building adaptive capacities of small inland fishermen for climate resilience and livelihood security under adaptation fund has been initiated in Madhya Pradesh as adaptive measures to address rainfall variability, building resilience, knowledge generation and management with focus on gender issues.

Further, NABARD is supporting projects many projects that have links with climate change adaptation and mitigation including forestry, agriculture, animal husbandry, land development, minor irrigation, etc., have components with emission reduction potential. Further, the support to farm mechanisation, SHGs, storage and market yards, etc have linkages with climate change adaptation. NABARD's programmes like Rural Infrastructure Development Fund, RID , Tribal Development Fund, TDF, Integrated Watershed Development Programme, IWDP, Umbrella Programme on Natural Resource Management, UPNRM, NABARD Infrastructure Development Assistance, NIDA, have components with emission reduction potential including forestry, biogas digesters, rural energy management, renewable energy and improving energy efficiency. All the above mentioned programmes have been yielding substantial benefits to the society.

Acknowledgement

The author is grateful to CGM, DEAR, NABARD for renewed support and financial assistance.

References/Suggested Readings

APRO (2017). Doubling of Farm Income through Innovative Agribusiness, APRO, NABARD, Hyderabad.

APRO (2018). Andhra Pradesh State focus Paper 2018-19, APRO, NABARD, Hyderabad.

BIRD (2017). Institutional analysis of Microfinance Institutions (MFIs), Bankers Institute of Rural Development, Lucknow.

NABARD (2017). Annual Report 2016-17, NABARD, Mumbai. www.nabard.org NABARD's website

Overall impact on agricultural economy: Cost of climate change estimated at around 1.5% of GDP from Agriculture every year (ICAR –NICRA project).

Palanisamy, K. (2016). Experiences in building climate changeresilience in fisheries and aquaculture sector through international and domestic funding mechanisms in India, National Bank for Agriculture and Rural Development (NABARD)**(Footnotes)**

Projections (medium-term - 2010-2039) : Yield reduction by 4.5 to 9% - posing challenge to food security and livelihood. q

The negative impact on crop productivity : Experienced in wheat and paddy due to increased temperature, reduction in rainy days and consequential water stress.q

31

Climate Change and Sorghum Productivity in India

V.M. Sandeep[1] and V.U.M. Rao[2]

Central Research Institute for Dryland Agriculture, Santosh Nagar, Saidabad PO, Hyderabad - 500059, India
[1]Research Associate, AICRPAM, India
[2]Project Co-ordinator (Retd), AICRPAM, India

Sorghum (Sorghum bicolour (L.) Moench) is one of the most nutritional and fifth most important cereal crop after maize, wheat, rice and barley as well as a major dietary staple food crop of more than 500 million people in various countries (Ashok Kumar *et al*., 2011). It is considered as the "King of Millets" that extensively grown in Semi-Arid Tropics (SAT). As a coarse grain, it is used not only for human food, but also for fodder and feed for animals, and building material etc. Its grain is used to make bread, biscuits, sugar, starch, syrups, and alcohol, beer and malt products. It is a good source of energy, protein, vitamin, minerals, and trace elements. Sorghum cultivation in India probably started in the first millennium BC (Dogget, 1988). In India, sorghum is commonly known as *jowar* and accounts 14% of total world's sorghum cultivated area and 8% production. India stands first in area of cultivation and second major producer until the end of 20th century. Currently (2012-16). India stands second in cultivated area after Sudan and fourth in production after USA, Nigeria and Mexico. It is the fifth major cereal crop in India after rice, wheat, maize, and pearl millet (bajra), and accounts 6% of total cereal cultivated area and 3% of total cereal production (DES, 2016). Sorghum is cultivated in India majorly during *kharif* (rainy) season as a rainfed crop and in rabi (post-rainy) season under residual soil moisture/limited-irrigated conditions. Sorghum is gaining importance along the coastal belt of Andhra Pradesh as rice fallows during *rabi* season in place of maize due to water scarcity in recent years. Maize requires 4 to 6 irrigations while 2 to 3 irrigations in case of sorghum during the crop season for potential yields under field conditions. In this chapter,a glimpse of sorghum growing regions in India, interrelationship between climatic changes and yield of sorghum and adaptation strategies to sustain sorghum productivity in India during future climatic periods are discussed in detail.

31.1 Climatic regimes of sorghum

Sorghum is well adapted to drought environments compared to other cereals, making it suitable for the semi-arid tropical agricultural production systems. Sorghum requires warm environment but it can be cultivated under a broad range of conditions. It is also extensively cultivated in temperate regions and at altitudes of up to 2300 m in the Tropics. It can withstand high temperature during its life cycle compared to any other crop. The climatic conditions suitable for sorghum crop along with the climatic regimes across *kharif* and *rabi* sorghum regions in India are discussed below. The beginning of the crop season is defined separately for both *kharif* and *rabi* season.The date of sowing during *kharif* season has been set immediate after the commencement of southwest monsoon rainfall over each district (Rana *et al*., 1999). The *rabi* sowing window was set as 25 September to 15 October and a minimum of 9 mm rain within 5 days was required to trigger sowing (Kholova*et al*., 2013). The average sorghum growing duration in India is 110 days characterized by Rao *et al*. (2004) and used the present study for both seasons.

31.1.1 Rainfall

Sorghum is best adapted to areas having an average annual rainfall between 400 to 700 mm. Even though it can respond to good moisture supplies, it is one of the best drought tolerant crops also as it can withstand waterlogging situation than any other cereal crop, except rice. The rainfall deficit of more than 50% can cause crop failure (Srivastava *et al*., 2010). Mastrorilli *et al*. (1995) reported that water stress at flowering reduces final biomass by 52%, number of seed per panicle by 58%, and grain yield by 61%. Rainfall across the *kharif* sorghum tract in India is observed below 250 mm across North-western Rajasthan as well as South-eastern Tamil Nadu and it exceeds 1500 mm across few parts of Coastal Karnataka and South-western Maharashtra. It is noticed that, the *kharif* sorghum regions spread across major parts of Tamil Nadu, Karnataka, Southwestern Maharashtra, Southern Andhra Pradesh, Central Gujarat, Central Rajasthan and Haryana, receive 250 to 500 mm rainfall, while thecropregions spread across Central Maharashtra, Southern and Eastern Gujarat, Southeastern Rajasthan, Western Uttar Pradesh, Eastern Andhra Pradesh, Southern Telangana, Northeastern Karnataka, and some parts of Northwestern Madhya Pradesh, receive 500 to 750 mm rainfall. The regions over Eastern Uttar Pradesh, Central Madhya Pradesh, Central/Eastern Maharashtra, Southeastern Gujarat, Central Telangana and few regions of Northeastern Andhra Pradesh, receive 750 to 1000 mm rainfall, while the crop regions across South-western Madhya Pradesh, far eastern Maharashtra, Northern Telangana and few regions of Western Maharashtra and Western Karnataka, receive 1000 to 1250 mm rainfall (Fig. 31.1). Few pockets of *kharif* sorghum regions across Western parts of

Karnataka and Maharashtra receives more than 1250 mm rainfall but the share of those regions in India is trivial, while the northwestern arid regions of Rajasthan and few regions of Southeastern Tamil Nadu, receive less than 250 mm rainfall during the season. It is clear that the yield is high across the regions with good rainfall. The higher (greater than 1250 kg ha^{-1}) yield over the regions of Northern Telangana, Eastern Maharashtra and Central/Eastern Madhya Pradesh was attributed due to good amount of rainfall (> 1000 mm). As per the data sourced from http://www.data.gov.in, the total percentage area under irrigation for sorghum across different states are as follows, Andhra Pradesh (29.3%), Gujarat (19.8%), Tamil Nadu (13.5%), Karnataka (10%), Maharashtra (9.5%), Telangana (3.4%), Madhya Pradesh (0.06%), Uttar Pradesh (0.04%) and Rajasthan (0.01%). The higher yield over Coastal Andhra Pradesh, Tamil Nadu, Eastern Gujarat and Central Karnataka could be attributed to improved irrigation facilities across sorghum regions. The yield is low (< 750 kg ha^{-1}) across Haryana and Northwestern Rajasthan, which is attributed by the receipt of limited rainfall (< 500 mm). In the case of *rabi* sorghum, the seasonal rainfall is below 50 mm in Gujarat, gradually increases towards South-eastern direction and exceeds 400 mm across major parts of Tamil Nadu and some parts of South-eastern Andhra Pradesh due to the receipt of Northeast monsoon rainfall. The *rabi* sorghum area spread across major parts of Maharashtra and Karnataka, few regions of Western Telangana, receive 50 to 100 mm rainfall, while the regions of Northwestern Maharashtra and Gujarat, receive less than 50 mm rainfall. The regions across Eastern Maharashtra, major parts of Telangana, Southern Karnataka and Southwestern Andhra Pradesh, receive 100 to 200 mm rainfall, while the regions across major parts of Tamil Nadu and Southeastern Andhra Pradesh accounts more than 400 mm rainfall from northeast monsoon. It is clear that the *rabi* sorghum yield is also significantly influenced by rainfall because the yield is high (> 1000 kg ha^{-1}) over the regions of Southeastern Andhra Pradesh and major parts of Tamil Nadu, which receive good amount of rainfall through Northeast Monsoon (more than 400 mm). The *rabi* productivity is high in the districts over coastal Andhra Pradesh, Eastern Gujarat and Central Karnataka which can be attributed to improved irrigation facilities, while the *rabi* productivity is low (<750 kg ha^{-1}) in major parts of Maharashtra as the crop is rain fed across those regions.

31.1.2 Temperature

Sorghum requires about 26-30°C temperature for good growth and can tolerate temperature up to 45°C, but lower temperature less than 8°C can mutilate flowering and pollination(Peacock, 1982). The lowest cardinal temperature for the germination of sorghum seed is 7 to 10°C (Rao, 2005). The optimum temperature range for sorghum is 21 to 35°C for seed germination, 26 to 34°C for vegetative growth and 25 to 28°C for reproductive growths (Prasad *et al.*, 2008). Germination will not happen in cool soil circumstances. Hence, winter cultivation is not possible in the subtropics and temperate regions. So *rabi* cultivation is not practiced in North India. Exceptionally high temperatures can cause retarded flower initiation and decreased yield. Under the environment of seasonal mean daily temperatures greater than 20 °C, early grain cultivars take 90 to 110 days and medium-duration cultivars, 110 to 140 days to mature, while the average daily temperature is below 20 °C, there is an extension of about 10 to 20 days in the growing season for each 0.5 °C decrease in temperature, depending on cultivar (Wani*et al.*, 2012). In India, the average seasonal maximum temperature for *kharif* sorghum is ranged between 27°C across the regions in South-western Karnataka to 39°C across the arid regions in North-western Rajasthan and for *rabi* it is ranged between 27°C in Southern Karnataka to 34°C in South-western Andhra Pradesh (Fig 31.1). The average seasonal minimum temperature for *kharif* sorghum is ranged between 17°C in South-western Karnataka to 28°C in North-western Rajasthan and for *rabi* it is ranged between 13°C in North-western Maharashtra to 25°C in South-eastern Tamil Nadu. The state wise sorghum growing environments for *kharif* and *rabi* sorghum in India are furnished in Table 31.1.

31.1 Climate variability and sorghum productivity

Earlier studies employing A2A scenario and the HADCM3 global climate model for 2020, 2050 and 2080 indicated that sorghum yield decreases of 3–76% for the rainy season and 7–32% for the winter crop due to climate change (Srivastava*et al.*, 2010). Talwar*et al.* (2010) noticed that climate change impact on *rabi* sorghum is projected to reduce the yield up to 7 % by 2020, up to 11 % by 2050 and up to 30 % by 2080. In the present study, we investigated the trends in seasonal maximum (T_{max}), minimum (T_{mix}) temperatures and rainfall in *kharif* and *rabi* crop growing seasons during baseline period (1981-2010) and future climates viz., near future (2021-2050) and far future (2051-2080) periods over sorghum growing regions in India. The area weighted values of rainfall, T_{max}, T_{min} and sorghum yield of the districts where continuous yield data is available during the period 1981-2010 for *kharif* and *rabi* sorghum growing seasons were considered in this study. The major sorghum growing district of Rajasthan, Gujarat, Madhya Pradesh, Maharashtra, Karnataka, Telangana, and Andhra Pradesh states for *kharif* and Gujarat, Maharashtra, Karnataka, Telangana and Andhra Pradesh states for *rabi* seasons were considered to prepare the area weighted values of the sorghum growing regions in the country. 152 districts for *kharif* and 71 districts for *rabi* season were selected based on the above criteria and the study area consists of 92% of *kharif* and 94% of *rabi* sorghum growing regions in India. For analysis, the time domain is divided

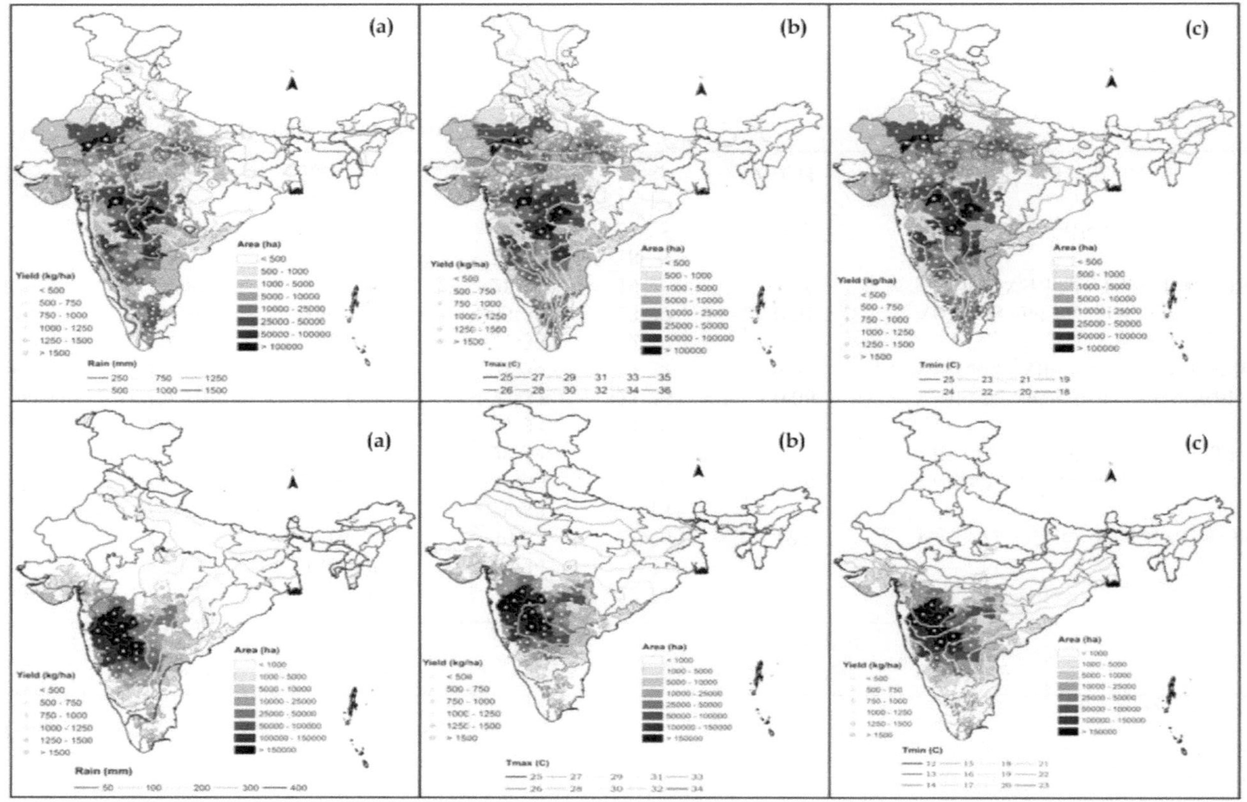

Fig 31.1: Distribution of seasonal average (a) rainfall, (b) maximum and (c) minimum temperatures across *kharif* (upper pane) and *rabi* (lower pane) sorghum growing regions in India.*(Maps prepared using IMD temperature (1° x 1°) and rainfall (0.25° x 0.25°) data for the baseline period 1981-2010)*

Table 31.1. State wise sorghum growing environment in India

State	Season	Rainfall (mm)	T_{max}(°C)	T_{min}(°C)	Area under irrigation (%)	Soil Type
Andhra Pradesh	*Kharif*	300-700	32-36	22-27	29.3	Vertisols, Alfisols and Ultisols
	Rabi	100-400	29-34	16-23		
Gujarat	*Kharif*	400-1200	30-35	22-26	19.8	Vertisols, Inceptisols and Aridisols
	Rabi	10-25	29-32	13-18		
Haryana	*Kharif*	350-600	33-36	21-26	77.6	Aridisols and Alfisols
Karnataka	*Kharif*	300-2300	28-34	17-23	10	Alfisols, Vertisols, Ultisols and Inceptisols
	Rabi	75-150	27-33	14-20		
Madhya Pradesh	*Kharif*	650-1100	31-37	21-26	0.06	Vertisols, Inceptisols and Alfisols
Maharashtra	*Kharif*	500-1500	28-34	21-24	9.5	Vertisols , Alfisols, Ultisols and Inceptisols
	Rabi	50-100	27-33	12-18		
Rajasthan	*Kharif*	150-650	33-39	23-27	0.01	Aridisols and Inceptisols
Tamil Nadu	*Kharif*	150-350	28-36	19-26	13.5	Alfisols, Vertisols, Ultisols and Inceptisols
	Rabi	250-800	29-32	18-23		
Telangana	*Kharif*	500-1000	31-34	21-25	3.4	Alfisols and Vertisols
	Rabi	50-100	28-33	15-19		
Uttar Pradesh	*Kharif*	600-1000	31-36	21-25	0.04	Inceptisols and Alfisols

(T_{max}- Maximum Temperature, T_{min}- Minimum Temperature)

into three periods*viz.*, baseline (1981–2010), near future (2021-2050) and far future (2051-2080). Daily data on T_{max} and T_{min} at 1°x1°, and rainfall at 0.25° x 0.25°resolution for baseline period are sourced from National Data Center (NDC), India Meteorological Department (IMD), Pune. Rainfall, T_{max} and T_{min} for 2021-2050 and 2051-2080 has been sourced from the selected model NorESM-1M of Coupled Model Inter comparison Project Phase 5 (CMIP5) in Representative Concentration Pathways 4.5 (RCP 4.5) after evaluation with IMD data.

The time series inter-annual variability of average sorghum yield with average seasonal rainfall, T_{max} and T_{min} for both *kharif* and *rabi* sorghum seasons during the period 1981-2010 over the study area are depicted in Figs. 31.2 and 31.3. Even though there is a gradual increase in sorghum yield due to technological advancement and improved crop management practices, it is observed that the higher temperature and decreased rainfall negatively influenced the sorghum productivity. It is clear that, the decline in *kharif* yield during the years 1982 (-14%), 1987 (-18%), 1992 (-21%), 1995 (-8%) and 2009 (-10%) compared to the previous years were attributed by below normal rainfall *viz.,* 1982 (-16%), 1987 (-23%), 1992 (-11%), 1995 (-9%) and 2009 (-14%) as well as higher T_{max} and T_{min} *viz.,* 1982 (2% and 1%), 1987 (5% and 3%), 1992 (2% and 1%), 1995 (1% and 1%) and 2009 (4% and 3%). It is also observed that decreased *rabi* sorghum productivity across the study area during the years 1983 (-7%), 1987 (-9%), 1998 (-34%) and 2004 (-32%) are attributed to higher temperatures in 1983 (-7%), 1987 (-9%), 1998 (-34%) and 2004 (-32%) and also it is clear that decreased *rabi* sorghum yield is influenced by rainfall deficit during the years 1986, 1987, 1992, 1998 and 2004.It is noticed that during baseline period the *kharif* sorghum seasonal rainfall experienced an increasing trend in at the rate of 63 mm 30 yr^{-1}, whereas seasonal rainfall for *rabi* sorghum was exhibited a marginal increase at the rate of 7.6 mm 30 yr^{-1}.The maximum temperature across *kharif* sorghum regions in India exhibited a marginal increasing trend at the rate of 0.07 °C30 yr^{-1}, while *rabi* sorghum experienced a sharp increasing trend at the rate of 0.71°C30 yr^{-1} during this period. The minimum temperature across *kharif*and *rabi*sorghum regions in India exhibited increasing trend at the rate of 0.21 °C30 yr^{-1} and of 0.65 °C30 yr^{-1}, respectively during baseline period.

31.2.1 Rainfall and Temperature in future climates

It is projected that the frequency of below normal rainfall years are more prominent between the period 2035-55 for *kharif* rainfall (Fig 31.4). The projected driest year is 2048 (512 mm, which is 30% less than baseline value), followed by 2038 (561 mm, which is 24% less than baseline value), 2042

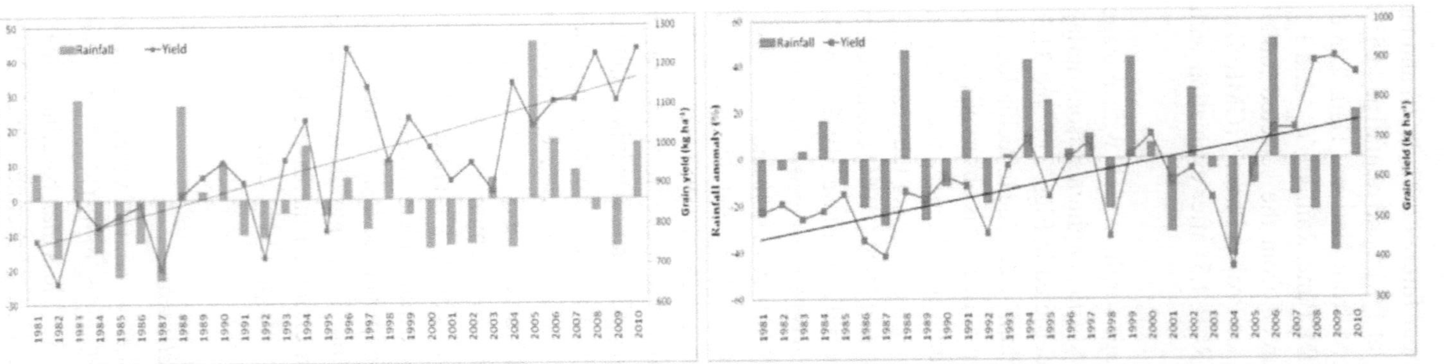

Fig. 31.2: Variability in rainfall and yield for kharif (left pane) and rabi (right pane) sorghum in India during 1981-2010

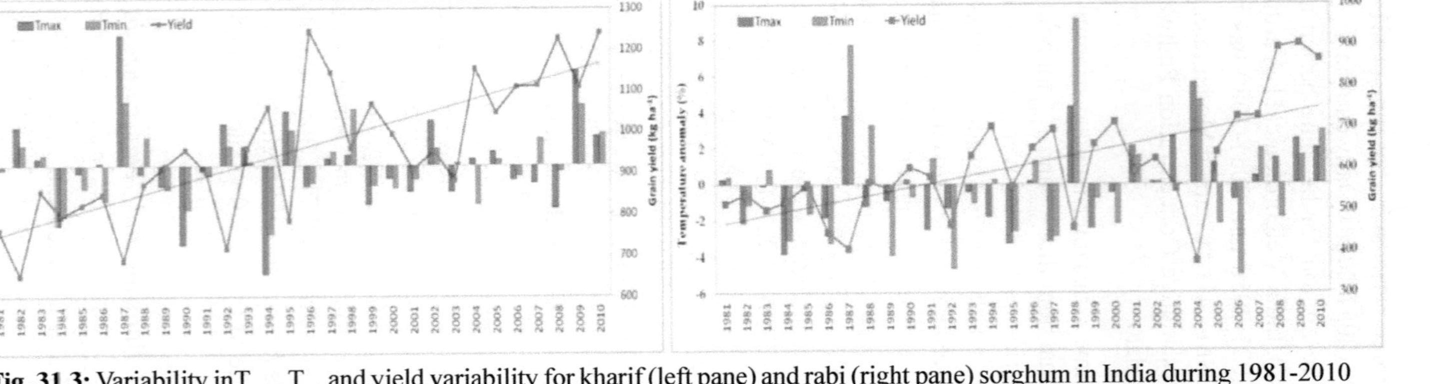

Fig. 31.3: Variability in T_{max}, T_{min} and yield variability for kharif (left pane) and rabi (right pane) sorghum in India during 1981-2010

(604 mm, which is 18% less than baseline value) and 2053 (631 mm, which is 15% less than baseline value). The projected wettest year is 2061 (1165 mm, which is 57% higher than baseline value), followed by 2021 (1077 mm, which is 45% higher than baseline value), 2056 (1028 mm, which is 38% higher than baseline value) and 2049 (1027 mm, which is 38% higher than baseline value). The *kharif* rainfall is projected to experience a sharp decreasing trend at the rate of -98.4 mm 30 yr^{-1} during 2021-50. During 2051-80 a sharp increasing trend at the rate of 87.8 mm 30 yr^{-1} is noticed in *kharif* sorghum seasonal rainfall. While considering the period as a whole (2021-80), it is observed that the *kharif* sorghum seasonal rainfall is experiencing a gradual increasing trend at the rate of 41.7 mm 60 yr^{-1} towards the end of the century. The *rabi* sorghum seasonal rainfall also projected a gradual increasing trend at the rate of 34.2 mm 60 yr^{-1} between the period 2021-2080. If we bisect this period as 2021-50 and 2051-80, both experience increasing trend at the rate of 35.1 and 36.9 mm 30 yr^{-1}, respectively

It is observed that, except couple of years between the period 2025and 2035, all the years in future climates may experience higher average seasonal maximum temperature for *kharif* sorghum (Fig 31.5). The increase in *kharif* seasonal average maximum temperature across India is observed nearly 0.36 °C 30 yr^{-1} between the period 2021-80. It is ranged between 31.3°C (2031) and 34.2°C (2044) during this period. During 2021-50, a sharp increase of 0.79 °C 30 yr^{-1} is noticed in seasonal average maximum temperature. It is found to be higher than baseline value throughout the period 2035-80. During 2051-80, a gradual decrease is noticed, nearly -0.29 °C 30 yr^{-1}, in maximum temperature. The *rabi* seasonal average maximum temperature shows a gradual increase, nearly about 1.14 °C 60 yr^{-1}, between the period 2021-2080 (Fig 31.5). It is ranged between 29.1 (2022) and 31.8 (2069) during this period. The rate of increase is observed in *rabi* maximum temperature during 2021-50 is 0.72 °C 30 yr^{-1} and during 2051-2080 is 0.32 °C 30 yr^{-1}. The year 2068 is found to be the warmest year followed by 2053, 2077 and 2058. The number of the years with above normal *rabi* seasonal maximum temperature is higher 2051-2080 compared to 2021-2050.

The *kharif* minimum temperature is projected to vary between 24 to 26°C between the period 2021-80. It is noticed that the *kharif* minimum temperature keeps maintaining near to baseline normal and show a sharp increase after 2030s until 2050s, and likely to decrease gradually afterwards (Fig 31.6). The highest value is noticed during the year 2045 (25.8 °C), followed by 2044 (25.5 °C). The *kharif* minimum temperature is exhibiting a sharp increasing at the rate of 0.71 °C 30 yr^{-1}during 2021-2050 and a gradual decrease at the rate of 0.24 °C 30 yr^{-1} during 2051-2080. However, an increase in *kharif* minimum

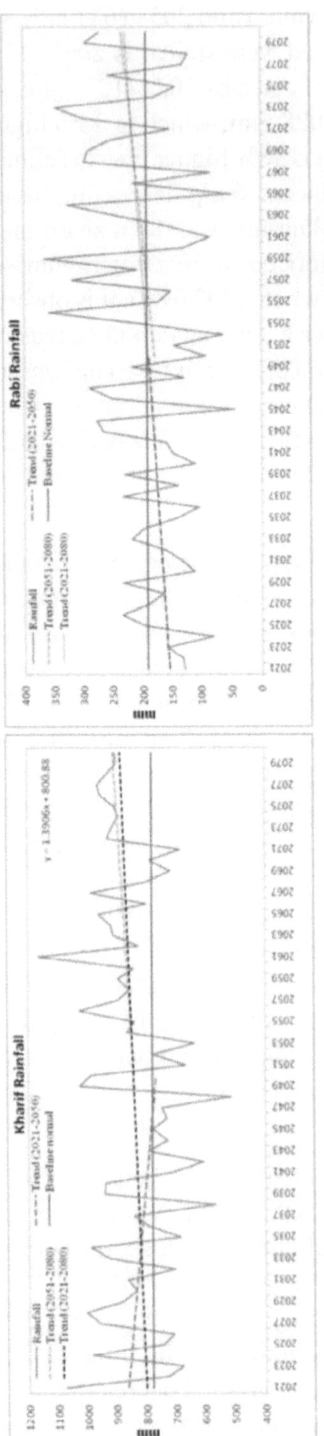

Fig. 31.4: Projected variability in *kharif* and *rabi* sorghum seasonal rainfall across India during 2021-2080

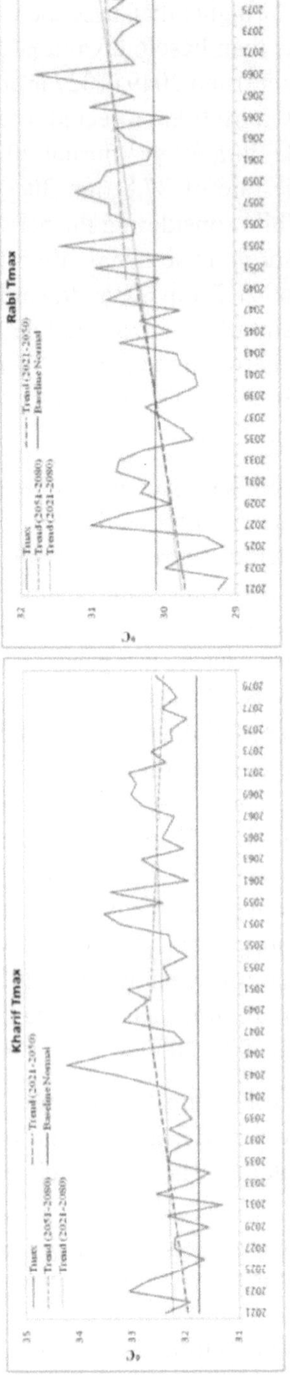

Fig. 31.5: Projected variability in *kharif* and *rabi* sorghum seasonal T_{max} across India during 2021-2080

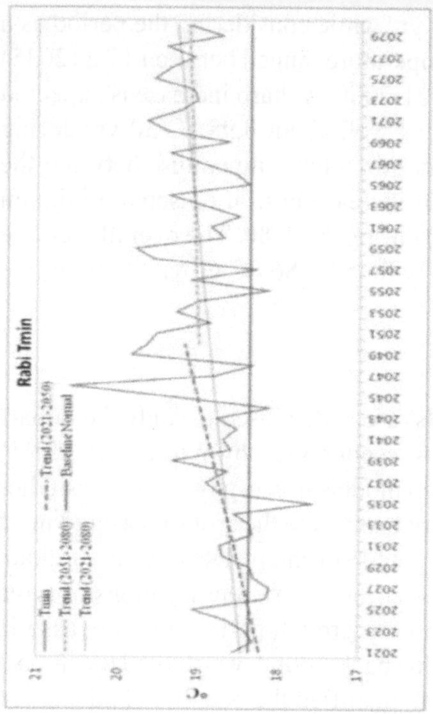

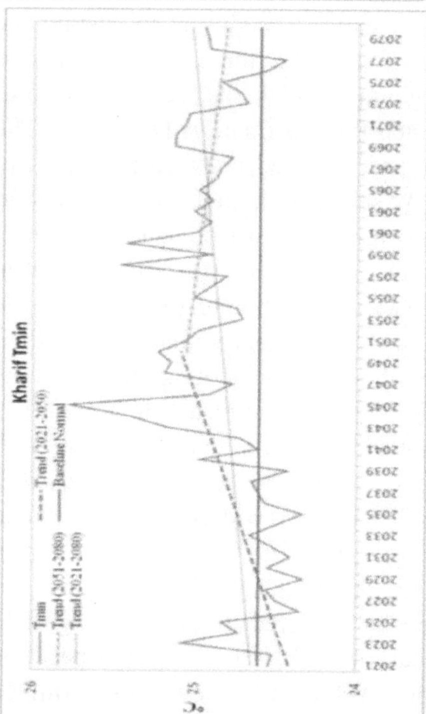

Fig. 31.6: Projected variability in *kharif*and *rabi*sorghum seasonal T_{min} across India during 2021-2080

temperature is noticed about 0.36 °C 30 yr^{-1} while considering the period as a whole (2021-80). The *rabi* minimum temperature ranged between 17.5 (2035) to 20.6 °C (2046) between the period 2021-2080. A sharp increase is expected in *rabi* seasonal minimum temperature as well about 0.98 °C 30 yr^{-1} during 2021-2050 due to the sudden increase in minimum temperature between the period 2035-2050 (Fig 31.6). The rate of increase in *rabi* seasonal minimum temperature is marginal (0.13 °C 30 yr^{-1}) during 2051-80. The overall increase in *rabi* seasonal minimum temperature is found 0.86 °C 60 yr^{-1} between the period 2021-80.

31.3 Adaptation strategies

Crop growth simulation models (CGSMs) are widely used to study the impact of climate change on crop productivity. A recent review by White *et al.*,(2011) on the use of more than 70 crop simulation models in eco-physiological studies pointed out that rice, wheat, maize and soybean were the crops measured most and very hardly any studies (<4%) targeted sorghum. Decline in sorghum productivity in future climate change scenarios in different locations of India was primarily attributed to reduction in crop growth period with increase in temperature (Boomiraj *et al.*, 2012). In this background, we studied the impact of climate change on sorghum yield and quantified the benefits of adaptation strategies like changing sowing time, irrigation scheduling to sustain the productivity of sorghum for *kharif* and *rabi* seasons in future climate scenarios (2025, 2050 and 2075) using DSSAT-Sorghum model. In the present study, four representative locations were selected each for *kharif* and *rabi* seasons. The major kharif sorghum growing regions are extended between the states of Rajasthan (North) to Karnataka (South). We selected Akola (Central), Dharwad (South), Surat (West) & Udaipur (North) for kharif season. The major *rabi* sorghum growing regions are confined to Deccan Plateau, Western and Central India. We selected Rahuri, Solapur, Dharwad andBijapur (South) for *rabi* seasons. These locations are selected on the basis of data availability and their climate/soil regimes. The soil/experimental details and geographical/seasonal climatic characteristics of the selected locations are furnished in Table 31.2. Site wise daily weather data on maximum and minimum temperature, rainfall and sunshine hours apart from crop and soil data were collected and used in DSSAT-Sorghum model simulations. Replicated trials widely cultivated Indian sorghum genotypes (CSH-9, CSH-16) during the *kharif* and (M-35-1) during the *rabi* were conducted at all the locations.The genotypic coefficients for various sorghum cultivars were chosen by repeated iterations using field experimental data for the different locations until a close match between simulated and observed yield was attained. Model efficiency was determined by comparing observed and simulated yields using statistical tools like root mean square error (RMSE) and D-index.

Table 31.2: Geographic, agro-climatic features, Soil and experimental details at different sites

Season	Location	Latitude	Longitude	Altitude (AMSL)	Seasonal rainfall (mm)	Seasonal average Tmax (C)	Seasonal average Tmin (C)	Crop growing period	Soil type	Soil pH	Selected cultivars
Kharif	Akola	20.7	77.02	296	712	33.2	24.1	1-10 July to 15-25 October	Clay Loam	7.8	CSH-9
	Dharwad	15.27	75.05	731	478	29.5	21.1	1-8 July to 15-25 October	Medium deep black (Clay)	7.8	CSH-16
	Surat	21.1	72.49	12	1393	33.2	25.4	15-30 June to 1-10 October	Deep black	7.8	CSH-16
	Udaipur	24.35	73.42	573	603	32.6	23.2	1-8 July to 15-25 October	Sandy clay loam	7.9	CSH-16
Rabi	Bijapur	16.49	75.42	614	186	31.2	18.3	20-30 September to 1-10 February	Clay	8.3	M-35-1
	Dharwad	15.27	75.05	731	204	30.3	17.8	1-15 October to 1-10 February	Medium deep black (Clay)	7.8	M-35-1
	Rahuri	19.23	74.38	516	156	31.2	14.4	20-30 September to 1-10 February	Clay	8.4	M-35-1
	Solapur	17.39	75.54	474	121	32.1	16.4	15-25 October to 15-25 February	Shallow and medium	8.6	M-35-1

In the present study, three crop management options, independently and in combination were tested as adaptation strategies to sustain sorghum productivity for kharif and rabi seasons in future climates. Six sowing time options (15, 10, 5 days ahead and 5, 10, 15 days delayed relative to normal sowing time) is the first low-cost option tried and to choose the one which provides the highest yield in climate change scenarios. Subsequently, we simulated the impacts of one additional irrigation of 50 mm at 45-60 days after sowing (DAS) and change in sowing time combined with additional irrigation. Series of sowing time and irrigation were input into CERES-Sorghum model and simulations were carried out for climate change scenarios to identify most suitable adaptation combination. The net effect of each adaptation strategy was assessed in terms of relative yield as

$$\text{Relative yield} = \left(\frac{Y_{fa} - Y_b}{Y_b}\right) * 100$$

Where Y_{fa} indicates future yield due to adaptation and Y_b indicates baseline yield without any adaptation. The combination(s), which gave the highest relative yield at each site in each scenario, was considered as the best suitable adaptation option. While simulating crop yields in future climates technological advancements in crop management, as well as impacts of pest and diseases on crop production, are not considered.

31.3.1 Calibration of CERES-Sorghum

The CERES-Sorghum model was calibrated for different sorghum cultivars (CSH-9, CHS 16) for the kharif and (M-35-1) during the *rabi* seasons based on data from field experiments carried out at respective experimental locations. Cultivar specific genotypic coefficients were estimated by the GENCALC software embedded in DSSAT v4.6. Thecalibrated genetic coefficients of various sorghum cultivars at respective sites used in the present analysis are presented in Table 31.3. The CERES-Sorghum model satisfactorily simulated the phenology and grain yield of sorghum cultivars in close match with observed values at different locations for both *kharif* and *rabi* seasons (Table 31.4). The differences with measured values are well within acceptable error level as indicated by a very high value of D-Index and acceptable RMSE values.

Table 31.3: Genotypic coefficients used for selected sorghum cultivars

Calibrated Genetic coefficients for rainy Season- DSSAT

Station	Variety	P1	P2	P2O	P2R	PANTH	P3	P4	P5	PHINT	G1	G2
Akola	CSH-9	680	100	16.5	125	617.5	152.5	81.5	500	49	20	6
Dharwad	CSH-16	300.5	102	16	270	617.5	152.5	81.5	700	49	18	4.5
Surat	CSH-16	470.5	102	16	170	617.5	152.5	81.5	700	49	22	6
Udaipur	CSH-16	505	102	17	295	617.5	152.5	81.5	600	49	22	6

Calibrated Genetic coefficients for Post rainy Season-DSSAT

Station	Variety	P1	P2	P2O	P2R	PANTH	P3	P4	P5	PHINT	G1	G2
Bijapur	M-35-1	236	102	17	95.6	617.5	152.5	81.5	540	49	12	6
Dharwad	M-35-1	310	102	16	195.6	617.5	152.5	81.5	540	49	22	6
Rahuri	M-35-1	305	102	13	55.6	617.5	152.5	81.5	446	49	22	6
Solapur	M-35-1	280	102	15	155.6	617.5	152.5	81.5	540	49	13	6

P1-Thermal time from seedling emergence to the end of the juvenile phase (expressed in degree days above TBASE during which the plant is not responsive to changes in photoperiod, P2-Thermal time from the end of the juvenile stage to tassel initiation under short days (degree days above TBASE), P2O-Critical photoperiod or the longest day length (in hours) at which development occurs at a maximum rate, P2R- Extent to which phasic development leading to panicle initiation (expressed in degree days) is delayed for each hour increase in photoperiod above P2O, PANTH-Thermal time from the end of tassel initiation to anthesis (degree days above TBASE), P3-Thermal time from to end of flag leaf expansion to anthesis (degree days above TBASE), P4-Thermal time from anthesis to beginning grain filling (degree days above TBASE), P5-Thermal time from beginning of grain filling to physiological maturity (degree days above TBASE), PHINT-Phylochron interval; the interval in thermal time between successive leaf tip appearances (degree days), G1-Scaler for relative leaf size, G2- Scaler for partitioning of assimilates to the panicle (head).

Table 31.4: Observed and simulated crop phenology and grain yields during baseline period

Season	Station	Parameter	Mean		Standard Deviation		D-Index	RMSE
			Observed	Estimated	Observed	Estimated		
Rainy	Akola	Anthesis day	74	75	3.57	2.07	0.87	1.84
	Dharwad		71	73	1.3	1.6	0.7	1.8
	Surat		64	64	2	2.07	0.8	1.5
	Udaipur		74	76	6.44	4.39	0.86	3.8
	Akola	Maturity day	112	108	4.32	3.3	0.59	4.1
	Dharwad		111	112	2.3	3.67	0.78	2.4
	Surat		105	103	1.3	2.38	0.68	2.7
	Udaipur		114	113	1.92	4.5	0.79	2.5
	Akola	Yield kg/ha	4143	3672	980	699	0.81	734.2
	Dharwad		4354	4150	946	614	0.77	646.5
	Surat		3493	3291	310	256	0.65	320.2
	Udaipur		4602	4406	1356	668	0.74	650.6
Post Rainy	Bijapur	Anthesis day	62	61	1.58	1.67	0.66	2.16
	Dharwad		65	66	2.65	2.06	0.77	1.94
	Rahuri		64	65	2.58	3.2	0.85	1.8
	Solapur		59	56	3.7	1.8	0.62	4.7
	Bijapur	Maturity day	105	102	2.56	2.73	0.67	3.82
	Dharwad		109	112	3.11	3.81	0.77	3.4
	Rahuri		109	108	3.16	2.55	0.75	2.55
	Solapur		98	97	3.38	3.2	0.59	4.2
	Bijapur	Yield kg/ha	1378	1573	507	567	0.72	441.8
	Dharwad		2832	2692	1304	618	0.8	754.7
	Rahuri		2088	2346	386	356	0.61	436.4
	Solapur		1034	1376	221	143	0.51	243.2

31.3.2 Climate change projections at different sites

The average percentage change in T_{max}, T_{min} and rainfallprojected by three models over baseline during the sorghum growing seasons (kharif and rabi) in 2025, 2050 and 2075 at four respective experimental sites are summarized in Table 31.5. The rise in T_{min} more severe in comparisonwith T_{max} over all the locations in future climatic periods during both (kharif and rabi) crop growing seasons. Rainfall (*kharif* and *rabi* seasons) is projected to be below normal during 2025 and expected to increase towards 2075 at all locations. Sorghum crop is likely to receive more rainfall at Dharwad (+18% & +28% for *kharif&rabi* crop seasons respectively) by 2075. The *kharif* season sorghum crop experience warmest day and night temperatures at Dharwad (+11% and +14%) and Udaipur (+10% and 17%), while *rabi* season sorghum crop may experience warmest daytime temperature at Bijapur (+10%) followed by Solapur (+10%) and warmest nights at Dharwad (+20%) followed by Bijapur (+18%) towards the end of 21st century.

31.3.3 Climate change projections

To the future climatic scenario datasets for the years 2025, 2050 and 2075 over the selected locations in this study are sourced from selected models of coupled model inter comparison project-5 (CMIP-5) in RCP 4.5.The projected CO_2 concentrations for the RCP 4.5 were 440, 530 and 570 ppm by 2025, 2050 and 2075s, respectively, and were used in simulating the future sorghum yields using CERES-Sorghum, while measured concentration was used for the baseline period. Garg *et al*., (2015) identified five best models (CCSM4, GFDL-ESM2M, MIROC5, NorESM1-M and NorESM1-ME)for Indian conditions based on the bias, temporal and spatial correlations, and coefficient of variation in the models output and observed data. Among these models, the average of three (GFDL-ESM2M, MIROC5 and NorESM1-M) are used in the present study and sourced from MarkSim DSSAT weather file generator. MarkSim is a tool specifically intended for use in the tropics that generates simulated daily weather data (rainfall, maximum and minimum temperatures and solar radiation) from GCMs and converts the downscaled weather data to DSSAT weather input file format.

31.4 Simulated future sorghum yield without any adaptation strategies

The future sorghum yields for *kharif* and *rabi* seasons simulated with the existing fertilizer (100% NPK) at each experimental site without any adaptation strategies indicate that during the first quarter of the 21st century (2025) the mean relative yield is achieved at Akola and Dharwad with existing crop management practices, while by middle of the 21st century (2050) it is confined only to Akola (Table 6). By the end of this century (2075) kharif sorghum yields are declinedat all the selected locations. From Table 31.5 it is evident that the

Table 31.5: Mean percentage change from the three selected models in T_{max}, T_{min} and rainfall over baseline during sorghum growing season in future climatic scenarios

Kharif Season						Rabi Season					
Location	Weather Variable	Normal Value	2025	2050	2075	Location	Weather Variable	Normal Value	2025	2050	2075
Akola	Rainfall	592	-2	4	3	Bijapur	Rainfall	153.35	-15	10	15
	Tmax	31.77	0	1	5		Tmax	30.49	10	0	11
	Tmin	23.39	3	3	9		Tmin	15.37	11	7	18
	Days to Anthesis	74	2	-4	-4		Days to Anthesis	62	-4	-1	-4
	Days to Maturity	112	6	-2	-5		Days to Maturity	105	-7	2	-9
Dharwad	Rainfall	456.4	-6	2	18	Dharwad	Rainfall	198.67	-21	-5	28
	Tmax	27.39	-4	7	11		Tmax	28.81	5	6	8
	Tmin	20.49	3	12	14		Tmin	15.72	12	17	20
	Days to Anthesis	71	2	-3	-5		Days to Anthesis	65	-4	-6	-6
	Days to Maturity	111	5	-7	-8		LGP	109	-8	-9	-9
Surat	Rainfall	965	0	-3	5	Rahuri	Rainfall	122	-21	-15	16
	Tmax	31.01	4	3	4		Tmax	30.69	9	6	8
	Tmin	26.57	6	8	9		Tmin	13.34	11	6	10
	Days to Anthesis	64	0	-4	-4		Days to Anthesis	64	-2	0	-2
	Days to Maturity	105	-2	-7	-7		Days to Maturity	109	-4	1	-5
Udaipur	Rainfall	569	-7	-5	7	Solapur	Rainfall	101.76	-10	-9	18
	Tmax	33.12	7	7	10		Tmax	31.59	8	9	10
	Tmin	23.75	10	11	17		Tmin	16.46	11	14	13
	Days to Anthesis	74	-2	-1	-2		Days to Anthesis	59	-1	-1	-2
	Days to Maturity	114	-4	-2	-7		Days to Maturity	98	-4	-2	-5

temperature is above normal over all the locations except during 2025, 2050 at Akola and during 2025 at Dharwad. Due to the increase in temperature as discussed in earlier section, the length of growing period is declining resulting in reduced yields. A decline of 4 days and 2 days in crop duration is simulated at Udaipurand Surat respectively and it causes a drop in yield about 10% at Udaipur and 4% at Surat in 2025. In 2050 the crop duration decreased up to 2 days with reduced yield of 13% at Udaipur, 7 days with reduced yield about 6% at Dharwad and Surat. Further a marked decline in growing period is simulated at Dharwad (8 days),followed by Udaipur (7 days), Surat (7 days) and Akola (5 days) and in 2075. A decline in yield is noticed about at 3% Akola, 14% at Dharwad, 3% at Surat and 13% at Udaipur in 2075.The R*abi* season sorghum yield Table (31.6) and length of growing period Table (31.5) also shows signs of decreasing trend in future climatic scenarios over all the locations except Bijapur and Rahuri during 2050. The average temperature during the crop growing seasons is above normal in future climatic scenarios over all the locations. A marked decrease of crop duration at Dharwad (8 days) followed by Bijapur (7 days), Rahuri (4 days) and Solapur (4 days) in 2025 is projected. A reduction in yield about 20% at Bijapur, 16% at Dharwad, 10% at Rahuri and 13% at Solapur is simulated in 2025. The crop duration is further decreased by 9 days and 2 days at Dharwad and Solapur respectively in 2050 with a decline in yield by 22% at Dharwad and 4% at Solapur. A further decrease in season length is simulated up to 9 days at DharwadandBijapuras well as5 days at Rahuriand Solapur in 2075. Yield reductionsare simulated about 22% at Bijapur, 23% at Dharwad, 7% at Rahuri and 14% at Solapur in 2075.

31.4.1 Impact of adaptation strategies

The average magnitude of improvement (%) in grain yield (mean of the three selected projected scenarios) by the various adaptation strategies in future climatic periods for *kharif* and *rabi* sorghum are furnished in Table 31.6. The influence of delayed sowing is found negatively on kharif and rabi yield over all the locations in future climatic scenarios, so the results are not provided here.

Kharif crop: The impact of six different sowing dates (three early and three late sowings at an interval of five days) with respect to the normal date of sowing (DOS) at each experimental site was assessed. The application of additional irrigation with normal DOSis not found beneficial adaptation strategy to alleviate the impact of climate change by attaining the targeted yield in future climatic scenarios as the yield were below targeted (Table 31.6). Among the four sites, significant improvement and targeted*kharif* sorghum yield was attained at Surat during all the three future periods and over Akola in 2075 when sowing was advanced compared to the normalDOS. The changes in mean temperature (average of maximum and minimum temperature) in 2025 in all the scenarios are +2, 0, +5, +9 per cent over Akola, Dharwad, Surat and Udaipur respectively. While it is +2, +10, +5, +9 percent in 2050 and +7, +13, +6, +14 per cent in 2075 over Akola, Dharwad, Surat and Udaipur respectively. Due to the relatively small change in mean temperature the targeted yield is attained with normal DOS without any adaptation strategies at Akola in 2025 & 2050 and at Dharwad in 2025. The advancement of sowing date is able to achieve the targeted yield by alleviating temperature stress and ensure the availability of moisture in rainfed condition at Akola in 2075 and at Surat in all the future periods. Due to the comparatively steep rise in temperature over Udaipur in all the future periods and at Dharwad in 2050 & 2075, the advancement of sowing time combined with an extra irrigation of 50 mm (45-60 DAS) is required to attain the targeted productivity.

Rabi crop: The average temperature projections at Bijapur for *rabi* season predicts +11, +4 and +15 percent increase in 2025, 2050 and 2075 respectively. While it is +9, +12, +14 percent at Dharwad, +10, +6 and +9 at Rahuri and +10, +11 and +11 percent at Solapur in 2025, 2050 and 2075 respectively. Due to comparatively lower temperature stress at Bijapur and Rahuri in 2050 the targeted yield is attained in normal DOS without any adaptation strategies The application of additional irrigation with normal DOSis found not beneficial adaptation strategy to alleviate the impact of climate change in attaining the targeted yield in future climatic scenarios over all the locations. The advancement of sowing date reduces the impact of temperature stress and facilitates to attain the targeted yield at Rahuri in 2075. Due to the higher rate of temperature stress advancement of sowing date combined with an extra irrigation of 50 mm (45-60 DAS) is required to sustain the targeted yield at Rahuri in 2025, at Bijapur in 2025 & 2075 and in all the future periods at Solapur and Dharwad (Table 31.6). Though adjusting sowing time is the most widely studied strategy of adaptation (White *et al.*, 2011), the impact by adjusting sowing time appears to be location specific as observed in the present study as the response is not uniform across the sites studied. The differential rise in day and night temperatures at different sites could be another reason for this response.

Summary and Conclusions

The current annual sorghum productivity in India is 864 kg ha^{-1},which is far low compared to the global average of 1481 kg ha^{-1}. India ranked 34 in terms of productivity out of 42 countries where sorghum produced is 1 lakh tonnes. The yield gap analysis by Murthy *et al.* (2007) shows that, the rainfed potential yield is ranged from 3210 to 3410 kg ha^{-1} for *kharif* sorghum, 1000 to 1360 kg ha^{-1} for *rabi* sorghum. Total yield gap in India is ranged from 2130 to 2560 kg ha^{-1} for *kharif* sorghum and 280 to 830 kg ha^{-1}for *rabi* sorghum. This is pointing

Table 31.6: Various adaptation strategies of *kharif* and *rabi* sorghum with average magnitude of advantage (% grain yield) for different climate periods.

Season	Location	Normal Date of Sowing			Normal Date of Sowing with additional irrigation (45-60 DAS)			Early Sowing			Early Sowing with additional irrigation (45-60 DAS)		
		2025	2050	2075	2025	2050	2075	2025	2050	2075	2025	2050	2075
Kharif	Akola	5	2	-3	8	3	-1	3	4	2	13	9	3
	Dharwad	7	-6	-14	8	-3	-11	-2	-5	-7	4	8	9
	Surat	-4	-6	-3	-3	-8	-7	5	6	3	8	9	16
	Udaipur	-10	-13	-13	-9	-7	-9	0	-4	-2	11	14	7
Rabi	Bijapur	-20	12	-22	-15	18	-15	-8	14	-11	7	18	4
	Dharwad	-16	-22	-23	-11	-13	-16	-7	-8	-10	4	9	7
	Rahuri	-10	4	-7	-3	9	-5	-6	9	7	5	13	16
	Solapur	-13	-6	-14	-13	-3	-15	-5	-8	-10	5	9	4

out that the yield of *kharif* sorghum can be improved 3.0 to 4.0 times and *rabi* sorghum 1.4 to 2.7 times from their current levels. For *kharif* sorghum, among biotic stresses shoot fly is an important constraint across all the regions in India. Stem borer is a major stress across the northern states *viz.,* Uttar Pradesh, Haryana and Rajasthan. Head bug is a main constraint across Karnataka, Gujarat, Maharashtra, Telangana, Andhra Pradesh and Tamil Nadu. Among the diseases, grain mold is a severe threat in Andhra Pradesh, Gujarat, Telangana and Tamil Nadu. Ergot is major threat in Andhra Pradesh, Gujarat, Maharashtra and Telangana. For fodder sorghum, anthracnose is a major constraint. Among the abiotic stress, mid season drought is a major constraint for sorghum productivity and the frequency of drought is found higher across Maharashtra, Karnataka, Andhra Pradesh, Telangana, Rajasthan, Tamil Nadu and Uttar Pradesh. Apart of these, the frequency of occurrence of extreme weather events like high rainfall episodes are found increasing across all the regions in recent decades and also projected to continue in the future can cause significant damage for the rainfed crops like sorghum. For Rabi sorghum, among the biotic stresses, shoot fly and head bugs are the major pests and stalk rot is a major disease across Karnataka, Southern Maharashtra, Gujarat and Southern Andhra Pradesh. Among the abiotic stresses for *rabi* sorghum, terminal drought caused by variability in rainfall is a major constraint as the crop dependent on stored moisture. The sorghum area and production are found diminishing over India during the past few decades as the farmers are shifting to other crops due to less profitability and the lack of value added products from sorghum. Higher temperature and decreased rainfall are negatively influencing the productivity for both *kharif* and *rabi* sorghum in India. It is noticed that the *kharif* sorghum growing regions in India experienced an increasing trend in seasonal rainfall at the rate of 63 mm 30 yr^{-1} during 1981-2010, while it is projected to decrease at the rate of 97.4 mm 30 yr^{-1} during 2021-50, and likely to increase at the rate of 98.6 mm 30 yr^{-1} during 2051-80.The seasonal rainfall for *rabi* sorghum in India was exhibited a marginal increase at the rate of 7.6 mm 30 yr^{-1} during 1981-2010, while it is projected to increase at the rate of 34.2 mm 30 yr^{-1} during 2021-50 and at the rate of 35.6 mm 30 yr^{-1} during 2051-80. The maximum temperature across *kharif* sorghum regions in India exhibited a marginal increasing trend at the rate of 0.07 °C30 yr^{-1} during baseline period, while in future climates it is projected a sharp increasing trend at the rate of 0.79 °C30 yr^{-1} during 2021-50 and gradual decreasing trend at the rate of 0.29 °C30 yr^{-1} during 2051-80.The maximum temperature across *rabi* sorghum regions in India exhibited an increasing trend at the rate of 0.71°C30 yr^{-1} during 1981-2010, while in future climates it is projected an increasing trend at the rate of 0.72 °C30 yr^{-1} and0.32 °C30 yr^{-1} during 2021-50 and 2051-80 respectively. The minimum temperature across *kharif* sorghum regions in India exhibited a marginal increasing trend at the rate of 0.21 °C30 yr^{-1} during 1981-2010, while in future climates it is projected an increasing trend at the rate of 0.71 °C30 yr^{-1} during 2021-50 and gradual decreasing trend at the rate of 0.21 °C30 yr^{-1} during 2051-80.The minimum temperature across *rabi* sorghum regions in India exhibited an increasing trend at the rate of 0.65 °C30 yr^{-1} during 1981-2010.A sharp increase is expected in *rabi* seasonal minimum temperature as well about 0.98 °C 30 yr^{-1} during 2021-50 whereas the rate of increase is marginal (0.13 °C 30 yr^{-1}) during 2051-80.The study pointing out that the sorghum production in India is likely to experience low moisture and high temperature stress during the period 2021-50. Both *kharif* and *rabi* sorghum yield is exhibiting a decreasing trend towards the end of the 21st century. Negative influences of climatic change over sorghum growing areas of India can be minimized by adapting one or a combination of management practices like adjusting sowing time, application of extra irrigation which are location specific as obtained from the present study.Looking at the rapid growth of population in the country, sorghum has to play a significant role considering the vast scope of increasing productivity with value added products and their multiple uses to make them more profitable for the farmers. In formulating any adaptation strategy involving selection of suitable variety, the ability to identify the one with most robust characteristics for withstanding climate change is crucial. Apart from cultivar selection, large scale adaptation of soil moisture conservation should form part of integrated approach to mitigate climate change effects and sustain sorghum yields in India.

References/Suggested Readings

Ashok Kumar, A., Reddy, B.V.S., Ramaiah, B. and Sharma, R. (2011). Heterosis in white-grained grain mold resistant sorghum hybrids. *Journal of SAT Agricultural Research*, 9: 1-6.

Boomiraj, K., Marimuthu, S., Wani, S.P., Ravikumar, S., Manikandan, N. and Tani, S. (2012). Vulnerability assessment of kharif rainfed sorghum to climate change in SAT regions of India. *Journal of Agrometeorology*, 14(1): 2-8.

Doggett, H. (1988). Sorghum, 2nd edition, Longman Scientific & Technical, London, United Kingdom, 512p.

Kholova, J., McLean, G., Vadez, V., Craufurd, P. and Hammer, G.L. (2013). Drought stress characterization of post-rainy season (rabi) sorghum in India. *Field Crops Research*, 141; 38–46.

Mastrorilli, M., Katerji, N. and Rana, G. (1995). Water efficiency and stress on grain sorghum at different reproductive stages. *Agricultural Water Management*, 28 (1): 23–34.

Peacock, J.M. and Heinrich, G.M. (1984). Light and temperature responses in sorghum. In: "Agrometeorology of Sorghum and Millet in the Semi-arid Tropics". Proceedings. of the International Symposium, ICRISAT, India, 15–20 November 1982. Patancheru, A.P., India.

Prasad, P.V.V., Pisipati, S.R., Mutava, R.N. and Tunistra, M.R. (2008) Sensitivity of grain sorghum to high temperatures stress during reproductive development. *Crop Science*, 48: 1911-1917.

Rana, B.S., Rao, M.H., Indira, S., Singh, B.U., Appaji, Chari. andTonapi, V. (1999). "Technology for Improving sorghum production and value addition". National Research Centre for Sorghum, Indian Council of Agricultural Research, Rajendranagar, Hyderabad A.P., India.

Rao, G.S.L.H.V.P. (2005). Agricultural Meteorology. Second Edition. Kerala Agricultural University, Thrissur, Kerala, India. 326 p.

Rao, S.S., Seetharama, N., Kiran Kumar, K.A. and Vanderlip, R.L. (2004). Characterization of sorghum growth stages. NRCS Bulletin Series no. 14. National Research Centre for Sorghum, Rajendranagar, Hyderabad - 500 030, India. 20p.

Srivastava, A., Naresh Kumar, S. and Aggarwal, P.K. (2010). Assessment on vulnerability of sorghum to climate change in India. *Agriculture Ecosystem and Environment*, 138 (3-4): 160–169.

Talwar, H.S., Elangovan, P.M., Kumali, A., Rao, S.S., Mishra, J.S. and Patil, J.V. (2010). Strategies to improve post flowering drought tolerance in rabi sorghum for predicted climate change scenario. *Crop Improvement*, 37 (2): 93-99.

Wani, S.P., Albrizio, R. and Rao, V.N. (2012). Sorghum. In: "Crop yield response to water". (Eds. Steduto P., Hsiao T.C., Fereres E. and Raes, D.). pp. 144-151. Food and Agricultural Organization, Rome, Italy.

White, J.W., Hoogenboom, G., Kimball, B.A. and Wall, G.W. (2011). Methodologies for simulating impacts of climate change on crop production. *Field Crop Research*, 124: 357-368.

32

Impacts of Climate Change on Indian Wheat Production

V.P. Pramod and B. Bapuji Rao

All India Coordinated Research Project on Agrometeorology
ICAR-Central Research Institute for Dryland Agriculture
Santoshnagar, Hyderabad, India

Wheat (*T. aestivum*) is a staple food of the vast majority of the human population in most part of the world; Europe, West Asia, South East Asia, Americas and North Africa. It provides more nourishment than any other food crop and leads all other food crops-rice, maize and potato on a global basis. The wide varieties of food products prepared from it make the crop popular and also a reason for its expansion even to non-traditional wheat cultivated areas. As a leading cereal crop, it is cultivated widely under varied agro-ecological regions and cropping systems throughout the world. It is one of the first plants domesticated and cultivated by humans, which played a critical role in the transition from hunter–gatherer groups to stabilized societies with an agrarian foundation. Wheat is harvested every month of the year somewhere in the world, from as far south as Argentina and as far north as Finland, and it is grown over a substantial area in almost every country of North and South America, Europe, and Asia (Fig. 32.1). Even though, the normal planting period of wheat falls in the same month across different countries, the harvesting period is different (Fig. 32.2). This is due to the delay in the pheno-phase of the crop by four days for every 122 m of altitude, one degree of latitude northward, and five degrees of longitude eastward in spring and early summer, and reverse in late summer and fall in the Northern hemisphere (Hopkins, 1919). Wheat, a C3 crop, is considered a cool-season grass and hence, it is mainly cultivated in areas between 30° N and 60° N latitude and 25° S and 40° S latitude (Briggle and Curtis, 1987). In India, the cultivation of wheat dates back to more than 5000 years during the era of Indus valley civilization. The spurt of Green Revolution, which started in late 1960s, makes India a wheat surplus country from a deficient one. Today, India stands first in area of cultivation and second after China in terms of wheat production. With increasing pressure on cultivable land, there is hardly any scope to expand the area under wheat. In addition, global warming and climate change is a threat to wheat production globally. Increase in concentration of the greenhouse gases (GHG's) due to human activities changes the Earth's climate on both global and regional scales. According to World Meteorological Organization (WMO) (2017), the global average concentrations of carbon dioxide (CO_2) in the atmosphere rose to 403 parts per million (ppm) in 2016, its greatest level in 800,000 years and 145% of pre-industrial levels (278 ppm). The impacts of climate change and variability are already visible on observed temperature and rainfall. The difference between maximum and minimum temperature is narrowing, which could be detrimental for agriculture (Bapuji Rao *et al.*, 2014). At the regional scale, precipitation declined while temperature increased over majority regions in India since last few decades, causing increased frequency of droughts and reduction in soil moisture that can hamper the crop production (Mishra *et al.*, 2014b). This will increase the gap between crop production and consumption in changing climate (Fischer *et al.*, 2005) and the impacts of climate change on food security could be even more, than previously thought (Schmidhuber and Tubiello, 2007).

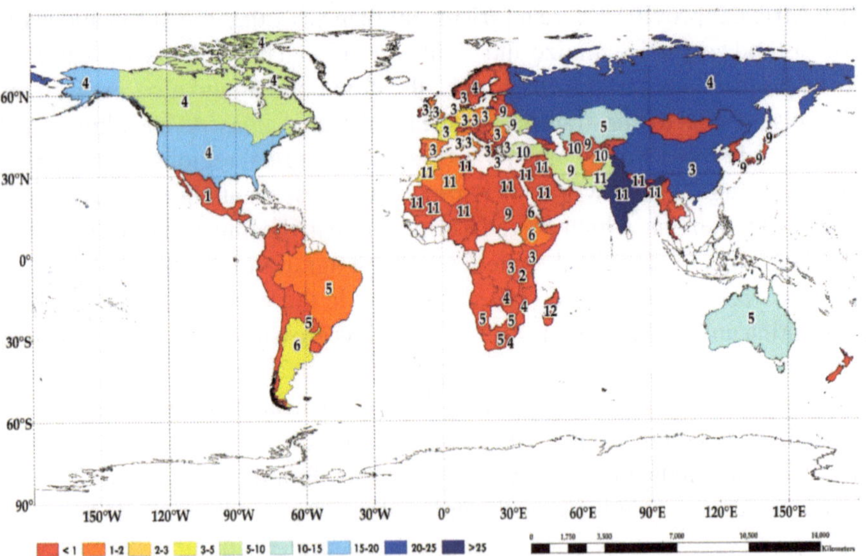

Fig. 32.1: Countries with average wheat growing areas (Mha) in the last decade (2007-16) and normal planting period across the globe. *(Number within the country indicates normal month of planting); (Data Source: FAO/GIEWS)*

32.1 Climates

Wheat crop has wide adaptability and hence it grows in all types of climatic conditions i.e. tropical, subtropical and temperate zones, and even in the cold tracts of far north, even beyond 60 °N latitude. With respect to altitude, it can be cultivated from sea level to as high as 3300 meters. Since wheat can tolerate severe cold and snow, it can resume growth with the onset of warm weather in spring. The best wheat is produced in areas favoured with cool, moist weather during the major portion of the growing period followed by dry, warm weather to enable the grain to ripen properly. The climatic conditions suitable for growing wheat are given below:

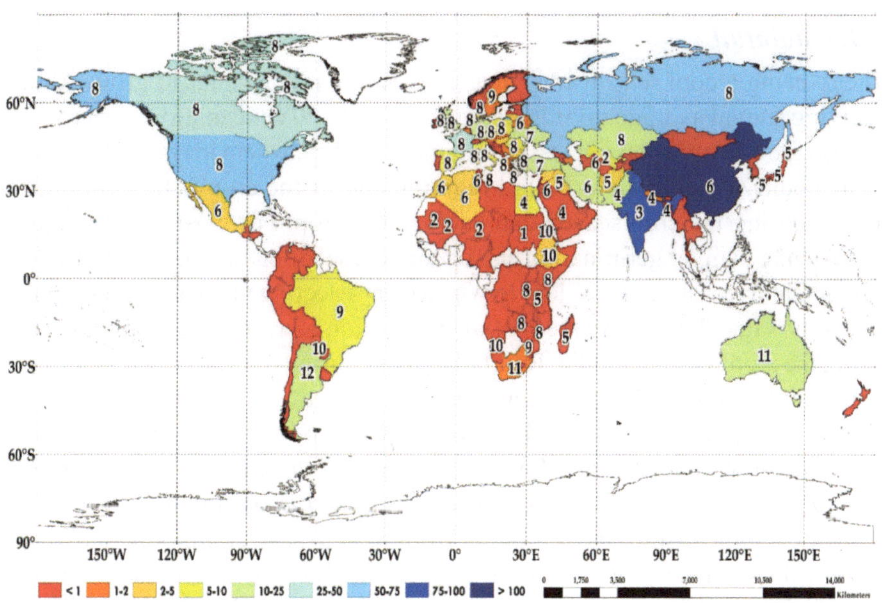

Fig. 32.2: Countries with average wheat production (MT) in the last decade (2007-16) and normal harvesting period across the globe. *(Number within the country indicates normal month of harvesting); (Data Source: FAO/GIEWS)*

32.1.1 Temperature

Even though seeds can germinate in the temperature range of 3.5 to 35 °C, the optimum range is 20-25 °C. Excessively high or low temperatures and drought are harmful to the crop, particularly during the heading and flowering stages. Cloudy weather, with high humidity and low temperatures is conducive for rust attack. Wheat plant requires about 14-15 °C optimum average temperature at the time of ripening (grain filling). Temperatures above 25 °C during the time of grain filling can reduce the grain weight and eventually the yield. When temperatures are high, too much energy is lost through the process of transpiration by the plants and the reduced residual energy results in poorer grain formation and lower yields. The high temperature at both ends of the crop season determine the duration available for cultivation. In India, wheat is mainly taken as a *rabi* (November-March) crop that normally sown during autumn and harvest in summer. The spatial pattern of average maximum and minimum temperature during the crop season over the country is shown in Figs. 32.3a and 32.3b, respectively. Temperature (maximum and minimum) distribution over the country shows a marked decrease from south to north. In North India, the minimum and maximum temperature during this season is 4 and 16 °C, respectively and in South India, it is about 18 and 32 °C, respectively.

32.1.2 Rainfall

Wheat, being a cool season crop does not require much water to grow. Areas receiving annual average rainfall between 30 and 100 cm can grow the crop. It can also grow in areas having lesser amount of rainfall, where irrigation facilities are available. Even though rainfall requirement for the crop is less, untimely rainfall during the crop season can damage the crop. For example, rains just after sowing hamper germination and encourage seedling blight, whereas at the time of maturity cause severe loss to wheat crop affecting yield and seed quality adversely. Fig. 32.3c shows the spatial distribution of average rainfall received during the crop season in India. Rainfall activity during the season was confined to the south peninsula, northern and eastern parts of the country. Over these regions, rainfall generally exceeded 200 mm. Extreme northern, coastal regions of southeastern peninsula and few eastern parts of the country received more than 500 mm of rainfall. The northwestern parts of the country are invaded by shallow cyclonic disturbances moving from west to east, commonly known as western disturbance, brings rainfall in Punjab, Kashmir and plains of Uttar Pradesh. These showers are very good and effective for the wheat over these regions due to less runoff and evaporation.

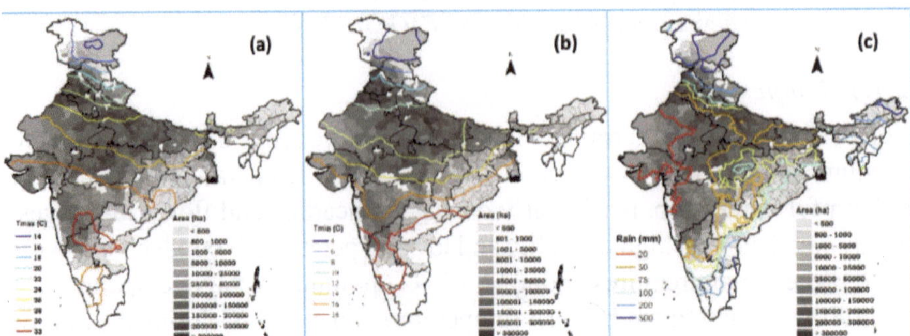

Fig. 32.3: Average crop seasonal (a) maximum temperature; (b) minimum temperature and (c) rainfall superimposed over wheat growing areas (Mha) *(Maps prepared using IMD* temperature (1° x 1°) and rainfall (0.25° x 0.25°) data for the baseline period 1981-10)

32.2 Wheat growing zones in India

Based on soil characterization, rainfall, temperature and terrain, six main agro climatic zones for wheat cultivation in India have been identified (Fig. 32.4). The major wheat growing zones in India are classified as

1. Northern Hill Zone (NHZ)
2. North Western Plain Zone (NWPZ)
3. North Eastern Plain Zone (NEPZ)
4. Central Zone (CZ)
5. Peninsular Zone (PZ)
6. Southern Hill Zone (SHZ)

Among the six wheat growing zones, NWPZ is the major wheat producing zone of the country. A short description regarding the climatic, soil and agronomic details of the six different wheat growing zones of the country is shown in Table 32.1.

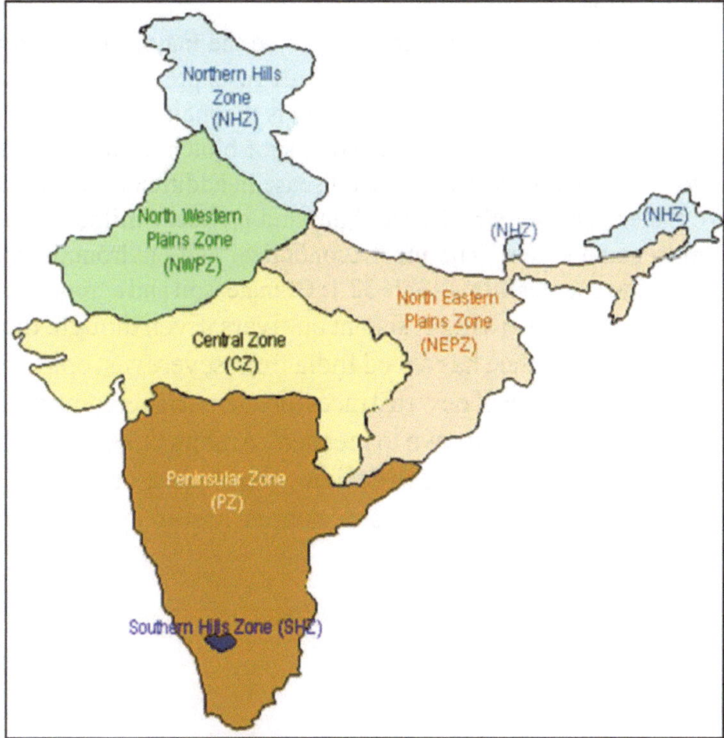

Fig. 32.4: Major wheat growing zones of India
(*Source*: http://krishisewa.com/krishi/Azone.html)

32.3 Indian wheat production - Challenges and constraints

Wheat is the second important food grain after rice in India and its products play an increasingly important role in managing country's food economy. The projected demand for wheat by the year 2025 will be high for an estimated population of 1.25 billion and the country has to produce 105 MT by 2025, demanding an average growth rate of 4% per annum (Prasad, 2011). This challenge will probably be further compounded by increased temperature due to global warming (Fischer *et al.*, 2005; Bapuji Rao *et al.*, 2014; 2015), dwindling water supplies for irrigation (Shiklomanov and Rodda, 2003), new races of pathogen and pest besides decrease in public sector investment in agriculture and rural affairs (Falcon and Naylor, 2005), changes in urbanization patterns, and demand for better quality wheat. The growing incidence and severity of heat/cold waves, droughts, floods, hail storms, frost and other extreme weather events severely affects the wheat production in the country. The yield gap between the potential and harvested yield in the eastern Gangetic Plains also cries out for solutions.

32.3.1 Biotic stress

Amongst the biotic stresses, rusts continue to be the major threat throughout most of India (Singh *et al.*, 2006). Out of three rusts prevalent in Indian sub-continent, leaf rust that caused by *P. triticina* is the major disease that affects almost whole of the India. In eastern India, spot blotch caused by *Bipolaris sorokiniana* is the most important wheat disease in addition to rust (Pandey *et al.*, 2005). Other diseases *viz.,* Karnal bunt and powdery mildew also affect wheat crop to some extent. The most economical and environmentally safe control measure to counter this is Table 32.1: Climate, soil and agronomic details of wheat growing zones growing resistant cultivars. Even though, the use of race specific rust resistance has saved India from severe rust epidemics for many years, the emergence of new rust races threatens to overcome the limited number of resistance genes known to breeders. Among other biotic stresses, weeds also play an important role in deciding the productivity since they are more resistant to abiotic stresses and their nutrient absorption capacity is also more than the wheat crop.

Table 32.1: Climate, soil and agronomic details of wheat growing zones

Environment	Wheat growing zones					
	NWPZ	NEPZ	CZ	PZ	NHZ	SHZ
Recommended wheat varieties	HD 3053, HD 2967, DBW 71, HD 2987, DPW 621-50	Raj 4120, K 0307, HD 2985,	MP 1202, MP 3336, MP 3288 HPW 349	UAS 428, NIAW 1415	HPW 249, HPW 349	HUW 318, HW 1085, HW 2044
Climate						
T_{max} range (°C)	18.3-31.2	23.3- 30.1	26.6-31.6	28.1-32.3	16.2-26.3	27.9-33.7; 28.6-31.9
T_{min} range (°C)	8.3-15.0	10.5-18.6	10.3-18.4	15.0-22.4	4.9-13.2	19.6-24.0; 17.6-22.2
Rainfall (mm)	14-394	0-384	0-79	12-788	149-821	85-765; 102-1032
Soil Type	Alluvial and Arid	Alluvial and Red	Alluvial, Black and Red	Red and Black	Forest and Mountainous	Forest and Mountainous
Sowing	1st FN of Nov	Mid Nov	Mid Nov	1st FN of Nov	1st FN of Nov	May; Oct

Contd.

Agronomic practices

Fertilizer dosage &	150:60:40 kg NPK ha^{-1}(1/3) N and P + K at sowing and(2/3) at 35-40 DAS		120:60:40 kg NPK ha^{-1}(1/3) N and P + K at sowing and (2/3) at 35-40 DAS			
Cropping system	Maize-Wheat; Rice-Wheat; Sorghum-Wheat; Bajra-Wheat; Cotton-Wheat; Sugarcane-Wheat	Rice-Wheat; Maize-Wheat; Sesame-Wheat	Rice-Wheat; Cotton-Wheat; Sorghum-Wheat; Soybean-Wheat	Rice-Wheat; Cotton-Wheat; Bajra-Wheat	Rice-Wheat; Maize-Wheat	Fallow-Wheat
Major diseases and insect-pests	Karnal bunt, yellow & brown rust	Leaf blight & brown rust	Leaf & stem rust, termites	Leaf & brown rust, aphid	Yellow & brown rust	Termite & black rust
	Harvesting	Apr	Mar-Apr	Mar	Mar	May-Jun Sep; Apr
Area (Mha)	12.0	8.3	6.1	1.5	0.8	0.2
Production (MT)	46.3	20.0	12.8	3.0	1.3	0.2
Yield (kg ha^{-1})	3900	2400	2100	2000	1600	1000

NWPZ (North Western Plain Zone) - Delhi, Jammu and Kathua districts of J & K, Una district and Paonta valley of HP, Tarai region of Uttarakhand, Haryana, Punjab, Rajasthan (except Kota and Udaipur divisions) and Western UP (except Jhansi division); **NEPZ** (North Eastern Plain Zone) - Eastern UP, Bihar, Jharkhand, WB, Odisha, Assam, Sikkim and plains of far eastern states; **CZ** (Central Zone) - Gujarat, South-eastern Rajasthan (Kota and Udaipur divisions), Jhansi division of UP, MP, and Chhattisgarh; **PZ** (Peninsular Zone) - Maharashtra, Andhra Pradesh, Karnataka, Goa and Plains of Tamil Nadu; **NHZ** (Northern Hill Zone) - Himalayan regions of J&K (except Jammu and Kathua district), HP (except Una and Paonta valley), Uttaranchal (except Tarai area), Sikkim and hills of WB and NE States; **SHZ** (Southern Hill Zone) - Hilly areas of TN and Kerala along with Nilgiri and Palani hills of southern Plateau; **FN**-Fortnight; **DAS**-Days after sowing

32.3.2 Abiotic stress

Even though production of food crops increased many folds due to the introduction of semi dwarf high yielding varieties, natural resources resulted in depletion due to their exhaustive exploitation and ultimately caused imbalance to the ecology thus, inviting problems of abiotic stresses. The most common abiotic stresses are high temperature, low temperature, drought, flood, salinity etc. According to world estimates, 50% of worldwide yield loss of major crops every year is due to abiotic factors like high temperature (20%), low temperature (7%), salinity (10%), drought (9%) and other forms of stresses (4%). In a developing country like India, the multitude combination of several abiotic stresses poses a serious threat to the agricultural production. Existence of high temperature at different growth stages of the crop decreases the duration of that particular stage and thus reduces the accumulation of photosynthates necessary for getting optimum yield. The crop completes its life cycle much faster under high temperature, than under normal temperature conditions. For instance, high temperature at flowering and grain filling stage reduce the duration of grain filling resulting in early maturity, thus reducing the crop yield. Sinha and Swaminathan (1991) observed a decrease in duration of the wheat crop by seven days due to an increase of 0.5 °C, that eventually reduce the yield by 500 kg ha^{-1} in north India. In India, wheat is exposing to high temperatures (> 35 °C) during grain filling stage because of the proximity to the equator and the popular cropping systems, which involve late sowing of wheat (Rane *et al.*, 2000). The northwestern plains are highly productive, but are prone to high temperature stress caused by dry winds from the Thar Desert whereas, the wheat grown in the northeastern plains is exposed to high temperature coupled with high humidity. Heat stress, particularly at the terminal stage, is seen as a serious reason why India has fallen well short of anticipated production gains (Nagarajan, 2005). Due to delayed planting, caused mostly by long duration paddy varieties in the rice-wheat cropping areas, a substantial wheat area is heat stressed (Joshi *et al.*, 2007a). A very recent study by Bapuji Rao *et al.* (2015) reported that Indian wheat yields were observed to be prone to continual heat stress and especially to short-term temperature extremes. They observed a mean yield decline of 204 kg ha^{-1} (approximately 7% Indian wheat yields) with 1°C rise in minimum temperature. Exposure to continual minimum temperature exceeding 12 °C for 6 days and terminal heat stress with maximum temperature exceeding 34 °C for 7 days during post-anthesis period is the other thermal constraints in achieving high productivity.

Like high temperature, low temperature can also have detrimental effects on crop growth, development and yield particularly at critical phenophases such as anthesis, especially in sub-tropical and temperate grain crops (Thakur *et. al.*, 2010). Since wheat is grown as a winter crop in India its flowering and anthesis time coincide during cold weather (December-January), which can cause sterility as high as 98% (Chakrabarti *et al.*, 2011). Temperature as low as freezing can severely injure wheat and greatly reduces grain yields at its reproductive stage as well. Among the abiotic stresses, drought and low soil fertility are the major significant problem affecting crop production. In arid and semi-arid regions, water stress is the most influential factor affecting crop yield particularly in irrigated agriculture. Therefore, the judicious use of water resource is very important in a developing country like India, in which wheat is grown on an area of about 30 Mha and nearly 54% of this area is irrigated. Only one-third receives full irrigation; the remainder is cultivated under partial irrigation and receives one to two irrigations over a cropping season. More than half of the wheat is cultivated under partially irrigated conditions, except for the NWPZ. In India, water will become a limiting factor for sustained production of different crops, including those of the water rich Indo Gangetic Plains. Various reports indicate declining water supplies for irrigation due to over mining of ground water in intensive rice-wheat cultivation and comparatively less water recharge from monsoon rains. Hence, apart from water conservation technologies, breeding for varieties that tolerate low-water conditions will become an increasingly higher priority.

32.3.3 Extreme weather events

Increasing frequency of extreme weather events due to the change in climate poses a serious threat to agriculture and food security of any country. Hot days, hot nights and heat waves have become more frequent, while cold days, cold nights and frost events have become less frequent. The rate of change of average production and yield of wheat for both global and India in every five years from 1961-2015 is shown in Fig. 32.5. Over the period, the average rate of change for both production and yield shows a decreasing trend. The production and yield of wheat is observed to be very low during the 2001-05 and 2006-10 period, when compared to other pentads. Since the start of modern measurements in 1850, the world experienced unprecedented high-impact climate extremes during the 2001-2010 decade (Plate 32.1). According to WMO (2010), more national temperature records were reported broken than in any previous decade. Argentina's 2002-03 wheat production declines by 13% compared to previous year due to heavy rains, whereas in Australia, it drops to half the normal value and in Canada, the lowest wheat production since 1974 was recorded due to extended dryness. Several extreme heat waves and drought adversely affected much of Europe during summer 2003, July and August months of 2003 in Africa. Over three quarters of the soft wheat output lost was estimated at the European Union level, attributable to the heat wave in France. France, the largest wheat producer of European Union, recorded a decline in wheat production by 21.5%, compared to 2002 (Agra Europe, 2003). The combination of December frost and February ice crusting during 2003-04 destroys the winter sown crop in Ukraine and Russia. Production in the European Union has fallen during 2005-06 because of excessive rainfall before and during the harvest period in both Germany and France. Australia wheat production trimmed down in 2006-07 following one of the worst drought country faced in the recent history. Heat and dryness during grain-fill stage has significantly reduced wheat production prospects in Germany in 2006-07. The 2007 frost occurred in USA during flowering period caused widespread devastation of crops. Wheat production declined to 20% in 2008-09 owing to drought in Middle East. The 2009 drought in Northern China, the worst in 50 years, threatens over half the wheat crop in eight provinces.

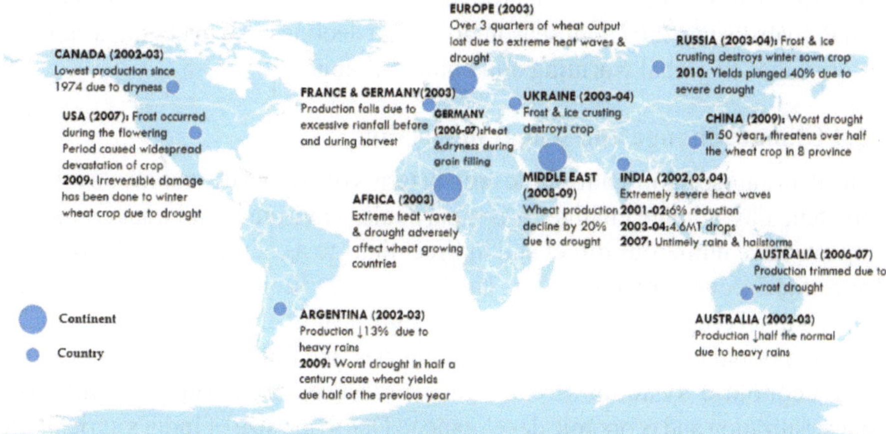

Plate 32.1: Weather extremes in major wheat growing countries during 2001-10

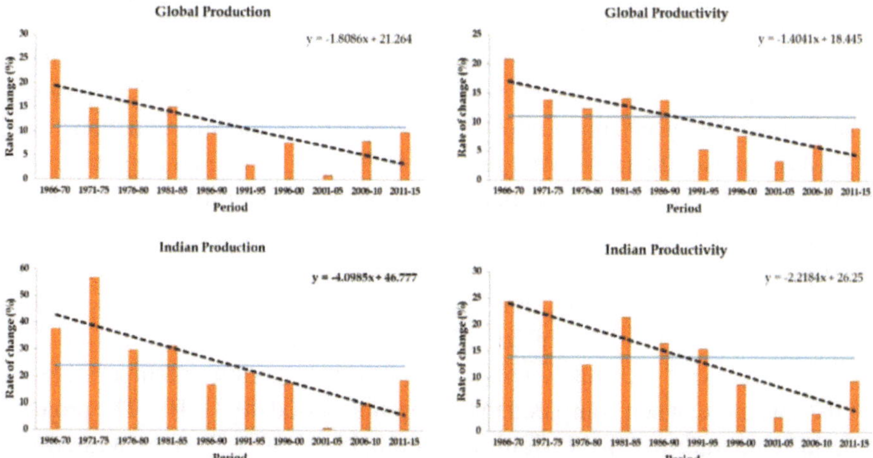

Fig. 32.5: Rate of change of average production and yield of global and Indian wheat for each pentad from 1961-2015 *(Blue line indicates the normal rate of change)*

In India, extremely severe heat waves hit continuously in 2002, 2003 and 2004. The largest year-to-year wheat production decrease occurred in the 2001-02 season with a reduction of six percent below the previous season. The year 2002 was a drought year in India and the seriousness of the drought and its effect on wheat production was underestimated. A significant drop in wheat productivity (4.6 MT) reported in 2003-04, due to intense heat wave conditions that prevailed over major part of northern western region of India. The wheat crop experienced terminal heat stress, which led to early maturity leading to smaller grain and lower 1000 grain weight. An average yield loss of 217 kg ha^{-1} was noticed during a heat wave year, whereas a gain of 356 kg ha^{-1} during a cold wave year (Samra and Singh, 2004; Samra *et al*., 2012). Untimely rains and hailstorms destroyed wheat crop of 15,000 ha over Uttar Pradesh, Haryana and Punjab during 2007 *rabi* season. Regional declines in wheat yields can also be attributes to global warming and climate change.

32.4 Climate change and wheat projection

Wheat, the most significant single crop in terms of human consumption is being threatened by climate change. It is a particular environmental concern, as the impacts have numerous direct and indirect links to agricultural production and hence food security of any region. By the end of 21st century, global production of the annual crops is expected to reduce significantly due to climate change. It is an additional burden in developing countries like India, since ecological and socioeconomic systems already face pressures from rapid population, industrialization and economic development. Due to the size of India's agricultural sector, the study of climate change impacts on Indian agriculture has gained global attention. Significant damage has already caused to our present crop profile due to climate change and threatens to bring even more consequences that are serious in the future. Wheat is highly sensitive to climate change especially temperature changes experienced in the later phase of crop season. The average change in temperature predicted by various global circulation models (GCMs) across India is to be in the range of 2.3-4.8 °C, even though the output from each GCMs differs considerably. In order to determine the changes in temperature and their association with future change in climate, many research works have been carried out on Indian context. A very recent study by Pramod (2018) projects an increasing trend in mean seasonal maximum temperature of the order of 0.91 °C 30 yr^{-1} in 2021-50 climatic periods over the major wheat growing areas of the country. This seasonal rise in maximum temperature will be 0.22 °C more, when compared to the baseline period (1981-10). The projected rise in temperature will be significant (@0.95 °C 30 yr^{-1}) over western and northwestern parts of the country, the major wheat growing zones of the country (Fig. 32.6a). In the case of minimum temperature, the trend signals a significant increasing trend over northeastern part of Uttar Pradesh, Bihar, West Bengal, Chhattisgarh and Punjab states @ 1.02 °C 30 yr^{-1} (Fig. 32.6b). One of the most important constraints limiting wheat productivity is the heat stress following anthesis, terminal heat stress, can significantly alter the normal physiological processes such as photosynthesis, respiration, membrane stability and stomatal responses. Over most of the wheat growing regions in India the crop reaches anthesis stage by the end of January and crop duration after January is considered as post-anthesis period. The trends in temperature (maximum and minimum) projected during the future climatic period 2021-50 is positive for the entire crop growing area in February. This significant rise in both maximum and minimum temperature projected during the post-anthesis period can negatively affect the yield of wheat crop.

Conclusion

The pace of growth of Indian wheat production has slowed down in recent years since 1995 due to the presence of abiotic stress *viz.,* high temperature, low temperature, drought, flood, salinity, and reduced water availability for irrigation. The increasing frequency of extreme weather events like cold and heat waves depending upon the crop phenology also pose a serious threat to wheat production in addition to several other constraints like socio-economic factors. The future projection of temperature across the major wheat growing regions of the country also looks not fruitful. Both the maximum and minimum temperatures are likely to increase significantly, in particular during the post-

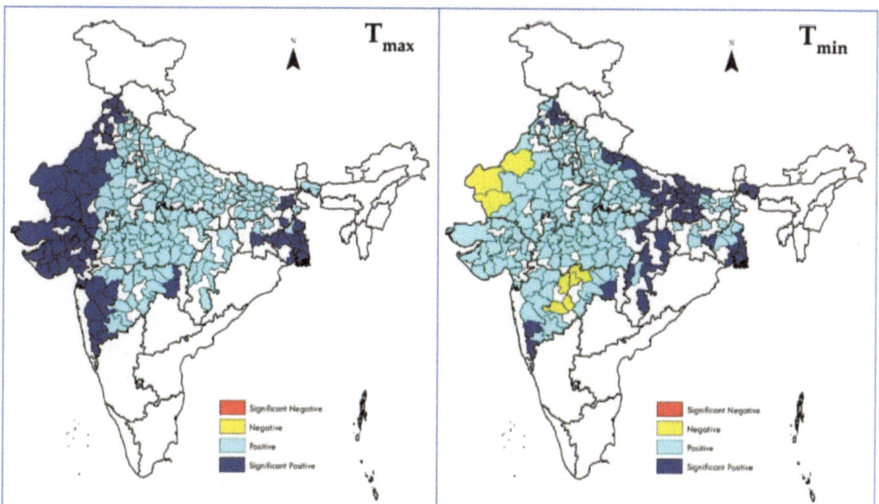

Fig. 32.6: Projected trends in maximum temperature and minimum temperature during 2021-50

anthesis stage, over major crop growing parts of the country. These show the need to develop better adaptation measures such as adjusting the sowing time, inclusion of early maturing, high yielding and heat tolerant wheat cultivars in the breeding program under Indian conditions.

References/Suggested Readings

Bapuji Rao, B., Santhibhushan Chowdary, P., Sandeep, V. M., Rao, V. U. M. and Venkateswarlu, B. (2014). Rising minimum temperature trends over India in recent decades: Implications for agricultural production. *Global Planet. Change*, 117: 1–8.

Bapuji Rao, B., Santhibhushan Chowdary, P., Sandeep, V. M., Pramod, V. P. and Rao, V. U. M. (2015). Spatial analysis of the sensitivity of wheat yields to temperature in India. *Agric. For. Meteorol.*, 200: 192–202.

Briggle, L. W. and Curtis, B.C. (1987). Wheat worldwide. In: "Wheat and Wheat Improvement". (Ed. E. G. Heyne). pp. 1-32. (Madison, Wisconsin).

Chakrabarti, B., Singh, S.D., Nagarajan, S. and Aggarwal, P. K. (2011). Impact of temperature on phenology and pollen sterility of wheat varieties. *Aust. J. Crop Sci.* 5 (8): 1039-1043.

Falcon, W. P. and Naylor, R. L. (2005). Rethinking food security for the 21st century. *Am. J. Agric. Econ.*, 87 (5): 1113–1127.

Fischer, G., Shah, M., Tubiello, F. N. and Van Velhuizen, H. (2005). Socio-economic and climate change impacts on agriculture: An integrated assessment, 1990–2080. *Phil. Trans. R. Soc. B.*, 360 (1463): 2067–2083.

Hopkins, A. D. (1919). The bioclimatic law as applied to entomological research and farm practise. *Sci. Mon.*, 8 (6): 496-513.

Joshi, A. K., Chand, R., Arun, B., Singh, R. P. and Ortiz, R. (2007a). Breeding crops for reduced-tillage management in the intensive, rice-wheat systems of South Asia. *Euphytica,* 153: 135-151.

Mishra, V., Shah, R. and Thrasher, B. (2014b). Soil moisture droughts under the retrospective and projected climate in India. *J. Hydrometeorol.*, 15: 2267-2292.

Nagarajan, S. (2005). Can India produce enough wheat even by 2020? *Curr. Sci.,* 89: 1467–1471.

Pandey, S. P., Kumar, S., Kumar, U., Chand, R. and Joshi, A. K. (2005). Sources of inoculum and reappearance of spot blotch of wheat in rice-wheat cropping system in eastern India. *Eur. J. Plant Pathol.,* 111: 47–55.

Pramod, V. P. (2018). "Thermal and Moisture Regimes of Wheat Under Projected Climate Change Scenario in India". Andhra University. 188. pp. (Unpublished).

Prasad, R. (2011). Nitrogen and food grain production in India. *Ind. J. Fert.,* 7 (12): 66-76.

Rane, J., Shoran, J. and Nagarajan, S. (2000). Heat stress environments and impact on wheat productivity in India: Guestimate of losses. *Ind. Wheat News Lett.,* 6 (1): 5- 6.

Samra, J. S. and Singh, G. (2004). "Heat Wave of March 2004: Impact on agriculture". Natural Resource Management Division, ICAR, New Delhi.

Samra, J. S., Kaur, P. and Mahal, A. K. (2012). Spectral density analysis of the cold wave (2010–11 and 2011–2012) and its impact on wheat productivity in Indian Punjab. In: 3rd International Agronomy Congress on "Agriculture Diversification, Climate Change Management and Livelihoods", 26-30 November, New Delhi.

Schmidhuber, J. and Tubiello, F. N. (2007). Global food security under climate change. *Proc. Natl. Acad. Sci.,* 104 (50): 19703–19708.

Shiklomanov, A. and Rodda, J. C. (2003). "World Water Resources at the Beginning of the 21st Century". (Cambridge University Press, New York). pp. 435.

Singh, R. P., Hodson, D. P., Jin, Y., Huerta-Espino, J., Kinyua, M. G., Wanyera, R., Njau, P. and Ward, R. W. (2006). Current status, likely migration and strategies to mitigate the threat to wheat production from race Ug99 (TTKS) of stem rust pathogen. *CAB Reviews: Perspectives in Agriculture, Veterinary Science, Nutrition and Natural Resources*, 1: 1–13.

Sinha, S. K. and Swaminathan, M. S. (1991). Deforestation, climate change and sustainable nutrition security- A case study of India. *Clim. Change*, 19: 201-209.

Thakur, P., Kumar, S., Malik, J. A., Berger, J. D. and Nayyar, H. (2010). Cold stress effects on reproductive development in grain crops: An overview. *Environ. Exp. Bot.,* 67: 429-443.

WMO (World Meteorological Organization). (2010). "The Global Climate 2001-2010". WMO, Geneva.

WMO (World Meteorological Organization). (2017). "WMO Greenhouse Gas Bulletin". WMO, Geneva.